ELECTRIC MACHINERY

McGRAW-HILL ELECTRICAL AND ELECTRONIC ENGINEERING SERIES

Frederick Emmons Terman, *Consulting Editor*
W. W. Harman, J. G. Truxal, and R. A. Rohrer
Associate Consulting Editors

TERMAN · Electronic and Radio Engineering
TERMAN AND PETTIT · Electronic Measurements
TOU · Digital and Sampled-data Control Systems
TOU · Modern Control Theory
TRUXAL · Automatic Feedback Control System Synthesis
TUTTLE · Electric Networks: Analysis and Synthesis
VALDES · The Physical Theory of Transistors
VAN BLADEL · Electromagnetic Fields
WEEKS · Antenna Engineering
WEINBERG · Network Analysis and Synthesis

ELECTRIC MACHINERY
Third Edition

The Processes, Devices, and Systems of Electromechanical Energy Conversion

A. E. FITZGERALD
Vice President for Academic Affairs
Dean of the Faculty
Northeastern University

CHARLES KINGSLEY, JR.
Associate Professor of Electrical Engineering
Massachusetts Institute of Technology

ALEXANDER KUSKO
Alexander Kusko, Inc., Consulting Engineers
Lecturer in Electrical Engineering
Massachusetts Institute of Technology

McGraw-Hill Book Company
New York
St. Louis
San Francisco
Düsseldorf
Johannesburg
Kuala Lumpur
London
Mexico
Montreal
New Delhi
Panama
Rio de Janeiro
Singapore
Sydney
Toronto

ELECTRIC MACHINERY

Library of Congress Catalog Card Number
70-137126
07-021140-X

13 14 15 16   KPKP   7832109

Contents

Contents

Preface

This book, like its two earlier editions, is intended to present a coordinated treatment of the processes, devices, and systems associated with electromechanical energy conversion. Emphasis is given to the physical concepts of voltage and torque, or force, production, for we continue to believe that these concepts underlie creative engineering and become the most valuable and permanent part of a student's background. Special attention is also given to performance features, limitations, and potentialities, and to the machine as one element of a sophisticated system. To accomplish these ends in the context of the engineering world of today and tomorrow, we have reorganized the book, rewritten many parts, and added a considerable amount of new material.

Probably the most significant development of recent years in this area is the use of power semiconductors with machines. As a result, earlier economic and technical limitations on control, performance, and applications have been extended or removed. Potentialities for the further development of such combinations during the next decade are great. Accordingly, we have added a considerable amount of material on solid-state drives.   We have also brought in a third author both to contribute expertise in this area and to reflect the viewpoint of a practicing engineer with wide experience (a rather rare commodity in today's educational world) throughout the book.

To provide better matching with the modern engineering curriculum, we have added an introduction to magnetic circuits to the first chapter. It is hoped that this material, in association with the treatment of iron-core transformers, will provide the background necessary to the study of energy-conversion devices.   We have also added an appendix on 3-phase circuits to highlight some of the simpler aspects of circuit theory which are of importance to energy processing.

The book obviously presents more material than can be treated in the time available in the typical curriculum.   Moreover, the detailed content and order of subject matter in specific courses are naturally governed by local circumstances and the desires and enthusiasms of individual instructors.   For these reasons, we have given particular attention to flexibility of use without loss of continuity.   Browsing in the book for an hour or two will enable an instructor to outline a variety of courses with differing content and sequence.

During the planning and writing of the book, we have incurred indebtedness to many people.   We especially wish to acknowledge the valuable advice and suggestions of many instructors who used the earlier editions.

A. E. FITZGERALD

CHARLES KINGSLEY, JR.

ALEXANDER KUSKO

ELECTRIC MACHINERY

1
magnetic circuits and transformers

The object of this book is to study the devices used in the interconversion of electrical and mechanical energy. Emphasis is given to the very common and very important electromagnetic rotating machines—the motors and generators which provide the power on which industrialized societies depend. Attention is also devoted to broader aspects of electro-mechanical energy conversion, not only because of the importance of mechanisms other than the rotating machine, but also to gain proper perspective.

Attention is likewise given to the transformer, which, although not an electromechanical energy-conversion device, is an important auxiliary in the overall problem of energy conversion. Moreover, in many respects its analytic details are closely related to those of motors and generators. The concepts of transformer behavior thus have the added feature of serving as an adjunct to the study of machines.

Practically all transformers and electric machinery utilize magnetic material for shaping the magnetic fields which act as the medium for transferring and converting energy. The relationships between the

magnetic-field quantities and the electric circuits with which they inter-
act play an important part in describing the operation of the various types
of equipment included in this book.   The magnetic material determines
the size of the equipment, its capability, and introduces limitations
because of saturation and loss on the performance.   In this chapter, we
shall proceed from the physical laws governing magnetic fields to relatively simple structures, describe the materials themselves, and then
treat transformers in some detail.   In the later chapters, we will use continuously the background material on magnetic circuits that is covered
in this chapter.

1-1
MAGNETIC CIRCUITS

The complete behavior of the magnetic field is described by Maxwell's
equations supplemented by constituent relationships which introduce the
parameters of the various materials occupied by the field.   In the case of
the electric machines and transformers treated in this book the frequencies
and sizes are such that the displacement current terms of Maxwell's
equations can be neglected and the *quasi-static* form used.   By this term
we mean that the magnetic fields under time-varying conditions are the
same as for static conditions at the same electrical levels.   From a practical standpoint, it means that we can solve all of our magnetic-circuit
problems under static conditions, then impart any time variation after-
ward.   Otherwise, the solution of magnetic-circuit problems for the
configurations of actual machines would be an extremely difficult task.

The basic law governing the relationship between electric current and
magnetic field is Ampere's law

$$\int_s J \cdot da = \oint \mathcal{3C} \cdot dl \tag{1-1}$$

where J is current density and $\mathcal{3C}$ is magnetic-field intensity.   In mks
units, J is given as amperes per square meter and $\mathcal{3C}$ as ampere-turns per
meter.   Equation 1-1 states that the line integral of $\mathcal{3C}$ around a surface
through which the current density J passes is equal to the total current
enclosed.   When Eq. 1-1 is applied to the simple core of Fig. 1-1, the left-
hand side merely becomes the product Ni of the turns and current; the
right-hand side becomes the product $\mathcal{3C}_c l_c$ of the intensity and the mean
core length.   We will show later why we can assume that $\mathcal{3C}_c$ is constant

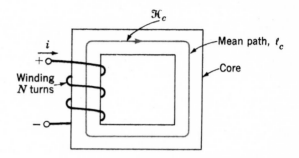

Fig. 1-1. Simple magnetic circuit.

along the path measured by l_c.   The relationship becomes

$$Ni = \mathcal{H}_c l_c \qquad (1\text{-}2)$$

The ampere turns Ni can be produced by one winding or more than one, where the total of all windings is Ni.   The direction of $\mathcal{H}_c$ with respect to Ni is given mathematically in Eq. 1-1, but practically by the right-hand rule.

The magnetic-field intensity $\mathcal{H}$ produces a magnetic-flux density $\mathcal{B}$ everywhere it exists, of value

$$\mathcal{B} = \mu\mathcal{H} \qquad (1\text{-}3)$$

The units of $\mathcal{B}$ are webers per square meter, where 1 weber $= 10^8$ lines of magnetic field.   The term μ is the permeability and is a property of the material.   In mks units the permeability of free space is $\mu_0 = 4\pi \times 10^{-7}$ webers/amp-turn m.   The permeability of ferromagnetic materials is usually expressed as a relative permeability μ_r to the free-space value, or $\mu = \mu_r\mu_0$.   Typical values of μ_r are in the 2000-to-6000 range for materials used in machines.   Equation 1-1 must be satisfied for every path in space which links the winding in Fig. 1-1.   The values of $\mathcal{H}$ along the paths are independent of whether they cross the material or free space.   However, the flux density $\mathcal{B}$ produced by the $\mathcal{H}$ is negligibly small everywhere except in the iron core.   When magnetic circuits are analyzed to determine the flux and flux density in the main magnetic paths through the core, the magnetic field outside the core and its gaps is usually neglected.   However, when two or more windings are placed on a magnetic circuit, as in a transformer or in a rotating machine, the fields outside the core, called leakage fields, are extremely important in determining the coupling between the windings.

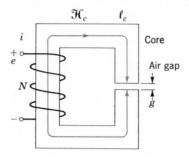

Fig. 1-2.   Magnetic circuit with air gap.

Transformers are wound on closed cores as shown in Fig. 1-1. Energy-conversion devices which incorporate a moving element must have air gaps in the cores. A magnetic circuit with a gap is shown in Fig. 1-2. Equation 1-1 applied to the circuit yields

$$Ni = \mathcal{3C}_c l_c + \mathcal{3C}_g g \qquad (1\text{-}4)$$

where $\mathcal{3C}_g$ is the intensity in the air gap. Around the magnetic circuit, the magnetic flux ϕ is continuous. The flux is defined as

$$\phi = \int_s \mathcal{B} \cdot da \qquad (1\text{-}5)$$

Within a core, the flux density can be assumed uniform across a cross-sectional area A_c, so that the flux is

$$\phi = \mathcal{B}_c A_c \qquad (1\text{-}6)$$

Equation 1-4 can be written first in terms of core and gap flux densities, then in terms of total flux.

$$Ni = \frac{\mathcal{B}_c}{\mu_c} l_c + \frac{\mathcal{B}_g}{\mu_0} g \qquad (1\text{-}7)$$

$$Ni = \frac{\phi l_c}{A_c \mu_c} + \frac{\phi}{A_g \mu_0} g \qquad (1\text{-}8)$$

The Ni term is designated $\mathcal{F}$ and termed *magnetomotive force* or *mmf*. The coefficients of the right-hand terms are designated *permeance* $\mathcal{P}$ or *reluctance* $\mathcal{R}$ and defined as

$$\mathcal{R}_c = \frac{1}{\mathcal{P}_c} = \frac{l_c}{\mu_c A_c} \tag{1-9}$$

Equation 1-8 becomes

$$\mathcal{F} = \phi(\mathcal{R}_c + \mathcal{R}_g) \tag{1-10}$$

The difficulty of solving Eq. 1-10 for ϕ for given $\mathcal{F}$ and geometry is that the permeability μ_c is not a constant, but depends on $\mathcal{B}_c$. The equation can be written as

$$\phi = \frac{\mathcal{F}/\mathcal{R}_g}{1 + \mathcal{R}_c/\mathcal{R}_g} = \frac{\mathcal{F}/\mathcal{R}_g}{1 + (\mu_0/\mu_c)(l_c/g)(A_g/A_c)} \tag{1-11}$$

In typical magnetic circuits, the term $(\mu_0 l_c A_g/\mu_c g A_c)$ is much less than unity, so that the behavior of the circuit is determined by the reluctance of the air gap alone. That reluctance, because it depends on μ_0, is independent of the flux density.

EXAMPLE 1-1

A magnetic circuit as shown in Fig. 1-2 has dimensions $A_c = 9$ cm²; $A_g = 9$ cm²; $g = 0.050$ cm; $l_c = 30$ cm; $N = 500$ turns. Assume the value $\mu_r = 5,000$ for the iron. Find the following: (*a*) Current i for $\mathcal{B}_c = 1$ weber/m². (*b*) Flux ϕ and flux linkage $\lambda = N\phi$.

Solution

a. From Eq. 1-4, the ampere turns for the circuit are

$$Ni = \frac{\mathcal{B}_c l_c}{\mu_r \mu_0} + \frac{\mathcal{B}_g g}{\mu_0}$$

Since $\phi = \mathcal{B}_c A_c = \mathcal{B}_g A_g$, the current is

$$i = \frac{\mathcal{B}_c}{\mu_0 N}\left(\frac{l_c}{\mu_r} + g\right)$$

$$= \frac{1}{4\pi \times 10^{-7} \times 500}(0.6 + 5.0)10^{-4}$$

$$= 0.89 \text{ amp}$$

Note that the reluctance of the iron path of 30 cm is only $0.6/5.0 = 0.12$ of the reluctance of the 0.050-cm air gap.

b. Equation 1-6 gives

$$\phi = \mathfrak{B}_c A_c = 1 \times 9 \times 10^{-4} = 9 \times 10^{-4} \text{ weber}$$

$$\lambda = N\phi = 500 \times 9 \times 10^{-4} = 0.45 \text{ weber-turn}$$

In detailed calculations of magnetic circuits, the tendency of the magnetic-field lines to bulge as they cross the air gap and so reduce the flux density $\mathfrak{B}_g$ relative to $\mathfrak{B}_c$ in the core is handled by a correction for fringing. The usual correction is to add the gap dimension g to each dimension of the gap. For example, if the core dimensions are 3 by 3 cm, the corrected gap area would be $A_g = (3 + 0.05)(3 + 0.05) = 9.3 \text{ cm}^2$, making the gap reluctance about 3 percent less than for the uncorrected gap.

EXAMPLE 1-2

The magnetic circuit of Fig. 1-3 has two parallel paths which link the winding. Find the flux and flux density in each of the legs of the magnetic circuit. Neglect fringing at the air gaps and any leakage fields. Assume that the relative permeability of the iron is so high that the ampere-turns of the winding are all consumed in the air gaps.

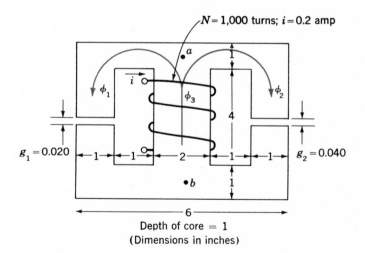

Depth of core $= 1$
(Dimensions in inches)

Fig. 1-3. Parallel-path magnetic circuit.

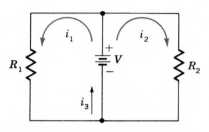

Fig. 1-4. Electric-circuit analog of magnetic circuit of Fig. 1-3.

Solution

The electric-circuit analog of the magnetic circuit is shown in Fig. 1-4. The battery voltage represents the winding mmf; the resistances R_1 and R_2 represent the reluctances $\mathcal{R}_1$ and $\mathcal{R}_2$ of the air gaps; the currents i_1 and i_2 represent the fluxes ϕ_1 and ϕ_2.

The area of the air gaps is

$$A_{g1} = A_{g2} = 1 \text{ in.}^2 \times \frac{1 \text{ m}^2}{(39.4 \text{ in.})^2} = 6.45 \times 10^{-4} \text{ m}^2$$

The lengths of the air gaps are

$$g_1 = 0.020 \text{ in.} \times \frac{1 \text{ m}}{39.4 \text{ in.}} = 5.08 \times 10^{-4} \text{ m}$$

$$g_2 = 0.040 \text{ in.} \times \frac{1 \text{ m}}{39.4 \text{ in.}} = 10.2 \times 10^{-4} \text{ m}$$

The reluctances of the air gaps are given by Eq. 1-9 as

$$\mathcal{R}_1 = \frac{g_1}{\mu_0 A_{g1}} = \frac{5.08 \times 10^{-4}}{(4\pi \times 10^{-7})(6.45 \times 10^{-4})} = 6.26 \times 10^5 \text{ amp-turns/weber}$$

$$\mathcal{R}_2 = \frac{10.2 \times 10^{-4}}{(4\pi \times 10^{-7})(6.45 \times 10^{-4})} = 12.5 \times 10^5 \text{ amp-turns/weber}$$

The fluxes in the legs are found from Eq. 1-10 as

$$\phi_1 = \frac{\mathcal{F}}{\mathcal{R}_1} = \frac{1000 \times 0.2}{6.26 \times 10^5} = 3.2 \times 10^{-4} \text{ weber}$$

$$\phi_2 = \frac{1000 \times 0.2}{12.5 \times 10^5} = 1.6 \times 10^{-4} \text{ weber}$$

$$\phi_3 = \phi_1 + \phi_2 = 4.8 \times 10^{-4} \text{ weber}$$

The areas of the air gap are the same as the legs,

$$A_{c1} = A_{c2} = 6.45 \times 10^{-4} \text{ m}^2$$

The flux densities are found from Eq. 1-6 as

$$\mathcal{B}_1 = \frac{\phi_1}{A_{c1}} = \frac{3.2 \times 10^{-4}}{6.45 \times 10^{-4}} = 0.495 \text{ weber/m}^2$$

$$\mathcal{B}_2 = \frac{\phi_2}{A_{c2}} = \frac{1.6 \times 10^{-4}}{6.45 \times 10^{-4}} = 0.248 \text{ weber/m}^2$$

$$\mathcal{B}_3 = \frac{\phi_3}{A_{c3}} = \frac{4.8 \times 10^{-4}}{12.9 \times 10^{-4}} = 0.372 \text{ weber/m}^2$$

1-2
AC OPERATION

When the magnetic field is allowed to vary with time, an electric field $\mathcal{E}$ is produced in space in accordance with Faraday's law

$$\oint \mathcal{E} \cdot dl = -\frac{d}{dt} \int_s \mathcal{B} \cdot da \qquad (1\text{-}12)$$

where the line integral is taken around the surface through which $\mathcal{B}$ passes. The units of $\mathcal{E}$ are volts per meter. In magnetic structures with windings, such as Fig. 1-2, the varying magnetic field in the core produces an emf e at the terminals, of value

$$e = N\frac{d\varphi}{dt} = \frac{d\lambda}{dt} \qquad (1\text{-}13)$$

where $\lambda = N\varphi$, termed *flux linkage*, in units of weber-turns.   The symbol φ is used to indicate the instantaneous value of a time-varying flux.

For a magnetic circuit which has a linear relationship between $\mathfrak{B}$ and $\mathfrak{K}$, because of material of constant permeability, or a dominating air gap, we can define the λ-i relationship by the *inductance L* as

$$L = \frac{\lambda}{i} \qquad\qquad (1\text{-}14)$$

The inductance can also be expressed in field quantities as

$$L = \frac{N^2 \mathfrak{B} A}{\mathfrak{K} l} = N^2 \mu \frac{A}{l} = N^2 \mathcal{P} \qquad\qquad (1\text{-}15)$$

which shows the geometrical form.   The inductance is measured in henrys or weber-turns per ampere.   Substituting Eq. 1-14 in 1-13 yields

$$e = \frac{d}{dt}(Li) \qquad\qquad (1\text{-}16)$$

For static magnetic circuits, the inductance is fixed and the equation reduces to the well-known circuit form of $L\, di/dt$.   However, in machines, the inductance may be time varying and the equation must be expressed as

$$e = L\frac{di}{dt} + i\frac{dL}{dt} \qquad\qquad (1\text{-}17)$$

The power at the terminals of the winding on a magnetic circuit is a measure of the *rate of energy flow* into the circuit through that particular winding.   The power at the terminals in Fig. 1-1 is

$$p = ie = i\frac{d\lambda}{dt} \qquad\qquad (1\text{-}18)$$

the units are watts, or joules per sec.   The change in energy in the magnetic circuit in the time interval t_1 to t_2 is

$$W_{\text{fld}} = \int_{t_1}^{t_2} p\, dt = \int_{\lambda_1}^{\lambda_2} i\, d\lambda \qquad\qquad (1\text{-}19)$$

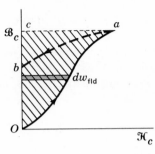

Fig. 1-5. $\mathcal{B}_c = \mathcal{H}_c$ character-istic and magnetic-field energy.

Equation 1-19 in field terms is

$$W_{\text{fld}} = \int_{\mathcal{B}_1}^{\mathcal{B}_2} \left(\frac{\mathcal{H}_c l_c}{N}\right) (A_c N) \, d\mathcal{B}_c = A_c l_c \int_{\mathcal{B}_1}^{\mathcal{B}_2} \mathcal{H}_c \, d\mathcal{B}_c \qquad (1\text{-}20)$$

The term $A_c l_c$ is recognized as the volume of the core; hence, $\mathcal{H}_c \, d\mathcal{B}_c$ is the magnetic-energy density in the core. If the magnetic circuit contains ferromagnetic material, the relation will be more or less nonlinear, as in the rising curve Oa of Fig. 1-5, and the integrations in Eq. 1-20 must then be performed graphically. Because of hysteresis and eddy currents, the relation between $\mathcal{H}_c$ and $\mathcal{B}_c$ is not single valued.

The falling curve is indicated by the broken line ab in Fig. 1-5. When the $\mathcal{H}_c$ is reduced to zero, only a part of the energy that was absorbed by the field during the build-up process is returned to the circuit, the energy returned being given by area abc. Some energy remains stored in the

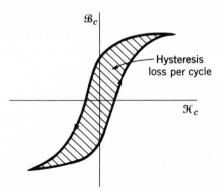

Fig. 1-6. Symmetrical hysteresis loop.

kinetic energy associated with the spinning electrons producing the residual flux, and some has been dissipated in the core losses caused by hysteresis and eddy currents.   The net energy absorbed during the build-up and build-down process Oab is the area $OabO$ in Fig. 1-5.   If the time rate of change is sufficiently slow, so that the effects of eddy currents can be neglected, the flux-mmf characteristics Oa and ab become the rising and falling magnetization curves.   Under cyclic conditions, the hysteresis loss can be found from the area of a closed hysteresis loop, as shown in Fig. 1-6.

EXAMPLE 1-3

For the magnetic circuit of Example 1-1 and Fig. 1-2, find the following: (*a*) Emf e for $\mathcal{B}_c = 1 \sin 377t$ webers/m². (*b*) Reluctances $\mathcal{R}_c$ and $\mathcal{R}_g$. (*c*) Inductance L. (*d*) Energy at $\mathcal{B}_c = 1$ weber/m².

Solution

a. The value of λ was previously found as 0.45 weber-turn for $\mathcal{B}_c = 1$ weber/m². Hence, for the sinusoidal variation of $\mathcal{B}_c$, the flux linkage is

$$\lambda = 0.45 \sin 377t \text{ weber-turns}$$

The emf is given by Eq. 1-13 as

$$e = \frac{d\lambda}{dt} = 170 \cos 377t \text{ volts}$$

b. The reluctances can be determined from Eq. 1-9 as

$$\mathcal{R}_c = \frac{l_c}{\mu_r \mu_0 A_c} = \frac{0.3}{5000 \times 4\pi \times 10^{-7} \times 9 \times 10^{-4}} = 5.3 \times 10^4 \\ \text{amp-turns/weber}$$

$$\mathcal{R}_g = \frac{g}{\mu_0 A_g} = \frac{5 \times 10^{-4}}{4\pi \times 10^{-7} \times 9 \times 10^{-4}} = 44.2 \times 10^4 \\ \text{amp-turns/weber}$$

c. The inductance from Eq. 1-14 is

$$L = \frac{\lambda}{i} = \frac{0.45}{0.89} = 0.51 \text{ henry}$$

It can be determined from Eq. 1-15 as follows:

$$L = N^2 \mathcal{P} = N^2 \frac{1}{\mathcal{R}} = \frac{N^2}{\mathcal{R}_c + \mathcal{R}_g}$$

$$= \frac{(500)^2}{(5.3 + 44.2)10^4} = 0.51 \text{ henry}$$

d. The energy from Eq. 1-19 is

$$W_{\text{fld}} = \int_0^\lambda i \, d\lambda = \int_0^\lambda \frac{\lambda}{L} \, d\lambda = \frac{1}{2} \frac{\lambda^2}{L}$$

$$= \frac{(0.45)^2}{2 \times 0.51} = 0.20 \text{ joule}$$

1-3

PROPERTIES OF MATERIALS

Magnetic materials are used in machines and transformers in a variety of
sizes and shapes ranging from stampings of thin silicon-steel sheet to
solid members of iron for synchronous-alternator rotors and dc machine
field poles. All ferromagnetic material used for machines is characterized by a high relative permeability and both a nonlinear and multivalued
relationship between $\mathcal{B}$ and $\mathcal{3C}$. The characteristics of the material cannot be described by a few numerical constants but must be described by
sets of curves relating pertinent variables with other variables such as
thickness and frequency as parameters.

The basic measure of the magnetic properties is the $\mathcal{B}$-$\mathcal{3C}$ or hysteresis
loop. The loop shows the instantaneous relationship between the flux
density $\mathcal{B}$ and the magnetic intensity $\mathcal{3C}$ over a complete cycle of operation.
For each value of maximum flux density, at the tips of the loop, the mate-
rial has a different $\mathcal{B}$-$\mathcal{3C}$ loop. A loop for a typical steel used in equipment, M-19 grade, fully processed, is shown in Fig. 1-7 for maximum flux
densities of 1.0 and 1.5 webers/m². The loop shows that the relationship
between $\mathcal{B}$ and $\mathcal{3C}$ is nonlinear and multivalued. The $\mathcal{B}$-$\mathcal{3C}$ loop demon-
strates the physics of operation but is of little engineering use. The
pertinent information of the loop is the relationship between the maximum
values of $\mathcal{B}$ and $\mathcal{3C}$ at the tip of the loop. This information is presented as
a dc or *normal magnetization curve*, which passes through the tips of a

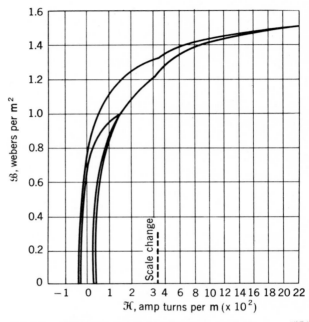

Fig. 1-7. $\mathcal{B}$-$\mathcal{H}$ loops for M-19 fully processed, 29-gage steel for $\mathcal{B}_{\max} = 1.0$ webers/m² and 1.5 webers/m².

succession of $\mathcal{B}$-$\mathcal{H}$ loops covering the range of flux density.   The dc magnetization curve for M-19, 29-gage, fully processed steel is shown in Fig. 1-8.   The presentation of ac magnetization data will be discussed in a later paragraph.   The dc magnetization curve can be used for more accurate calculations of magnetic circuits than shown in Examples 1-1 and 1-2.

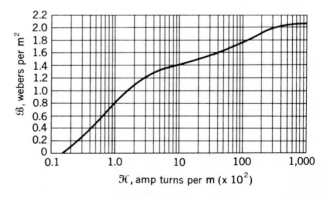

Fig. 1-8.   DC magnetization curve for M-19 fully processed, 29-gage steel.

EXAMPLE 1-4

Assume that the material in Fig. 1-2 has the dc magnetization curve of Fig. 1-8.   Find the current i for $\mathcal{B}_c = 1$ weber/m².

Solution

The value of $\mathcal{H}_c$ for $\mathcal{B}_c = 1$ weber/m² is read from Fig. 1-8 as

$$\mathcal{H}_c = 1.6 \times 10^2 \text{ amp-turns/m}$$

The mmf for the core path is

$$\mathcal{F}_c = \mathcal{H}_c l_c = 1.6 \times 10^2 \times 0.3 = 48 \text{ amp-turns}$$

The mmf for the air gap is

$$\mathcal{F}_g = \mathcal{H}_g g = \frac{\mathcal{B}_g g}{\mu_0} = \frac{5 \times 10^{-4}}{4\pi \times 10^{-7}} = 396 \text{ amp-turns}$$

The current is

$$i = \frac{\mathcal{F}_c + \mathcal{F}_g}{N} = \frac{444}{500} = 0.89 \text{ amp}$$

Note that the relative permeability agrees with the value assumed in Example 1-1,

$$\mu_r = \frac{\mathcal{B}_c}{\mu_0 \mathcal{H}_c} = \frac{1}{4\pi \times 10^{-7} \times 1.6 \times 10^2} = 5{,}000$$

EXAMPLE 1-5

The magnetic-circuit problem of Example 1-2 was solved under the assumption that the mmf of the iron parts of the magnetic paths could be neglected.   Assume that the core material is described by the characteristic of Fig. 1-8.   Find the flux densities in the three legs of the circuit.

Solution

The problem can be solved either graphically or by successive approximation because the magnetization curve is not described by an analytic function.   Let us use successive approximation with the solution of Example 1-2 as the first approximation.

The three magnetic-path lengths between points a and b are

$$l_1 = 10 \text{ in.} = 0.254 \text{ m}$$

$$l_2 = 10 \text{ in.} = 0.254 \text{ m}$$

$$l_3 = 5 \text{ in.} = 0.127 \text{ m}$$

The values of $\mathcal{H}_c$ from Fig. 1-8 for the first approximation are

for $\mathcal{B}_1 = 0.495 \text{ weber/m}^2$ $\mathcal{H}_1 = 0.53 \times 10^2 \text{ amp-turns/m}$

$\mathcal{B}_2 = 0.248$ $\mathcal{H}_2 = 0.30 \times 10^2$

$\mathcal{B}_3 = 0.372$ $\mathcal{H}_3 = 0.40 \times 10^2$

The mmfs for the paths are

$$\mathcal{F}_1 = \mathcal{H}_1 l_1 = 0.254 \times 0.53 \times 10^2 = 13.5 \text{ amp-turns}$$

$$\mathcal{F}_2 = \mathcal{H}_2 l_2 = 0.254 \times 0.30 \times 10^2 = 7.6$$

$$\mathcal{F}_3 = \mathcal{H}_3 l_3 = 0.127 \times 0.40 \times 10^2 = 5.1$$

For the second approximation of the flux densities, we will reduce the winding ampere-turns by the mmfs of the core calculated above and apply the resultant to the gaps,

$$\mathcal{B}_1 = \frac{\mu_0}{g_1} (Ni - \mathcal{F}_1 - \mathcal{F}_3)$$

$$= \frac{4\pi \times 10^{-7}}{5.08 \times 10^{-4}} (200 - 18.6) = 0.45 \text{ weber/m}^2$$

$$\mathcal{B}_2 = \frac{\mu_0}{g_2} (Ni - \mathcal{F}_2 - \mathcal{F}_3)$$

$$= \frac{4\pi \times 10^{-7}}{10.2 \times 10^{-4}} (200 - 12.7) = 0.233 \text{ weber/m}^2$$

The flux ϕ_3 in the center leg is given by

$$\phi_3 = \mathcal{B}_1 A_1 + \mathcal{B}_2 A_2$$

and $\mathcal{B}_3 = \dfrac{\phi_3}{A_3} = \dfrac{\mathcal{B}_1 A_1 + \mathcal{B}_2 A_2}{A_3} = \dfrac{\mathcal{B}_1 + \mathcal{B}_2}{2} = 0.342 \text{ weber/m}^2$

The new values of $\mathcal{3C}$ are read from Fig. 1-8 as

$$\mathcal{3C}_1 = 0.48 \times 10^2 \text{ amp-turns/m}$$

$$\mathcal{3C}_2 = 0.30 \times 10^2$$

$$\mathcal{3C}_3 = 0.38 \times 10^2$$

When this second set of $\mathcal{3C}$ values is used to calculate a third-round approximation to the $\mathcal{B}$ values, the results are

$$\mathcal{B}_1 = 0.45 \frac{200 - 17.0}{200 - 18.6} = 0.454 \text{ weber/m}^2$$

$$\mathcal{B}_2 = 0.233 \frac{200 - 12.4}{200 - 12.7} = 0.234 \text{ weber/m}^2$$

$$\mathcal{B}_3 = \frac{\mathcal{B}_1 + \mathcal{B}_2}{2} = 0.344 \text{ weber/m}^2$$

The values of $\mathcal{B}$ converge rapidly to the final values. Even so, the values above are within 10 percent of the values calculated in Example 1-2 with the mmf of the core material neglected.

As shown in Art. 1-2, magnetic material undergoes a loss of energy each time the material is subjected to one cycle of its $\mathcal{B}$-$\mathcal{3C}$ loop. The loss is separated into eddy-current loss and hysteresis loss; the first component is caused by the I^2R loss of currents that circulate in the material tending to oppose the change of flux density; the second component is caused by the energy expended in orienting the magnetic domains of the material into line with the direction of the field. The losses depend upon the metallurgy of the material, particularly the percent silicon; the frequency; the thickness of the material in a plane normal to the field; and the maximum flux density. In electrical equipment, the core loss must be dissipated as heat and also contributes to the total losses of the equipment itself. Information on core loss is presented as shown in Fig. 1-9 as watts per pound vs. flux density at a specific frequency. It is further supplemented by curves of core loss vs. frequency with the flux density as a parameter.

The *ac* excitation characteristics of magnetic material are usually employed in terms of the *volt-amperes per pound* P_a of material as a function of flux density rather than a magnetization curve relating $\mathcal{B}$ and $\mathcal{3C}$.

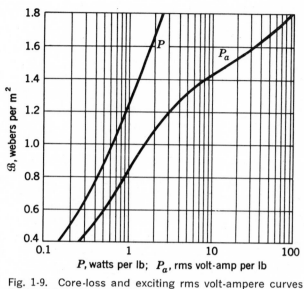

Fig. 1-9. Core-loss and exciting rms volt-ampere curves for M-19 fully processed, 29-gage steel at 60 Hz.

Particularly in transformers and air-gap reactors, and to a lesser extent in rotating machinery, the magnetic material throughout the structure operates at the same flux density; the excitation volt-amperes for the material is then found as the product of P_a and the weight.   The basis for P_a can be seen by applying a voltage to the winding of the magnetic circuit of Fig. 1-1

$$v = \sqrt{2}\, V \sin 2\pi ft \qquad\qquad (1\text{-}21)$$

where V is the rms voltage.   The flux density in the core is found from Eq. 1-13 as

$$\mathcal{B}_c = \frac{1}{N A_c} \int v\, dt$$

$$\mathcal{B}_c = \frac{-\sqrt{2}\, V}{N A_c 2\pi f} \cos 2\pi ft$$

$$\mathcal{B}_c = -\mathcal{B}_{max} \cos 2\pi ft \qquad\qquad (1\text{-}22)$$

where

$$\mathcal{B}_{max} = \frac{\sqrt{2}\, V}{N A_c 2\pi f} = \frac{V}{4.44 N A_c f} \qquad\qquad (1\text{-}23)$$

The rms voltage V is then related to the peak flux density $\mathcal{B}_{max}$

$$V = 4.44\mathcal{B}_{max}A_cNf \qquad (1\text{-}24)$$

As we shall see later in this chapter, the waveform of $\mathcal{H}_c$ is highly non-sinusoidal when $\mathcal{B}_c$ is sinusoidal. We can define an rms value $\mathcal{H}_{rms}$ of $\mathcal{H}_c$ and an rms value of current I such that

$$I = \frac{\mathcal{H}_{rms}l_c}{N} \qquad (1\text{-}25)$$

The product VI is the rms volt-ampere input to the winding of Fig. 1-1 to provide the excitation of the magnetic material to the peak flux density $\mathcal{B}_{max}$

$$VI = 4.44f(\mathcal{B}_{max}\mathcal{H}_{rms})(A_cl_c) \qquad (1\text{-}26)$$

For a magnetic material of density ρ_c, the weight is $A_cl_c\rho_c$ and the volt-amperes per pound is given by

$$P_a = \frac{VI}{A_cl_c\rho_c} = \frac{4.44f}{\rho_c}(\mathcal{B}_{max}\mathcal{H}_{rms}) \qquad (1\text{-}27)$$

The characteristic P_a at a given frequency f is dependent only on $\mathcal{B}_{max}$ since $\mathcal{H}_{rms}$ is a function of $\mathcal{B}_{max}$, and is independent of turns and geometry. The characteristic for M-19 material is shown in Fig. 1-9.

Nearly all transformers and certain sections of electric machines use sheet-steel material that has highly favorable directions of magnetization along which the core loss is low and the permeability is high. The material is termed *grain-oriented steel*. The reason for this property lies in the atomic structure of the simple crystal of the silicon-iron alloy, which is a body-centered cube; each cube has an atom at each corner as well as one in the center of the cube. In the cube, the easiest axis of magnetization is the cube edge; the diagonal across the cube face is more difficult, and the diagonal through the cube is the most difficult. By suitable manufacturing technique, the majority of the cube edges are aligned in the rolling direction to make it the favorable direction of magnetization. The behavior in this direction is superior in core loss and required $\mathcal{H}$ to nonoriented steels, so that the oriented steels can be operated at higher flux densities than the nonoriented grades. A set of magnetization curves at various angles from the favorable direction are shown in Fig. 1-10 to show the marked effect.

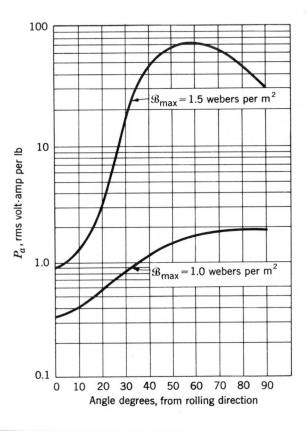

P_a, rms volt-amp per lb

$\mathcal{B}_{max} = 1.5$ webers per m^2

$\mathcal{B}_{max} = 1.0$ webers per m^2

Angle degrees, from rolling direction

Fig. 1-10. Exciting rms volt-ampere curves for M-6, 0.014-in., grain-oriented steel at 60 Hz as a function of the angle to the rolling direction. $\mathcal{B}_{max} = 1.0$ and 1.5 webers/m^2.

EXAMPLE 1-6

The magnetic core in Fig. 1-11 is made from laminations of M-19, 29-gage, electrical steel. The winding is excited with a voltage to produce a flux density in the steel of $\mathcal{B} = 1.5 \sin 377t$ webers/m^2. The steel occupies 0.94 of the gross core volume. The density of the steel is 7.65 gm/cm^3. Find: (a) the applied voltage; (b) the peak current; (c) the rms current; (d) the core loss; (e) the rms current if the material is replaced with M-6, 0.014-in. (29-gage) material oriented in the favorable direction with the field.

Solution

a. From Eq. 1-13, the voltage is

$$e = N \frac{d\varphi}{dt} = NA_c \frac{d\mathcal{B}}{dt} = 200 \times 4 \text{ in.}^2 \times 0.94 \times \frac{1 \text{ m}^2}{(39.4)^2 \text{ in.}^2}$$

$$\times 1.5 \times 377 \cos 377t = 275 \cos 377t \text{ volts}$$

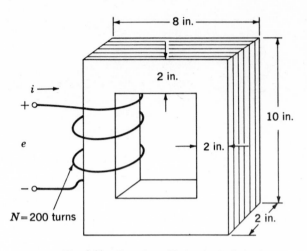

Fig. 1-11.   Reactor with laminated steel core.

b. The magnetic intensity corresponding to $\mathcal{B}_{max}$ = 1.5 webers/m² is given in Fig. 1-8 as $\mathcal{H}$ = 22 × 10² amp-turns/m

$$l_c = (6 + 6 + 8 + 8) \text{ in.} \times \frac{1 \text{ m}}{39.4 \text{ in.}} = 0.71 \text{ m}$$

The peak current is

$$I = \frac{22 \times 10^2 \times 0.71}{200} = 7.8 \text{ amp}$$

c. The rms current is obtained from the value of P_a of Fig. 1-9 for $\mathcal{B}_{max}$ = 1.5 webers/m²

$$P_a = 18 \text{ volt-amp/lb}$$

The core volume and weight are

$$V_c = 4 \text{ in.}^2 \times 0.94 \times 28 \text{ in.} = 105.5 \text{ in.}^3$$

$$W_c = 105.5 \text{ in.}^3 \times \frac{7.65 \text{ gm}}{1 \text{ cm}^3} \times \frac{(2.54)^3 \text{ cm}^3}{1 \text{ in.}^3} \times \frac{2.2 \text{ lb}}{1,000 \text{ gm}} = 29.1 \text{ lb}$$

The total volt-amperes and current are

$$P_a = 18 \frac{\text{va}}{\text{lb}} \times 29.1 \text{ lb} = 525 \text{ va}$$

$$I = \frac{P_a}{V} = \frac{525}{275 \times 0.707} = 2.7 \text{ amp}$$

d. The core-loss density is obtained from Fig. 1-9 as $P = 1.6$ watts/lb.   The total core loss is

$$P = 1.6 \times 29.1 \text{ lb} = 46.5 \text{ watts}$$

e. For the M-6 grain-oriented material, the volt-amperes per pound at 0°, 1.5 webers/m² is given by Fig. 1-10 as

$$P_a = 0.86 \text{ volt-amp/lb}$$

The rms current is

$$I = \frac{0.86 \times 29.1 \text{ lb}}{275 \times 0.707} = 0.128 \text{ amp}$$

The M-6 material could be operated at over 1.8 webers/m² to reach the exciting current of 2.7 amp obtained with the M-19 material at 1.5 webers/m², an increase of 20 percent.

1-4

MAGNETICALLY COUPLED CIRCUITS

Before proceeding with a study of electric machinery, it is desirable to discuss certain aspects of the theory of magnetically coupled circuits, with emphasis on transformer action.   Although the static transformer is not an energy-conversion device, it is an indispensable component in many energy-conversion systems.   It is one of the principal reasons for the widespread use of ac power systems, for it makes possible electric generation at the most economical generator voltage, power transfer at the most economical transmission voltage, and power utilization at the most suitable voltage for the particular utilization device.   The transformer is also widely used in low-power low-current electronic and control

circuits for performing such functions as matching the impedances of a
source and its load for maximum power transfer, insulating one circuit
from another, or isolating direct current while maintaining ac continuity
between two circuits.

Moreover, the transformer is one of the simpler devices comprising
two or more electric circuits coupled by a common magnetic circuit, and
its analysis involves many of the principles essential to the study of
electric machinery.

Essentially, a transformer consists of 2 or more windings interlinked
by a mutual magnetic field.   If one of these windings, the *primary*, is connected to an alternating-voltage source, an alternating flux will be produced whose amplitude will depend on the primary voltage and number
of turns.   The mutual flux will link the other winding, the *secondary*, and
will induce a voltage in it whose value will depend on the number of
secondary turns.   By properly proportioning the numbers of primary
and secondary turns, almost any desired voltage ratio, or *ratio of transformation*, can be obtained.

Transformer action evidently demands only the existence of alternating mutual flux linking the 2 windings and is simply utilization of the
mutual-inductance concept.   Such action will be obtained if an air core
is used, but it will be obtained much more effectively with a core of iron or

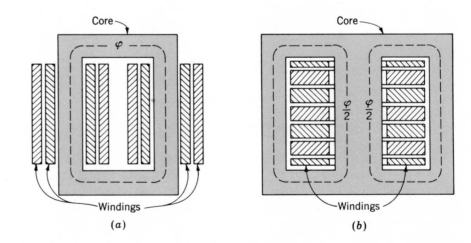

Fig. 1-12.   *(a)* Core-type and *(b)* shell-type transformer.

other ferromagnetic material, because most of the flux is then confined to a definite path linking both windings and having a much higher permeability than that of air.   Such a transformer is commonly called an *iron-core transformer*.   The majority of transformers are of this type.   The following discussion will be concerned almost wholly with iron-core transformers.

In order to reduce the losses caused by eddy currents in the core, the magnetic circuit usually consists of a stack of thin laminations, two common types of construction being shown in Fig. 1-12.   In the *core type* (Fig. 1-12a) the windings are wound around two legs of a rectangular magnetic core, while in the *shell type* (Fig. 1-12b) the windings are wound around the center leg of a three-legged core.   Silicon-steel laminations

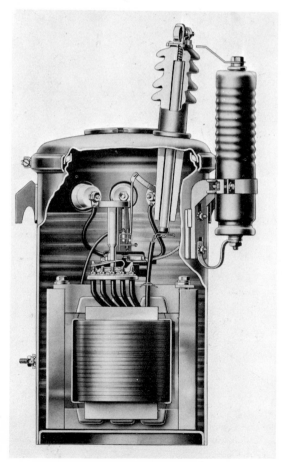

Fig. 1-13.   Cutaway view of self-protected transformer typical of sizes 2 to 25 kva, 7,200:240/120 volts.   Only one high-voltage insulator and lightning arrester is needed because one side of the 7,200-volt line and one side of the primary are grounded.   *(General Electric Company.)*

0.014 in. thick are generally used for transformers operating at frequencies below a few hundred hertz. Silicon steel has the desirable properties of low cost, low core loss, and high permeability at high flux densities (65 to 90 kilolines/in.²). The cores of small transformers used in communication circuits at high frequencies and low energy levels are sometimes made of compressed powdered ferromagnetic alloys such as permalloy.

Most of the flux is confined to the core and therefore links both windings. Although leakage flux which links one winding without linking the other is a small fraction of the total flux, it has an important effect on the behavior of the transformer. Leakage is reduced by subdividing the windings into sections placed as close together as possible. In the core-type construction, each winding consists of two sections, one section on

Fig. 1-14. A 660-Mva 3-phase 50-Hz transformer used to step up generator voltage of 20 kv to transmission voltage of 405 kv. (*CEM Le Havre, French Member of the Brown Boveri Corporation.*)

each of the two legs of the core, the primary and secondary windings being concentric coils. In the shell-type construction, variations of the concentric-winding arrangement may be used, or the windings may consist of a number of thin "pancake" coils assembled in a stack with primary and secondary coils interleaved.

Figure 1-13 illustrates the internal construction of a *distribution transformer* such as is used in public-utility systems to provide the appropriate voltage at the consumer's premises. A large power transformer is shown in Fig. 1-14.

1-5
NO-LOAD CONDITIONS

Consider the transformer shown in Fig. 1-15 with its secondary circuit open and an alternating voltage v_1 applied to its primary terminals. A small steady-state current i_φ, called the *exciting current*, exists in the primary and establishes an alternating flux in the magnetic circuit. This flux induces an emf in the primary equal to

$$e_1 = \frac{d\lambda_1}{dt} = N_1 \frac{d\varphi}{dt} \tag{1-28}$$

where λ_1 is the flux linkage with the primary, φ the flux (here assumed all confined to the core), and N_1 the number of turns in the primary winding. The voltage e_1 is in volts when φ is in webers. Lenz' law shows it to be a counter emf having the polarity relative to v_1 shown by the $+$ and $-$ signs in Fig. 1-15. This counter emf together with the drop in the pri-

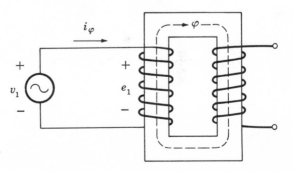

Fig. 1-15. Transformer with open secondary.

mary resistance r_1 must balance the applied voltage v_1; thus

$$v_1 = r_1 i_\varphi + e_1 \qquad (1\text{-}29)$$

In most power apparatus the no-load resistance drop is very small indeed, and the induced emf e_1 very nearly equals the applied voltage v_1. Furthermore, the waveforms of voltage and flux are very nearly sinusoidal. The analysis can then be greatly simplified. Thus, if the instantaneous flux is

$$\varphi = \phi_{max} \sin \omega t \qquad (1\text{-}30)$$

the induced voltage is

$$e_1 = N_1 \frac{d\varphi}{dt} = \omega N_1 \phi_{max} \cos \omega t \qquad (1\text{-}31)$$

where ϕ_{max} is the maximum value of the flux and $\omega = 2\pi f$, the frequency being f Hz. For the positive directions shown in Fig. 1-15, the induced emf leads the flux by 90°. The rms value of the induced emf is

$$E_1 = \frac{2\pi}{\sqrt{2}} f N_1 \phi_{max} = 4.44 f N_1 \phi_{max} \qquad (1\text{-}32)$$

If the resistance drop is negligible, the counter emf equals the applied voltage. Under these conditions, if a sinusoidal voltage is applied to a winding, a sinusoidally varying core flux must be established whose maximum value ϕ_{max} satisfies the requirement that E_1 in Eq. 1-32 equals the rms value V_1 of the applied voltage, thus

$$\phi_{max} = \frac{V_1}{4.44 f N_1} \qquad (1\text{-}33)$$

The flux is determined solely by the applied voltage, its frequency, and the number of turns in the winding. This important relation applies not only to transformers but also to any device operated with sinusoidal alternating impressed voltage, so long as the resistance drop is negligible. The magnetic properties of the core determine the exciting current. It must adjust itself so as to produce the mmf required to create the flux demanded by Eq. 1-33.

Because of the nonlinear magnetic properties of iron, the waveform of the exciting current differs from the waveform of the flux. A curve of the exciting current as a function of time can be found graphically from

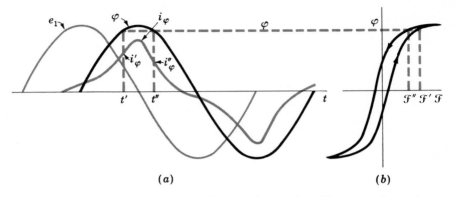

Fig. 1-16. Excitation phenomena. *(a)* Voltage, flux, and exciting-current waveforms and *(b)* corresponding flux-mmf loop.

the magnetic characteristics of the core material in the manner illustrated in Fig. 1-16. Sine waves of voltage e_1 and flux φ in accordance with Eqs. 1-30 to 1-33 are shown in Fig. 1-16a. The corresponding flux-mmf loop for the core is shown in Fig. 1-16b. Values of the mmf corresponding to various values of the flux can be found from this hysteresis loop. For example, at time t' the instantaneous flux is φ' and the flux is increasing; the corresponding value of the mmf is $\mathcal{F}'$ read from the increasing-flux portion of the hysteresis loop. The corresponding value i'_φ of the exciting current is plotted at time t' in Fig. 1-16a. At time t'' the flux also has the instantaneous value φ', but it is decreasing, and the corresponding values of mmf and current are $\mathcal{F}''$ and i''_φ. In this manner the complete curve of exciting current i_φ can be plotted, as shown in Fig. 1-16a.

If the exciting current is analyzed by Fourier-series methods, it will be found to comprise a fundamental and a family of odd harmonics. The fundamental can, in turn, be resolved into two components, one in phase with the counter emf and the other lagging the counter emf by 90°. The fundamental in-phase component accounts for the power absorbed by hysteresis and eddy-current losses in the core. It is called the *core-loss component* of the exciting current. When the core-loss component is subtracted from the total exciting current, the remainder is called the *magnetizing current*. It comprises a fundamental component lagging the counter emf by 90°, together with all the harmonics. The principal harmonic is the third. For typical power transformers, the third harmonic usually is about 40 percent of the exciting current.

Except in problems concerned directly with the effects of harmonics, the peculiarities of the exciting-current waveform usually need not be

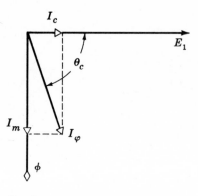

Fig. 1-17. No-load phasor diagram

taken into account, because the exciting current itself is small. For example, the exciting current of a typical power transformer is about 5 percent of full-load current. Consequently the effects of the harmonics usually are swamped out by the sinusoidal-current requirements of other linear elements in the circuit. The exciting current may then be represented by its *equivalent sine wave*, which has the same effective value and frequency, and produces the same average power as the actual wave. Such representation is essential to the construction of a phasor diagram. In Fig. 1-17, the phasors E_1 and ϕ, respectively, represent the induced emf and the flux. The phasor I_φ represents the equivalent sinusoidal exciting current. It lags the induced emf E_1 by a phase angle θ_c such that

$$P_c = E_1 I_\varphi \cos \theta_c \qquad (1\text{-}34)$$

where P_c is the core loss. The component I_c in phase with E_1 represents the core-loss current. The component I_m in phase with the flux represents an equivalent sine wave having the same rms value as the magnetizing current. Typical core-loss and exciting volt-ampere characteristics of two grades of high-quality silicon steel used for power- and distribution-transformer laminations are shown in Figs. 1-9 and 1-10.

EXAMPLE 1-7

In Example 1-6 the core loss and exciting volt-amperes for the core of Fig. 1-11 at $\mathcal{B}_{max} = 1.5$ webers/m² and 60 Hz were found to be

$$P_c = 46.5 \text{ watts} \qquad (VI)_{rms} = 525 \text{ va}$$

and the induced voltage was $275/\sqrt{2} = 194$ volts rms when the winding had 200 turns.

Find the power factor, the core-loss current I_c, and the magnetizing current I_m.

Solution

$$\text{Power factor } \cos \theta_c = \frac{46.5}{525} = 0.089$$

$$\theta_c = 84.9° \qquad \sin \theta_c = 0.996$$

$$\text{Exciting current } I_\varphi = \frac{525}{194} = 2.71 \text{ amp rms}$$

$$\text{Core-loss component } I_c = \frac{46.5}{194} = 0.24 \text{ amp rms}$$

Magnetizing component $I_m = I_\varphi \sin \theta_c = 2.71$ amp rms. Because θ_c is nearly 90°, there is negligible difference between the rms values of the exciting current I_φ and its magnetizing component I_m.

1-6
EFFECT OF SECONDARY CURRENT.   IDEAL TRANSFORMER

As a first approximation to a quantitative theory, consider a transformer with a primary winding of N_1 turns and a secondary winding of N_2 turns, as shown schematically in Fig. 1-18. Let the properties of this transformer be idealized in that the winding resistances are negligible, all of the flux is confined to the core and links both windings, core losses are negligible, and the permeability of the core is so high that only a negligible

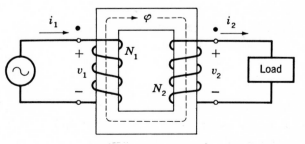

Fig. 1-18.   Ideal transformer and load.

exciting current is required to establish the flux. These properties are closely approached but never actually attained in practical transformers. A hypothetical transformer having these properties is often called an *ideal transformer.*

When a time-varying voltage v_1 is impressed on the primary terminals, a core flux φ must be established such that the counter emf e_1 equals the impressed voltage when winding resistance is negligible. Thus

$$v_1 = e_1 = N_1 \frac{d\varphi}{dt} \qquad (1\text{-}35)$$

The core flux also links the secondary and produces an induced emf e_2 and an equal secondary terminal voltage v_2 given by

$$v_2 = e_2 = N_2 \frac{d\varphi}{dt} \qquad (1\text{-}36)$$

From the ratio of Eqs. 1-35 and 1-36,

$$\frac{v_1}{v_2} = \frac{N_1}{N_2} \qquad (1\text{-}37)$$

Thus an ideal transformer changes voltages in the direct ratio of the turns in its windings.

Now, let a load be connected to the secondary. A current i_2 and an mmf $N_2 i_2$ are then present in the secondary. Unless this secondary mmf is counteracted in the primary, the core flux will be radically changed and the balance between impressed voltage and counter emf in the primary will be disturbed. Hence, a compensating primary mmf and current i_1 must be called into being such that

$$N_1 i_1 = N_2 i_2 \qquad (1\text{-}38)$$

This is the means by which the primary knows of the presence of current in the secondary. Note that for the reference directions shown in Fig. 1-18 the mmfs of i_1 and i_2 are in opposite directions and therefore compensate. The net mmf acting on the core therefore is zero, in accordance with the assumption that the exciting current of an ideal transformer is zero. From Eq. 1-38,

$$\frac{i_1}{i_2} = \frac{N_2}{N_1} \qquad (1\text{-}39)$$

Thus an ideal transformer changes currents in the inverse ratio of the turns in its windings.   Also notice from Eqs. 1-37 and 1-39 that

$$v_1 i_1 = v_2 i_2 \qquad\qquad (1\text{-}40)$$

that is, instantaneous power input equals instantaneous power output, a necessary condition because all causes of active- and reactive-power losses in the transformer have been neglected.

For further study, consider the case of a sinusoidal applied voltage and an impedance load.   Phasor symbolism can then be used.   The circuit is shown in simplified form in Fig. 1-19a, in which the dot-marked terminals of the transformer correspond to the similarly marked terminals in Fig. 1-18.   The dot markings indicate terminals of corresponding polarity, i.e., if one follows through the primary and secondary windings of Fig. 1-18 beginning at their dot-marked terminals, one will find that both windings encircle the core in the same direction with respect to the flux.   Therefore, if one compares the voltages of the two windings, the voltages from a dot-marked to an unmarked terminal will be of the same instantaneous polarity for primary and secondary.   In other words, the voltages V_1 and V_2 in Fig. 1-19a are in phase.   Also, the currents I_1 and

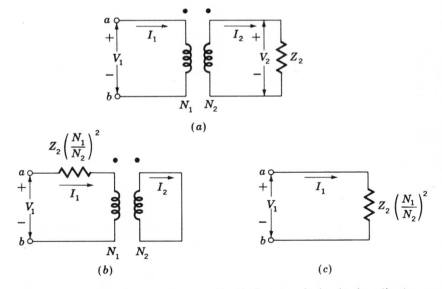

Fig. 1-19.   Three circuits which are identical at terminals ab when the transformer is ideal.

I_2 are in phase. The fact that their mmfs must balance is accounted for by their being in opposite directions through the windings.

In phasor form, Eqs. 1-37 and 1-39 can be expressed as

$$V_1 = \frac{N_1}{N_2} V_2 \quad \text{and} \quad V_2 = \frac{N_2}{N_1} V_1 \qquad (1\text{-}41)$$

$$I_1 = \frac{N_2}{N_1} I_2 \quad \text{and} \quad I_2 = \frac{N_1}{N_2} I_1 \qquad (1\text{-}42)$$

From these equations,

$$\frac{V_1}{I_1} = \left(\frac{N_1}{N_2}\right)^2 \frac{V_2}{I_2} = \left(\frac{N_1}{N_2}\right)^2 Z_2 \qquad (1\text{-}43)$$

where Z_2 is the complex impedance of the load. Consequently, as far as its effect is concerned, an impedance Z_2 in the secondary circuit may be replaced by an equivalent impedance Z_1 in the primary circuit, provided that

$$Z_1 = \left(\frac{N_1}{N_2}\right)^2 Z_2 \qquad (1\text{-}44)$$

Thus, the three circuits of Fig. 1-19 are indistinguishable as far as their performance viewed from terminals ab is concerned. Transferring an impedance from one side of a transformer to the other in this fashion is called *referring the impedance* to the other side. In a similar manner, voltages and currents may be *referred* to one side or the other by using Eqs. 1-41 and 1-42 to evaluate the equivalent voltage and current on that side.

To sum up, *in an ideal transformer voltages are transformed in the direct ratio of turns, currents in the inverse ratio, and impedances in the direct ratio squared; and power and volt-amperes are unchanged.*

1-7

TRANSFORMER REACTANCES AND EQUIVALENT CIRCUITS

The departures in an actual transformer from the ideal properties assumed in Art. 1-6 must be included to a greater or lesser degree in most analyses of transformer performance. A more complete theory must take into account the effects of winding resistances, magnetic leakage, and exciting

current.   Sometimes the capacitances of the windings also have important effects, notably in problems involving transformer behavior at frequencies above the audio range or during rapidly changing transient
conditions such as those encountered in pulse transformers and in power-
system transformers as a result of voltage surges caused by lightning or
switching transients.   The analysis of these high-frequency problems is
beyond the scope of the present treatment, however, and accordingly the
capacitances of the windings are neglected in the following analyses.

Two methods of analysis by which the departures from the ideal can
be taken into account are (1) an equivalent-circuit technique based on
physical reasoning and (2) a mathematical attack based on the classical
theory of magnetically coupled circuits.   Both methods are in everyday
use, and both have very close parallels in the theories of rotating machines.
Because it offers an excellent example of the thought process involved in
translating physical concepts into a quantitative theory, the equivalent-
circuit technique is presented here.

The total flux linking the primary winding may be divided into two
components: the resultant mutual flux, confined essentially to the iron
core and produced by the combined effect of the primary and secondary
currents; and the primary leakage flux, which links only the primary.
These components are identified in the elementary transformer shown in
Fig. 1-20, where for simplicity the primary and secondary windings are
shown on opposite legs of the core.   In an actual transformer with inter-

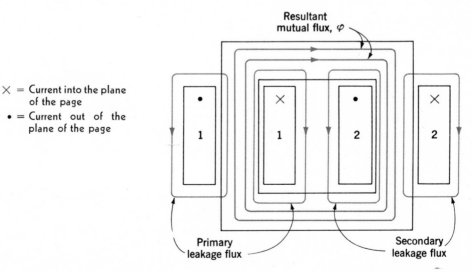

Fig. 1-20.   Component fluxes.

leaved windings, the details of the flux map are more complicated, but the essential features remain the same.   Because the leakage path is largely in air, the leakage flux and the voltage induced by it vary linearly with primary current I_1.   The effect on the primary circuit is the same as that of flux linkages anywhere in the circuit leading up to the transformer primary and may be simulated by assigning to the primary a *leakage inductance* (equal to the leakage-flux linkages with the primary per unit of primary current) or *leakage reactance* x_{l1} (equal to $2\pi f$ times leakage inductance).   In addition there will be a voltage drop in the primary effective resistance r_1.

The impressed voltage V_1 is then opposed by three phasor voltages: the $I_1 r_1$ drop in the primary resistance, the $I_1 x_{l1}$ drop arising from primary leakage flux, and the counter emf E_1 induced in the primary by the resultant mutual flux.   All these voltages are appropriately included in the equivalent circuit of Fig. 1-21a.

The resultant mutual flux links both the primary and secondary windings and is created by their combined mmfs.   It is convenient to treat these mmfs by considering that the primary current must meet two requirements of the magnetic circuit: it must not only (1) counteract the demagnetizing effect of the secondary current but also (2) produce sufficient mmf to create the resultant mutual flux.   According to this physical picture, it is convenient to resolve the primary current into two components, a load component and an exciting component.   The *load component* I_2' is defined as the component current in the primary which would exactly counteract the mmf of the secondary current I_2.   Thus, for opposing currents,

$$I_2' = \frac{N_2}{N_1} I_2 \tag{1-45}$$

It equals the secondary current referred to the primary as in an ideal transformer.   The *exciting component* i_φ is defined as the additional primary current required to produce the resultant mutual flux.   It is a nonsinusoidal current of the nature described in Art. 1-5.

The exciting current can be treated as an equivalent sinusoidal current I_φ, in the manner described in Art. 1-5, and can be resolved into a core-loss component I_c in phase with the counter emf E_1 and a magnetizing component I_m lagging E_1 by $90°$.   In the equivalent circuit (Fig. 1-21b) the equivalent sinusoidal exciting current is accounted for by means of a shunt branch connected across E_1, comprising a noninductive resistance whose conductance is g_c in parallel with a lossless inductance whose susceptance is b_m.   Alternatively, a series combination of resistance and reactance can be connected across E_1.   In the parallel combination (Fig.

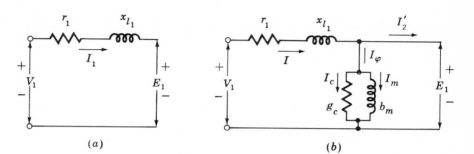

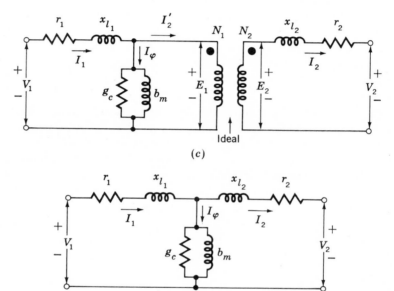

Fig. 1-21.   Steps in development of the transformer equivalent circuit.

1-21b) the power $E_1^2 g_c$ accounts for the core loss due to the resultant mutual flux.   When g_c is assumed constant, the core loss is thereby assumed to vary as E_1^2 or (for sine waves) as $\phi_{max}^2 f^2$, where ϕ_{max} is the maximum value of the resultant mutual flux.   The magnetizing susceptance b_m varies with the saturation of the iron.   When the inductance corresponding to b_m is assumed constant, the magnetizing current is thereby assumed to be independent of frequency and directly proportional to the resultant mutual flux.   Both g_c and b_m are usually determined at rated voltage and frequency; they are then assumed to remain constant for the small departures from rated values associated with normal operation.

The resultant mutual flux φ induces an emf E_2 in the secondary, and since this flux links both windings, the induced-voltage ratio is

$$\frac{E_1}{E_2} = \frac{N_1}{N_2} \tag{1-46}$$

just as in an ideal transformer. This voltage transformation and the current transformation of Eq. 1-45 can be accounted for by introducing an ideal transformer in the equivalent circuit, as in Fig. 1-21c. The emf E_2 is not the secondary terminal voltage, however, because of the secondary resistance and because the secondary current I_2 creates *secondary leakage flux* (see Fig. 1-20). The secondary terminal voltage V_2 differs from the induced voltage E_2 by the voltage drops due to secondary resistance r_2 and *secondary leakage reactance x_{l2}*, as in the portion of the equivalent circuit (Fig. 1-21c) to the right of E_2.

The actual transformer therefore is equivalent to an ideal transformer plus external impedances. By referring all quantities to the primary or secondary, the ideal transformer in Fig. 1-21c may be moved out to the right or left, respectively, of the equivalent circuit. This is almost invariably done, and the equivalent circuit is usually drawn as in Fig. 1-21d with the ideal transformer not shown and all voltages, currents, and impedances referred to the same side. In order to avoid a complicated notation, the same symbols have been used for the *referred* values in Fig. 1-21d as were used for the *actual* values in Fig. 1-21c. In what follows we shall almost always deal with the referred values. One simply keeps in mind the side of the transformer to which all quantities have been referred. The circuit of Fig. 1-21d is often called the *T circuit* for a transformer.

EXAMPLE 1-8

A 50-kva 2,400:240-volt 60-Hz distribution transformer has a leakage impedance of $0.72 + j0.92$ ohm in the high-voltage winding and $0.0070 + j0.0090$ ohm in the low-voltage winding. At rated voltage and frequency the admittance Y_φ of the shunt branch accounting for the exciting current is $(0.324 - j2.24) \times 10^{-2}$ mho when viewed from the low-voltage side.

Draw the equivalent circuit (*a*) referred to the high-voltage side and (*b*) referred to the low-voltage side, and label the impedances numerically.

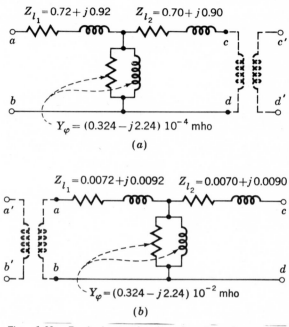

$Z_{l_1} = 0.72 + j0.92$ $Z_{l_2} = 0.70 + j0.90$

$Y_\varphi = (0.324 - j2.24)\ 10^{-4}$ mho

(a)

$Z_{l_1} = 0.0072 + j0.0092$ $Z_{l_2} = 0.0070 + j0.0090$

$Y_\varphi = (0.324 - j2.24)\ 10^{-2}$ mho

(b)

Fig. 1-22. Equivalent circuits for transformer of Example 1-8.

Solution

The circuits are given in Fig. 1-22a and b, respectively, with the high-voltage side numbered 1 and the low-voltage side numbered 2. The voltages given on the nameplate of a power-system transformer are based on the turns ratio and neglect the small leakage-impedance voltage drops under load.   Since this is a 10-to-1 transformer, imped-ances are referred by multiplying or dividing by 100.   The value of an impedance referred to the high-voltage side is greater than its value referred to the low-voltage side.   Since admittance is the reciprocal of impedance, an admittance is referred from one side to the other by use of the reciprocal of the referring factor for impedance. The value of an admittance referred to the high-voltage side is smaller than its value referred to the low-voltage side.

The ideal transformer may be explicitly drawn, as shown dotted in Fig. 1-22, or it may be omitted in the diagram and remembered mentally, making the unprimed letters the terminals.

1-8

ENGINEERING ASPECTS OF TRANSFORMER ANALYSIS

In engineering analyses involving the transformer as a circuit element, it is customary to adopt one of several approximate forms of the equivalent circuit of Fig. 1-21 rather than the full circuit. The approximations chosen in a particular case depend largely on physical reasoning based on orders of magnitude of the neglected quantities. The more common approximations are presented in this article. In addition, test methods are given for determining the transformer constants.

a. Approximate Equivalent Circuits—Power Transformers

The approximate equivalent circuits commonly used for constant-frequency power-transformer analyses are summarized for comparison in Fig. 1-23. All quantities in these circuits are referred to either the primary or the secondary, and the ideal transformer is not shown.

The computational labor involved often can be appreciably reduced by moving the shunt branch representing the exciting current out from the middle of the T circuit to either the primary or the secondary terminals, as in Fig. 1-23a and b. These are *cantilever circuits*. The series branch is the combined resistance and leakage reactance referred to the same side. This impedance is sometimes called the *equivalent impedance*

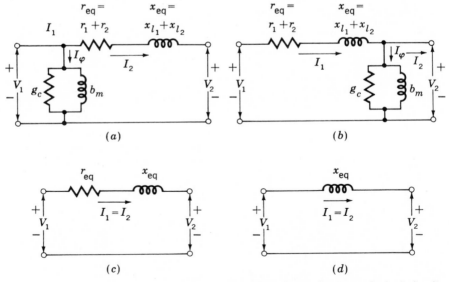

Fig. 1-23. Approximate equivalent circuits.

and its components the *equivalent resistance* r_{eq} and *equivalent reactance* x_{eq}, as shown in Fig. 1-23a and b.   Error is introduced by neglect of the voltage drop in the primary or secondary leakage impedance caused by the exciting current, but this error is insignificant in most problems involving power-system transformers.

Further simplification results from neglecting the exciting current entirely, as in Fig. 1-23c, in which the transformer is represented as an equivalent series impedance.   If the transformer is large (several hundred kva or over), the equivalent resistance r_{eq} is small compared with the equivalent reactance x_{eq} and may frequently be neglected, giving Fig. 1-23d.   The circuits of Fig. 1-23c and d are sufficiently accurate for most ordinary power-system problems.   Finally, in situations where the currents and voltages are determined almost wholly by the circuits external to the transformer or when a high degree of accuracy is not required, the entire transformer impedance may be neglected and the transformer may be considered to be ideal as in Art. 1-6.

The circuits of Fig. 1-23 have the additional advantage that the total equivalent resistance r_{eq} and equivalent reactance x_{eq} can be found from a very simple test, whereas measurement of the values of the component leakage reactances x_{l1} and x_{l2} is a difficult experimental task.

EXAMPLE 1-9

The 50-kva 2,400:240-volt transformer whose constants are given in Example 1-8 is used to step down the voltage at the load end of a feeder whose impedance is $0.30 + j1.60$ ohms.   The voltage V_s at the sending end of the feeder is 2,400 volts.

Find the voltage at the secondary terminals of the transformer when the load connected to its secondary draws rated current from the transformer and the power factor of the load is 0.80 lagging. Neglect the voltage drops in the transformer and feeder caused by the exciting current.

Solution

The circuit with all quantities referred to the high-voltage (primary) side of the transformer is shown in Fig. 1-24a, wherein the transformer is represented by its equivalent impedance, as in Fig. 1-23c.   From Fig. 1-22a, the value of the equivalent impedance is $Z_{eq} = 1.42 + j1.82$ ohms, and the combined impedance of the feeder and transformer in series is $Z = 1.72 + j3.42$ ohms.   From the transformer rating, the load current referred to the high-voltage side is $I = 50{,}000/2{,}400 = 20.8$ amp.

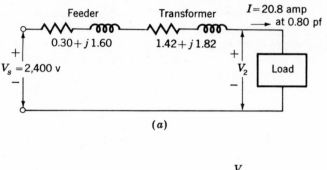

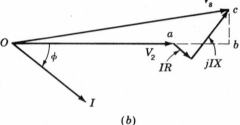

(b)

Fig. 1-24.   Equivalent circuit and phasor diagram, Example 1-9.

The phasor diagram referred to the high-voltage side is shown
in Fig. 1-24b, from which

$$Ob = \sqrt{V_s^2 - (bc)^2}$$

and $V_2 = Ob - ab$

Note that

$$bc = IX \cos \phi - IR \sin \phi$$
$$ab = IR \cos \phi + IX \sin \phi$$

where R and X are the combined resistance and reactance, respectively.   Thus

$$bc = (20.8)(3.42)(0.80) - (20.8)(1.72)(0.60) = 35.5 \text{ volts}$$
$$ab = (20.8)(1.72)(0.80) + (20.8)(3.42)(0.60) = 71.4 \text{ volts}$$

Substitution of numerical values shows that Ob very nearly equals
V_s, or 2,400 volts.   Then $V_2 = 2,329$ volts referred to the high-

voltage side.   The actual voltage at the secondary terminals is 2,329/10, or

$$V_2 = 233 \text{ volts}$$

---

b. Approximate Equivalent Circuits—Variable-frequency Transformers

Small iron-core transformers operating in the audio-frequency range (hence called *audio-frequency transformers*) are often used as coupling devices in electronic circuits for communications, measurements, and control.   Their principal functions are either to step up voltage, thereby contributing to the overall voltage gain in amplifiers, or to act as impedance-transforming devices bringing about the optimum relation between the apparent impedance of a load and the impedance of a source.   They may also serve other auxiliary functions, such as providing a path for direct current through the primary while keeping it out of the secondary circuit.

Application of transformers for impedance matching makes direct use of the impedance-transforming property shown in Eq. 1-44.   Oscillators and amplifiers, for example, give optimum performance when working into a definite order of magnitude of load impedance, and transformer coupling may be used to change the apparent impedance of the actual load to this optimum.   A transformer so used is called an *output transformer*.

When the frequency varies over a wide range, it is important that the output voltage be as closely as possible instantaneously proportional to the input voltage.   Ideally, this means that voltages should be amplified equally and phase shift should be zero for all frequencies.   The *amplitude-frequency characteristic* (often abbreviated to *frequency characteristic*) is a curve of the ratio of the load voltage on the secondary side to the internal source voltage on the primary side plotted as a function of frequency, a flat characteristic being the most desirable.   The *phase characteristic* is a curve of the phase angle of the load voltage relative to the source voltage plotted as a function of frequency, a small phase angle being desirable. These characteristics are dependent not only on the transformer but also on the constants of the entire primary and secondary circuits.

As an example of the use of engineering approximations, consider an amplifier coupled to its load through an output transformer.   The amplifier is considered as equivalent to a source of voltage E_G in series with an internal resistance r_G, and the load is considered as a resistance r_L, as shown in Fig. 1-25a, wherein the transformer is represented by the equivalent circuit of Fig. 1-21c, with core loss neglected.   Sometimes the stray capacitances of the windings must be taken into account at

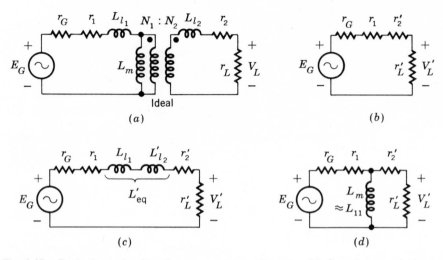

Fig. 1-25. Equivalent circuits of an output transformer. *(a)* Complete equivalent circuit, *(b)* approximate equivalent in the middle range of audio frequencies, *(c)* high-frequency equivalent, *(d)* low-frequency equivalent.

high audio frequencies, especially when the source impedance is higher than a few thousand ohms.

The analysis of a properly designed circuit breaks down into three frequency ranges:

1 At intermediate frequencies (around 500 Hz) none of the inductances is important, and the equivalent circuit reduces to a network of resistances, as shown in Fig. 1-25*b*, wherein all quantities have been referred to the primary, as indicated by the prime superscripts. Analysis of this circuit shows that the ratio of load voltage V_L to source voltage E_G is

$$\frac{V_L}{E_G} = \frac{N_2}{N_1} \frac{r'_L}{R'_{se}} \tag{1-47}$$

where
$$R'_{se} = r_G + r_1 + r'_2 + r'_L \tag{1-48}$$

In this middle range (which usually extends over several octaves) the voltage ratio is very nearly constant; i.e., the amplitude characteristic is flat and the phase shift is zero.

2 As the frequency is increased, however, the leakage reactances of the transformer become increasingly important. The equivalent circuit in the high audio range is shown in Fig. 1-25*c*. Analysis of this circuit shows that at high frequencies

$$\frac{V_L}{E_G} = \frac{N_2}{N_1} \frac{r'_L}{R'_{se}} \frac{1}{\sqrt{1 + (\omega L'_{eq}/R'_{se})^2}} \qquad (1\text{-}49)$$

where L'_{eq} is the equivalent leakage inductance.   The voltage ratio relative to its midrange value is

$$\text{Relative voltage ratio} = \frac{1}{\sqrt{1 + (\omega L'_{eq}/R'_{se})^2}} \qquad (1\text{-}50)$$

The phase angle by which the load voltage lags the source voltage is

$$\phi = \tan^{-1} \frac{\omega L'_{eq}}{R'_{se}} \qquad (1\text{-}51)$$

Curves of the relative voltage ratio and phase angle as functions of the reactance-to-resistance ratio $\omega L'_{eq}/R'_{se}$ are shown in the right-hand half of Fig. 1-26.

3 At low frequencies, the leakage reactances are negligible, but the shunting effect of the magnetizing branch becomes increasingly important as its reactance decreases.   The inductance of the mag-

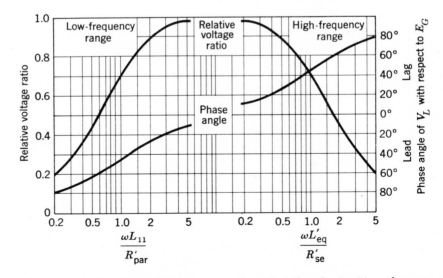

Fig. 1-26.   Normalized frequency characteristics of output transformers.

netizing branch very nearly equals the self-inductance L_{11} of the primary. The equivalent circuit at low frequencies is shown in Fig. 1-25d, from which

$$\frac{V_L}{E_G} = \frac{N_2}{N_1} \frac{r'_L}{R'_{se}} \frac{1}{\sqrt{1 + R'_{par}/\omega L_{11})^2}} \tag{1-52}$$

where
$$R'_{par} = \frac{(r_G + r_1)(r'_2 + r'_L)}{r_G + r_1 + r'_2 + r'_L} \tag{1-53}$$

The voltage ratio relative to its midrange value is

$$\text{Relative voltage ratio} = \frac{1}{\sqrt{1 + (R'_{par}/\omega L_{11})^2}} \tag{1-54}$$

and the phase angle by which the load voltage leads the source voltage is

$$\phi = \tan^{-1} \frac{R'_{par}}{\omega L_{11}} \tag{1-55}$$

Curves of the relative voltage ratio and phase angle as functions of the reactance-to-resistance ratio $\omega L_{11}/R'_{par}$ are shown in the left-hand half of Fig. 1-26.

The points at which the relative voltage ratio is 0.707 are called the *half-power points*. From Eq. 1-50, the upper half-power point occurs at a frequency f_h for which the equivalent leakage reactance $\omega_h L'_{eq}$ equals the series resistance R'_{se}, or

$$f_h = \frac{R'_{se}}{2\pi L'_{eq}} \tag{1-56}$$

and from Eq. 1-54 the lower half-power point occurs at a frequency f_l for which the self-reactance ωL_{11} equals the parallel resistance R'_{par}, or

$$f_l = \frac{R'_{par}}{2\pi L_{11}} \tag{1-57}$$

The bandwidth is describable in terms of the ratio

$$\frac{f_h}{f_l} = \frac{R'_{se}}{R'_{par}} \frac{L_{11}}{L'_{eq}} \tag{1-58}$$

A broad bandwidth requires a high ratio of self-inductance to leakage inductance, or a coefficient of coupling as close as possible to unity.

Stray capacitances of the windings may have a significant effect on the frequency characteristics at high audio frequencies.   This effect tends to produce a peak in the voltage ratio in the high-frequency range at the frequency when the capacitances are in resonance with the leakage inductances.   The height of the peak depends on the circuit impedances as well as the transformer inductances and stray capacitances.   These capacitances are difficult to calculate accurately.   The frequency-response characteristics of a specific circuit can usually best be determined by tests.   The foregoing analysis should be regarded more as an explanation of the effects of the transformer inductances than as an accurate analysis.

c. Pulse Transformers—Qualitative Explanation[1]

Many of the circuits found in applications such as radar, television, and digital computers are called *pulse* or *digital* circuits because the voltage and current waveforms are pulses.   The transformers used in such circuits are *pulse transformers*.   They are inserted for the same general reasons that they appear in other circuits—to change the amplitude of a voltage pulse, to couple successive stages of pulse amplifiers, to change impedance levels, or to isolate direct current from a circuit element.

To meet these requirements, it is important that the transformer reproduce the input pulse as faithfully as possible at its secondary terminals.   Figure 1-27a shows a square-wave input pulse.   The pulse width will usually range from a fraction of a microsecond to about 20

---

[1] For a quantitative analysis, see Reuben Lee, "Electronic Transformers and Circuits," John Wiley & Sons, Inc., New York, 1947.

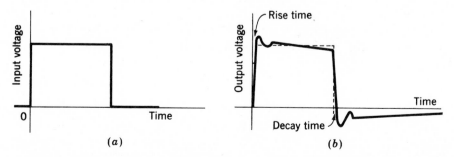

Fig. 1-27.   *(a)* Square-wave input voltage to a pulse transformer and *(b)* the corresponding output voltage waveform.

μsec, and a relatively long time will elapse before the pulse repeats. To determine the output waveform requires a transient rather than a steady-state examination. A typical result is that shown in Fig. 1-27*b*.

The response to the leading edge of the pulse is determined by the high-frequency equivalent circuit including stray capacitance. Because of the presence of leakage inductance, an appreciable time is required for the output voltage to build up to the desired value. This is called the *rise time*. Because of the stray capacitance, there will usually be damped oscillations. Leakage inductance is kept to a minimum for the shortest rise time.

The response to the flat-top portion of the input pulse is determined by the low-frequency equivalent circuit of Fig. 1-25*d*. The output volt-age cannot remain flat, for that would be the equivalent of transmitting steady direct current through the transformer. Instead, the waveform shows a downward tilt, or drop-off of voltage (see Fig. 1-27*b*). In time, of course, the voltage would become zero, but the duration of the pulse is short compared with this time. The tilt of the pulse top is kept within allowable limits by having high magnetizing inductance, that is, by con-structing the core of high-permeability material.

When the input voltage is removed at the termination of the pulse, there is an appreciable decay time before the secondary voltage reaches zero. There is also a significant backswing associated with a damped oscillation and a long-duration negative overshoot of the voltage. In effect, during this period, the magnetizing inductance is discharging the energy of the decaying magnetic field through the stray capacitance and circuit resistances. The resultant waveform is that of a parallel *RLC* circuit.

Pulse transformers are of a small physical size and have relatively few turns in order to minimize leakage inductance. The cores are con-structed of ferrites or of wound strips of high-permeability alloys such as Hipersil (a special high-permeability silicon steel) or permalloy. Since the time interval between pulses is long compared with the pulse duration, the load-carrying duty on the transformer is light. As a result, a very small transformer can handle surprisingly high pulse-power levels.

d. Short-circuit and Open-circuit Tests

Two very simple tests serve to determine the constants of the equivalent circuits of Fig. 1-23 and the power losses in a transformer. These consist in measuring the input voltage, current, and power to the primary, first with the secondary short-circuited, and then with the secondary open-circuited.

With the secondary short-circuited, a primary voltage of only 2 to 12 percent of the rated value need be impressed to obtain full-load current. For convenience, the high-voltage side is usually taken as the primary in this test.   If V_{sc}, I_{sc}, and P_{sc} are the impressed voltage, primary current, and power input, the short-circuit impedance Z_{sc} and its resistance and reactance components R_{sc} and X_{sc} referred to the primary are

$$Z_{sc} = \frac{V_{sc}}{I_{sc}} \tag{1-59}$$

$$R_{sc} = \frac{P_{sc}}{I_{sc}^2} \tag{1-60}$$

$$X_{sc} = \sqrt{Z_{sc}^2 - R_{sc}^2} \tag{1-61}$$

The equivalent circuit with the secondary terminals short-circuited is shown in Fig. 1-28.   The voltage induced in the secondary by the resultant core flux equals the secondary leakage-impedance voltage drop, and at rated current this voltage is only about 1 to 6 percent of rated voltage. At the correspondingly low value of core flux, the exciting current and core losses are entirely negligible.   The exciting admittance, shown dashed in Fig. 1-28, then can be omitted, and the primary and secondary currents are very nearly equal when referred to the same side.   The power input very nearly equals the total copper loss in the primary and secondary windings, and the impressed voltage equals the drop in the combined primary and secondary leakage impedance Z_{eq}.   The equivalent resistance and reactance referred to the primary very nearly equal the short-circuited resistance and reactance of Eqs. 1-60 and 1-61, respectively.   The equivalent impedance can, of course, be referred from one side to the other in the usual manner.   On the rare occasions when the equivalent T circuit of Fig. 1-21d must be resorted to, approximate values of the individual primary and secondary resistances and leakage

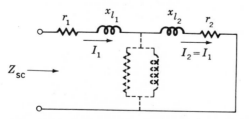

Fig. 1-28. Equivalent circuit with short-circuited secondary.

reactances can be obtained by assuming that $r_1 = r_2 = 0.5r_{eq}$ and $x_{l1} = x_{l2} = 0.5x_{eq}$ when all impedances are referred to the same side.

With the secondary open-circuited and rated voltage impressed on the primary, an exciting current of only 2 to 6 percent of full-load current is obtained. If the transformer is to be used at other than its rated voltage, the test should be taken at that voltage. For convenience, the low-voltage side is usually taken as the primary in this test. The voltage drop in the primary leakage impedance caused by the small exciting current is entirely negligible, and the primary impressed voltage V_1 very nearly equals the emf E_1 induced by the resultant core flux. Also, the primary copper loss caused by the exciting current is entirely negligible, so that the power input P_1 very nearly equals the core loss P_c. Thus the exciting admittance $Y_\varphi = g_c - jb_m$ in Fig. 1-21d very nearly equals the open-circuit admittance $Y_{oc} = g_{oc} - jb_{oc}$ determined from the impressed voltage V_1, exciting current I_φ, and power input P_1 measured in the primary with the secondary open-circuited; thus the exciting admittance and its conductance and susceptance components are very nearly

$$Y_\varphi = Y_{oc} = \frac{I_\varphi}{V_1} \tag{1-62}$$

$$g_c = g_{oc} = \frac{P_1}{V_1^2} \tag{1-63}$$

$$b_m = b_{oc} = \sqrt{Y_{oc}^2 - g_{oc}^2} \tag{1-64}$$

The values so obtained are, of course, referred to the side which was used as the primary in this test. When the approximate equivalent circuits of Fig. 1-23c and d are used, the open-circuit test is used only to obtain core loss for efficiency computations and to check the magnitude of the exciting current. Sometimes the voltage at the terminals of the open-circuited secondary is measured as a check on the turns ratio.

EXAMPLE 1-10

With the instruments located in the high-voltage side and the low-voltage side short-circuited, the short-circuit test readings for the 50-kva 2,400:240-volt transformer of Example 1-8 are 48 volts, 20.8 amp, and 617 watts. An open-circuit test with the low-voltage side energized gives instrument readings on that side of 240 volts, 5.41 amp, and 186 watts.

Determine the efficiency and the voltage regulation at full load, 0.80 power factor lagging.

Solution

From the short-circuit test, the equivalent impedance, resistance, and reactance of the transformer (referred to the high-voltage side as denoted by the subscript H) are

$$Z_{eqH} = \frac{48}{20.8} = 2.31 \text{ ohms}$$

$$r_{eqH} = \frac{617}{(20.8)^2} = 1.42 \text{ ohms}$$

$$x_{eqH} = \sqrt{(2.31)^2 - (1.42)^2} = 1.82 \text{ ohms}$$

Full-load high-tension current is

$$I_H = \frac{50,000}{2,400} = 20.8 \text{ amp}$$

Copper loss $= I_H^2 r_{eqH} = (20.8)^2(1.42) = \quad$ 617 watts
Core loss $\qquad\qquad\qquad\qquad\qquad = \qquad$ 186
Total losses at full load $\qquad\qquad\quad = \qquad$ 803
Output $= (50,000)(0.80) \qquad\qquad\; = $ 40,000
Input $\qquad\qquad\qquad\qquad\qquad\quad = $ 40,803 watts

$$\frac{\text{Losses}}{\text{Input}} = \frac{803}{40,803} = 0.0197$$

The *efficiency* of any power-transmitting device is

$$\text{Efficiency} = \frac{\text{power output}}{\text{power input}}$$

which can also be expressed as

$$\text{Efficiency} = \frac{\text{input} - \text{losses}}{\text{input}} = 1 - \frac{\text{losses}}{\text{input}}$$

For the specified load, the efficiency of this transformer is

$$\text{Efficiency} = 1 - 0.0197 = 0.980$$

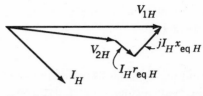

Fig. 1-29. Phasor diagram, Example 1-10.

The *voltage regulation* of a transformer is the change in secondary terminal voltage from no load to full load and is usually expressed as a percentage of the full-load value. The equivalent circuit of Fig. 1-23c will be used with everything still referred to the high-voltage side. The primary voltage is assumed to be adjusted so that the secondary terminal voltage has its rated value at full load, or $V_{2H} = 2{,}400$ volts referred to the high-voltage side. The required value of the primary voltage V_{1H} can be computed from the phasor diagram shown in Fig. 1-29.

$$V_{1H} = V_{2H} + I_H(r_{eqH} + jx_{eqH})$$

$$= 2{,}400 + (20.8)(0.80 - j0.60)(1.42 + j1.82)$$

$$= 2{,}446 + j13$$

The magnitude of V_{1H} is 2,446 volts. If this voltage were held constant and the load removed, the secondary voltage on open circuit would rise to 2,446 volts referred to the high-voltage side. Then,

$$\text{Regulation} = \frac{2{,}446 - 2{,}400}{2{,}400}(100) = 1.92\%$$

1-9
TRANSFORMERS IN THREE–PHASE CIRCUITS

Three single-phase transformers may be connected to form a 3-phase bank in any of the four ways shown in Fig. 1-30. In all four parts of this figure, the windings at the left are the primaries, those at the right are the secondaries, and any primary winding is mated in one transformer with the secondary winding drawn parallel to it. Also shown are the voltages and currents resulting from balanced impressed primary line-to-line voltages V and line currents I when the ratio of primary to secondary turns N_1/N_2

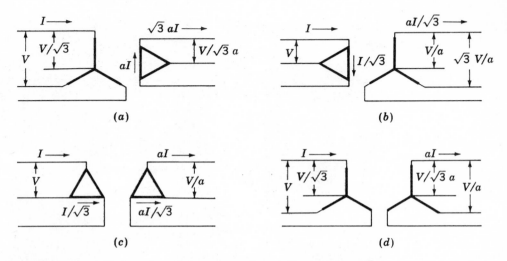

Fig. 1-30. Common 3-phase transformer connections.   The transformer windings are indicated by the heavy lines.

is a and ideal transformers are assumed.   It will be noted that, for fixed line-to-line voltages and total kva, the kva rating of each transformer is one-third of the kva rating of the bank, regardless of the connections used, but that the voltage and current ratings of the individual transformers depend on the connections.

The Y-Δ connection is commonly used in stepping down from a high voltage to a medium or low voltage.   One of the reasons is that a neutral is thereby provided for grounding on the high-voltage side, a procedure which may be shown to be desirable in most cases.   Conversely, the Δ-Y connection is commonly used for stepping up to a high voltage.   The Δ-Δ connection has the advantage that one transformer may be removed for repair or maintenance while the remaining two continue to function as a 3-phase bank with, however, the rating reduced to 58 percent of that of the original bank; this is known as the *open-delta*, or V, *connection*. The Y-Y connection is seldom used, because of difficulties with exciting-current phenomena.

Instead of three single-phase transformers, a 3-phase bank may consist of one 3-phase transformer having all 6 windings on a common core and contained within a common tank.   Advantages of 3-phase transformers are that they cost less, weigh less, require less floor space, and have somewhat higher efficiency.   A photograph of the internal parts of a large 3-phase transformer is shown in Fig. 1-31.

Circuit computations involving 3-phase transformer banks under bal-

Fig. 1-31. A 200-Mva 3-phase 50-Hz 3-winding 210/80/10.2 kv transformer removed from its tank. The 210-kv winding has an onload tap changer for adjustment of the voltage. (*Brown Boveri Corporation.*)

anced conditions can be made by dealing with only one of the transformers or phases and recognizing that conditions are the same in the other two phases except for the phase displacements associated with a 3-phase system.   It is usually convenient to carry out the computations on a per-phase-Y line-to-neutral basis, since transformer impedances can then be added directly in series with transmission-line impedances.   The imped-ances of transmission lines can be referred from one side of the transformer bank to the other by use of the square of the ideal line-to-line voltage ratio of the bank.   In dealing with Y-Δ or Δ-Y banks, all quantities can be referred to the Y-connected side.   In dealing with Δ-Δ banks in series with transmission lines, it is convenient to replace the Δ-connected imped-ances of the transformers by equivalent Y-connected impedances.   It is well known that a balanced Δ-connected circuit of Z_Δ ohms per phase is equivalent to a balanced Y-connected circuit of Z_Y ohms per phase if

$$Z_Y = \frac{1}{3}Z_\Delta \qquad\qquad (1\text{-}65)$$

EXAMPLE 1-11

Three single-phase 50-kva 2,400:240-volt transformers identical with that of Example 1-10 are connected Y-Δ in a 3-phase 150-kva bank to step down the voltage at the load end of a feeder whose impedance is $0.15 + j1.00$ ohm per phase.   The voltage at the sending end of the feeder is 4,160 volts, line to line.   On their secondary sides the transformers supply a balanced 3-phase load through a feeder whose impedance is $0.0005 + j0.0020$ ohm per phase.

Find the line-to-line voltage at the load when the load draws rated current from the transformers at a power factor of 0.80 lagging.

Solution

The computations can be made on a per-phase-Y basis by referring everything to the high-voltage Y-connected side of the transformer bank.   The voltage at the sending end of the feeder is equivalent to a source voltage V_s of

$$V_s = \frac{4,160}{\sqrt{3}} = 2,400 \text{ volts to neutral}$$

From the transformer rating, the rated current on the high-voltage side is 20.8 amp per phase Y.   The low-voltage feeder impedance referred to the high-voltage side by means of the square of the ideal line-to-line voltage ratio of the bank is

$$\left(\frac{4,160}{240}\right)^2 (0.0005 + j0.0020) = 0.15 + j0.60 \text{ ohm}$$

and the combined series impedance of the high-voltage and low-voltage feeders referred to the high-voltage side is

$$Z_{\text{feeder}} = 0.30 + j1.60 \text{ ohms per phase Y}$$

From Example 1-10, the equivalent impedance of the transformer bank referred to its high-voltage Y-connected side is

$$Z_{eqH} = 1.42 + j1.82 \text{ ohms per phase Y}$$

The equivalent circuit for 1 phase referred to the Y-connected primary side then is exactly the same as Fig. 1-24a, and the solution on a per-phase basis is exactly the same as the solution of Example 1-9, whence the load voltage referred to the high-voltage side is 2,329 volts to neutral.   The actual load voltage is

$$V_{\text{load}} = 233 \text{ volts, line to line}$$

This is the line-to-line voltage because the secondaries are Δ-connected.

EXAMPLE 1-12

The three transformers of Example 1-9 are connected Δ-Δ and supplied with power through a 2,400-volt (line-to-line) 3-phase feeder whose reactance is 0.80 ohm per phase.   The equivalent reactance of each transformer referred to its high-voltage side is 1.82 ohms. All circuit resistances will be neglected.   At its sending end the feeder is connected to the secondary terminals of a 3-phase Y-Δ connected transformer whose 3-phase rating is 500 kva, 24,000:2,400 volts (line-to-line voltages).   The equivalent reactance of the sending-end transformers is 2.76 ohms per phase Δ referred to the 2,400-volt side.   The voltage applied to the primary terminals is 24,000 volts, line to line.

A 3-phase shortcircuit occurs at the 240-volt terminals of the receiving-end transformers.   Compute the steady-state short-circuit current in the 2,400-volt feeder wires, in the primary and secondary windings of the receiving-end transformers, and at the 240-volt terminals.

Solution

The computations will be made on an equivalent line-to-neutral basis with everything referred to the 2,400-volt feeder.   The source voltage then is

$$\frac{2,400}{\sqrt{3}} = 1,385 \text{ volts to neutral}$$

Sending-end transformers

$$X_{\text{eq}} = \frac{2.76}{3} = 0.92 \text{ equivalent ohm per phase Y}$$

Receiving-end transformers

$$X_{eq} = \frac{1.82}{3} = 0.61 \text{ equivalent ohm per phase Y}$$

Feeder X = 0.80 ohm per phase Y

Total reactance = sum = 2.33 ohms

$$\text{Current in 2,400-volt feeder} = \frac{1,385}{2.33} = 594 \text{ amp}$$

$$\text{Current in 2,400-volt windings} = \frac{594}{\sqrt{3}} = 342 \text{ amp}$$

Current in 240-volt windings = 3,420 amp

Current at the 240-volt terminals = 5,940 amp

<hr>

1-10
THE PER–UNIT SYSTEM

Very often, computations relating to machines, transformers, and systems of machines are carried out in per-unit form—i.e., with all pertinent quantities expressed as decimal fractions of appropriately chosen base values.   All the usual computations are then carried out in these per-unit values instead of the familiar volts, amperes, ohms, etc.

There are two advantages to the system.   One is that the constants of machines and transformers lie in a reasonably narrow numerical range when expressed in a per-unit system related to their rating.   The correctness of their values is thus subject to a rapid approximate check.   The other is that the analyst is relieved of the worry of referring circuit quantities to one side or the other of transformers.   For complicated systems involving many transformers of different turns ratios, this advantage is a significant one in that a possible cause of serious mistakes is removed. The per-unit system is also very useful in simulating machine systems on analog and digital computers for transient and dynamic analyses.

Quantities such as voltage V, current I, power P, reactive power Q, volt-amperes VA, resistance R, reactance X, impedance Z, conductance G, susceptance B, and admittance Y can be translated to and from per-unit form as follows:

$$\text{Quantity in per unit} = \frac{\text{actual quantity}}{\text{base value of quantity}} \qquad (1\text{-}66)$$

where *actual quantity* refers to the value in volts, amperes, ohms, etc. To a certain extent, base values may be chosen arbitrarily, but certain relations among them must be observed for the normal electrical laws to hold in the per-unit system. Thus, for a single-phase system,

$$P_{\text{base}}, Q_{\text{base}}, VA_{\text{base}} = V_{\text{base}}I_{\text{base}} \qquad (1\text{-}67)$$

$$R_{\text{base}}, X_{\text{base}}, Z_{\text{base}} = \frac{V_{\text{base}}}{I_{\text{base}}} \qquad (1\text{-}68)$$

$$G_{\text{base}}, B_{\text{base}}, Y_{\text{base}} = \frac{I_{\text{base}}}{V_{\text{base}}} \qquad (1\text{-}69)$$

In normal usage, values of VA_{base} and V_{base} are chosen first; values of I_{base} and all other quantities in Eqs. 1-67 to 1-69 are thereby established.

The value of VA_{base} must be the same over the entire system concerned. When a transformer is encountered, the values of V_{base} are different on each side and must be in the same ratio as are the turns on the transformer. Usually, the rated or nominal voltages of the respective sides are chosen. The process of referring quantities to one side of the transformer is then taken care of automatically by the use of Eqs. 1-66 to 1-69 in finding and interpreting per-unit values. The procedure thus becomes one of translating all quantities to per-unit values, using these values in all the customary circuit-analysis techniques, and translating the end results back to the more usual forms.

When only one electrical device, such as a transformer, is involved, the device's own rating is generally used for the volt-ampere base. When expressed in per unit on the rating as a base, the characteristics of power and distribution transformers do not vary much over a wide range of ratings. For example, the exciting current usually is between 0.02 and 0.06 per unit, the equivalent resistance usually is between 0.005 and 0.02 per unit (the smaller values applying to large transformers), and the equivalent reactance usually is between 0.015 and 0.10 per unit (the larger values applying to large high-voltage transformers). Similarly, the per-unit values of synchronous- and induction-machine constants fall within a relatively narrow range. When several devices are involved, however, an arbitrary choice of volt-ampere base must usually be made in order that the same base be used for the overall system. Per-unit values may be changed from one volt-ampere base to another with the same voltage base by the relations

$$(P,Q,VA,G,B,Y)_{\text{pu on base 2}} = (P,Q,VA,G,B,Y)_{\text{pu on base 1}} \frac{(VA)_{\text{base 1}}}{(VA)_{\text{base 2}}} \quad (1\text{-}70)$$

$$(R,X,Z)_{\text{pu on base 2}} = (R,X,Z)_{\text{pu on base 1}} \frac{(VA)_{\text{base 2}}}{(VA)_{\text{base 1}}} \quad (1\text{-}71)$$

EXAMPLE 1-13

The exciting current measured on the low-voltage side of a 50-kva 2,400:240-volt transformer is 5.41 amp. Its equivalent impedance referred to the high-voltage side is $1.42 + j1.82$ ohms. Take the transformer rating as a base.

a. Express the exciting current in per unit on the low-voltage side and also on the high-voltage side.

b. Express the equivalent impedance in per unit on the high-voltage side and also on the low-voltage side.

Solution

The base values of voltages and currents are

$$V_{\text{base }H} = 2,400 \text{ volts} \qquad V_{\text{base }X} = 240 \text{ volts}$$

$$I_{\text{base }H} = 20.8 \text{ amp} \qquad I_{\text{base }X} = 208 \text{ amp}$$

where subscripts H and X indicate the high- and low-voltage sides, respectively.

From Eq. 1-68,

$$Z_{\text{base }H} = \frac{2,400}{20.8} = 115.2 \text{ ohms}$$

$$Z_{\text{base }X} = \frac{240}{208} = 1.152 \text{ ohms}$$

a. From Eq. 1-66,

$$I_{\varphi X} = \frac{5.41}{208} = 0.0260 \text{ per unit}$$

The exciting current referred to the high-voltage side is 0.541 amp.
Its per-unit value is

$$I_{\varphi H} = \frac{0.541}{20.8} = 0.0260 \text{ per unit}$$

The per-unit values are the same referred to either side.   The
turns ratios required to refer currents in amperes from one side of
the transformer to the other are taken care of in the per-unit system by the base values for currents on the two sides when the
volt-ampere base is the same on both sides and the voltage bases
are in the ratio of the turns.

b. From Eq. 1-66 and the value for $Z_{\text{base } H}$,

$$Z_{eqH} = \frac{1.42 + j1.82}{115.2} = 0.0123 + j0.0158 \text{ per unit}$$

The equivalent impedance referred to the low-voltage side is
$0.0142 + j0.0182$ ohm.   Its per-unit value is

$$Z_{eqX} = \frac{0.0142 + j0.0182}{1.152} = 0.0123 + j0.0158 \text{ per unit}$$

The per-unit values are the same, the referring factors being taken
care of in per unit by the base values.

When applied to 3-phase problems, the base values for the per-unit
system are chosen so that the relations for a balanced 3-phase system
hold among them:

$$(P_{\text{base}}, Q_{\text{base}}, VA_{\text{base}}) \text{ 3-phase} = 3 VA_{\text{base per phase}} \tag{1-72}$$

$$V_{\text{base (line to line)}} = \sqrt{3} \, V_{\text{base (line to neutral)}} \tag{1-73}$$

$$I_{\text{base (per phase } \Delta)} = \frac{1}{\sqrt{3}} I_{\text{base (per phase Y)}} \tag{1-74}$$

In dealing with 3-phase systems the 3-phase kva base and the line-to-line
voltage base are usually chosen first.   The base values for phase voltages
and currents then follow from Eqs. 1-72, 1-73, and 1-74.   Equations 1-67,
1-68, and 1-69 still apply to the base values per phase.   For example, the

base value for Y-connected impedances is given by Eq. 1-68 with V_{base} taken as the base voltage to neutral and I_{base} taken as the base current per phase Y; the base value for Δ-connected impedances is also given by Eq. 1-68 but with V_{base} taken as the base line-to-line voltage and I_{base} taken as the base current per phase Δ.   Division of Eq. 1-73 by Eq. 1-74 shows that

$$Z_{base \text{ (per phase } \Delta)} = 3Z_{base \text{ (per phase Y)}} \tag{1-75}$$

The factors of $\sqrt{3}$ and 3 relating Δ and Y quantities in volts, amperes, and ohms in a balanced 3-phase system are thus automatically taken care of in per unit by the base values.   Such 3-phase problems can be solved in per unit as if they were single-phase problems, without paying any attention to the details of the transformer connections except in translating volt-ampere-ohm values into and out of the per-unit system.

EXAMPLE 1-14

Solve Example 1-12 in per unit on a 3-phase 150-kva rated-voltage base.   The reactance of the 3-phase 500-kva sending-end transformer is 0.08 per unit on its rating as a base.   The reactance of the 2,400-volt feeder is 0.80 ohm per phase.   The reactance of the receiving-end transformers is 0.0158 per unit as computed in Example 1-13.

Solution

Convert reactance of sending-end transformer to 150-kva base.

$$X_{sending\ end} = 0.08\,\frac{150}{500} = 0.024 \text{ per unit}$$

For the 2,400-volt feeder

$$V_{base} = \frac{2,400}{\sqrt{3}} = 1,385 \text{ volts, line to neutral}$$

$$I_{base} = \frac{50,000}{1,385} = 36.1 \text{ amp per phase Y}$$

$$Z_{base} = \frac{1,385}{36.1} = 38.3 \text{ ohms per phase Y}$$

Therefore $$X_{feeder} = \frac{0.80}{38.3} = 0.021 \text{ per unit}$$

For the receiving-end transformers

$$X_{\text{receiving end}} = 0.0158 \text{ per unit}$$

$$\text{Total reactance} = \text{sum} = 0.0608 \text{ per unit}$$

$$\text{Short-circuit current} = \frac{1.00}{0.0608} = 16.4 \text{ per unit}$$

The currents in various parts of the circuit can now be computed. For example, the current in the 2,400-volt feeder wires is $(16.4)(36.1) = 594$ amp. Compare with the result in Example 1-12.

(Note: It can readily be shown that the reactance of 0.08 per unit for the sending-end transformers is consistent with the value of 2.76 ohms per phase given in Example 1-12.)

1-11
AUTOTRANSFORMERS. MULTICIRCUIT TRANSFORMERS

The principles discussed in the foregoing three articles have been developed with specific reference to 2-winding transformers. They are also generally applicable to transformers with other than 2 separate windings. Aspects relating to autotransformers and multiwinding transformers are considered in this article.

a. Autotransformers

Viewed from the terminals, substantially the same transformation effect on voltages, currents, and impedances can be obtained with the connections of Fig. 1-32a as in the normal transformer with 2 separate windings

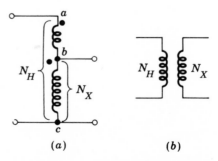

(a)   (b)

Fig. 1-32. (a) Autotransformer compared with (b) 2-winding transformer.

shown in Fig. 1-32b. In Fig. 1-32a, the winding bc is common to both the primary and secondary circuits. This type of transformer is called an *autotransformer*. It is really nothing but a normal transformer connected in a special way. The only difference structurally is that winding ab must be provided with extra insulation. The performance of an autotransformer is governed by the same fundamental considerations already discussed for transformers having 2 separate windings. Autotransformers have lower leakage reactances, lower losses, and smaller exciting current and cost less than 2-winding transformers when the voltage ratio does not differ too greatly from 1 to 1. A disadvantage is the direct copper connection between the high- and low-voltage sides.

EXAMPLE 1-15

The 2,400:240-volt 50-kva transformer of Examples 1-10 and 1-11 is connected as an autotransformer as shown in Fig. 1-33, in which ab is the 240-volt winding and bc is the 2,400-volt winding. (It is assumed that the 240-volt winding has sufficient insulation so that it can withstand a voltage of 2,640 volts to ground.)

a. Compute the voltage ratings V_H and V_X of the high-tension and low-tension sides, respectively, when the transformer is connected as an autotransformer.

b. Compute the kva rating as an autotransformer.

c. Data with respect to the losses are given in Example 1-10. Compute the full-load efficiency as an autotransformer at 0.80 power factor.

Solution

a. Since the 2,400-volt winding bc is connected to the low-tension circuit, $V_X = 2,400$ volts.

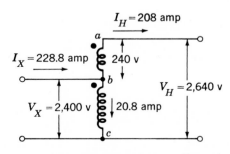

Fig. 1-33. Autotransformer, Example 1-15.

When $V_{bc} = 2{,}400$ volts, a voltage $V_{ab} = 240$ volts in phase with V_{bc} will be induced in winding ab (leakage-impedance voltage drops being neglected). The voltage of the high-tension side therefore is

$$V_H = V_{ab} + V_{bc} = 2{,}640 \text{ volts}$$

b. From the rating of 50 kva as a normal 2-winding transformer, the rated current of the 240-volt winding is $50{,}000/240$, or 208 amp. Since the 240-volt winding is in series with the high-tension circuit, the rated current of this winding is the rated current I_H on the high-tension side as an autotransformer. The kva rating as an autotransformer therefore is

$$\frac{V_H I_H}{1{,}000} = \frac{(2{,}640)(208)}{1{,}000} = 550 \text{ kva}$$

The rating can also be computed on the low-tension side in a manner which highlights the current-transforming properties. Thus, if the current in the 240-volt winding has its rated value of 208 amp, the current in the 2,400-volt winding must produce an equal and opposite mmf (exciting current being neglected) and therefore must be 20.8 amp in the arrow direction (Fig. 1-33). The current I_X on the low-tension side as an autotransformer therefore is

$$I_X = 208 + 20.8 = 228.8 \text{ amp}$$

and the kva rating is

$$\frac{V_X I_X}{1{,}000} = \frac{(2{,}400)(228.8)}{1{,}000} = 550 \text{ kva}$$

Note that this transformer, whose rating as a normal 2-winding transformer is 50 kva, is capable of handling 550 kva as an autotransformer. The higher rating as an autotransformer is a consequence of the fact that all of the 550 kva does not have to be transformed by electromagnetic induction. In fact, all that the transformer has to do is to boost a current of 208 amp through a potential rise of 240 volts, corresponding to a rating of 50 kva.

c. When connected as an autotransformer with the currents and voltages shown in Fig. 1-33, the losses are the same as in Example 1-10, namely, 803 watts. But the output as an autotransformer at 0.80 power factor is $(0.80)(550,000) = 440,000$ watts. The efficiency therefore is

$$1 - \frac{803}{440,803} = 0.9982$$

The efficiency is so high because the losses are those incident to transforming only 50 kva.

b. Multicircuit Transformers

Transformers having 3 or more windings, known as *multicircuit*, or *multi-winding*, *transformers*, are often used to interconnect three or more circuits which may have different voltages. For these purposes a multi-circuit transformer costs less and is more efficient than an equivalent number of two-circuit transformers. A transformer having a primary and two secondaries is generally used to supply power to electronic units. The distribution transformers used to supply power for domestic purposes usually have two 120-volt secondaries connected in series. Lighting cir-cuits are connected across each of the 120-volt windings, while electric ranges, domestic hot-water heaters, and other similar loads are supplied with 240-volt power from the series-connected secondaries. A large dis-tribution system may be supplied through a 3-phase bank of multicircuit transformers from two or more transmission systems having different voltages. The 3-phase transformer banks used to interconnect two trans-mission systems of different voltages often have a third, or *tertiary*, set of windings to provide voltage for auxiliary power purposes in the substation or to supply a local distribution system. Static capacitors, or synchron-ous condensers, may be connected to the tertiary windings for purposes of power-factor correction or voltage regulation. Sometimes Δ-connected tertiary windings are put on 3-phase banks to provide a circuit for the third harmonics of the exciting current.

Some of the problems arising in the use of multicircuit transformers concern the effects of leakage impedances on voltage regulation, short-circuit currents, and division of load among circuits. These problems can be solved by an equivalent-circuit technique similar to that used in dealing with two-circuit transformers.

The equivalent circuits of multiwinding transformers are more com-plicated than in the 2-winding case because they must take into account the leakage impedances associated with each pair of windings. In these equivalent circuits all quantities are referred to a common base, either by use of the appropriate turns ratios as referring factors or by expressing all quantities in per unit. The exciting current usually is neglected.

The following discussion will be confined to three-circuit transformers. A three-circuit transformer is shown schematically in Fig. 1-34a, in which 1-1' is the primary winding. Secondary and tertiary quantities are referred to the primary, and windings 2-2' and 3-3' indicate the equivalent secondary and tertiary on the basis of a 1-to-1-to-1 turns

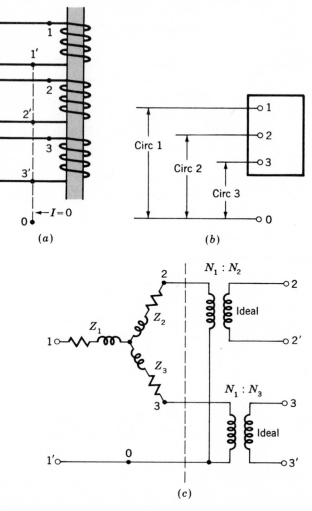

Fig. 1-34. (a) Elementary 3-winding transformer. (b) and (c) Steps in the development of its equivalent circuit.

ratio.   Terminals $1'$, $2'$, and $3'$ are of like polarity.   These terminals can be considered to be connected together to a common terminal 0, as shown by the dashed lines.   The three external circuits can then be considered to be connected between terminals 1, 2, and 3, respectively, and the common terminal 0.   If the exciting current is neglected, the phasor sum of the currents I_1, I_2, and I_3 is zero.   The current in the connection between terminal 0 and the junction of terminals $1'$, $2'$, $3'$ therefore is zero, and this connection can be omitted without disturbing the currents. The transformer then is equivalent to the arrangement shown in Fig. 1-34b, in which the box is a network with three terminals and contains impedances accounting for the leakage-impedance voltage drops.   Insofar as its effects on the external circuits are concerned, such a three-terminal network is equivalent to three impedances connected either in Δ or in Y.   The Y arrangement usually is more convenient.   The transformer therefore is equivalent to the circuit shown to the left of the dashed line in Fig. 1-34c, in which the impedances Z_1, Z_2, and Z_3 account for the effects of the leakage impedances among the three pairs of windings.   If desired, the factors referring all quantities to, say, the primary can be shown explicitly by means of ideal transformers as shown to the right of the dashed line in Fig. 1-34c.   The terminals 2-$2'$ and 3-$3'$ to the right of these ideal transformers then are equivalent to the actual secondary and tertiary terminals.   When the ideal transformers are included, the equivalent circuit does not require a conductive connection among the three circuits.   Usually the ideal transformers can be omitted, however, and the external circuits can be considered to be connected between the common point 0 and terminals 1, 2, and 3, respectively, to the left of the dashed line.   One then simply remembers that all quantities are referred to a common base.

This equivalent circuit represents the impedance phenomena associated with 3 windings on a common magnetic core.   It applies equally well to the external behavior of autotransformers and of transformers having separate windings, although the internal phenomena differ.

The impedances of Fig. 1-34c can readily be determined from the results of three simple short-circuit tests.   Thus, if Z_{12} is the short-circuit impedance of circuits 1 and 2 with circuit 3 open, inspection of the equivalent circuit (Fig. 1-34c) shows that

$$Z_{12} = Z_1 + Z_2 \qquad (1\text{-}76)$$

Similarly
$$Z_{13} = Z_1 + Z_3 \qquad (1\text{-}77)$$

$$Z_{23} = Z_2 + Z_3 \qquad (1\text{-}78)$$

where Z_{13} is the short-circuit impedance of circuits 1 and 3 with circuit 2 open and Z_{23} is the short-circuit impedance of circuits 2 and 3 with circuit 1 open. These short-circuit impedances are the values referred to a common base. Solution of Eqs. 1-76, 1-77, and 1-78 then gives

$$Z_1 = \tfrac{1}{2}(Z_{12} + Z_{13} - Z_{23}) \tag{1-79}$$

$$Z_2 = \tfrac{1}{2}(Z_{23} + Z_{12} - Z_{13}) \tag{1-80}$$

$$Z_3 = \tfrac{1}{2}(Z_{13} + Z_{23} - Z_{12}) \tag{1-81}$$

EXAMPLE 1-16

The results of three short-circuit tests on a 7,960:2,400:600-volt 60-Hz single-phase transformer are as follows:

Test	Winding excited	Winding short-circuited	Applied voltage, volts	Current in excited winding, amp
1	1	2	252	62.7
2	1	3	770	62.7
3	2	3	217	208

Resistances may be neglected. The rating of the 7,960-volt primary winding is 1,000 kva, of the 2,400-volt secondary is 500 kva, and of the 600-volt tertiary is 500 kva.

a. Compute the per-unit values of the equivalent-circuit impedances of this transformer on a 1,000-kva rated-voltage base.

b. Three of these transformers are used in a 3,000-kva Y-Δ-Δ 3-phase bank to supply 2,400-volt and 600-volt auxiliary power circuits in a generating station. The Y-connected primaries are connected to the 13,800-volt main bus. Compute the per-unit values of the steady-state short-circuit currents and of the voltage at the terminals of the secondary windings if a 3-phase shortcircuit occurs at the terminals of the tertiary windings with 13,800 volts maintained at the primary line terminals. Use a 3,000-kva 3-phase rated-voltage base.

Solution

a. First convert the short-circuit data to per unit on 1,000 kva per phase.

For primary: $V_{\text{base}} = 7,960$ volts

$$I_{\text{base}} = \frac{1,000}{7.96} = 125.4 \text{ amp}$$

For secondary: $V_{\text{base}} = 2,400$ volts

$$I_{\text{base}} = \frac{1,000}{2.4} = 416 \text{ amp}$$

Conversion of the test data to per unit then gives

Test	Windings	V	I
1	1 and 2	0.0316	0.500
2	1 and 3	0.0967	0.500
3	2 and 3	0.0905	0.500

From test 1, the short-circuit impedance Z_{12} is

$$Z_{12} = \frac{0.0316}{0.500} = 0.0632 \text{ per unit}$$

Similarly, from tests 2 and 3,

$$Z_{13} = \frac{0.0967}{0.500} = 0.1934 \text{ per unit}$$

$$Z_{23} = \frac{0.0905}{0.500} = 0.1810 \text{ per unit}$$

From Eqs. 1-79, 1-80, and 1-81, the equivalent-circuit constants are

$$Z_1 = jX_1 = j0.0378 \text{ per unit}$$
$$Z_2 = jX_2 = j0.0254 \text{ per unit}$$
$$Z_3 = jX_3 = j0.1556 \text{ per unit}$$

b. Base line-to-line voltage for the Y-connected primaries is $\sqrt{3}$ (7,960) = 13,800 volts, or the bus voltage is 1.00 per unit. From the equivalent circuit with a shortcircuit on the tertiaries,

$$I_{sc} = \frac{V_1}{Z_1 + Z_3} = \frac{V_1}{Z_{13}} = \frac{1.00}{0.1934} = 5.18 \text{ per unit}$$

(Note, however, that this current is 10.36 per unit on the rating of the tertiaries.)   If the voltage drops caused by the secondary load current are neglected in comparison with those due to the short-circuit current, the secondary terminal voltage equals the voltage at the junction of the three impedances Z_1, Z_2, and Z_3 in Fig. 1-34c, whence

$$V_2 = I_{sc}Z_3 = (5.18)(0.1556) = 0.805 \text{ per unit}$$

1-12
RÉSUMÉ

Electromechanical devices of the magnetic-field type use ferromagnetic material for guiding and concentrating the fields.   Because the permeability of the ferromagnetic material is several thousand times that of the surrounding space, most of the flux is confined to fairly definite paths and the field reduces to a magnetic circuit.   The frequencies are sufficiently low so that the field can be considered to be quasi-static.   The analogy to a resistive electric circuit is useful for its conceptual insight.   Analogous quantities are shown in Table 1-1.   For the magnetic circuit the

TABLE 1-1
ANALOGOUS QUANTITIES IN MAGNETIC AND ELECTRIC CIRCUITS

Mmf $\mathfrak{F} = Ni$	Emf E
Flux ϕ	Current I
Reluctance $\mathfrak{R}$	Resistance R
Permeance $\mathcal{P} = 1/\mathfrak{R}$	Conductance $G = 1/R$
Flux density $B = \phi/A$	Current density $J = I/A$
Magnetizing force H	Potential gradient ε
Permeability μ	Conductivity σ

analogies to Kirchhoff's current and voltage laws state that: (1) the sum of the fluxes $\Sigma\phi$ at a node equals zero, and (2) the sum of the reluctance drops $\Sigma\phi\mathfrak{R}$ around a loop equals the sum of the mmfs ΣNi.

Computationally, magnetic circuits differ from resistive electric circuits in two important respects: (1) leakage is greater, and (2) the reluctance of a magnetic-circuit branch is a function of the flux in that branch. As the flux increases, the circuit tends to saturate. Both the permeability and reluctance may change from one operating condition to another. Hence, direct numerical application of the electric-circuit analogy is rare. Its principal value lies in the guidance it provides to the thought process. For quantitative analysis, graphical methods based on the *B-H* characteristics of the core material are generally used.

The electric-circuit analog breaks down completely when power dissipation analogous to I^2R in an electric circuit is associated with a non-time-varying flux in a magnetic circuit. The physical phenomena of conduction and ferromagnetism are entirely different in nature. Rather, hysteresis and eddy-current losses occur in a ferromagnetic core only when the flux is time varying.

Ferromagnetic materials are available having a wide variety of characteristics. Numerous alloys of iron with other metals are used when special magnetic properties are desired. The material commonly used in the construction of transformers, reactors, solenoids, and rotating machines is an alloy of iron and a small amount (2 to 4 percent) of silicon. It is available commercially in thin laminations (0.014 to 0.025 in. thick) for use in parts subjected to ac fields. The magnetic properties of two grades of sheet steels are shown in Figs. 1-9 and 1-10. The grain-oriented sheets, Fig. 1-10, have superior magnetic properties when the magnetic field is aligned in the direction of rolling of the sheets. This material is commonly used for transformer cores.

Although no electromechanical energy conversion is involved in a static transformer, its performance has many points of similarity with the electrical behavior of ac rotating machines. The following discussion of these points will therefore constitute a brief preview of the thought processes involved in ac machine theory as presented in subsequent chapters.

In both transformers and rotating machines, a magnetic field is created by the combined action of the currents in the windings. In an iron-core transformer most of this flux is confined to the core and links all the windings. This resultant mutual flux induces voltages in the windings proportional to their numbers of turns and provides the voltage-changing property. In rotating machines most of the flux crosses the air gap, like the core flux in a transformer, and links all the windings on both stator and rotor. The voltages induced in the windings by this resultant mutual air-gap flux are similar to those induced by the resultant core flux in a transformer. The difference is that mechanical motion together with electromechanical energy conversion is involved in rotating machines. The torque associated with this energy-conversion process is created by the interaction of the air-gap flux with the magnetic field of the rotor currents.

In addition to the useful mutual fluxes, in both transformers and rotating machines there are leakage fluxes which link one winding without linking the other. Although the detailed picture of the leakage fluxes in rotating machines is more complicated than in transformers, their effects are essentially the same. In both, the leakage fluxes induce voltages in ac windings which are accounted for as leakage-reactance voltage drops. In both, the leakage-flux paths are mostly in air, and the leakage fluxes are nearly linearly proportional to the currents producing them. The leakage reactances therefore are often assumed to be constant, independent of the degree of saturation of the main magnetic circuit.

From the viewpoint of the winding, the induced-voltage phenomena in transformers and rotating machines are essentially the same, although the internal phenomena causing the time variations in flux linkages are different. In a rotating machine the time variation in flux linkages is caused by relative motion of the field and the winding, and the induced voltage is sometimes referred to as a speed voltage. Speed voltages accompanied by mechanical motion are a necessary counterpart of electro-mechanical energy conversion. In a static transformer, however, the time variation of flux linkages is caused by the growth and decay of a stationary magnetic field, no mechanical motion is involved, and no electromechanical energy conversion takes place.

The resultant core flux in a transformer induces a counter emf in the primary which, together with the primary resistance and leakage-reactance voltage drops, must balance the applied voltage. Since the resistance and leakage-reactance voltage drops usually are small, the counter emf must approximately equal the applied voltage and the core flux must adjust itself accordingly. Exactly similar phenomena must take place in the armature windings of an ac motor—the resultant air-gap flux wave must adjust itself to generate a counter emf approximately equal to the applied voltage. In both transformers and rotating machines, the net mmf of all the currents must accordingly adjust itself to create the resultant flux required by this voltage balance. In any ac electromagnetic device in which the resistance and leakage-reactance voltage drops are small, the resultant flux is very nearly determined by the applied voltage and frequency, and the currents must adjust themselves accordingly so as to produce the mmf required to create this flux.

In a transformer, the secondary current is determined by the voltage induced in the secondary, the secondary leakage impedance, and the electrical load. In an induction motor, the secondary (rotor) current is determined by the voltage induced in the secondary, the secondary leakage impedance, and the mechanical load on its shaft. Essentially the same phenomena take place in the primary winding of the transformer and in the armature (stator) windings of induction and synchronous motors. In all three, the primary, or armature, current must adjust itself so that the combined mmf of all currents creates the flux required by the applied voltage.

Further examples of these basic similarities can be cited. Except for friction and windage, the losses in transformers and rotating machines are essentially the same. Tests for determining the losses and equivalent-circuit constants are essentially the same: an open-circuit, or no-load, test gives information regarding the excitation requirements and core losses (and friction and windage in rotating machines), while a short-circuit test together with dc resistance measurements gives information regarding leakage reactances and copper losses. The handling of the effects of magnetic saturation is another example: in both transformers and ac rotating machines, the leakage reactances are usually assumed to be unaffected by saturation, and the saturation of the main magnetic circuit is assumed to be determined by the resultant mutual or air-gap flux.

PROBLEMS

1-1. If the coil in Fig. 1-35 is excited with direct current, determine the coil current required to produce a flux of 7.5×10^{-4} weber in the center leg. Neglect magnetic leakage, fringing, and iron reluctance.

M-19 grade steel
29 gage laminations
Stacking height = 1.00 in.
Stacking factor = 0.94
Coil = 1,000 turns

Gap = 0.010

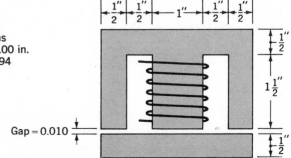

Fig. 1-35. Air-gap reactor.

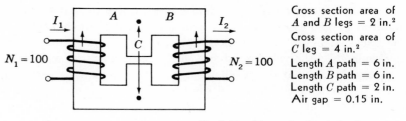

Cross section area of
A and B legs $= 2$ in.^{2}
Cross section area of
C leg $= 4$ in.^{2}
Length A path $= 6$ in.
Length B path $= 6$ in.
Length C path $= 2$ in.
Air gap $= 0.15$ in.

Fig. 1-36.   Magnetic circuit for Prob. 1-2.

If the coil is excited with alternating current instead of the direct current of part a, determine the core loss and the rms value of the magnetizing current for a sinusoidal applied voltage of 120 volts rms at 60 Hz. Neglect the effects of harmonics and the resistance drop in the coil.

1-2. The coils of the magnetic-circuit device shown in Fig. 1-36 are connected in series so that the mmf of paths A and B both tend to set up flux in the center leg C in the same direction.   The material is M-19 grade, 29-gage steel, stacking factor 0.94.   Neglect fringing and leakage.

 a. How many amperes are required to set up a flux density of 0.6 weber/m^2.

 b. How many joules of energy are stored in the magnetic field in the air gap?

1-3. A square voltage wave having a fundamental frequency of 60 Hz and equal positive and negative half cycles of amplitude E volts is impressed on a resistanceless winding of 1,000 turns surrounding a closed iron core of 10^{-3} m^2 cross section.

 a. Sketch curves of voltage and flux as functions of time.

 b. Find the maximum permissible value of E if the maximum flux density is not to exceed 1.00 weber/m^2.

1-4. Data for the top half of a symmetrical hysteresis loop for the core of Prob. 1-3 are given below:

$\mathcal{B}$, webers/m^2	0	0.2	0.4	0.6	0.7	0.8	0.9	1.0	0.95	0.9	0.8	0.7	0.6	0.4	0.2	0
$\mathcal{H}$, amp-turns/m	48	52	58	73	85	103	135	193	80	42	2	−18	−29	−40	−45	−48

The mean length of the flux paths in the core is 0.30 m.

Find graphically the hysteresis loss in watts for a maximum flux density of 1.00 weber/m² at a frequency of 60 Hz.

1-5. Figure 1-37 shows an inductor wound on a high-permeability laminated iron core of rectangular cross section. Assume that the permeability of the iron is infinite. Neglect magnetic leakage and fringing in the air gap g. The winding is insulated copper wire whose resistivity is ρ ohm-m. Assume that the fraction k_w of the winding space is available for copper, the rest of the space being used for insulation.

a. Estimate the mean length l of a turn of the winding.
b. Derive an expression for the electric power input to the coil for a specified steady flux density $\mathcal{B}$. This expression should be in terms of $\mathcal{B}$, ρ, μ_0, l, k_w, and the given dimensions. Note that the expression is independent of the number of turns if the winding factor k_w is assumed to be independent of the turns.
c. Derive an expression for the magnetic stored energy in terms of $\mathcal{B}$ and the given dimensions.
d. From (b) and (c) derive an expression for the time constant L/r of the coil.

1-6. The inductor of Fig. 1-37 has the following dimensions:

$$a = h = w = 1 \text{ cm}$$

$$b = 2 \text{ cm} \qquad g = 0.2 \text{ cm}$$

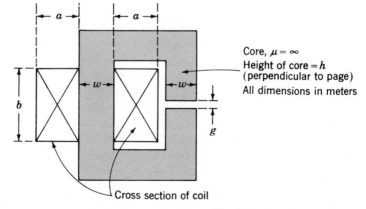

Core, $\mu = \infty$
Height of core $= h$
(perpendicular to page)
All dimensions in meters

Cross section of coil

Fig. 1-37. Iron-core inductor.

The winding space factor $k_w = 0.7$. The resistivity of copper $= 1.73$ microhm-cm.

The coil is to be operated with a constant applied voltage of 50 volts, and the air-gap flux density is to be 1.0 weber/m².

Find the power input to the coil, the coil current, the number of turns, the coil resistance, the inductance, the time constant, and the wire size to the nearest standard size.

1-7. A 500-kva 60-Hz transformer with an 11,000-volt primary winding takes 3.35 amp and 2,960 watts at no load, rated voltage and frequency. Another transformer has a core with all its linear dimensions $\sqrt{2}$ times as large as the corresponding dimensions of the first transformer. Core material and lamination thickness are the same in both transformers.

If the primary windings of both transformers have the same number of turns, what no-load current and power will the second transformer take with 22,000 volts at 60 Hz impressed on its primary?

1-8. The flux density and core loss of a transformer operating on a voltage of 6,600 volts at 60 Hz are, respectively, 70 kilolines/in.² and 2,500 watts. Suppose that all the linear dimensions of the transformer core are doubled, the numbers of turns in the primary and secondary windings are halved, and the new transformer is operated on a voltage of 13,200 volts at 60 Hz. The same grade of iron and the same thickness of laminations are used for both transformers.

What are the values of flux density and core loss for the new transformer?

1-9. The resistances and leakage reactances of a 10-kva 60-Hz 2,400:240-volt distribution transformer are as follows:

$$r_1 = 4.20 \text{ ohms} \qquad r_2 = 0.0420 \text{ ohm}$$

$$x_{l1} = 5.50 \qquad x_{l2} = 0.0550$$

where subscript 1 denotes the 2,400-volt winding, and subscript 2 the 240-volt winding. Each quantity is referred to its own side of the transformer.

a. Find the equivalent impedance referred to the high-voltage side and referred to the low-voltage side.

b. Consider the transformer to deliver its rated kva at 0.80 power factor lagging to a load on the low-tension side with 240 volts across the load. Find the high-tension terminal voltage.

1-10. A single-phase load is supplied through a 33,000-volt feeder whose impedance is $105 + j360$ ohms and a 33,000:2,400-volt transformer whose equivalent impedance is $0.26 + j1.08$ ohms referred to its low-voltage side.   The load is 180 kw at 0.85 leading power factor and 2,250 volts.

    a. Compute the voltage at the sending end of the feeder.

    b. Compute the voltage at the primary terminals of the transformer.

    c. Compute the power and reactive-power input at the sending end of the feeder.

1-11. When a 50-kva 2,300:230-volt 60-Hz transformer is operated at no load on rated voltage, the input is 200 watts at 0.15 power factor. When it is operating at rated load, the voltage drops in the total resistance and leakage reactances are, respectively, 1.2 and 1.8 percent of rated voltage.

Determine the input power and power factor when the transformer delivers 30 kw at 0.80 power factor lagging and 230 volts to a load on the low-voltage side.

1-12. A source which may be represented by a constant voltage of 5 volts rms in series with an internal resistance of 2,000 ohms is connected to a 50-ohm load resistance through an ideal transformer.   Plot the power in milliwatts supplied to the load as a function of the transformer ratio, covering ratios ranging from 0.1 to 10.0.

1-13. An audio-frequency output transformer has a primary-to-secondary turns ratio of 31.6.   Its primary inductance measured with the secondary open is 19.6 henrys and measured with the secondary short-circuited is 0.207 henry.   The winding resistances are negligible.

This transformer is used to connect an 8-ohm resistance load to a source which may be represented by a variable-frequency internal emf in series with an internal impedance of 5,000 ohms resistance.   Compute the following that relate to the frequency characteristics of the circuit:

    a. The upper half-power frequency

    b. The lower half-power frequency

    c. The geometric mean of these frequencies

    d. The ratio of load voltage to source voltage at the frequency of (*c*)

1-14. An audio-frequency output transformer, having a turns ratio of 17.32, is to be used to match a source, having an internal resistance of

3,000 ohms, to a resistance load of 10 ohms. The upper and lower half-power frequencies are to be 50 and 10,000 Hz. Neglect core loss and winding resistances. Specify:

a. The primary self-inductance
b. The equivalent leakage inductance referred to the primary

1-15. The following data were obtained for a 20-kva 60-Hz 2,400: 240-volt distribution transformer tested at 60 Hz:

	Voltage, volts	Current, amp	Power, watts
With high-voltage winding open-circuited	240	1.066	126.6
With low-voltage terminals short-circuited	57.5	8.34	284

a. Compute the efficiency at full-load current and rated terminal voltage at 0.8 power factor.
b. Assume that the load power factor is varied while the load current and secondary terminal voltage are held constant. By means of a phasor diagram, determine the load power factor for which the regulation is greatest. What is this regulation?

1-16. a. Show that the maximum efficiency of a transformer operating at a constant output voltage and power factor occurs at that kva load for which the copper losses equal the core losses. In doing so, recall that the core losses remain constant, while the copper losses vary as the square of the kva load.

b. For the transformer of Prob. 1-15, determine the kva output at maximum efficiency.

1-17. The high-voltage terminals of a 3-phase bank of three single-phase transformers are connected to a 3-wire 3-phase 13,800-volt (line to line) system. The low-voltage terminals are connected to a 3-wire 3-phase substation load rated at 1,500 kva and 2,300 volts line to line.
Specify the voltage, current, and kva ratings of each transformer (both high- and low-voltage windings) for the following connections:

a. High-voltage windings Y, low-voltage windings Δ
b. High-voltage windings Δ, low-voltage windings Y
c. High-voltage windings Y, low-voltage windings Y
d. High-voltage windings Δ, low-voltage windings Δ

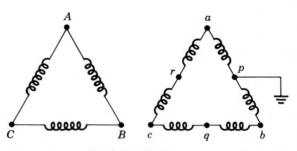

Fig. 1-38. Transformer bank, Prob. 1-18.

1-18. Figure 1-38 shows a Δ-Δ bank of 2,400:240-volt transformers. The secondaries ab, bc, ca have center taps p, q, r.   Neglect leakage-impedance voltage drops, and assume rated primary impressed voltage. With secondary voltage V_{ab} as reference phasor, draw a phasor diagram showing voltages ab, bc, ca, pq, qr, rp, ap, bp, cp.   Find the magnitudes of these voltages.

1-19. A Δ-Y-connected bank of three identical 100-kva 2,400:120-volt 60-Hz transformers is supplied with power through a feeder whose impedance is $0.80 + j0.30$ ohm per phase.   The voltage at the sending end of the feeder is held constant at 2,400 volts line to line.   The results of a single-phase short-circuit test on one of the transformers with its low-voltage terminals short-circuited are

$$V_H = 52.0 \text{ volts} \qquad f = 60 \text{ Hz}$$
$$I_H = 41.6 \text{ amp} \qquad P = 950 \text{ watts}$$

a. Determine the secondary line-to-line voltage when the bank delivers rated current to a balanced 3-phase 1.00-power-factor load.
b. Compute the currents in the transformer primary and secondary windings and in the feeder wires if a solid 3-phase shortcircuit occurs at the secondary line terminals.

1-20. A 480:120-volt 5-kva 2-winding transformer is to be used as an autotransformer to supply a 480-volt circuit from a 600-volt source. When tested as a 2-winding transformer at rated load, 0.80 power factor lagging, its efficiency is 0.965.

a. Show a diagram of connections as an autotransformer.
b. Determine its kva rating as an autotransformer.
c. Find its efficiency as an autotransformer at full load, 0.80 power factor lagging.

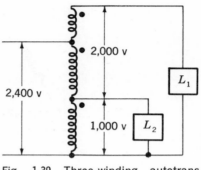

Fig. 1-39. Three-winding autotransformer, Prob. 1-21.

1-21. Figure 1-39 shows a 3-winding autotransformer supplying two loads L_1 and L_2. Voltage drops and exciting current may be neglected. Find the currents in the 3 windings for the following load conditions:

a. $L_1 = 360$ kva, $L_2 = 0$
b. $L_1 = 0$, $L_2 = 120$ kva
c. $L_1 = 360$ kva, $L_2 = 120$ kva at same power factor

1-22. A 3-phase bank consisting of three single-phase 3-winding transformers is used to step down the voltage of a 3-phase 110-kv transmission line. The following data apply to one of the transformers:

Ratings:
Primary 1:10,000 kva, 63,500 volts
Secondary 2:5,000 kva, 11,000 volts
Tertiary 3:5,000 kva, 7,580 volts

Short-circuit reactances on 5,000-kva base:
$X_{12} = 0.071$ per unit
$X_{23} = 0.054$ per unit
$X_{13} = 0.092$ per unit

Resistances are negligible.
The transformers are connected Y-Δ-Y. The Δ-connected secondaries supply their rated current to a balanced load at 0.80 power factor.

The tertiaries supply their rated current to a balanced load at 1.00 power factor.

    a. Compute the primary line-to-line voltage to maintain rated voltage at the secondary terminals.

    b. For the conditions of part *a*, compute the line-to-line voltage at the tertiary terminals.

    c. If the primary voltage is held constant as in part *a*, to what value will the tertiary voltage rise if the secondary load is removed? Consider that the tertiary load behaves as a constant resistance.

2

electromechanical-energy-conversion principles

Energy is converted to electrical form because of the ease of its transmission and processing; it is seldom available or used in electrical form, but is converted into that form at the input to a system and back to non-electrical form at the output of a system. A typical example is the processing of energy from a nuclear reactor; it is converted to electrical form at the power plant, carried by transmission and distribution lines, and converted to mechanical energy in an electric motor at the point of use. A second example is the conversion of the energy in sound pressure waves, and the transmission in electrical form from the talker to the listener in a telephone system.

We are concerned here with the electromechanical-energy-conversion process. The process takes place through the medium of the electric or the magnetic field of the conversion device. Although the various conversion devices operate on similar physical principles, the structures of the devices depend on their function. Devices for measurement and control are frequently referred to as transducers; they generally operate under linear input-output conditions and with relatively small signals. The many examples include torque motors, microphones, pickups, and loudspeakers.   A second category of devices encompasses force-producing devices and includes solenoid actuators, relays, and electromagnets. The third category of devices includes continuous energy-conversion equipment such as motors and generators.

As in many branches of engineering, energy-conversion devices were developed and used before the details of their operation were fully understood.   This chapter is devoted to the principles of electromechanical energy conversion and the analysis of such devices.   The purpose for analyzing devices is first to assist us in understanding how energy conversion takes place; the second purpose is to provide us with the techniques to design and optimize the devices for specific requirements; the third purpose is to show us how to develop models for the devices whereby their performance in systems can be treated.   The devices treated in this chapter are the signal transducers and force-producing devices; continuous energy-conversion equipment is treated in the rest of the book.

In this chapter we will start from the concept of energy balance and develop the expressions for forces of field origin.   Then we will incorporate the peripheral electrical and mechanical circuit elements and express their effect by differential equations.   Methods for solving these equations for special cases will be described and models will be developed for describing the behavior of devices in systems.

2-1
ENERGY BALANCE

A general principle applicable to all physical systems in which mass is neither created nor destroyed is the *principle of conservation of energy,* which states that energy then is neither created nor destroyed; it is merely changed in form.   This principle, together with the laws of electric and magnetic fields, electric circuits, and Newtonian mechanics, is a convenient means for finding the characteristic relationships of electromechanical coupling.   Because the frequencies and velocities involved are relatively low, quasi-static field conditions prevail and electromagnetic radiation is entirely negligible.   Electromechanical energy conversion then involves energy in four forms, and conservation of energy leads to the following relation among these forms:

$$\begin{pmatrix} \text{Energy input} \\ \text{from electrical} \\ \text{source} \end{pmatrix} = \begin{pmatrix} \text{mechanical} \\ \text{energy} \\ \text{output} \end{pmatrix} + \begin{pmatrix} \text{increase in} \\ \text{energy stored} \\ \text{in coupling} \\ \text{field} \end{pmatrix} + \begin{pmatrix} \text{energy} \\ \text{converted} \\ \text{to heat} \end{pmatrix}$$

$$(2\text{-}1)$$

Equation 2-1 is applicable to all conversion devices; it is written so that the electrical and mechanical energy terms have positive values for motor action. The equation applies equally well to generator action: the electrical and mechanical energy terms then have negative values.

Irreversible conversion of energy to heat arises from three causes: part of the electrical energy is converted directly to heat in the resistances of the current paths, part of the mechanical energy developed within the device is absorbed in friction and windage and converted to heat, and part of the energy absorbed by the coupling field is converted to heat in magnetic core loss (for magnetic coupling) or dielectric loss (for electric coupling). If the energy losses in the electrical system, the mechanical system, and the coupling field are grouped with the corresponding terms in Eq. 2-1, the energy balance may be written in the following form:

$$\begin{pmatrix} \text{Electrical} \\ \text{energy input} \\ \text{minus resis-} \\ \text{tance losses} \end{pmatrix} = \begin{pmatrix} \text{mechanical energy} \\ \text{output plus} \\ \text{friction and} \\ \text{windage losses} \end{pmatrix} + \begin{pmatrix} \text{increase in energy} \\ \text{stored in} \\ \text{coupling field plus} \\ \text{associated losses} \end{pmatrix}$$

$$(2\text{-}2)$$

The left-hand side of Eq. 2-2 can be expressed in terms of the currents and voltages in the electric circuits of the coupling device. Consider, for example, the energy-conversion device shown schematically in Fig. 2-1. The differential energy input from the electrical source in time dt is $v_t i \, dt$, where v_t is the instantaneous terminal voltage and i is the instantaneous current. The energy loss in the resistance of the device is $i^2 r \, dt$, where r is the resistance. Hence, the left-hand side of Eq. 2-2 is

$$dW_{\text{elec}} = v_t i \, dt - i^2 r \, dt \tag{2-3}$$

$$dW_{\text{elec}} = (v_t - ir) i \, dt \tag{2-4}$$

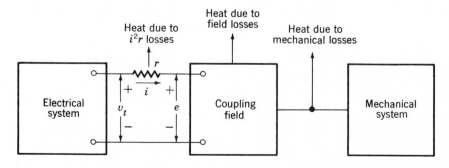

Fig. 2-1. General representation of electromechanical energy conversion.

where dW_{elec} is the net electrical energy input to the coupling device after resistance losses have been taken into account.   For the coupling device to absorb energy from the electric circuit, the coupling field must produce a reaction in the circuit.   This reaction is the emf indicated by the voltage e in Fig. 2-1.   In electromagnetic devices, for example, it is a voltage induced by the magnetic field.   Reaction on the input is an essential part of the process of transfer of energy between an electric circuit and any other medium.   The coupling field may be likened to a reservoir of energy, releasing energy to the output system and being replenished through the reaction of the field on the input system.   For the voltages to balance properly, the emf e must be

$$e = v_t - ir \tag{2-5}$$

Substitution of Eq. 2-5 in 2-4 gives

$$dW_{elec} = ei \, dt \tag{2-6}$$

If electrical energy is supplied to the coupling field from more than one circuit, the total electrical energy input is the sum of terms of the form of Eq. 2-6.

The first term on the right-hand side of Eq. 2-2 is the total energy converted to mechanical form.   It differs from the useful mechanical energy by the mechanical friction and windage losses caused by the motion of the mechanical parts of the energy-conversion device.   The second term on the right-hand side of Eq. 2-2 is the total energy absorbed by the coupling field, including both stored energy and losses.

From the foregoing discussion, it should be evident that the resistances of the electric circuits and the friction and windage of the mechanical system, though always present, play no basic parts in the energy-conversion process.   They can be accounted for as losses in the electrical and mechanical systems on the two sides of the coupling element, as indicated in block-diagram form in Fig. 2-1.   The basic energy-conversion process is one involving the coupling field and its action and reaction on the electrical and mechanical systems.   For motor action, the sum of the energy absorbed by the coupling field and the internal energy converted to mechanical form can always be equated to the internal electrical energy associated with the flow of charge against the emf e caused by the coupling field.   In differential form, Eq. 2-2 may be written

$$dW_{elec} = ei \, dt = dW_{fld} + dW_{mech} \tag{2-7}$$

where dW_mech is the differential energy converted to mechanical form and dW_fld is the differential energy absorbed by the coupling field.

Equation 2-7, together with Faraday's law for induced voltage, is the fundamental basis for analysis of energy-conversion devices.

ENERGY IN SINGLY EXCITED MAGNETIC SYSTEMS

In Chap. 1 we were concerned primarily with the closed-core magnetic circuits used for transformers. Energy in those devices is stored in the leakage fields and to some extent in the core itself; the stored energy does not enter directly into the transformation process. In this chapter we are dealing with energy conversion; the magnetic circuits have air gaps between the stationary and moving members in which considerable energy is stored in the magnetic field. The field acts as the energy-conversion medium and its energy is the reservoir between the electrical and mechanical systems.

a. Induced Voltage and Electric Input

Consider first a magnetic system comprising a single exciting coil and its associated magnetic field, as indicated schematically in Fig. 2-2. The circuit is described by

$$v_t = ir + e \tag{2-8}$$

If the flux is increasing, the emf e induced in the coil is in a direction to oppose the current, as shown by the $+$ and $-$ signs associated with the emf e in Fig. 2-2. When the flux is increasing, its time derivative is positive and the instantaneous value of the counter emf e is given by

$$e = +\frac{d\lambda}{dt} \tag{2-9}$$

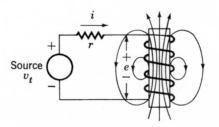

Fig. 2-2. Elementary singly excited magnetic system.

where λ is the instantaneous value of the flux linkage with the circuit and t is time.

For a winding in which all of the flux φ links all N turns of the winding, the flux linkage λ is $N\varphi$ and Eq. 2-8 may be written

$$v_t = ir + \frac{d\lambda}{dt} = ir + N\frac{d\varphi}{dt} \qquad (2\text{-}10)$$

Actually, part of the magnetic field, usually a small portion in practical devices, is distributed throughout the space occupied by the turns of the winding and therefore links only a fraction of the turns. In computing the total linkage, proper account must be taken of the actual flux linking each turn. The effect of the partial linkages can be taken into account by defining the flux as $\varphi \equiv \lambda/N$, where φ is the equivalent flux linking all N turns. In most practical devices having ferromagnetic cores the effect of the partial linkages is relatively slight, because most of the flux is confined to the core and therefore links all the turns.

As shown in Eq. 2-6, the differential energy dW_{elec} supplied by the electrical source in time dt (after heat loss $i^2r\,dt$ in the coil has been accounted for) is

$$dW_{elec} = ei\,dt \qquad (2\text{-}11)$$

From Eqs. 2-11 and 2-9,

$$dW_{elec} = i\,d\lambda = Ni\,d\varphi = \mathfrak{F}\,d\varphi \qquad (2\text{-}12)$$

where $\mathfrak{F} \equiv Ni$ is defined as the *magnetomotive force*, or *mmf*, of the coil.

Equation 2-12 shows that a change in flux linking a circuit is associated with flow of energy in the circuit. The change in flux may be caused by a change in excitation, or by mechanical motion, or by both. For example, in the relay shown in Fig. 2-3 mechanical forces are created by

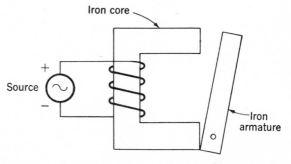

Fig. 2-3. Magnetic relay.

the magnetic field tending to shorten the air gap, and if the armature is allowed to move, the flux will change. The electrical energy input is always given by Eq. 2-12 when all factors causing the change in flux are taken into account.

b. Energy in the Magnetic Field

In the energy-balance equation 2-7, the energy associated with the magnetic field is determined by the mmf of the coil and the configuration of the magnetic material and coil. Mechanical forces are created by the field, but if there is no mechanical motion, no mechanical work will be done. For example, if the armature of the relay of Fig. 2-3 is held in a fixed position, Eq. 2-7 reduces to the special case

$$dW_{elec} = dW_{fld} + 0 \qquad (2\text{-}13)$$

The field energy for a specified configuration can then be found from the energy supplied by the source in establishing the field *with the configuration considered to be fixed.* Substitution of Eq. 2-12 in Eq. 2-13 then gives

$$dW_{fld} = i \, d\lambda = \mathfrak{F} \, d\varphi \qquad (2\text{-}14)$$

That is, for a fixed configuration, the electrical energy input $i \, d\lambda$ associated with a change in flux is absorbed by the field.

The energy absorbed by the field in changing the flux linkage from λ_1 to λ_2, or the flux from φ_1 to φ_2, is

$$\Delta W_{fld} = \int_{\lambda_1}^{\lambda_2} i(\lambda) \, d\lambda = \int_{\varphi_1}^{\varphi_2} \mathfrak{F}(\varphi) \, d\varphi \qquad (2\text{-}15)$$

where the functional notation $i(\lambda)$ and $\mathfrak{F}(\varphi)$ is used to emphasize the fact that λ and φ are the variables of integration. If the initial flux is zero, the energy absorbed by the field when flux linkage λ, or flux φ, is established is

$$W_{fld} = \int_{0}^{\lambda} i(\lambda) \, d\lambda = \int_{0}^{\varphi} \mathfrak{F}(\varphi) \, d\varphi \qquad (2\text{-}16)$$

In these equations, the mmf is a function of the flux, the relation between them depending on the geometry of the coil and magnetic circuit and on the magnetic properties of the core material.

The relay of Fig. 2-3 has two sources of electrical energy losses: the resistance of the winding and the eddy current and hysteresis losses of the iron. When the flux φ is raised from φ_1 to φ_2 then returned to φ_1 by adjusting the current i, the energy delivered by the source is not all returned because of these losses. If we assume that the elements causing the losses are separated and included in the electrical source, then the relay will return all of the energy to the source after traversing a path of flux φ_1 to φ_2 and return to φ_1. The relay is now termed a *conservative* system.

A property of a conservative system is that its energy is a function only of its *state*, as described by the independent variables which describe the state. In the case of the relay, the independent variable φ describes the state as shown in Eq. 2-16. Hence, the energy at any value of φ is independent of how the relay was brought to that value of φ. The same is true for $N\varphi$ or λ. Keep in mind that the relay is assumed lossless and its mechanical position is fixed when we describe its state by the single independent variable φ. We will introduce mechanical motion in a later article.

Electromagnetic-energy-conversion devices are built with air gaps in the magnetic circuit to separate the fixed and moving parts. Most of the mmf of the windings is required to overcome the air-gap reluctance so that most of the energy is then stored in the air gap and is returned to the electric source when the field is reduced. Because of the simplicity of the resulting relations, magnetic nonlinearity and core losses are often neglected in the analysis of practical devices. The final results of such approximate analyses can, if necessary, be corrected for the effects of these neglected factors by semiempirical methods. Consequently analyses are carried out under the assumption that the flux and mmf are directly proportional, as in air, for the entire magnetic circuit. The relationship between flux φ and mmf $\mathfrak{F}$ is given by the *reluctance* $\mathfrak{R}$ and the *permeance* $\mathcal{P}$, defined as

$$\mathfrak{R} = \frac{\mathfrak{F}}{\varphi} \tag{2-17}$$

$$\mathcal{P} = \frac{\varphi}{\mathfrak{F}} = \frac{1}{\mathfrak{R}} \tag{2-18}$$

With nonlinearity and hysteresis neglected, the reluctance and permeance are constant with φ and $\mathfrak{F}$; from Eqs. 2-16 to 2-18, we can then specify the energy as

$$W_{\text{fld}} = \tfrac{1}{2}i\lambda = \tfrac{1}{2}\mathfrak{F}\varphi = \tfrac{1}{2}\mathfrak{R}\varphi^2 \tag{2-19}$$

Under the same conditions of linearity, the *self-inductance* L of the coil in henrys is defined as the flux linkage per ampere, or

$$L = \frac{\lambda}{i} = \frac{N\varphi}{i} = N^2\mathcal{P} \qquad (2\text{-}20)$$

Substitution of this relation in Eq. 2-19 gives the expression for the energy stored in the field at constant inductance,

$$W_{\text{fld}} = \frac{1}{2}\frac{\lambda^2}{L} \qquad (2\text{-}21)$$

The energy associated with the field is distributed throughout the space occupied by the field. For a magnetic medium with no losses and *constant permeability*, the energy density is

$$w_{\text{fld}} = \frac{1}{2}\mathcal{H}\mathcal{B} = \frac{1}{2}\frac{\mathcal{B}^2}{\mu} \qquad (2\text{-}22)$$

where w_{fld} is expressed in joules per cubic meter; the magnetic-field intensity is $\mathcal{H}$ in ampere-turns per meter, the flux density is $\mathcal{B}$ in webers per square meter, and the permeability is μ in rationalized mks units.

In the foregoing discussion the field-energy relations have been expressed in three ways, and it is appropriate to comment briefly on the three viewpoints. In Eq. 2-22 the magnetic stored energy is expressed in terms of the specific or per-unit-volume properties of the magnetic field. This viewpoint is that of the designer. He thinks in terms of properties of materials and field intensities, stress intensities, flow densities, and like concepts. He then builds up the geometrical form and arrangement of any specific device from a knowledge of what he can do with unit volume of the available materials. In Eq. 2-21, the field energy is expressed in terms of flux linkage and inductance, familiar and useful concepts, particularly when nonlinearity is unimportant. The viewpoint here is that of the circuit analyst. Except for difficulties in taking into account nonlinearity, the theory of the operating characteristics of most electromagnetic-energy-conversion devices can be developed on the basis of assuming the device to be a circuit element with time-varying inductance parameters. This viewpoint, however, gives very little insight into the internal phenomena and gives no conception of physical size. In Eq. 2-19, the field energy is expressed in terms of the whole field. The viewpoint here is somewhere between the other two. The expressions readily can be translated into the language of either the designer or the circuit analyst. In this text all three viewpoints will be taken at various times.

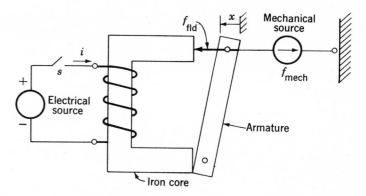

Fig. 2-4.   Relay restrained by external mechanical force.

2-3

MECHANICAL FORCE AND ENERGY

The relay of Fig. 2-3, as an example of an electromagnetic-energy-conversion device, is shown in Fig. 2-4 with an external source of mechanical force f_{mech} acting to maintain the armature at a particular displacement x. Previously, we had merely assumed that the position of the armature was fixed in dealing with the magnetic-field energy introduced from the electrical terminals, without showing the force holding it fixed.   The force produced by the magnetic field f_{fld} is also shown in Fig. 2-4 acting to move the armature in the direction x.   In the static case, and assuming no friction forces, the field force and the mechanical force are in balance,

$$f_{fld} = f_{mech} \qquad\qquad (2\text{-}23)$$

If the armature is allowed to move a distance dx, then the field does work on the armature dW_{mech} because the force and displacement are in the same direction,

$$dW_{mech} = f_{fld}\, dx \qquad\qquad (2\text{-}24)$$

However, the mechanical force f_{mech} and displacement are in opposite directions so that the work done by the mechanical source is negative, that is, the mechanical source *absorbs* the energy,

$$dW_{mech} = f_{mech}\, dx = f_{fld}\, dx \qquad\qquad (2\text{-}25)$$

If the armature of Fig. 2-4 were accelerating or loaded mechanically, then the forces of the field and the mechanical source would not be equal; Eq. 2-23 would have to be replaced by a differential equation that accounted for all of the force components.

We can now return to Eqs. 2-7 and 2-12, and substitute Eq. 2-25 to obtain

$$dW_{elec} = i \, d\lambda = dW_{fld} + f_{mech} \, dx \tag{2-26}$$

Equation 2-26 is more general than Eq. 2-14 for the change in the magnetic-field energy of the singly excited device. If the armature is assumed stationary, then $dx = 0$ and the change in field energy all comes from the electrical source, as in Eq. 2-14,

$$dW_{fld} = i \, d\lambda \tag{2-27}$$

If the flux linkages are assumed stationary, $d\lambda = 0$, then the change in magnetic-field energy all comes from the mechanical source,

$$dW_{fld} = - f_{mech} \, dx \tag{2-28}$$

The condition $d\lambda = 0$ results from imposing zero voltage at the terminals of the resistanceless winding on the relay of Fig. 2-4. However, we must have a finite λ to insure a force f_{fld} to transfer energy into the field from the mechanical terminals. For example, we can start a current i in the winding at some fixed displacement x of the armature and then short-circuit the winding terminals. From Eq. 2-22 we see that the energy density w_{fld} in the air gap is constant at constant $\mathcal{B}$ so that the total field energy W_{fld} is then proportional to the air-gap volume. If x is allowed to increase, the volume is reduced and energy flows from the magnetic field to the mechanical source. If x is reduced under the pull of the mechanical source, energy flows from the mechanical source to the magnetic field in accordance with Eq. 2-28.

Equations 2-27 and 2-28 show the true nature of the magnetic field as the coupling medium between the electrical and mechanical systems. Energy can be made to flow into or out of the field from either system depending upon the conditions imposed on the independent variables λ and x. In some machines and devices, the magnetic-field energy remains constant during the energy-conversion process; the electrical and mechanical rates of energy flow remain equal. In other machines, such as in single-phase ac motors, the process is much more complicated, with interchanges taking place during each revolution of the shaft.

STATE FUNCTIONS, VARIABLES, COENERGY

Energy is a *state function* of a conservative system.   Equation 2-26 as rearranged in the following form

$$dW_{\text{fld}}(\lambda,x) = i\, d\lambda - f_{\text{mech}}\, dx \qquad (2\text{-}26)$$

shows the magnetic-field energy of the singly excited lossless device as a function of the two independent variables λ and x.   The differential of the energy $dW_{\text{fld}}(\lambda,x)$ can be expressed mathematically in terms of the partial derivatives as

$$dW_{\text{fld}}(\lambda,x) = \frac{\partial W_{\text{fld}}}{\partial \lambda}\, d\lambda + \frac{\partial W_{\text{fld}}}{\partial x}\, dx \qquad (2\text{-}29)$$

Since the variables λ and x are independent, the coefficients of the terms of Eqs. 2-26 and 2-29 must also be independently equal, leading to the parametric equations

$$i = \frac{\partial W_{\text{fld}}(\lambda,x)}{\partial \lambda} \qquad (2\text{-}30)$$

$$f_{\text{mech}} = f_{\text{fld}} = -\frac{\partial W_{\text{fld}}(\lambda,x)}{\partial x} \qquad (2\text{-}31)$$

Equation 2-30 for the current i corresponds to Eq. 2-27 for the field energy when the armature is restrained, $dx = 0$.   Equation 2-31 for the mechanical force f_{mech}, which also equals the field force f_{fld}, corresponds to Eq. 2-28 for the field energy when the flux linkage is restrained, $d\lambda = 0$.

Note that the energy always has been expressed until now in terms of the electrical variable λ.   The force in Eq. 2-31 is obtained as a function of the flux linkage λ.   Once we obtain the expression for the force, we can use the equations for the connected system to find the force in terms of the current i, if we so desire.   The alternative to obtaining the force as a function of λ is to use a different state function than energy, namely, the *coenergy*, to obtain the force as a function of current.   The selection of the state function is a matter of convenience; it depends upon the desired variables in the result and the initial description of the system being analyzed.

The coenergy W_{fld}' is defined as a function of i and x such that

$$W_{\text{fld}}'(i,x) = i\lambda - W_{\text{fld}}(\lambda,x) \qquad (2\text{-}32)$$

and can be obtained from the energy expression of Eq. 2-26. The transformation is carried out by using the differential of $i\lambda$

$$d(i\lambda) = i\,d\lambda + \lambda\,di \qquad\qquad \cdot \qquad\qquad (2\text{-}33)$$

and the differential of $dW_{\text{fld}}(\lambda,x)$ from Eq. 2-26. The differential of the coenergy is

$$dW'_{\text{fld}}(i,x) = d(i\lambda) - dW_{\text{fld}}(\lambda,x) \qquad\qquad (2\text{-}34a)$$

We can substitute Eqs. 2-33 and 2-26 in Eq. 2-34a to obtain

$$dW'_{\text{fld}}(i,x) = i\,d\lambda + \lambda\,di - i\,d\lambda + f_{\text{mech}}\,dx \qquad\qquad (2\text{-}34b)$$

which reduces simply to

$$dW'_{\text{fld}}(i,x) = \lambda\,di + f_{\text{mech}}\,dx \qquad\qquad (2\text{-}34c)$$

As in the case of the energy $W_{\text{fld}}(\lambda,x)$ in Eq. 2-29, the differential of the coenergy $W'_{\text{fld}}(i,x)$ can be expressed as

$$dW'_{\text{fld}}(i,x) = \frac{\partial W'_{\text{fld}}}{\partial i}\,di + \frac{\partial W'_{\text{fld}}}{\partial x}\,dx \qquad\qquad (2\text{-}34d)$$

Since the variables i and x are now independent, the coefficients of the terms of Eqs. 2-34c and 2-34d must be independently equal, resulting in the parametric equations

$$\lambda = \frac{\partial W'_{\text{fld}}(i,x)}{\partial i} \qquad\qquad (2\text{-}35)$$

$$f_{\text{mech}} = f_{\text{fld}} = + \frac{\partial W'_{\text{fld}}(i,x)}{\partial x} \qquad\qquad (2\text{-}36)$$

Compare Eqs. 2-31 and 2-36. The first gives the force in terms of the flux linkage λ; the second gives the force in terms of the current i. The selection of energy or coenergy to determine the force usually depends on the variables required for the resulting expressions.

The coenergy for a singly excited system when the armature position is fixed, $dx = 0$, is obtained from Eq. 2-34c as

$$W'_{\text{fld}} = \int_0^i \lambda\,di \qquad\qquad (2\text{-}37)$$

For a linear system in which λ is proportional to i, that is, constant inductance for change in λ, or constant μ for change in $\mathcal{B}$, the coenergy in circuit form is

$$W'_{\text{fld}} = \int_0^i Li\ di = \tfrac{1}{2}Li^2 \tag{2-38}$$

In coenergy density form it is

$$w'_{\text{fld}} = \int_0^{\mathcal{H}} \mathcal{B}\ d\mathcal{H} = \int_0^{\mathcal{H}} \mu\mathcal{H}\ d\mathcal{H} = \tfrac{1}{2}\mu\mathcal{H}^2 \tag{2-39}$$

And, in total coenergy field form it is

$$W'_{\text{fld}} = \tfrac{1}{2}\lambda i = \tfrac{1}{2}\mathcal{F}\varphi = \tfrac{1}{2}\mathcal{P}\mathcal{F}^2 \tag{2-40}$$

These expressions are recognized as counterparts to Eqs. 2-21, 2-22, and 2-19, respectively, for the energy function.

For a linear system, the energy and coenergy are numerically equal; for example, $\tfrac{1}{2}Li^2 = \tfrac{1}{2}\lambda^2/L$; $\tfrac{1}{2}\mu\mathcal{H}^2 = \tfrac{1}{2}\mathcal{B}^2/\mu$; and $\tfrac{1}{2}\mathcal{P}\mathcal{F}^2 = \tfrac{1}{2}\mathcal{R}\varphi^2$. For a nonlinear system in which λ and i or $\mathcal{B}$ and $\mathcal{H}$ are not proportional, the two functions are not even numerically equal. A graphical interpretation of the energy and coenergy for a nonlinear system is shown in Fig. 2-5. The area between the λ-i curve and the vertical axis given by the integral of $i\ d\lambda$ is the energy. The area to the horizontal axis given by the integral of $\lambda\ di$ is the coenergy. The sum for the singly excited system is by definition

$$W_{\text{fld}} + W'_{\text{fld}} = \lambda i \tag{2-41}$$

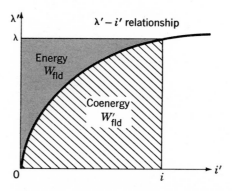

Fig. 2-5. Graphical interpretation of energy and coenergy in singly excited system.

The force of the field in the device of Fig. 2-4 for some particular value of x and i or λ must be independent of whether it is calculated from the energy or coenergy. A graphical illustration will demonstrate this point. Assume that the armature is at position x so that the device is operating at point a in Fig. 2-6a. The partial derivative of Eq. 2-31 can be interpreted as the limit of $-\Delta W_{\mathrm{fld}}/\Delta x$ with λ constant as $\Delta x \to 0$. If we allow a change Δx, the change $-\Delta W_{\mathrm{fld}}$ is shown by the shaded area in Fig. 2-6a. Hence, the force $f_{\mathrm{fld}} = $ (shaded area)$/\Delta x$ as $\Delta x \to 0$. On the other hand, the partial derivative of Eq. 2-36 can be interpreted as the limit of $\Delta W'_{\mathrm{fld}}/\Delta x$ with i constant as $\Delta x \to 0$. This perturbation of the device is shown in Fig. 2-6b; the force $f_{\mathrm{fld}} = $ (shaded area)$/\Delta x$ as $\Delta x \to 0$. The shaded areas only differ by the small triangle abc of sides Δi and $\Delta \lambda$, so that in the limit the shaded areas resulting from Δx at constant λ or at constant i are equal. Thus the force of the field is independent of whether the determination is made with energy or coenergy.

Equations 2-31 and 2-36 express the mechanical force of electrical origin in terms of partial derivatives of the energy and coenergy functions $W_{\mathrm{fld}}(\lambda,x)$ and $W'_{\mathrm{fld}}(i,x)$. It is important to note two things about them: (1) the variables in terms of which they must be expressed and (2) their algebraic signs. Physically, of course, the force depends on the dimension x and the magnetic field. The field can be specified in terms of flux linkage λ, or current i, or related variables. Mathematically, however, it

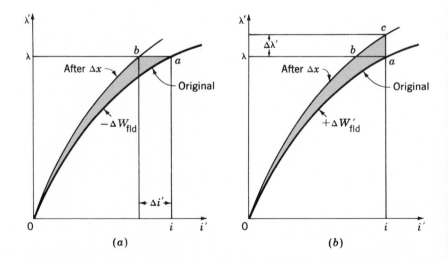

Fig. 2-6. Effect of Δx on energy and coenergy of singly excited device. (a) Change of energy, λ held constant; (b) change of coenergy, i held constant.

makes a great deal of difference which energy function is used and the variables in terms of which it is expressed. The selection of the energy or coenergy function as a basis for analysis is a matter of convenience; the choice depends upon the initial description of the system and the desired variables in the result.

The algebraic signs in Eqs. 2-31 and 2-36 show that the field force acts in a direction to decrease the magnetic-field stored energy at constant flux or to increase the coenergy at constant current. In a singly excited device the force acts to increase the inductance by pulling on members so as to reduce the reluctance of the magnetic path linking the winding.

EXAMPLE 2-1

The λ-i characteristics of a magnetic circuit are frequently described with straight-line segments as shown in Fig. 2-7. The circuit is considered linear to point a and in saturation from a to b. Find the energy W_{fld} and coenergy W'_{fld} for the magnetic circuit at point a and at point b.

Solution

The energy is given by Eq. 2-16 as

$$W_{\text{fld}} = \int_0^{\lambda} i \, d\lambda$$

The energy at point a (note that the equation of curve $0a$ is $\lambda = i$) is

$$W_{\text{fld}} = \int_0^{1.0} \lambda \, d\lambda = 0.5 \text{ joule}$$

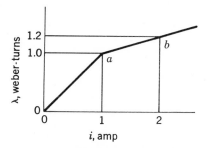

Fig. 2-7. Straight-line approximation of λ–i characteristics.

The energy at point b includes the additional energy

$$\Delta W_{fld} = \int_{1.0}^{1.2} (5\lambda - 4) \, d\lambda = 0.3 \text{ joule}$$

The total energy at point b is thus 0.8 joule.
The coenergy is given by Eq. 2-37 as

$$W'_{fld} = \int_0^i \lambda \, di$$

The coenergy at point a, $i = 1$ amp, is

$$W'_{fld} = \int_0^{1.0} i \, di = 0.5 \text{ joule}$$

The energy and coenergy at point a are seen to be equal.
The coenergy at point b includes the additional coenergy,

$$\Delta W'_{fld} = \int_{1.0}^{2.0} (0.2i + 0.8) \, di = 1.1 \text{ joules}$$

The total coenergy at point b is 1.6 joules and is larger than the energy of 0.8 joule at the same point.

EXAMPLE 2-2

The magnetic circuit shown in Fig. 2-8 is made of cast steel. The rotor is free to turn about a vertical axis. The dimensions are shown in the figure.

a. Derive an expression in mks rationalized units for the torque acting on the rotor in terms of the dimensions and the magnetic field in the two air gaps. Neglect the effects of fringing.
b. The maximum flux density in the overlapping portions of the air gaps is limited to approximately 130 kilolines/in.^{2}, because of saturation in the steel. Compute the maximum torque in inch-pounds for the following dimensions: $r_1 = 1.00$ in.; $h = 1.00$ in.; $g = 0.10$ in.

Solution

The torque T can be obtained from the derivative of field coenergy with respect to angle θ, as in Eq. 2-36.

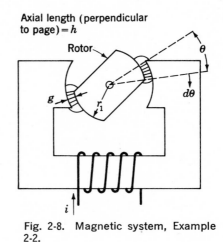

Fig. 2-8. Magnetic system, Example 2-2.

a. The air-gap field intensity $\mathcal{3C}_{ag}$ will be used as the independent variable because it is related to the terminal variable i by a constant, that is, $Ni = 2g\mathcal{3C}_{ag}$. The field coenergy density is $\mu_0\mathcal{3C}_{ag}^2/2$ (Eq. 2-39), and the volume of the two overlapping air gaps is $2gh(r_1 + 0.5g)\theta$. Consequently, the field coenergy is

$$W'_{ag} = \mu_0\mathcal{3C}_{ag}^2 gh(r_1 + 0.5g)\theta$$

$$T = \frac{\partial W'_{ag}(\mathcal{3C}_{ag},\theta)}{\partial\theta}$$

$$T = \mu_0\mathcal{3C}_{ag}^2 gh(r_1 + 0.5g) = \frac{\mathcal{B}_{ag}^2 gh(r_1 + 0.5g)}{\mu_0}$$

The torque acts in a direction to align the rotor with the stator pole faces.

b. Convert the flux density and dimensions to mks units.

$$\mathcal{B}_{ag} = \frac{130{,}000}{6.45} \times 10^4 \times 10^{-8} = 2.02 \text{ webers/m}^2$$

$$g = 0.1 \times 2.54 \times 10^{-2} = 0.00254 \text{ m}$$

$$h = r_1 = 1.00 \times 2.54 \times 10^{-2} = 0.0254 \text{ m}$$

$$\mu_0 = 4\pi \times 10^{-7}$$

Substitution of these numerical values gives

$$T = 5.56 \text{ newton-meters}$$
$$= 5.56 \times 0.738 \times 12 = 49.3 \text{ in.-lb}$$

Moving-iron devices are used in a wide variety of applications for producing mechanical force or torque. Some of them, such as lifting magnets and magnetic chucks, are required merely to hold a piece of ferromagnetic material; others, such as iron-core solenoids, relays, and contactors, are required to exert a force through a specified distance; still others, such as iron-vane instruments, are required to produce a rotational torque against a restraining spring so that the deflection of a pointer is indicative of the steady-state value of the current or voltage of the circuit to which they are connected; with others, such as moving-iron telephone receivers and the electromagnets used for controlling the operation of hydraulic motors in servomechanisms, the force or torque should be very nearly proportional to an electrical signal, and the dynamic response should be as rapid as possible. The dynamics of electromagnetically coupled systems are discussed in Art. 2-7.

2-5
SINGLY EXCITED ELECTRIC–FIELD SYSTEMS

An electric-field-energy-conversion system can be treated in an analogous manner to the magnetic-field counterpart to find the force produced by the electric field and the charge or voltage at the electrical terminals. The selection of independent variables determines whether the state function is energy or coenergy, and the form of the parametric equations.

a. Energy Balance

The representation of a singly excited electric-field device is shown in Fig. 2-9. The electrical terminals of the device are supplied from a current source I with a shunt loss element G. The movable plate is connected to an unspecified mechanical system. The flow of energy for any change in the system can be expressed as

$$dW_{\text{elec}} = dW_{\text{fld}} + dW_{\text{mech}} \tag{2-42}$$

With i and q representing the current and charge into the device, the electrical energy into the terminals is

$$dW_{\text{elec}} = vi \, dt = v \, dq \tag{2-43}$$

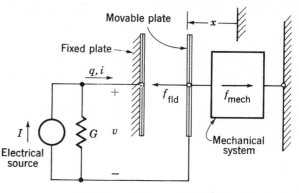

Fig. 2-9.   Singly excited electric-field device.

and the mechanical energy out of the device is

$$dW_{\text{mech}} = f_{\text{mech}} \, dx \tag{2-44}$$

Rewriting Eq. 2-42 in terms of Eqs. 2-43 and 2-44 yields the energy balance as

$$v \, dq = dW_{\text{fld}} + f_{\text{mech}} \, dx \tag{2-45}$$

The system described by Eq. 2-45 is assumed lossless; the loss element G is lumped with the source and the remainder of the system is assumed to have no dielectric loss, friction, or other losses.   The system is conservative; all energy supplied to the system is recoverable.

b. Energy

Rearrangement of Eq. 2-45 to

$$dW_{\text{fld}} = v \, dq - f_{\text{mech}} \, dx \tag{2-45a}$$

shows that the electric-field energy can be introduced through either the electrical or mechanical terminals.   Regardless of how it is introduced the expression for energy in terms of the independent variables q and x will correctly describe the energy because energy is a state function. Hence, for the energy introduced from the electrical terminals with the mechanical terminal fixed, $dx = 0$,

$$W_{\text{fld}} = \int_0^q v \, dq \tag{2-46}$$

and, if the electrical terminal is *opened* to maintain $dq = 0$, the energy is

$$W_{fld} = -\int_0^x f_{mech}\, dx \qquad (2\text{-}47)$$

To evaluate these integrals, we must have the dependence of v on q and f_{mech} on x.

In a linear-electric-field device in which q is proportional to v, that is, permittivity ϵ is constant, the relationship between q and v is defined by capacitance,

$$C \equiv \frac{q}{v} \qquad (2\text{-}48)$$

and the circuit expression for energy from Eq. 2-46 is

$$W_{fld} = \frac{1}{2}\frac{q^2}{C} \qquad (2\text{-}49)$$

By substitution of C from Eq. 2-48, the total energy becomes

$$W_{fld} = \tfrac{1}{2}vq \qquad (2\text{-}50)$$

The analogy between the electric- and magnetic-field cases can be seen more fully by expressing the energy density as a function of the potential gradient $\mathcal{E}$ and the electric-field flux density $\mathcal{D} = \epsilon\mathcal{E}$. Thus

$$w_{fld} = \int_0^{\mathcal{D}} \mathcal{E}\, d\mathcal{D} = \frac{1}{2}\frac{\mathcal{D}^2}{\epsilon} \qquad (2\text{-}51)$$

Compare Eqs. 2-49, 2-50, and 2-51 with Eqs. 2-19, 2-21, and 2-22.

c. Variables, Coenergy

The electric-field energy is a function of the independent variables q and x. For example, consider Eq. 2-49 for a single parallel-plate capacitor of plate area A, spacing x, and permittivity ϵ_0; the capacitance $C = \epsilon_0 A/x$ and the energy $W_{fld} = \tfrac{1}{2}q^2x/\epsilon_0 A$, clearly a function of q and x. The consequence of the selection of energy as the state function to describe the system is that the force f_{mech} will be a function of q and x. If the analysis of the overall system would be easier with the force expressed as a function of v and x, then the coenergy should be selected as the state function. The coenergy $W'_{fld}(v,x)$ of the electric-field system is defined as

$vq - W_{\text{fld}}(q,x)$. Substituting Eq. 2-45a for the energy and expanding $d(vq)$ we obtain

$$dW'_{\text{fld}}(v,x) = d(vq) - dW_{\text{fld}}(q,x) = q\,dv + f_{\text{mech}}\,dx \qquad (2\text{-}52)$$

The coenergy for the system can be evaluated by holding the mechanical terminal fixed, $dx = 0$, and bringing the system up to the voltage v so that all of the coenergy enters through the electrical terminals

$$W'_{\text{fld}}(v,x) = \int_0^v q\,dv \qquad (2\text{-}53)$$

For a linear system where v is proportional to q, ϵ is constant and capacitance C can be defined as before. The circuit expression for coenergy is then

$$W'_{\text{fld}}(v,x) = \tfrac{1}{2}Cv^2 \qquad (2\text{-}54)$$

where capacitance C is a function of x. The coenergy density is now

$$w'_{\text{fld}} = \tfrac{1}{2}\epsilon\mathcal{E}^2 \qquad (2\text{-}55)$$

d. Force

The force produced by the electric field on the mechanical terminal can be found from either the energy or the coenergy of the system. The energy is a function of q and x; the differential of the energy is

$$dW_{\text{fld}}(q,x) = \frac{\partial W_{\text{fld}}}{\partial q}\,dq + \frac{\partial W_{\text{fld}}}{\partial x}\,dx \qquad (2\text{-}56)$$

The coefficients of each of the two terms of Eq. 2-56 must be independently equal to the coefficients of Eq. 2-45, which yields the parametric equations for voltage and force

$$v = \frac{\partial W_{\text{fld}}(q,x)}{\partial q} \qquad (2\text{-}57)$$

$$f_{\text{mech}} = f_{\text{fld}} = -\frac{\partial W_{\text{fld}}(q,x)}{\partial x} \qquad (2\text{-}58)$$

The coenergy is a function of v and x; the differential of coenergy is

$$dW'_{\text{fld}}(v,x) = \frac{\partial W'_{\text{fld}}}{\partial v}\,dv + \frac{\partial W'_{\text{fld}}}{\partial x}\,dx \qquad (2\text{-}59)$$

The coefficients of each of the two terms of Eq. 2-59 must be independently equal to the coefficients of Eq. 2-52, which yields the parametric equations for charge and force

$$q = \frac{\partial W'_{fld}(v,x)}{\partial v} \tag{2-60}$$

$$f_{mech} = f_{fld} = + \frac{\partial W'_{fld}(v,x)}{\partial x} \tag{2-61}$$

Equations 2-58 and 2-61 for the electric-field system correspond to Eqs. 2-31 and 2-36 for the magnetic-field system.

EXAMPLE 2-3

Find the force between two parallel plates each of area $A = 1$ m² and with an electric field between them at the nominal breakdown strength of air of 3×10^6 volts/m. Use both the energy and the coenergy.

Solution

Using the model of Fig. 2-9, the spacing between the plates can be taken as $x_0 - x$ and the capacitance as $C = A\epsilon_0/(x_0 - x)$. The energy is given by Eq. 2-49 as

$$W_{fld}(q,x) = \frac{1}{2}\frac{q^2}{C} = \frac{1}{2}\frac{q^2(x_0 - x)}{A\epsilon_0}$$

and the force by Eq. 2-58 as

$$f_{fld} = -\frac{\partial W_{fld}(q,x)}{\partial x} = \frac{1}{2}\frac{q^2}{A\epsilon_0}$$

The charge q is given by $q = \mathcal{D}A = \mathcal{E}\epsilon_0 A$, so that the force is

$$f_{fld} = \frac{1}{2}\mathcal{E}^2\epsilon_0 A = \frac{1}{2}(3 \times 10^6)^2 \left(\frac{1}{36\pi} \times 10^{-9}\right)(1) = \frac{1}{8\pi} \times 10^3 \text{ newtons}$$

The coenergy is given by Eq. 2-54 as

$$W'_{fld}(v,x) = \frac{1}{2}Cv^2 = \frac{1}{2}v^2\frac{A\epsilon_0}{(x_0 - x)}$$

and the force by Eq. 2-61 as

$$f_{\text{fld}} = \frac{\partial W'_{\text{fld}}(v,x)}{\partial x} = \frac{1}{2} v^2 \frac{A\epsilon_0}{(x_0 - x)^2}$$

The voltage v is given by $v = \mathcal{E}(x_0 - x)$, so that the force is

$$f_{\text{fld}} = \frac{1}{2} \mathcal{E}^2 \epsilon_0 A = \frac{1}{2} (3 \times 10^6)^2 \left(\frac{1}{36\pi} \times 10^{-9} \right) (1) = \frac{1}{8\pi} \times 10^3 \, \text{newtons}$$

Both state functions yield the same force, as they should. Note that the derivatives for the force are taken with the electrical variables of q and v explicitly shown; the numerical values for $\mathcal{E}$ are substituted *after* the force expression is obtained.

It is interesting to compare the force produced on a square meter of surface bounding a magnetic field with the value found in the example. The value of electric field in the example was taken as the nominal breakdown strength of air. The value of the magnetic field can be taken as $\mathcal{B} = 1.6$ webers/m^2, which is a typical saturation level for ferromagnetic material. The total energy in a volume of $A = 1$ m^2 area and $(x_0 - x)$ spacing is given by Eq. 2-22 as

$$W_{\text{fld}}(\mathcal{B},x) = \frac{1}{2} \frac{\mathcal{B}^2 A (x_0 - x)}{\mu_0}$$

and the force from Eq. 2-31 as

$$f_{\text{fld}} = - \frac{\partial W_{\text{fld}}(\mathcal{B},x)}{\partial x} = \frac{1}{2} \frac{\mathcal{B}^2 A}{\mu_0}$$

$$= \frac{1}{2} \frac{(1.6)^2}{4\pi \times 10^{-7}} = \frac{0.32}{\pi} \times 10^7 \, \text{newtons}$$

The force density on the bounding surfaces in the magnetic field is about 25,000 times as great as that in the electric field at the field strengths assumed. This example shows why practically all energy-conversion devices utilize the magnetic field rather than the electric field as the coupling medium.

The force acts on any members to pull them in a direction to increase the capacitance; for example, it acts to pull the plates or electrodes together and to pull dielectric material both into the interelectrode space and into the region of highest electric-field strength.

MULTIPLY EXCITED MAGNETIC–FIELD SYSTEMS

Singly excited devices are generally used to develop bulk and noncontrolled forces; examples are relays, solenoids, and force actuators of various kinds. To obtain forces proportional to electrical signals, and signals proportional to forces and velocities, devices must be used which have two or more paths for excitation and energy exchange with sources. Permanent magnets are frequently used as one of the excitation paths. In many devices, one excitation path sets the level of the electrical or magnetic field while the other handles signals. Examples of signal-handling devices are loudspeakers, torque motors, pickups, and tachometers. All of the known types of motors and generators with minor exceptions are examples of continuous energy-conversion power-handling devices.

The model of a simple system with two sets of electrical terminals and one mechanical terminal is shown in Fig. 2-10. The system must be described in terms of three independent variables; these can be the flux linkages λ_1, λ_2 and the mechanical angle θ, or the currents i_1, i_2 and the angle θ, or a hybrid set.[1] When the flux linkages are used, the para-

---

[1] See, for example, H. H. Woodson and J. R. Melcher, "Electromechanical Dynamics," Part 1, John Wiley & Sons, Inc., New York, 1968, Chap. 3.

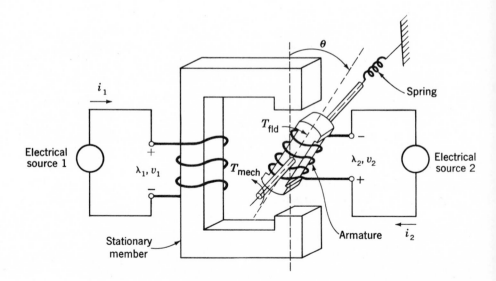

Fig. 2-10.   Elementary multiply excited magnetic system.

metric equations are extensions of Eqs. 2-30 and 2-31

$$i_1 = \frac{\partial W_{\text{fld}}(\lambda_1, \lambda_2, \theta)}{\partial \lambda_1} \tag{2-62}$$

$$i_2 = \frac{\partial W_{\text{fld}}(\lambda_1, \lambda_2, \theta)}{\partial \lambda_2} \tag{2-63}$$

$$T_{\text{fld}} = - \frac{\partial W_{\text{fld}}(\lambda_1, \lambda_2, \theta)}{\partial \theta} \tag{2-64}$$

where the energy is given by

$$W_{\text{fld}}(\lambda_1, \lambda_2, \theta) = \int_0^{\lambda_1} i_1 \, d\lambda_1 + \int_0^{\lambda_2} i_2 \, d\lambda_2 \tag{2-65}$$

and the system is held fixed at position θ as the integration is carried out. In a linear system, the relationships between λ and i are specified in terms of inductances

$$\lambda_1 = L_{11}i_1 + L_{12}i_2 \tag{2-66a}$$

$$\lambda_2 = L_{21}i_1 + L_{22}i_2 \tag{2-66b}$$

Solving Eqs. 2-66a and 2-66b for i_1 and i_2, substituting them in Eq. 2-65, and integrating yields

$$W_{\text{fld}}(\lambda_1, \lambda_2, \theta) = \tfrac{1}{2}\Gamma_{11}\lambda_1^2 + \Gamma_{12}\lambda_1\lambda_2 + \tfrac{1}{2}\Gamma_{22}\lambda_2^2 \tag{2-66c}$$

where

$$\Gamma_{11} = L_{22}/D$$

$$\Gamma_{22} = L_{11}/D$$

$$\Gamma_{12} = \Gamma_{21} = -L_{12}/D$$

$$D = L_{11}L_{22} - L_{12}^2$$

and are generally functions of angle θ.

When the currents are used to describe the state of the system, the parametric equations are extensions of Eqs. 2-35 and 2-36

$$\lambda_1 = \frac{\partial W'_{\text{fld}}(i_1, i_2, \theta)}{\partial i_1} \tag{2-67}$$

$$\lambda_2 = \frac{\partial W'_{\text{fld}}(i_1, i_2, \theta)}{\partial i_2} \tag{2-68}$$

$$T_{\text{fld}} = \frac{\partial W'_{\text{fld}}(i_1, i_2, \theta)}{\partial \theta} \tag{2-69}$$

where the coenergy is given by

$$W'_{\text{fld}}(i_1, i_2, \theta) = \int_0^{i_1} \lambda_1 \, di_1 + \int_0^{i_2} \lambda_2 \, di_2 \qquad (2\text{-}70)$$

and the system is held fixed at position θ as the integration is carried out. Using Eqs. 2-66a and 2-66b for a linear system in Eq. 2-70, we obtain

$$W'_{\text{fld}}(i_1, i_2, \theta) = \tfrac{1}{2} L_{11} i_1^2 + L_{12} i_1 i_2 + \tfrac{1}{2} L_{22} i_2^2 \qquad (2\text{-}71)$$

where the inductances are generally functions of angle θ.

Systems with more than two pairs of electrical terminals are handled by assigning additional independent variables to the terminals and proceeding in the same manner as for two pairs.

EXAMPLE 2-4

In the system shown in Fig. 2-10, the inductances in henrys are given as $L_{11} = (3 + \cos 2\theta) \times 10^{-3}$; $L_{12} = 0.1 \cos \theta$; $L_{22} = 30 + 10 \cos 2\theta$. Find the torque $T_{\text{fld}}(\theta)$ for the currents $i_1 = 1$ amp; $i_2 = 0.01$ amp.

Solution

Since the expression for torque should be a function of the currents i_1 and i_2, rather than the flux linkages, we will use the coenergy of the system as the state function. When the inductances are given, the coenergy of the system from Eq. 2-71 is

$$W'_{\text{fld}} = \tfrac{1}{2} L_{11} i_1^2 + L_{12} i_1 i_2 + \tfrac{1}{2} L_{22} i_2^2$$

The torque is given by Eq. 2-69 as

$$T_{\text{fld}} = + \frac{\partial W'_{\text{fld}}}{\partial \theta} = -1 \times 10^{-3} i_1^2 \sin 2\theta - 0.1 i_1 i_2 \sin \theta - 10 i_2^2 \sin 2\theta$$

At $i_1 = 1$ amp and $i_2 = 0.01$ amp, the torque is

$$T_{\text{fld}} = -2 \times 10^{-3} \sin 2\theta - 10^{-3} \sin \theta \text{ newton-meters}$$

The torque is plotted in Fig. 2-11.   The first term at spatial distribution $-\sin 2\theta$ is called the reluctance torque and is the torque tending to increase the inductances $L_{11}(\theta)$ and $L_{22}(\theta)$ by aligning the armature with

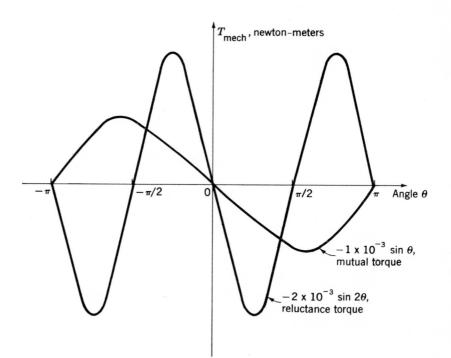

Fig. 2-11. Plot of torque components for multiply excited device.

the stationary member. The dependence on 2θ shows that the armature tends to align with the stationary member independently of "polarity." The second term at spatial distribution $-\sin\theta$ is the torque tending to increase the mutual inductance $L_{12}(\theta)$ and is polarity sensitive. If the self-inductance terms L_{11} and L_{22} are independent of angle θ, the reluctance torque disappears and the torque is produced only by the mutual term $L_{12}(\theta)$. The negative sign shows that the torque acts in a restoring direction.

2-7

DYNAMIC EQUATIONS

We have derived expressions for the forces produced in electromechanical-energy-conversion devices as functions of the electrical variables and the mechanical displacement. These expressions were derived for a conservative energy-conversion system; any losses were assigned to the electrical

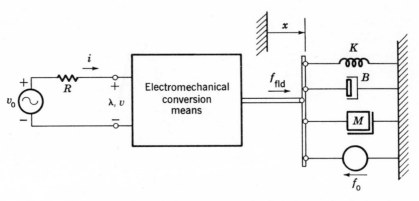

Fig. 2-12.   Model of singly excited electromechanical system.

and mechanical sources.   These conversion devices always operate as a coupling means between an electrical system and a mechanical system. Hence, we should be interested not only in how the conversion device itself operates, but in how the complete electromechanical system operates with the conversion device contained within it.

The model of a simple conversion system is shown in Fig. 2-12; it shows the basic portions, the details of which may vary from system to system.   The system shown consists of three parts: an external electrical portion, the energy-conversion means, and the external mechanical portion.   The electrical portion is merely represented by a voltage source v_0 and resistance R; the source could be represented by a current source and a parallel conductance G.   The electrical equation is thus

$$v_0 = iR + \frac{d\lambda}{dt} \qquad (2\text{-}72)$$

If the flux linkage λ can be expressed as $\lambda = L(x)i$, then the external equation becomes

$$v_0 = iR + L(x)\frac{di}{dt} + i\frac{dL(x)}{dx}\frac{dx}{dt} \qquad (2\text{-}73)$$

The term $L\,di/dt$ is the self-inductance voltage term; the term $i(dL/dx)(dx/dt)$ is the velocity voltage term which is responsible for the energy transfer between the external electrical system and the energy-conversion means.   For a multiply excited system, electrical equations corresponding to Eq. 2-72 are written for each input pair.   If the expressions for the

λ's are to be expanded in terms of inductances, as Eq. 2-73, then both self- and mutual-inductance terms will be required.

The mechanical portion of Fig. 2-12 shows symbols for the spring K, damping B, and mass M.   These symbols relate force to displacement, velocity, and acceleration as

$$f_K = K(x - x_0) \tag{2-74}$$

$$f_D = B \frac{dx}{dt} \tag{2-75}$$

$$f_M = M \frac{d^2x}{dt^2} \tag{2-76}$$

where x_0 is the value of x with the spring unstretched and the applied mechanical force $f_0 = 0$.   The force-balance equation for the mechanical portion is thus

$$f_{\text{fld}} = f_K + f_D + f_M + f_0 \tag{2-77}$$

$$f_{\text{fld}} = K(x - x_0) + B \frac{dx}{dt} + M \frac{d^2x}{dt^2} + f_0 \tag{2-78}$$

The force of electric origin f_{fld} is calculated from the derivatives of the energy or coenergy of the conversion medium as previously described.

The differential equations for the overall system of Fig. 2-12 for arbitrary inputs $v_0(t)$ and $f_0(t)$ are

$$v_0(t) = iR + L(x) \frac{di}{dt} + i \frac{dL(x)}{dx} \frac{dx}{dt} \tag{2-79}$$

$$-f_0(t) = M \frac{d^2x}{dt^2} + B \frac{dx}{dt} + K(x - x_0) - f_{\text{fld}}(x,i) \tag{2-80}$$

The functions $L(x)$ and $f_{\text{fld}}(x,i)$ depend upon the construction of the conversion means.

EXAMPLE 2-5

Figure 2-13 shows in cross section a cylindrical solenoid magnet in which the cylindrical plunger of mass M kg moves vertically in brass guide rings of thickness t and mean diameter d.   The permeability of brass is the same as that of free space and is $\mu_0 = 4\pi \times 10^{-7}$ rational-

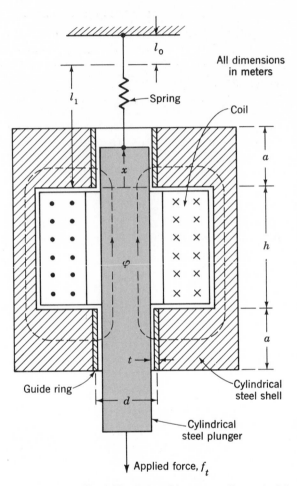

Fig. 2-13.   Solenoid magnet, Example 2-5.

ized mks units.   The plunger is supported by a spring whose elas-
tance is K newtons/m.   Its unstretched length is l_0.   A mechanical
load force f_t is applied to the plunger from the mechanical system
connected to it, as shown in Fig. 2-13.   Assume that frictional force
is linearly proportional to velocity and that the coefficient of friction
is B newton-sec/m.   The coil has N turns and resistance R ohms.
Its terminal voltage is v_t, and its current is i.   The effects of magnetic
leakage and reluctance of the steel are negligible.

Derive the dynamic equations of motion of the electromechanical system, i.e., the differential equations expressing the dependent variables i and x in terms of v_t, f_t, and the given constants and dimensions.

Solution

Express the inductance as a function of x. The coupling terms, i.e., magnetic force f_{fld} and induced emf e, can then be expressed in terms of x and i and these relations substituted in the equations for the mechanical and electrical systems.

The reluctance of the magnetic circuit is that of the two guide rings in series, with the flux directed radially through them, as shown by the dotted flux lines φ in Fig. 2-13. Because $t \ll d$, the flux density in the guide rings is very nearly constant with respect to radial distance. In a region where the flux density is constant the reluctance is

$$\frac{\text{Length of flux path in direction of field}}{\mu(\text{area perpendicular to field})}$$

The reluctance of the upper gap is

$$\mathcal{R}_1 = \frac{t}{\mu_0\pi \; dx}$$

in which it is assumed that the field is concentrated in the area between the upper end of the plunger and the lower end of the upper guide ring. Similarly the reluctance of the lower gap is

$$\mathcal{R}_2 = \frac{t}{\mu_0\pi \; da}$$

The total reluctance is

$$\mathcal{R} = \mathcal{R}_1 + \mathcal{R}_2 = \frac{t}{\mu_0\pi d}\left(\frac{1}{a} + \frac{1}{x}\right) = \frac{t}{\mu_0\pi \; da}\frac{a+x}{x}$$

The permeance is

$$\mathcal{P} = \frac{1}{\mathcal{R}} = \frac{\mu_0\pi \; da}{t}\frac{x}{a+x}$$

Hence, the inductance is

$$L = N^2 \mathcal{P} = \frac{\mu_0 \pi \, da N^2}{t} \frac{x}{a+x} = L' \frac{x}{a+x} \tag{2-81}$$

where

$$L' = \frac{\mu_0 \pi \, da N^2}{t} \tag{2-82}$$

The magnetic force acting upward on the plunger in the positive direction of x is

$$f_{\text{fld}} = \frac{+\partial W'_{\text{fld}}(i,x)}{\partial x} = +\frac{1}{2} i^2 \frac{dL}{dx} = +\frac{1}{2} L' \frac{ai^2}{(a+x)^2} \tag{2-83}$$

The counter emf induced in the coil is

$$e = \frac{d}{dt}(Li) = L\frac{di}{dt} + i\frac{dL}{dx}\frac{dx}{dt} \tag{2-84}$$

$$e = L' \frac{x}{a+x}\frac{di}{dt} + L' \frac{ai}{(a+x)^2}\frac{dx}{dt} \tag{2-85}$$

Substitution of the magnetic force in the differential equation of motion of the mechanical system gives

$$-f_t + \frac{1}{2} L' \frac{ai^2}{(a+x)^2} = M\frac{d^2x}{dt^2} + B\frac{dx}{dt} - K(l_1 - x) + Mg \tag{2-86}$$

The voltage equation for the electrical system is

$$v_t = Ri + e = Ri + L' \frac{x}{a+x}\frac{di}{dt} + L' \frac{ai}{(a+x)^2}\frac{dx}{dt} \tag{2-87}$$

Equations 2-86 and 2-87 are the desired results. They are valid only so long as the upper end of the plunger is well within the upper guide ring, say, between the limits $0.1a < x < 0.9a$. This is the normal working range of the solenoid.

2-8
ANALYTICAL TECHNIQUES

We have described relatively simple devices in this chapter. The devices have one or two pairs of electrical terminals and one mechanical terminal, which is usually constrained to incremental motion. More complicated devices capable of continuous energy conversion will be treated in the following chapters. The analytical techniques to be discussed here apply to the simple devices, but the principles are applicable to the more complicated devices as well.

a. Purposes for Analysis

Some of the devices described in this chapter are used with gross motion, such as in relays and solenoids, where the devices operate under essentially "on" and "off" conditions. Analysis on these devices is carried out to determine force as a function of displacement and reaction on the electrical source. Such calculations have already been made in this chapter. If the details of the motion are required, such as the displacement as a function of time after energizing the device, then nonlinear differential equations of the form of Eqs. 2-79 and 2-80 must be solved.

As compared with gross-motion devices, other devices are used with small motion, as in loudspeakers, pickups, and transducers of various kinds, for converting electrical signals into mechanical signals and vice versa. The relationship between the electrical and mechanical variables is made linear either by the design of the device or by restricting the excursion of the signals to a linear range. In such case the differential equations are linear and can be solved for transient response, frequency response, or equivalent-circuit representation as required.

b. Gross Motion

The differential equations for a singly excited device as derived in Example 2-5 are of the form

$$\frac{1}{2} L' \frac{ai^2}{(a + x)^2} = M \frac{d^2x}{dt^2} + B \frac{dx}{dt} + K(x - l_1) + f_t \tag{2-88}$$

$$v = iR + L' \frac{x}{a + x} \frac{di}{dt} + \frac{L'ai}{(a + x)^2} \frac{dx}{dt} \tag{2-89}$$

A typical problem using these differential equations is to find the excursion $x(t)$ when a prescribed voltage $v = V$ is applied at $t = 0$. An even simpler problem is to find the time required for the armature to move from its position $x(0)$ at $t = 0$ to a given displacement $x = X$ when a voltage $v = V$ is applied at $t = 0$. There is no general solution for these differential equations; they involve products of variables x and i and their derivatives and are termed nonlinear.   They can be solved using numerical step-by-step techniques or can be solved with a digital computer. Either a program can be prepared for the solution of the equations or a standard, already prepared, program can be called on and the coefficients for the particular problem inserted into that program.

In many cases the gross-motion problem can be simplified and a solution found by relatively simple methods.   For example, when the winding of the device is connected to the voltage source with a relatively large resistance, the iR term dominates on the right-hand side of Eq. 2-89 compared to the di/dt transformer-voltage term and the dx/dt velocity-voltage term.   The current i can then be assumed equal to V/R and inserted directly into Eq. 2-88.   The same assumption can be made when the winding is driven from a transistor, which acts as a current source to the winding.

With the assumption that $i = V/R$, two cases can be solved easily. First, we can handle those devices in which the dynamic motion is governed by damping rather than inertia; for example, relays having dashpots or dampers to slow down the motion, or devices purposely having low inertia.   In such case, the differential equation 2-88 reduces to

$$f(x) = \frac{1}{2} L' \frac{a}{(a + x)^2} \left(\frac{V}{R}\right)^2 - K(x - l_1) = B \frac{dx}{dt} \qquad (2\text{-}90)$$

The left-hand side, $f(x)$, is the difference between the force of electrical origin and the spring force in the device of Fig. 2-13.   The velocity at any value of x is merely $dx/dt = f(x)/B$; the time t to reach $x = X$ is given by

$$t = \int_0^X \frac{B}{f(x)} dx \qquad (2\text{-}91)$$

The operation of Eq. 2-91 can be carried out analytically or graphically.

The second case that can be solved easily for constant current drive of the winding is that where the dynamic motion is governed by the inertia rather than the damping.   Where additional damping is not introduced, the inertia usually governs the motion.   In such case, the differential equation 2-88 reduces to

$$f(x) = M \frac{d^2x}{dt^2} \qquad (2\text{-}92)$$

The acceleration at any value of x is $d^2x/dt^2 = f(x)/M$, and the velocity

$v(x)$ at any value of x is obtained from

$$f(x) = \frac{M}{2} \frac{d}{dx} \left(\frac{dx}{dt}\right)^2 \tag{2-93}$$

$$v(x) = \frac{dx}{dt} = \left[\frac{2}{M} \int_0^X f(x)\, dx\right]^{1/2} \tag{2-94}$$

The operation of Eq. 2-94 can be carried out analytically or graphically to find $v(x)$ and also to find the time t to reach any value of x.

c. Linearization

Devices which are characterized by nonlinear differential equations such as Eqs. 2-88 and 2-89 will yield nonlinear responses to input signals when used as transducers.   To obtain linear behavior, such devices must be restricted to small excursions of displacement and electrical quantities about an equilibrium position.   The equilibrium position is determined either by a bias mmf produced by current or a permanent magnet acting against a spring or by a pair of windings producing mmfs whose forces cancel at the equilibrium point.   The equilibrium point must be stable; the armature following a small disturbance should return to the equilibrium position.   The equilibrium condition is determined with the time derivatives set to zero in Eqs. 2-88 and 2-89; this occurs for

$$\frac{1}{2} L' \frac{a I_0^2}{(a + X_0)^2} = K(X_0 - l_1) + f_{t0} \tag{2-95}$$

$$V_0 = I_0 R \tag{2-96}$$

where $i = I_0$ and $x = X_0$ at equilibrium.

The incremental operation can be described by expressing the variables as $i = I_0 + i_1$ and $x = X_0 + x_1$ and canceling the products of increments as being of second order.   Equations 2-88 and 2-89 thus become

$$\frac{1}{2} \frac{L'a(I_0 + i_1)^2}{(a + X_0 + x_1)^2} = M \frac{d^2 x_1}{dt^2} + B \frac{dx_1}{dt} + K(X_0 + x_1 - l_1) + f_{t0} + f_1 \tag{2-97}$$

$$V_0 + v_1 = (I_0 + i_1)R + \frac{L'(X_0 + x_1)}{a + X_0 + x_1} \frac{di_1}{dt}$$

$$+ \frac{L'a(I_0 + i_1)}{(a + X_0 + x_1)^2} \frac{dx_1}{dt} \tag{2-98}$$

The equilibrium terms cancel out leaving a set of linear differential equations in terms of just the incremental variables of first order as

$$\frac{L'aI_0 i_1}{(a + X_0)^2} = M \frac{d^2 x_1}{dt^2} + B \frac{dx_1}{dt} + K' x_1 + f_1 \tag{2-99}$$

$$v_1 = i_1 R + \frac{L'(X_0)}{a + X_0} \frac{di_1}{dt} + \frac{L'aI_0}{(a + X_0)^2} \frac{dx_1}{dt} \tag{2-100}$$

The constant K' represents the effect of the spring force and the component of magnetic-field force proportional to x_1. The equations can be written in more compact form in terms of the self-inductance L_0 at the equilibrium point and a coefficient K_0 of energy conversion as

$$K_0 i_1 = M \frac{d^2 x_1}{dt^2} + B \frac{dx_1}{dt} + K' x_1 + f_1 \tag{2-101}$$

$$v_1 = i_1 R + L_0 \frac{di_1}{dt} + K_0 \frac{dx_1}{dt} \tag{2-102}$$

d. Transfer Functions and Block Diagrams

In the study of systems involving more than a few relatively simple equations an analytical attack by means of the equations themselves becomes cumbersome and confusing. It is then helpful to represent the system by means of either a block diagram or a signal-flow chart. This technique is useful in analytical studies of system characteristics such as stability, accuracy, and frequency response. The technique has been highly developed in connection with studies of feedback control systems. Closely related to the analytical technique based on block diagrams is the technique of programming a problem for solution on an analog computer.

The *block diagram* is a pictorial representation of the equations of the system. Each block represents a mathematical operation. The blocks are then interconnected in accordance with the dictates of the system. There is no need to solve simultaneous equations in setting up the block diagram. The block diagram itself is a chart of the procedure to be followed in combining the simultaneous equations. Often useful information can be obtained from the block diagram without making a complete analytical solution.

The symbols needed in block diagrams of linear systems are shown in Fig. 2-14. In Fig. 2-14a, X is an *input* variable, and Y is an *output* variable. They may be functions of time or complex-amplitude functions

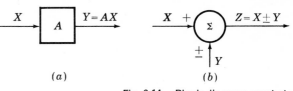

(a)   (b)

Fig. 2-14.   Block-diagram symbols.

of frequency.   The operator A is the *transfer function* representing the mathematical operation performed on X to obtain Y.   If X and Y are time functions, A is a differential operator $A(p)$, where p is the derivative operator d/dt.   If X and Y are complex amplitudes, A is a complex operator $A(s)$, where s is complex frequency.   For steady-state sinusoidal variables, X and Y are phasors and $s = j\omega$.   The symbol in Fig. 2-14b represents addition or subtraction.

EXAMPLE 2-6

Draw a block diagram in the complex-frequency domain for the circuit of Fig. 2-15 with input $e_1(t)$ and output $e_2(t)$.

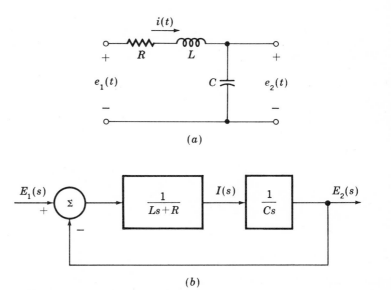

Fig. 2-15.   (a) Circuit diagram and (b) complex-frequency block diagram of its transfer function, Example 2-6.

Solution

The differential equations can be written in the form

$$e_1(t) - e_2(t) = Lpi + Ri \qquad e_2(t) = \frac{1}{Cp} i$$

When transformed to complex quantities and rearranged, the equations become

$$\frac{E_1(s) - E_2(s)}{Ls + R} = I(s) \qquad E_2(s) = \frac{1}{Cs} I(s)$$

The block diagram is shown in Fig. 2-15b.

In a linear system the response of an output variable to any one input can be found by considering all other inputs to be zero. The over-all response to several inputs can be found by superposition. The blocks can be combined by means of a few simple rules. Two basic ones are shown in Fig. 2-16. For example, in Fig. 2-16b,

$$Y = A(X \pm BY) \qquad \text{or} \qquad Y = \frac{A}{1 \mp AB} X \qquad (2\text{-}103)$$

2-9
RÉSUMÉ

Magnetic and electrostatic fields are seats of energy storage. Whenever the energy in the field is influenced by the configuration of the mechanical

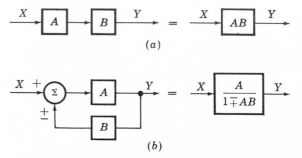

Fig. 2-16. Basic operations of block-diagram algebra.

parts constituting the boundaries of the field, mechanical forces are created which tend to move the mechanical elements so that energy is transmitted from the field to the mechanical system.

The singly excited magnetic system is considered first in Art. 2-2. When the electrical and mechanical loss elements are removed from the electromechanical-conversion device and combined with the external electrical and mechanical systems, the device can be considered as a conservative system. Its energy is thus a state function and is described by its independent variables of λ and x. The energy can be specified in terms of field quantities, electric-circuit quantities, and magnetic-circuit quantities.

In Art. 2-4, the state function is enlarged in concept to include the coenergy as a function of i and x. The force of field origin acting on the movable members is derived as the partial derivatives of the energy or coenergy expressed in the suitable independent variables. The force always acts to increase the self-inductance of the system, and the values of force are independent of the method of analysis. The energy or coenergy of the system can be introduced through both the electrical and mechanical terminals, depending upon the constraints placed upon the variables.

Expressions for force of field origin using the variables q and v are derived for electric-field systems in Art. 2-5. Both the energy and coenergy state functions are used; they are expressed in field and circuit terms. Example 2-3 shows that the force of magnetic-field origin at the typical saturation flux densities of magnetic materials far exceeds that of electric-field origin at the electric-field strength of air. The forces always act to increase the self-capacitance of the system either by moving metallic surfaces or dielectric bodies.

All rotating machinery and signal-handling transducers are built with multiple windings or excitation sources. The analysis is extended in Art. 2-6 for such magnetic-field systems in terms of energy and coenergy. The use of self- and mutual inductances for expressing the state functions is introduced.

Energy-conversion devices operate between electrical and mechanical systems. The behavior is described in Art. 2-7 by the differential equations which include the coupling terms between the systems. The equations are usually nonlinear and can be solved by computer or numerical methods if necessary. Usually, the solution is required for special conditions which can be handled as discussed in Art. 2-8. The most useful condition is that for small signal amplitudes, where the resulting equations are linear.

This chapter has been concerned with basic principles applying broadly to the electromechanical-energy-conversion process, with emphasis on magnetic-field coupling.   Basically, rotating machines and linear-motion transducers work in the same way.   The remainder of this text is devoted almost entirely to the rotating-machine aspects of electro-mechanical energy conversion, both the dynamic characteristics of the machines as system components and behavior under steady-state con-ditions.   The rotating machine is a fairly complicated assembly of elec-tric and magnetic circuits with a number of variations of machine types. For all of them, the electromechanical coupling terms—torque and induced voltage—can be found by the principles developed in this chapter.

PROBLEMS

2-1. An iron-clad plunger magnet is shown in Fig. 2-17.   For the purposes of this problem, neglect fringing and leakage fluxes, and assume that all the reluctance of the magnetic circuit is in the working gap g between the movable plunger and the center core.   The coil has an inductance of 1.00 henry when $g = 1.00$ cm.

If the coil is excited from a constant-current source of 1.00 amp, compute:

   $a.$ The energy stored in the magnetic field, in joules, when $g = 1.00$ cm

   $b.$ The magnetic force f on the plunger, in newtons, when $g = 1.00$ cm

   $c.$ The force f when $g = 0.50$ cm

   $d.$ The mechanical work done by the force f, in joules, when the plunger is allowed to move slowly from $g = 1.00$ cm to $g = 0.50$ cm

2-2. The cylindrical iron-clad solenoid magnet shown in Fig. 2-18 is used for tripping circuit breakers, for operating valves, and in other appli-

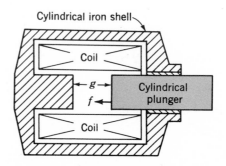

Fig. 2-17. Cylindrical plunger magnet, Prob. 2-1.

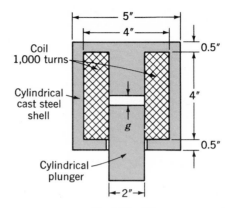

Fig. 2-18. Plunger magnet.

cations in which a relatively large force is applied to a member which moves a relatively short distance. When the coil current is zero, the plunger drops against a stop such that the gap g is 0.50 in. When the coil is energized by a direct current of sufficient magnitude, the plunger is raised until it hits another stop set so that g is 0.10 in. The plunger is supported so that it can move freely in the vertical direction. The air gap between the shell and the plunger can be assumed to be uniform and 0.01 in. long. For the purposes of this problem neglect magnetic leakage and fringing in the air gaps. The exciting coil has 1,000 turns and carries a constant current of 3.0 amp.

If the mmf in the iron is neglected:

a. Compute the flux densities, in webers per square meter, between the working faces of the center core and plunger for gaps g of 0.10, 0.20, and 0.50 in.

b. Compute the corresponding values of the energy stored in the magnetic field, in watt-seconds.

c. Compute the corresponding values of the coil inductance, in henrys.

2-3. Data for the magnetization curve of the iron portion of the magnetic circuit of the plunger magnet of Problem 2-2 are given below:

Flux, kilolines	100	150	200	240	250	260	270	275
Mmf, amp-turns	60	95	150	250	305	425	600	725

Plot magnetization curves for the complete magnetic circuit (flux in webers vs. total mmf in ampere-turns) for the following conditions:

a. Gap g = 0.50 in.
b. Gap g = 0.10 in.
c. From these curves find graphically the magnetic-field energy and coenergy for each of the gaps in (a) and (b) with 3.0-amp coil current.

2-4. The time constant L/r of the field winding of a 10-kw 1,150-rpm dc shunt generator is 0.15 sec. At normal operating conditions, the i^2r loss in its field winding is 350 watts. Compute the energy stored in its magnetic field, in watt-seconds, at normal operating conditions.

2-5. An iron-clad plunger magnet is shown in Fig. 2-18. The air gap between the shell and the plunger can be assumed to be uniform and 0.01 in. long. Magnetic leakage and fringing may be neglected. The coil has 1,000 turns and carries a constant current of 3.0 amp.

a. If the plunger is allowed to move slowly so as to reduce the gap g from 0.50 to 0.10 in., how much mechanical work will be done by the plunger, in joules?
b. For the conditions of (a), how much energy will be supplied by the electrical source (in excess of heat loss in the coil)?

2-6. For the plunger magnet of Prob. 2-5, find a numerical expression for the magnetic force f acting on the plunger, in newtons, as a function of the gap g in meters with a constant coil current of 3.0 amp. Plot a curve of f as a function of g. From the area under this force-displacement curve, compute the mechanical work done by the plunger when it moves slowly so as to reduce the gap g from 0.50 to 0.10 in. Compare with the result of Prob. 2-5a.

2-7. Data for the magnetization curve of the iron portion of the magnetic circuit of the plunger magnet are given in Prob. 2-3. Plot magnetization curves for the complete magnetic circuit (flux in webers vs. total mmf in ampere-turns) for the following conditions:

a. Gap g = 0.50 in.
b. Gap g = 0.10 in.

    c. From these curves find graphically the work done by the plunger
if it is allowed to move slowly from $g = 0.50$ in. to $g = 0.10$ in.
while the coil carries a constant current of 3.0 amp.   Compare
with the result of Prob. 2-5*a*.

    d. Also find the energy supplied by the electrical source (in excess
of heat loss in the coil).   Compare with the result of Prob. 2-5*b*.

    2-8. Consider the iron-cored solenoid shown in Fig. 2-19.   The iron
plunger of mass M is supported by a spring and guided vertically by non-
magnetic spacers of thickness t and permeability μ_0.   Assume the iron
to be infinitely permeable, and neglect magnetic fringing and leakage
fields.   All quantities are in rationalized mks units.

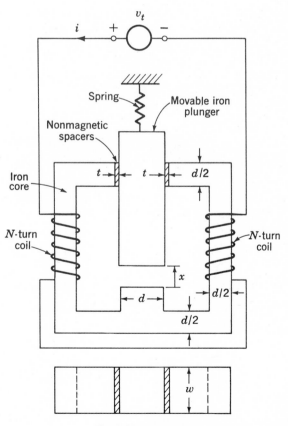

Fig. 2-19.   Solenoid magnet, Prob. 2-8.

a. Find the inductance of the coil as a function of the plunger position x.   Under what conditions is your expression fairly accurate?

b. Find an expression for the magnetic force acting on the plunger in terms of the coil current i.

c. Suppose the dimensions are

$$w = 5 \text{ cm} \qquad d = 4 \text{ cm} \qquad t = 0.1 \text{ cm}$$

Approximately what maximum force would you expect to get on the plunger without saturating the iron?

2-9. The armature and field structures of a simplified 2-pole dc machine are shown in the end view (*a*) and cross section (*b*) in Fig. 2-20. Because of an error in assembly, the armature core is displaced 0.5 in. in an axial direction from its correct position.   Other numerical data are as follows:

Length of each air gap = 0.10 in.
Diameter of armature = 10.0 in.
Air-gap flux density = 50 kilolines/in.²
Angle subtended by each pole shoe = 100°

The air-gap length can be considered constant under the pole shoes, and the armature can be considered as a smooth cylinder.

Find the axial force in pounds tending to center the armature.

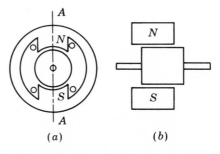

(*a*)   (*b*)

Fig.   2-20.   Simplified   dc   machine, Prob. 2-9.

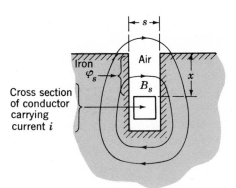

Fig. 2-21.   Conductor in a slot, Prob. 2-10.

2-10. Figure 2-21 shows the general nature of the slot-leakage flux produced by current i in a rectangular conductor embedded in a rectangular slot in iron.   Assume that the slot-leakage flux φ_s goes straight across the slot in the region between the top of the conductor and the top of the slot.

  a. Derive an expression for the flux density B_s in the region between the top of the conductor and the top of the slot.
  b. Derive an expression for the slot-leakage flux φ_s crossing the slot above the conductor, in terms of the height x of the slot above the conductor, the slot width s, and the embedded length l perpendicular to the paper.
  c. Derive an expression for the force f created by this magnetic field on a conductor of length l.   Use rationalized mks units.   In what direction does this force act on the conductor?
  d. Compute the force in pounds on a conductor 1.0 ft long in a slot 1.0 in. wide when the current in the conductor is 1,000 amp.

2-11. Two coils have self- and mutual inductances in henrys as functions of a displacement x in meters as follows:

$$L_{11} = 1 + x$$
$$L_{22} = 2(1 + x)$$
$$L_{12} = 1 - x$$

The resistances are negligible.

a. For constant currents $I_1 = +10.0$ amp and $I_2 = -5.0$ amp, compute the mechanical work done in increasing x from 0 to $+1.0$ m.

b. Does the force developed in part *(a)* tend to increase or decrease x?

c. During the motion of part *(a)*, how much energy is supplied by source 1? By source 2?

d. Compute the average value of the force developed for $x = 0.50$ m when coil 2 is short-circuited and a sinusoidal voltage 377 volts rms at 60 Hz is applied to coil 1.

2-12. The self-inductance L_{11} and L_{22} and the absolute value of the mutual inductance L_{12} of the device shown in Fig. 2-22 are given in the table below for two angular positions θ_0 of the rotor, where θ_0 is measured from a horizontal reference axis to the axis of the rotor:

θ_0	L_{11}	L_{22}	L_{12}
45°	0.60	1.10	0.30
75°	1.00	2.00	1.00

The inductances are given in henrys and may be assumed to vary linearly with θ_0 over the range $45° < \theta_0 < 75°$.

For each of the following cases, compute the electromagnetic torque in newton-meters when the rotor is stationary at an angular position $\theta_0 = 60°$ (approximately the position shown in Fig. 2-22), and state whether this torque tends to turn the rotor in a clockwise or counterclockwise direction.

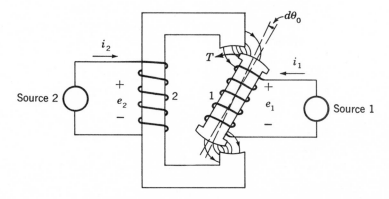

Fig. 2-22.   Elementary multiply excited magnetic system.

a. $i_1 = 10.0$ amp, $i_2 = 0$.

b. $i_1 = 0$, $i_2 = 10.0$ amp.

c. $i_1 = 10.0$ amp, and $i_2 = 10.0$ amp in arrow directions (Fig. 2-22).

d. $i_1 = 10.0$ amp in arrow direction, and $i_2 = 10.0$ amp in reverse direction.

e. $i_1 = 10.0$ amp rms value of sinusoidal alternating current, with coil 2 short-circuited. In this case the resistance of coil 2 may be neglected, and it is the time average of the torque which is wanted.

2-13. Two coils, one mounted on a stator and the other on a rotor, have self- and mutual inductances of

$$L_{11} = 0.20 \text{ mh}$$

$$L_{22} = 0.10 \text{ mh}$$

$$L_{12} = 0.05 \cos \theta \text{ mh}$$

where θ is the angle between the axes of the coils. The coils are connected in series and carry a current

$$i = \sqrt{2}\, I \sin \omega t$$

a. Derive an expression for the instantaneous torque T on the rotor as a function of the angular position θ.

b. Give an expression for the time-average torque T_{av} as a function of θ.

c. Compute the numerical value of T_{av} for $I = 10$ amp and $\theta = 90°$.

d. Sketch three curves of T_{av} versus θ for currents $I = 5$, 7.07, and 10 amp, respectively.

e. A helical restraining spring which tends to hold the rotor at $\theta = 90°$ is now attached to the rotor. The restraining torque of spring is proportional to the angular deflection from $\theta = 90°$ and is 0.004 newton-meter when the rotor is turned to $\theta = 0$. Show on the curves of part (d) how you could find the angular position of the rotor-plus-spring combination for coil currents $I = 5$, 7.07, and 10 amp, respectively. Sketch a curve showing θ versus I. A reasonable-looking sketch with estimated numerical values is all that is required.

Note: This problem illustrates the principles of the dynamometer-type ac ammeter.

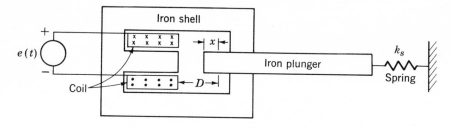

Fig. 2-23. Solenoid magnet, Prob. 2-14.

2-14. Figure 2-23 shows in cross section a cylindrical-plunger magnet. The plunger is constrained to move in the x direction and is restrained by a spring whose spring constant is k_s. The position of the plunger when the coil is unexcited is D. The mass of the plunger is M. Friction is negligible. The cross-sectional area of the working air gap is A. Neglect magnetic leakage and fringing. Also neglect the reluctance of the iron. The coil has N turns and negligible resistance.

The plunger is operating under steady-state sinusoidal conditions. The flux density in the air gap is

$$B(t) = B_m \sin \omega t$$

a. Find an expression for the magnetic force acting on the plunger in terms of B_m, ω, and t.

b. Write the equation for the coil voltage $e(t)$ in terms of B_m, ω, and t.

c. Write the differential equation of motion of the plunger in terms of B_m, ω, t, and x.

d. Assume that the system is in the steady state. Solve for the position of the plunger $x(t)$. Note that the solution must contain a constant part X_0 and a time-varying part $x_1(t)$.

2-15. Consider a solenoid magnet similar to that shown in Fig. 2-13, except that the length of the cylindrical steel plunger is $a + h$.

Derive the dynamic equations of motion of the system.

2-16. The following dimensions and data apply to the solenoid magnet of Prob. 2-15.

$$a = 2 \text{ cm} \qquad h = 6 \text{ cm} \qquad d = 2 \text{ cm} \qquad t = 0.05 \text{ cm}$$

Turns $N = 1,000$

Coil resistance $r = 100$ ohms

Mass of plunger $M = 0.2$ kg
Spring constant $K = 6.25$ newtons/cm
With $i = 0$, plunger is at rest at $x = 0.25$ cm
External applied force $= 0$
Friction negligible

The voltage applied to the coil has a quiescent value of 10 volts.

    a. Find the quiescent operating point.   Is it stable?
    b. Linearize the dynamic equations of motion for small incremental
        changes in the applied voltage.   Give numerical values of the
        parameters in mks units.

2-17. Two windings, one mounted on a stator and the other on a
rotor, having self- and mutual inductances as given below:

$$L_{11} = 2.20 \text{ henrys} \qquad L_{22} = 1.00 \text{ henry}$$

$$L_{12} = 1.414 \cos \theta_0 \text{ henrys}$$

where θ_0 is the angle between the axes of the windings.   The resistances
of the windings may be neglected.
    Winding 2 is short-circuited, and the current in winding 1 as a func-
tion of time is $i_1 = 14.14 \sin \omega t$.

    a. If the rotor is stationary, derive an expression for the numerical
        value, in newton-meters, of the instantaneous torque on the rotor
        in terms of the angle θ_0.
    b. Compute the average torque in newton-meters when $\theta_0 = 45°$.
    c. If the rotor is allowed to move, will it rotate continuously or will
        it tend to come to rest?   If the latter, at what value of θ_0?

2-18. Figure 2-24 shows a capacitor with one fixed and one movable
plate constrained to move in the x direction.   Its capacitance is given by

$$C = \frac{1}{x} \quad \mu\text{f}$$

with x in meters.   There are no electrical or mechanical losses.   Initially
$E = 200$ volts and $x = 0.01$ m.

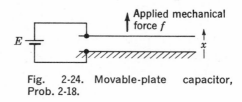

Fig.   2-24.   Movable-plate   capacitor,
Prob. 2-18.

The capacitor is then put through the following cycle:

a. With $E = 200$ volts, x is increased to 0.02 m.
b. With $x = 0.02$ m, E is decreased to 100 volts.
c. With $E = 100$ volts, x is decreased to 0.01 m.
d. With $x = 0.01$, E is increased to 200 volts, thereby closing the
cycle.

For this closed cycle, find the mechanical work done.   Is the net
energy flow into or out of the battery?

3
rotating machines:
basic concepts

The object of this chapter is to discuss the approximations involved in reducing a physical machine to a simple mathematical model, to apply the basic principles of Chap. 2 to the analysis of this model, and to give some simple concepts relating to the basic machine types.

ELEMENTARY CONCEPTS

Faraday's law, $e = d\lambda/dt$, describes quantitatively the induction of voltages by a time-varying magnetic field. Electromechanical energy conversion takes place when the change in flux is associated with mechanical motion. In rotating machines voltages are generated in windings or groups of coils by rotating these windings mechanically through a magnetic field, by mechanically rotating a magnetic field past the winding, or by designing the magnetic circuit so that the reluctance varies with rotation of the rotor. By any of these methods the flux linking a specific coil is changed cyclically, and a voltage is generated.   A group of such coils interconnected so that their generated voltages all make a positive contribution to the desired result is called an *armature winding*.   The armature of a dc machine is shown in Fig. 3-1; the armature is the rotating member, or *rotor*.   Figure 3-2 shows the armature of an ac generator, *alternator*, or *synchronous generator*.   Here the armature is the stationary member, or *stator*.

These coils are wound on iron cores in order that the flux path through them may be as effective as possible.   Because the armature iron is subjected to a varying magnetic flux, eddy currents will be induced in it; to minimize the eddy-current loss, the armature iron is built up of thin laminations as illustrated in Fig. 3-3 for the armature of an ac machine. The magnetic circuit is completed through the iron of the other machine member, and exciting coils, or *field windings*, are placed on that member to act as the primary sources of flux.   Permanent magnets may be used in small machines.

Fig. 3-1.   Armature of a dc motor.   *(General Electric Company.)*

Fig. 3-2. Stator of a 190-Mva 3-phase 12-kv 375-rpm hydroelectric generator. The conductors have hollow passages through which cooling water is circulated. *(Brown Boveri Corporation.)*

Fig. 3-3. Partially completed stator core for an ac motor. *(Westinghouse Electric Corporation.)*

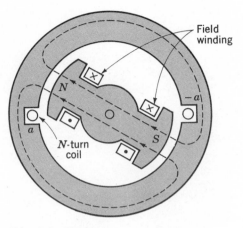

Fig. 3-4. Elementary synchronous generator.

a. Elementary Synchronous Machines

Preliminary ideas of generator action can be gained by discussing the
armature-induced voltage in the very much simplified ac synchronous
generator shown in Fig. 3-4.   With rare exceptions, the armature winding of a synchronous machine is on the stator, and the field winding is on
the rotor, as in Fig. 3-4.   The field winding is excited by direct current
conducted to it by means of carbon *brushes* bearing on *slip rings* or *collector
rings*.   Constructional factors usually dictate this orientation of the
2 windings: it is advantageous to have the low-power field winding on the
rotor.   The armature winding, consisting of a single coil of N turns, is
indicated in cross section by the two coil sides a and $-a$ placed in diametrically opposite narrow slots on the inner periphery of the stator of

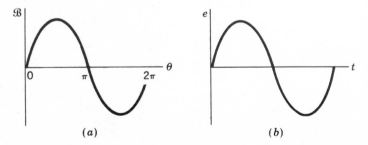

Fig. 3-5.   (a) Space distribution of flux density and (b) corresponding
waveform of the generated voltage.

Fig. 3-4.   The conductors forming these coil sides are parallel to the shaft
of the machine and are connected in series by end connections which are
not shown in the figure.   The rotor is turned at a constant speed by a
source of mechanical power connected to its shaft.   Flux paths are shown
by dotted lines in Fig. 3-4.

The radial distribution of the air-gap flux density $\mathcal{B}$ is shown in Fig.
3-5a as a function of the space angle θ around the air-gap periphery.   The
flux-density wave of practical machines can be made to approximate a
sinusoidal distribution by properly shaping the pole faces.   As the rotor
revolves, the flux waveform sweeps by the coil sides a and $-a$.   The
resulting coil voltage (Fig. 3-5b) is a time function having the same waveform as the spatial distribution $\mathcal{B}$.   The coil voltage passes through a
complete cycle of values for each revolution of the 2-pole machine of
Fig. 3-4.   Its frequency in cycles per second (hertz) is the same as the
speed of the rotor in revolutions per second; i.e., the electrical frequency
is synchronized with the mechanical speed, and this is the reason for the
designation *synchronous* machine.   Thus a 2-pole synchronous machine
must revolve at 3,600 rpm to produce a 60-Hz voltage.

A great many synchronous machines have more than 2 poles.   As a
specific example, Fig. 3-6 shows the most elementary 4-pole single-phase
alternator.   The field coils are connected so that the poles are of alternate
north and south polarity.   There are two complete wavelengths or cycles
in the flux distribution around the periphery, as shown in Fig. 3-7.   The
armature winding now consists of 2 coils a_1, $-a_1$ and a_2, $-a_2$ connected
in series by their end connections.   The span of each coil is one-half wave-

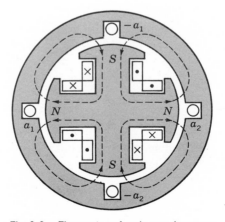

Fig. 3-6.   Elementary 4-pole synchronous
generator.

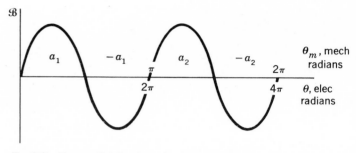

Fig. 3-7. Space distribution of flux density in a 4-pole synchronous generator.

length of flux.   The generated voltage now goes through two complete cycles per revolution of the rotor.   The frequency f in hertz then is twice the speed in revolutions per second.

When a machine has more than 2 poles, it is convenient to concentrate on a single pair of poles and to recognize that the electric, magnetic, and mechanical conditions associated with every other pole pair are repetitions of those for the pair under consideration.   For this reason it is convenient to express angles in *electrical degrees* or *electrical radians* rather than in mechanical units.   One pair of poles in a P-pole machine or one cycle of flux distribution equals 360 electrical degrees or 2π electrical radians.   Since there are $P/2$ complete wavelengths or cycles in one complete revolution, it follows that

$$\theta = \frac{P}{2}\,\theta_m \tag{3-1}$$

where θ is the angle in electrical units and θ_m is the mechanical angle. The coil voltage of a P-pole machine passes through a complete cycle every time a pair of poles sweeps by, or $P/2$ times each revolution.   The frequency of the voltage wave is therefore

$$f = \frac{P}{2}\frac{n}{60} \qquad \text{Hz} \tag{3-2}$$

where n is the mechanical speed in rpm and $n/60$ is the speed in revolutions per second.   The radian frequency ω of the voltage wave is

$$\omega = \frac{P}{2}\,\omega_m \tag{3-3}$$

where ω_m is the mechanical speed in radians per second.

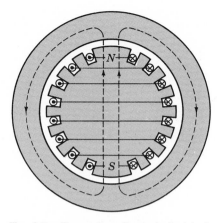

Fig. 3-8.   Elementary 2-pole cylindrical-rotor field winding.

The rotors shown in Figs. 3-4 and 3-6 have *salient*, or projecting, poles with *concentrated* windings.   Figure 3-8 shows diagrammatically a *non-salient-pole*, or *cylindrical, rotor*.   The field winding is a *distributed* winding placed in slots and arranged so as to produce an approximately sinusoidal 2-pole field.   The constructional reasons for some synchronous generators having salient-pole rotor structures and others having cylindrical rotors can be appreciated with the aid of Eq. 3-2.   Most power systems in the United States operate at a frequency of 60 Hz.   A salient-pole construction is characteristic of hydroelectric generators because hydraulic turbines operate at relatively low speeds and a relatively large number of poles are required to produce the desired frequency; the salient-pole construction is better adapted mechanically to this situation. The rotor of a large hydroelectric generator is shown in Fig. 3-9.   Steam turbines and gas turbines, on the other hand, operate best at relatively high speeds, and turbine-driven alternators or *turbine generators* are commonly 2- or 4-pole cylindrical-rotor machines.   The rotors are made from a single steel forging, or from several forgings, as shown in Figs. 3-10 and 3-11.

With very few exceptions, synchronous generators are 3-phase machines because of the advantages of 3-phase systems for generation, transmission, and heavy-power utilization.   For the production of a set of three voltages phase-displaced by 120 electrical degrees in time, it follows that a minimum of 3 coils phase-displaced 120 electrical degrees in space must be used.   An elementary 3-phase 2-pole machine with 1 coil per phase is shown in Fig. 3-12a.   The 3 phases are designated by

Fig. 3-9. Water-cooled rotor of the 190-Mva hydroelectric generator whose stator is shown in Fig. 3-2. *(Brown Boveri Corporation.)*

Fig. 3-10.   Rotor of a 2-pole 3,600-rpm turbine generator.   *(Westinghouse Electric Corpora-tion.)*

Fig. 3-11.   Parts of multipiece rotor for a 1333-Mva 3-phase 4-pole 1800-rpm turbine generator.   The separate forgings will be shrunk on the shaft before final machining and milling slots for the windings.   Total weight of the rotor is 435,000 pounds.   This is the same generator shown in Fig. 4-23.   *(Brown Boveri Corporation.)*

the letters a, b, and c. In an elementary 4-pole machine, a minimum of two such sets of coils must be used, as illustrated in Fig. 3-12b; in an elementary P-pole machine, $P/2$ such sets must be used. The 2 coils in each phase of Fig. 3-12b are connected in series so that their voltages add, and the 3 phases may then be either Y- or Δ-connected. Figure 3-12c shows how the coils are interconnected to form a Y connection.

When a synchronous generator supplies electrical power to a load, the armature current creates a component flux wave in the air gap which rotates at synchronous speed, as will be shown in Art. 3-4. This flux reacts with the flux created by the field current, and electromagnetic torque results from the tendency of the two magnetic fields to align themselves. In a generator this torque opposes rotation, and mechanical torque must be applied from the prime mover in order to sustain

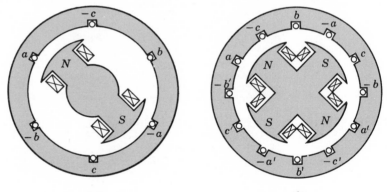

(a)   (b)

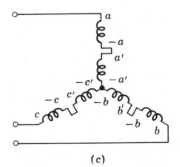

(c)

Fig. 3-12. Elementary 3-phase generators. (a) 2-pole; (b) 4-pole; (c) Y connection of the windings.

Fig. 3-13. Cutaway view of a high-speed synchronous motor. The exciter shown on the left end of the rotor is a small ac generator with a rotating semiconductor rectifier assembly. *(General Electric Company.)*

rotation. The electromagnetic torque is the mechanism through which greater electrical power output calls for greater mechanical power input.

The counterpart of the synchronous generator is the *synchronous motor*. A cutaway view of a 3-phase 60-Hz synchronous motor is shown in Fig. 3-13. Alternating current is supplied to the armature winding (usually the stator), and dc excitation is supplied to the field winding (usually the rotor). The magnetic field of the armature currents rotates at synchronous speed. In order to produce a steady electromagnetic torque, the magnetic fields of stator and rotor must be constant in amplitude and stationary with respect to each other. In a synchronous motor the steady-state speed is determined by the number of poles and the frequency of the armature current exactly as in Eq. 3-2 or 3-3 for a synchronous generator. Thus a synchronous motor operated from a constant-frequency ac source must run at a constant steady-state speed.

In a motor the electromagnetic torque is in the direction of rotation

Fig. 3-14.   A large dc motor showing brush rigging and commutator.   *(Westinghouse Electric Corporation.)*

and balances the opposing torque required to drive the mechanical load. The rotational or speed voltage then acts in opposition to the applied voltage and current.   In both generators and motors an electromagnetic torque and a rotational voltage are produced.   These are the essential phenomena for electromechanical energy conversion.   Quantitative analysis of these phenomena will be discussed in a preliminary way in Chap. 4, and in more detail in Chaps. 6 and 10.

b. Elementary DC Machines

The armature winding of a dc generator is on the rotor with current conducted from it by means of carbon brushes.   The field winding is on the stator and is excited by direct current.   A photograph of a large dc machine is shown in Fig. 3-14 and a cutaway view of a dc motor in Fig. 3-15.

A very elementary 2-pole dc generator is shown in Fig. 3-16.   The armature winding, consisting of a single coil of N turns, is indicated by the two coil sides a and $-a$ placed at diametrically opposite points on the rotor with the conductors parallel to the shaft.   The rotor is normally turned at a constant speed by a source of mechanical power connected to the shaft.   The air-gap flux distribution usually approximates a flat-topped wave rather than the sine wave found in ac machines and is shown

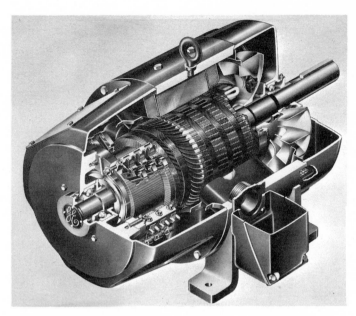

Fig. 3-15. Cutaway view of a typical integral-horsepower dc motor. *(General Electric Company.)*

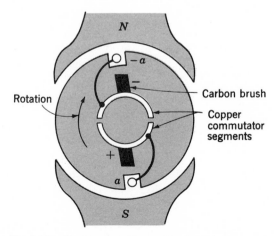

Fig. 3-16. Elementary dc machine with commutator.

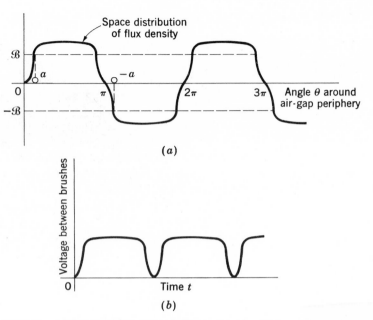

Fig. 3-17. (a) Space distribution of air-gap flux density in an elementary dc machine and (b) waveform of voltage between brushes.

in Fig. 3-17*a*. Rotation of the coil generates a coil voltage which is a time function having the same waveform as the spatial flux-density distribution.

Although the ultimate purpose is the generation of a direct voltage, the voltage induced in an individual armature coil is an alternating voltage. This voltage must therefore be rectified. Rectification is sometimes provided externally, for example, by means of semiconductor rectifiers. The machine then is an ac generator plus external rectifiers. In the conventional dc machine rectification is provided mechanically by means of a *commutator*, which is a cylinder formed of copper segments insulated from each other by mica and mounted on, but insulated from, the rotor shaft. Stationary carbon brushes held against the commutator surface connect the winding to the external armature terminals. The commutator and brushes may readily be seen in Figs. 3-14 and 3-15. The necessity for commutation is the reason why the armature windings of dc machines are placed on the rotor.

For the elementary dc generator the commutator takes the form shown in Fig. 3-16. For the direction of rotation shown, the commutator at all times connects the coil side which is under the south pole to the positive brush and that under the north pole to the negative brush.   The commutator provides full-wave rectification, transforming the voltage waveform between brushes to that of Fig. 3-17b and making available a unidirectional voltage to the external circuit.   The dc machine of Fig. 3-16 is, of course, simplified to the point of being unrealistic in the practical sense, and it will be essential later to examine the action of more realistic commutators.

If direct current flows in the external circuit connected to the brushes, torque is created by the interaction of the magnetic fields of stator and rotor.   If the machine is acting as a generator, this torque opposes rotation.   If it is acting as a motor, the electromagnetic torque acts in the direction of rotation.   Remarks similar to those already made concerning the roles played by the generated voltage and electromagnetic torque in the energy-conversion process in synchronous machines apply equally well to dc machines.   A wide variety of operating characteristics can be obtained, as will be shown in a preliminary way later in Chap. 4, and in more detail in Chaps. 5 and 9.

c. Elementary Induction Machines

A third variation on exciting stator and rotor windings is the *induction machine*, in which there are alternating currents in both stator and rotor windings.   The most common example is the induction motor in which alternating current is supplied directly to the stator and by induction (i.e., transformer action) to the rotor.   The induction machine may be regarded as a generalized transformer in which electric power is transformed between stator and rotor together with a change of frequency and a flow of mechanical power.   Although the induction motor is the most common of all motors, the induction machine is seldom used as a generator; its performance characteristics as a generator are unsatisfactory for most applications.   The induction machine may also be used as a frequency changer.

In the induction motor, the stator winding is essentially the same as that of a synchronous motor.   On the rotor, the winding is electrically closed on itself and very often has no external terminals; currents are induced in it by transformer action from the stator winding.   A cutaway view of an induction motor is shown in Fig. 3-18.   The usual induction-motor speed-torque characteristic is that the speed drops off slightly as load is added to its shaft.   These characteristics will be discussed in a preliminary way in Chap. 4, and in more detail in Chaps. 7 and 10.

Thus a variety of basic machine types is available, all operating on the same basic principles and each having its own special characteristics

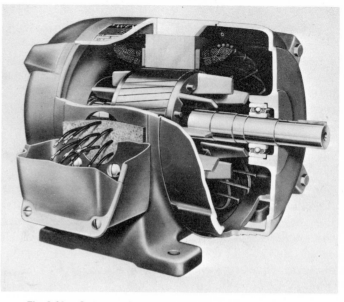

Fig. 3-18.   Cutaway view of an induction motor.   *(Westinghouse Electric Corporation.)*

and uses.   Obviously further study of the interacting magnetic fields is needed for an understanding of the basic theory.   Since the geometry of rotating machines is far too complicated to permit a direct solution by means of field theory, it will be necessary to develop a simplified mathematical model through making reasonable simplifying approximations.

3-2
GENERATED VOLTAGE

Study of the voltages induced in any of the armature windings of rotating machines resolves into study of the voltage induced in a single coil followed by addition of the individual coil voltages in the manner dictated by the specific interconnection of the coils forming the complete winding.   The general nature of the induced voltage has already been discussed in Art. 3-1.   The voltage magnitude will now be determined by Faraday's law.

a. AC Machines

An elementary ac machine is shown in cross section in Fig. 3-19.   The armature winding is a single N-turn coil whose coil sides are placed in diametrically opposite narrow slots on the inner periphery of the stator.

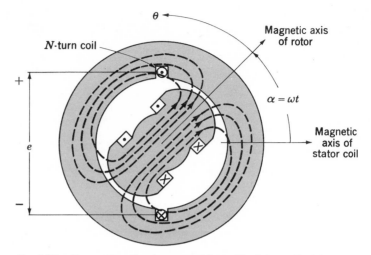

Fig. 3-19. Elementary 2-pole ac machine with stator coil of N turns.

The coil spans a full 180 electrical degrees or a complete pole pitch and is hence a *full-pitch* coil. For simplicity, a 2-pole machine is indicated, salient poles being chosen for diagrammatic symbolism only. The field winding on the rotor is assumed to produce a sinusoidal space wave of flux density $\mathcal{B}$ at the stator surface. The rotor is spinning at constant angular velocity ω electrical radians per second.

When the rotor poles are in line with the magnetic axis of the stator coil, the flux linkage with the stator coil is $N\Phi$, where Φ is the air-gap flux per pole. For the assumed sinusoidal flux-density wave

$$\mathcal{B} = B_{\text{peak}} \cos \theta \tag{3-4}$$

where B_{peak} is its peak value at the rotor pole center and θ is measured in electrical radians from the rotor pole axis. The air-gap flux per pole is the integral of the flux density over the pole area; thus, for a 2-pole machine

$$\Phi = \int_{-\pi/2}^{+\pi/2} B_{\text{peak}} \cos \theta \, lr \, d\theta = 2B_{\text{peak}} lr \tag{3-5}$$

where l is the axial length of the stator and r is its radius at the air gap. For a P-pole machine

$$\Phi = \frac{2}{P} 2B_{\text{peak}} lr \tag{3-6}$$

because the pole area is $2/P$ times that of a 2-pole machine of the same length and diameter.

As the rotor turns, the flux linkage varies as the cosine of the angle α between the magnetic axes of the stator coil and rotor. With the rotor spinning at constant angular velocity ω, the flux linkage with the stator coil is

$$\lambda = N\Phi \cos \omega t \qquad (3\text{-}7)$$

where time t is arbitrarily reckoned as zero when the peak of the flux-density wave coincides with the magnetic axis of the stator coil. By Faraday's law, the voltage induced in the stator coil is

$$e = -\frac{d\lambda}{dt} = \omega N\Phi \sin \omega t - N\frac{d\Phi}{dt} \cos \omega t \qquad (3\text{-}8)$$

The minus sign associated with Faraday's law in Eq. 3-8 implies generator reference directions as shown in Fig. 3-19. That is, while the flux linking the coil is decreasing, an emf will be induced in it in a direction to try to produce a current which would tend to prevent the flux linking it from decreasing.

The first term on the right-hand side of Eq. 3-8 is a speed voltage due to relative motion of the field and the coil. The second term is a transformer voltage and is only present when the amplitude of the flux-density wave changes with time. In the normal steady-state operation of most rotating machines the amplitude of the air-gap flux wave is constant and the induced voltage is simply the speed voltage

$$e = \omega N\Phi \sin \omega t \qquad (3\text{-}9)$$

Equation 3-9 was derived directly from Faraday's law. Alternatively, the voltage e induced in a conductor of length l moving with linear velocity v in a non-time-varying magnetic field of flux density B is given by the "cutting-of-flux" equation

$$e = Blv \qquad (3\text{-}10)$$

where B, l, and v are mutually perpendicular. When properly interpreted this equation also applies to rotating machines and gives a useful alternative basis for visualizing induced voltages. To show that Eq. 3-10 applies to rotating machines, we shall now derive it from Eq. 3-9.

Substitution of Eq. 3-6 in Eq. 3-9 gives

$$e = \omega N \frac{2}{P} 2B_{\text{peak}} lr \sin \omega t \tag{3-11}$$

$$e = \frac{2}{P} \omega r (2lN) B_{\text{peak}} \sin \omega t \tag{3-12}$$

Now $B_{\text{peak}} \sin \omega t$ is the flux density B_{coil} at the coil side a, Fig. 3-19, $2\omega/P$ is the mechanical angular velocity ω_m, and $2lN$ is the total active length of conductors in the two coil sides; thus, for a concentrated full-pitch coil

$$e = B_{\text{coil}}(2lN) r \omega_m = B_{\text{coil}}(2lN) v \tag{3-13}$$

where $v = r\omega_m$ is the linear velocity of the conductor relative to the field. Even though the conductors are embedded in slots, the voltage can be computed by the "cutting-of-flux" concept, Eq. 3-10, just as if the conductors were lifted out of the slots and placed directly in the air-gap field!

In the normal steady-state operation of ac machines, we are usually interested in the rms values of voltages and currents rather than their instantaneous values. From Eq. 3-9 the maximum value of the induced voltage is

$$E_{\text{max}} = \omega N \Phi = 2\pi f N \Phi \tag{3-14}$$

and its rms value is

$$E_{\text{rms}} = \frac{2\pi}{\sqrt{2}} f N \Phi = 4.44 f N \Phi \tag{3-15}$$

where f is the frequency in hertz. These equations are identical in form to the corresponding emf equations for a transformer. Relative motion of a coil and a constant-amplitude spatial flux-density wave in a rotating machine produces the same voltage effect as does a time-varying flux in association with stationary coils in a transformer. Rotation, in effect, introduces the time element and transforms a *space* distribution of flux density into a *time* variation of voltage.

The voltage induced is a single-phase voltage. For the production of a set of 3-phase voltages it follows that 3 coils displaced 120 electrical degrees in space must be used, as shown in elementary form in Fig. 3-12. Equation 3-15 then gives the rms voltage per phase when N is the total series turns per phase. All of these elementary windings are full-pitch concentrated windings because the two sides of any coil are 180 electrical degrees apart and all the turns of that coil are concentrated in one pair of slots.   In actual ac machine windings the armature coils of each phase are distributed in a number of slots, as can be seen in Fig. 3-2.   A distributed winding makes better use of the iron and copper and improves the waveform.   For distributed windings a reduction factor k_w must be applied because the emfs induced in the individual coils of any one-phase group are not in time phase.   Their phasor sum is then less than their numerical sum when they are connected in series.   (See Appendix B for details.   For most 3-phase windings, k_w is about 0.85 to 0.95.)   For distributed windings Eq. 3-15 becomes

$$E_{rms} = 4.44 f k_w N_{ph} \Phi \text{ rms volts per phase} \qquad (3\text{-}16)$$

where N_{ph} is the number of series turns per phase.

b. DC Machines

Even if the ultimate purpose is the generation of a direct voltage, it is evident that the speed voltage generated in an armature coil is an alternating voltage.   The alternating waveform must therefore be rectified. Mechanical rectification is provided by the commutator, the device already described in elementary form in Art. 3-1b.   For the single coil of Fig. 3-16, the commutator provides full-wave rectification.   With the continued assumption of sinusoidal flux distribution, the voltage waveform between brushes is transformed to that of Fig. 3-20.   The average, or dc, value of the voltage between brushes is

$$E_a = \frac{1}{\pi} \int_0^{\pi} \omega N \Phi \sin \omega t \, d(\omega t) = \frac{2}{\pi} \omega N \Phi \qquad (3\text{-}17)$$

For dc machines it is usually more convenient to express the voltage E_a in terms of the mechanical speed ω_m rad/sec or n rpm.   Substitution of

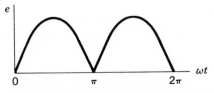

Fig. 3-20.   Voltage between brushes in an elementary dc machine.

Eq. 3-3 in Eq. 3-17 for a P-pole machine then yields

$$E_a = \frac{PN}{\pi} \Phi\omega_m = 2PN\Phi \frac{n}{60} \qquad (3\text{-}18)$$

The single-coil dc winding implied here is, of course, unrealistic in the practical sense, and it will be essential later to examine more carefully the action of commutators. Actually, Eq. 3-18 gives correct results for the more practical distributed dc armature windings as well, provided that N be taken as the total number of turns in series between armature terminals. Usually, the voltage is expressed in terms of the total number of active conductors Z_a and the number a of parallel paths through the armature winding. Because it takes two coil sides to make a turn and $1/a$ of these are connected in series, the number of series turns $N = Z_a/2a$. Substitution in Eq. 3-18 then gives

$$E_a = \frac{PZ_a}{2\pi a} \Phi\omega_m = \frac{PZ_a}{a} \Phi \frac{n}{60} \qquad (3\text{-}19)$$

We now have useful quantitative expressions for the rms voltage induced in ac machines, Eq. 3-16, and for the average induced voltage between brushes in dc machines, Eq. 3-19. We will now examine the magnetic fields produced by currents in the armature windings.

3-3
MMF OF DISTRIBUTED WINDINGS

Most armatures have distributed windings—i.e., windings which are spread over a number of slots around the air-gap periphery as in the photographs of Figs. 3-1 and 3-2. The individual coils are interconnected so that the result is a magnetic field having the same number of poles as the field winding.

As in the study of generated voltages, the study of the magnetic fields of distributed windings can be approached by study of the magnetic field of the single full-pitch N-turn coil shown in Fig. 3-21a. The dots and crosses indicate current toward and away from the reader, respectively. For simplicity, a concentric cylindrical rotor is shown. The general nature of the magnetic field produced by the current in the coil is shown by the dotted lines in Fig. 3-21a. Since the permeability of the armature and field iron is much greater than that of air, it is

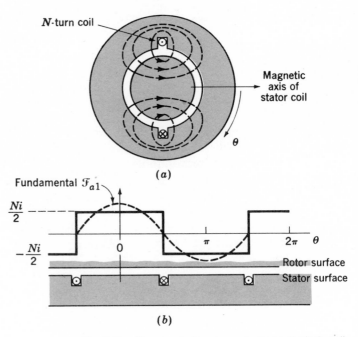

(a)

(b)

Fig. 3-21.   The mmf of a concentrated full-pitch coil.

sufficiently accurate for our present purposes to assume that all the reluctance of the magnetic circuit is in the two air gaps.   From symmetry of the structure it is evident that the magnetic-field intensity $\mathcal{3C}$ in the air gap at angle θ under one pole is the same in magnitude as that at $\theta + \pi$ under the opposite pole, but the fields are in the opposite direction.

Around any of the closed paths shown by the flux lines in Fig. 3-21a the mmf is Ni, or the mmf at each pole is $Ni/2$ ampere-turns per pole, where i is the coil current.   Figure 3-21b shows this winding in developed form, i.e., laid out flat.   The mmf wave is shown by the steplike distribution of amplitude $\pm Ni/2$.   On the assumption of narrow slot openings, the mmf wave jumps abruptly by Ni in crossing from one side to the other of a coil.

a. AC Machines

In the design of ac machines, serious efforts are made to distribute the winding so as to produce a close approximation to a sinusoidal space distribution of mmf.   We shall focus attention on the fundamental component.

The rectangular mmf wave of the concentrated full-pitch coil of Fig. 3-21b can be resolved into a Fourier series comprising a fundamental component and a series of odd harmonics. The fundamental component $\mathcal{F}_{a1}$ is

$$\mathcal{F}_{a1} = \frac{4}{\pi} \frac{Ni}{2} \cos \theta \tag{3-20}$$

where θ is measured from the axis of the stator coil, as shown by the dotted sinusoid in Fig. 3-21b. It is a sinusoidal space wave of amplitude

$$F_{1\text{peak}} = \frac{4}{\pi} \frac{Ni}{2} \tag{3-21}$$

with its peak aligned with the magnetic axis of the coil.

Now consider the effect of distributing a winding in several slots. For example, Fig. 3-22a shows phase a of the armature winding of a somewhat simplified 2-pole 3-phase ac machine. Phases b and c occupy the empty slots. The windings of the 3 phases are identical and are located with their magnetic axes 120 electrical degrees apart. We shall direct attention to the mmf of phase a alone, postponing the discussion of the effects of all 3 phases until Art. 3-4. The winding is arranged in two layers, each coil of n_c turns having one side in the top of a slot and the other coil side in the bottom of a slot a pole pitch away. The two-layer arrangement simplifies the geometrical problem of getting the end turns of the individual coils past each other.

Figure 3-22b shows one pole of this winding laid out flat. The mmf wave is a series of steps each of height $2n_c i_c$ equal to the ampere-conductors in the slot, where i_c is the coil current. Its space-fundamental component is shown by the sinusoid. It can be seen that the distributed winding produces a closer approximation to a sinusoidal mmf wave than the concentrated coil of Fig. 3-21.

The resultant fundamental mmf wave of a distributed winding is less than the sum of the fundamental components of the individual coils because the magnetic axes of the individual coils are not aligned with the resultant. The modified form of Eq. 3-20 for a distributed P-pole winding having N_{ph} series turns per phase is

$$\mathcal{F}_{a1} = \frac{4}{\pi} k_w \frac{N_{\text{ph}}}{P} i_a \cos \theta \tag{3-22}$$

in which the factor $4/\pi$ arises from the Fourier-series analysis of the rectangular mmf wave of a concentrated full-pitch coil, as in Eq. 3-20,

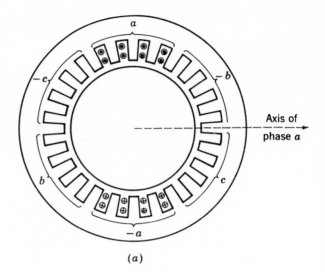

(a)

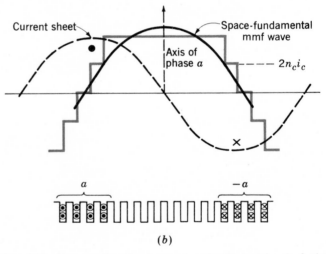

(b)

Fig. 3-22. The mmf of 1 phase of a distributed 2-pole 3-phase winding with full-pitch coils.

and the winding factor k_w takes into account the distribution of the winding.   (See Appendix B for details.)   Through the use of fractional-pitch coils and other artifices the effects of space harmonics in ac machines can be made small.

Equation 3-22 describes the space-fundamental component of the mmf wave produced by current in phase a.   It is equal to the mmf wave produced by a finely divided sinusoidally distributed current sheet placed on the inner periphery of the stator, as shown by the sine wave labeled "current sheet" in Fig. 3-22b.   The mmf is a standing wave whose *spatial* distribution around the periphery is described by $\cos \theta$. Its peak is along the magnetic axis of phase a and its peak amplitude is proportional to the instantaneous current i_a.   Accordingly, if the current $i_a = I_m \cos \omega t$, the *time* maximum of the peak is

$$F_{\max} = \frac{4}{\pi} k_w \frac{N_{\mathrm{ph}}}{P} I_m \qquad (3\text{-}23)$$

In Art. 3-4 we shall study the effect of 3-phase currents in all 3 phases.

b. DC Machines

Because of the restrictions imposed on the winding arrangement by the commutator, the mmf wave of a dc-machine armature approximates more nearly a triangular waveform rather than the sine wave of ac machines.   For example, Fig. 3-23 shows diagrammatically in cross section the armature of a 2-pole dc machine.   (In practice a larger number of slots would probably be used.)   The current directions are shown by dots and crosses.   The armature winding is equivalent to a coil wrapped around the armature and producing a magnetic field whose axis is vertical.   As the armature rotates, the coil connections to the external circuit are changed through commutator action, so that the magnetic field of the armature is always perpendicular to that of the field winding and a continuous unidirectional torque results.   Commutator action is described in Art. 4-3.

Figure 3-24a shows this winding laid out flat.   The mmf wave is shown in Fig. 3-24b.   On the assumption of narrow slots, it consists of a series of steps.   The height of each step equals the number of ampere-conductors $2n_c i_c$ in a slot, where n_c is the number of turns in each coil and i_c is the coil current, a two-layer winding and full-pitch coils being assumed.   The peak value of the mmf wave is along the magnetic axis of the armature, midway between the field poles.   This winding is equivalent to a coil of $12n_c i_c$ amp-turns distributed around the armature.   On the assumption of symmetry at each pole, the peak value of the mmf wave at each armature pole is $6n_c i_c$ amp-turns.

This mmf wave can be represented approximately by the triangular wave drawn in Fig. 3-24b and repeated in Fig. 3-24c.   For a more realistic winding with a larger number of armature slots per pole, the triangular distribution becomes a close approximation. It is the exact equivalent of the mmf wave of a uniformly distributed current sheet wrapped around the armature and carrying current in the dot and cross directions, as shown by the rectangular space distribution of current density in Fig. 3-24c.

For our preliminary study, it is convenient to resolve the mmf waves of distributed windings into a Fourier series. The fundamental component of the triangular mmf wave of Fig. 3-24c is shown by the sine wave. Its peak value is $8/\pi^2 = 0.81$ times the height of the triangular wave. The fundamental mmf wave is the exact equivalent of the mmf wave of a sinusoidally distributed current sheet wrapped around the armature whose peak value equals the fundamental component of the rectangular current sheet of Fig. 3-24c. This sinusoidally distributed current sheet is shown dotted in Fig. 3-24c.

It should be noted that the mmf wave depends only on the winding arrangement and symmetry of the magnetic structure at each pole. The

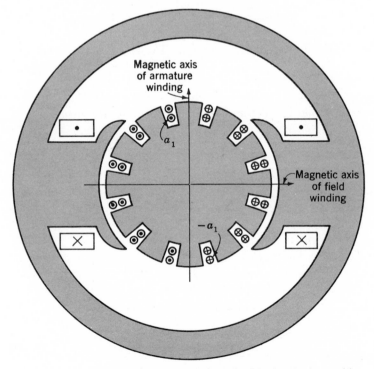

Fig. 3-23. Cross section of a 2-pole dc machine.

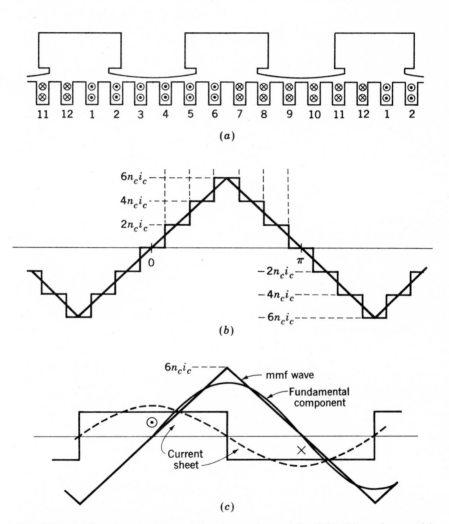

Fig. 3-24. (a) Developed sketch of the dc machine of Fig. 3-23; (b) mmf wave; (c) equivalent triangular mmf wave, its fundamental component, and equivalent rectangular current sheet.

flux-density wave, however, depends not only on the mmf but also on the magnetic boundary conditions, primarily the length of the air gap, the effect of the slot openings, and the shape of the pole face.   The designer takes these effects into account by means of flux maps, but these details need not concern us here.

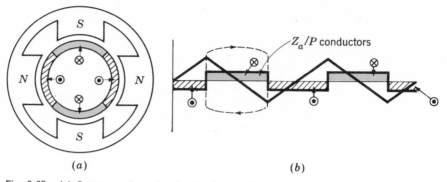

Fig. 3-25. (a) Cross section of a 4-pole dc machine and (b) development of current sheet and mmf wave.

Machines often have a magnetic structure with more than 2 poles. For example, Fig. 3-25a shows schematically a 4-pole dc machine. The field winding produces alternate north-south-north-south polarity, and the armature conductors are distributed in four belts of slots carrying currents alternately toward and away from the reader, as symbolized by the crosshatched areas. This machine is shown in laid-out form in Fig. 3-25b. The corresponding triangular armature-mmf wave is also shown. On the assumption of symmetry of the winding and field poles each successive pair of poles is like every other pair of poles. Magnetic conditions in the air gap can then be determined by examining any pair of adjacent poles—i.e., 360 electrical degrees.

The peak value of the triangular armature-mmf wave is

$$F_a = \frac{1}{2} \frac{Z_a}{P} \frac{i_a}{a} \qquad \text{amp-turns per pole} \qquad (3\text{-}24)$$

where Z_a is the total number of conductors in the armature winding, P is the number of poles, i_a is the armature current, and a is the number of parallel paths through the armature winding. Thus i_a/a is the current in each conductor. This equation comes directly from the line integral around the dotted closed path in Fig. 3-25b which crosses the air gap twice and encloses Z_a/P conductors each carrying current i_a/a in the same direction. In more compact form

$$F_a = \frac{N_a}{P} i_a \qquad (3\text{-}25)$$

where $N_a = Z_a/2a$ is the number of series armature turns.   For the triangular mmf wave of Fig. 3-25b the peak value of the space fundamental is $(8/\pi^2)(N_a/P)i_a$.

We shall base our preliminary investigations of both ac and dc machines on the assumption of sinusoidal space distributions of mmf. This model will be found to give very satisfactory results for most problems involving ac machines because their windings can be distributed so as to minimize the effects of harmonics.   Because of the restrictions placed on the winding arrangement by the commutator, the mmf waves of dc machines inherently approach more nearly a triangular waveform. Nevertheless the theory based on a sinusoidal model brings out the essential features of dc machine theory.   The results can readily be modified whenever necessary to account for any significant discrepancies.

We shall also use a 2-pole machine as our mathematical model.   The results can immediately be extrapolated to a P-pole machine when it is recalled that electrical angles θ and electrical angular velocities ω are related to mechanical angles θ_m and mechanical angular velocities ω_m through Eqs. 3-1 and 3-3.

For a preliminary study we shall further simplify our model by assuming that the stator and rotor air-gap surfaces are smooth, concentric cylinders.

3-4
ROTATING MAGNETIC FIELDS

To understand the theory of polyphase ac machines, it is necessary to study the nature of the magnetic field produced by a polyphase winding. In particular, we shall study the mmf patterns of a 3-phase winding such as those found on the stator of 3-phase induction and synchronous machines.   Once again attention will be focused on a 2-pole machine or one pair of poles of a P-pole winding.

In a 3-phase machine, the windings of the individual phases are displaced from each other by 120 electrical degrees in space around the air-gap circumference as shown by the coils a, $-a$, b, $-b$, and c, $-c$ in Fig. 3-26.   The concentrated full-pitch coils shown here may be considered to represent distributed windings producing sinusoidal mmf waves centered on the magnetic axes of the respective phases.   The three-component sinusoidal mmf waves are accordingly displaced 120 electrical degrees in *space*.   But each phase is excited by an alternating current which varies in magnitude sinusoidally with *time*.   Under balanced

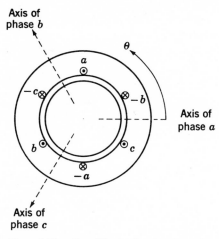

Fig. 3-26. Simplified 2-pole 3-phase stator winding.

3-phase conditions the instantaneous currents are

$$i_a = I_m \cos \omega t \tag{3-26}$$

$$i_b = I_m \cos (\omega t - 120°) \tag{3-27}$$

$$i_c = I_m \cos (\omega t - 240°) \tag{3-28}$$

where I_m is the maximum value of the current and the time origin is arbitrarily taken as the instant when the phase-a current is a positive maximum. The phase sequence is assumed to be abc. The instantaneous currents are shown in Fig. 3-27. The dots and crosses in the coil sides, Fig. 3-26, indicate the reference directions for positive phase currents.

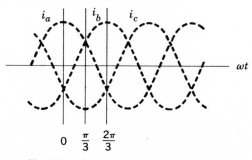

Fig. 3-27. Instantaneous 3-phase currents.

The corresponding component mmf waves vary sinusoidally with time. Each component is a stationary, pulsating sinusoidal distribution of mmf around the air gap with its peak located along the magnetic axis of its phase and its amplitude proportional to the instantaneous phase current—in other words, a standing space wave varying sinusoidally with time. Each component can be represented by an oscillating space vector drawn along the magnetic axis of its phase with length proportional to the instantaneous phase current. The resultant mmf is the sum of the components from all three phases.

a. Graphical Analysis

Now consider the state of affairs at $t = 0$, Fig. 3-27, the moment when the phase-a current is at its maximum value I_m. The mmf of phase a then has its maximum value F_{max}, given by Eq. 3-23 and shown by the vector $\mathbf{F}_a = \mathbf{F}_{max}$ drawn along the magnetic axis of phase a in Fig. 3-28a. At this moment the currents i_b and i_c are both $I_m/2$ in the negative direction, as shown by the dots and crosses in Fig. 3-28a indicating the *actual instantaneous* directions. The corresponding mmfs of phases b and c are shown by the vectors $\mathbf{F}_b$ and $\mathbf{F}_c$, both equal to $\mathbf{F}_{max}/2$ drawn in the negative direction along the magnetic axes of phases b and c, respectively. The resultant, obtained by adding the individual contributions of the 3 phases, is a vector $\mathbf{F} = \tfrac{3}{2}\mathbf{F}_{max}$ centered on the axis of phase a. It represents a sinusoidal space wave with its positive half wave centered on the axis of phase a and having an amplitude $\tfrac{3}{2}$ times that of the phase-a contribution alone.

At a later time $\omega t_1 = \pi/3$, Fig. 3-27, the currents in phases a and b are a positive half-maximum and that in phase c is a negative maximum. The individual mmf components and their resultant are now shown in Fig. 3-28b. The resultant has the same amplitude as at $t = 0$, but it has now rotated counterclockwise 60 electrical degrees in space. Similarly, at $\omega t_2 = 2\pi/3$ (when the phase-b current is a positive maximum and the phase-a and -c currents are a negative half-maximum) the same resultant mmf distribution is again obtained, but it has rotated counterclockwise 60 electrical degrees still farther and is now aligned with the magnetic axis of phase b (see Fig. 3-28c). As time passes, then, the resultant mmf wave retains its sinusoidal form and amplitude but shifts progressively around the air gap. This shift corresponds to a field rotating uniformly around the circumference of the air gap. Results consistent with this conclusion may be obtained by sketching the distribution at any arbitrary instant of time.

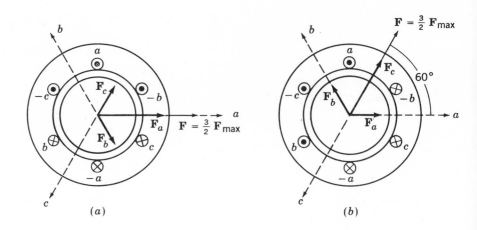

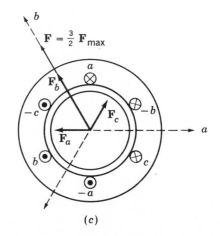

Fig. 3-28. Illustrating the production of a rotating magnetic field by means of 3-phase currents.

In one cycle the resultant mmf must be back in the position of Fig. 3-28a. The mmf wave therefore makes one revolution per cycle in a 2-pole machine. In a P-pole machine the wave travels 1 wavelength or $2/P$ revolutions per cycle.

b. Traveling Waves

To study the resultant field analytically, let the origin for angle θ around the air-gap periphery be placed at the axis of phase a (Fig. 3-26). At any time t, all 3 phases contribute to the air-gap mmf at any point θ.

The contribution from phase a is

$$F_{a(\text{peak})} \cos \theta$$

where $F_{a(\text{peak})}$ is the amplitude of the component mmf wave at time t. Similarly, the contributions from phases b and c are

$$F_{b(\text{peak})} \cos (\theta - 120°)$$

and
$$F_{c(\text{peak})} \cos (\theta - 240°)$$

respectively.   The 120° displacements appear because the machine is so wound that the axes of the 3 phases are 120 electrical degrees apart in space.   The resultant mmf at point θ is then

$$\mathfrak{F}(\theta) = F_{a(\text{peak})} \cos \theta + F_{b(\text{peak})} \cos (\theta - 120°)$$
$$+ F_{c(\text{peak})} \cos (\theta - 240°) \quad (3\text{-}29)$$

But the mmf amplitudes vary with time in accordance with the current variations.   Thus, with the time origin arbitrarily taken at the instant when the phase-a current is a positive maximum,

$$F_{a(\text{peak})} = F_{a(\text{max})} \cos \omega t \qquad (3\text{-}30)$$

$$F_{b(\text{peak})} = F_{b(\text{max})} \cos (\omega t - 120°) \qquad (3\text{-}31)$$

and
$$F_{c(\text{peak})} = F_{c(\text{max})} \cos (\omega t - 240°) \qquad (3\text{-}32)$$

The quantities $F_{a(\text{max})}$, $F_{b(\text{max})}$, and $F_{c(\text{max})}$ are, respectively, the time-maximum values of the amplitudes $F_{a(\text{peak})}$, $F_{b(\text{peak})}$, and $F_{c(\text{peak})}$.   The 120° displacements appear here because the three currents are 120° phase-displaced in time.   Since the currents in the 3 phases are balanced and therefore of equal amplitude, the three amplitudes $F_{a(\text{max})}$, $F_{b(\text{max})}$, and $F_{c(\text{max})}$ are also equal and the symbol F_{max} may be used for all three. Equation 3-29 accordingly becomes

$$\mathfrak{F}(\theta,t) = F_{\text{max}} \cos \theta \cos \omega t + F_{\text{max}} \cos (\theta - 120°) \cos (\omega t - 120°)$$
$$+ F_{\text{max}} \cos (\theta - 240°) \cos (\omega t - 240°) \quad (3\text{-}33)$$

Each of the three components on the right-hand side of Eq. 3-33 is a pulsating standing wave.   In each term the trigonometric function of θ defines the space distribution as a stationary sinusoid, and the trigonometric function of t indicates that the amplitudes pulsate with time.

The first of the three terms expresses the phase-a component; the second and third terms express, respectively, the phase-b and -c components.

By use of the trigonometric transformation

$$\cos \alpha \cos \beta = \tfrac{1}{2} \cos (\alpha - \beta) + \tfrac{1}{2} \cos (\alpha + \beta) \qquad (3\text{-}34)$$

each of the components in Eq. 3-33 can be expressed as cosine functions of sum and difference angles.   Thus,

$$\begin{aligned}
\mathcal{F}(\theta,t) = &\ \tfrac{1}{2}F_{max} \cos (\theta - \omega t) + \tfrac{1}{2}F_{max} \cos (\theta + \omega t) \\
&+ \tfrac{1}{2}F_{max} \cos (\theta - \omega t) + \tfrac{1}{2}F_{max} \cos (\theta + \omega t - 240°) \\
&+ \tfrac{1}{2}F_{max} \cos (\theta - \omega t) + \tfrac{1}{2}F_{max} \cos (\theta + \omega t - 480°) \quad (3\text{-}35)
\end{aligned}$$

Now the three cosine terms involving the angles $\theta + \omega t$, $\theta + \omega t - 240°$, and $\theta + \omega t - 480°$ are three equal sinusoids displaced in phase by 120°. (Note that a lag angle of 480° is equivalent to a lag angle of $480° - 360° = 120°$.)   Their sum is therefore zero, and Eq. 3-35 reduces to

$$\mathcal{F}(\theta,t) = \tfrac{3}{2}F_{max} \cos (\theta - \omega t) \qquad (3\text{-}36)$$

which is the desired expression for the resultant mmf wave.

The wave described by Eq. 3-36 is a sinusoidal function of the space angle θ.   It has a constant amplitude and a space-phase angle ωt which is a linear function of time.   The angle ωt provides rotation of the entire wave around the air gap at the constant angular velocity ω.   Thus, at a fixed time t_x, the wave is a sinusoid in space with its positive peak displaced ωt_x electrical radians from the fixed point on the winding which is the origin for θ; at a later instant t_y, the positive peak of the same wave is displaced ωt_y from the origin, and the wave has moved $\omega(t_y - t_x)$ around the gap.   At $t = 0$, the current in phase a is a maximum, and the positive peak of the resultant mmf wave is located at the axis of phase a; one-third of a cycle later, the current in phase b is a maximum, and the positive peak is located at the axis of phase b; and so on.   The angular velocity of the wave is $\omega = 2\pi f$ electrical radians per second.   For a P-pole machine the rotational speed is

$$\omega_m = \frac{2}{P} \omega \qquad \text{rad/sec} \qquad (3\text{-}37)$$

or

$$n = \frac{120f}{P} \qquad \text{rpm} \qquad (3\text{-}38)$$

results which are consistent with Eqs. 3-2 and 3-3.

In general it may be shown that a rotating field of constant amplitude will be produced by a q-phase winding excited by balanced q-phase currents when the respective phases are wound $2\pi/q$ electrical radians apart in space. The constant amplitude will be $q/2$ times the maximum contribution of any one phase, and the speed will be $\omega = 2\pi f$ electrical radians per second.

A polyphase winding excited by balanced polyphase currents is thus seen to produce the same general effect as spinning a permanent magnet about an axis perpendicular to the magnet or as the rotation of dc-excited field poles.

3-5

TORQUE IN NONSALIENT–POLE MACHINES

The behavior of any electromagnetic device as a component in an electromechanical system can be described in terms of its Kirchhoff-law voltage equations and its electromagnetic torque. The purpose of this article is to derive the voltage and torque equations for an idealized elementary machine—results which can readily be extended later to more complex machines. We shall derive these equations from two viewpoints and show that basically they stem from the same ideas.

The first viewpoint is essentially the same as that of Art. 2-6. The machine will be regarded as a circuit element whose inductances depend on the angular position of the rotor. The flux linkages λ and magnetic-field coenergy will be expressed in terms of the currents and inductances. The torque can then be found from the partial derivative of the magnetic-field energy or coenergy with respect to angle, and the terminal voltages from the sum of the resistance drops Ri and the Faraday-law voltages $d\lambda/dt$. The result will be a set of nonlinear differential equations describing the dynamic performance of the machine.

The second viewpoint regards the machine as two groups of windings producing magnetic fields in the air gap, one group on the stator and the other on the rotor. By making suitable assumptions regarding these fields, simple expressions can be derived for the flux linkages and magnetic field energy stored in the air gap in terms of the field quantities. The torque and generated voltage can then be found in terms of these quantities. Thus torque is expressed explicitly as the tendency for two magnetic-fields to line up in the same way as permanent magnets tend to align themselves, and generated voltage is expressed as the result of relative motion between a field and a winding. These expressions lead to a simple physical picture of the normal steady-state behavior of rotating machines.

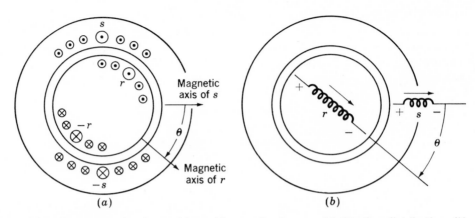

Fig. 3-29. Elementary 2-pole machine with smooth air gap. (a) Winding distribution; (b) schematic representation.

a. The Coupled-circuit Viewpoint

Consider the elementary machine of Fig. 3-29 with 1 winding on the stator and 1 on the rotor. These windings are distributed over a number of slots so that their mmf waves can be approximated by space sinusoids. In Fig. 3-29a the coil sides s, $-s$ and r, $-r$ mark the positions of the centers of the belts of conductors comprising the distributed windings. An alternative way of drawing these windings is shown in Fig. 3-29b, which also shows reference directions for voltages and currents. Here it is assumed that current in the arrow direction produces a magnetic field in the air gap in the arrow direction, so that a single arrow defines reference directions for both current and flux. The stator and rotor are concentric cylinders, and slot openings are neglected. Consequently, our elementary model does not include the effects of salient poles—effects which will be investigated in later chapters. We shall also assume that the reluctances of the stator and rotor iron are negligible.

On these assumptions the stator and rotor self-inductances L_{ss} and L_{rr} are constant, but the stator-rotor mutual inductance depends on the angle θ between the magnetic axes of the stator and rotor windings. The mutual inductance is a positive maximum when $\theta = 0$ or 2π, is zero when $\theta = \pm\pi/2$, and is a negative maximum when $\theta = \pm\pi$. On the assumption of sinusoidal mmf waves and a uniform air gap, the space distribution of the air-gap flux wave is sinusoidal, and the mutual inductance is

$$\mathcal{L}_{sr}(\theta) = L_{sr} \cos \theta \qquad\qquad (3\text{-}39)$$

where the script letter $\mathcal{L}$ is used to denote an inductance which is a function of the electrical angle θ. The italic capital letter L denotes a constant value. Thus L_{sr} is the value of the mutual inductance when the magnetic axes of stator and rotor are in line. In terms of the inductances, the stator and rotor flux linkages λ_s and λ_r are

$$\lambda_s = L_{ss}i_s + \mathcal{L}_{sr}(\theta)i_r = L_{ss}i_s + L_{sr}i_r \cos \theta \tag{3-40}$$

$$\lambda_r = \mathcal{L}_{sr}(\theta)i_s + L_{rr}i_r = L_{sr}i_s \cos \theta + L_{rr}i_r \tag{3-41}$$

In matrix notation

$$\begin{bmatrix} \lambda_s \\ \lambda_r \end{bmatrix} = \begin{bmatrix} L_{ss} & \mathcal{L}_{sr}(\theta) \\ \mathcal{L}_{sr}(\theta) & L_{rr} \end{bmatrix} \times \begin{bmatrix} i_s \\ i_r \end{bmatrix}$$

The terminal voltages v_s and v_r are

$$v_s = R_s i_s + p\lambda_s \tag{3-42}$$

$$v_r = R_r i_r + p\lambda_r \tag{3-43}$$

where R_s and R_r are the winding resistances and p is the time-derivative operator d/dt. When the rotor is revolving, θ must be treated as a variable. Differentiation of Eqs. 3-40 and 3-41 and substitution of the results in Eqs. 3-42 and 3-43 then give

$$v_s = R_s i_s + L_{ss}p i_s + L_{sr} \cos \theta \, (p i_r) - L_{sr}i_r \sin \theta \, (p\theta) \tag{3-44}$$

$$v_r = R_r i_r + L_{rr}p i_r + L_{sr} \cos \theta \, (p i_s) - L_{sr}i_s \sin \theta \, (p\theta) \tag{3-45}$$

where $p\theta$ is the instantaneous speed ω in *electrical* radians per second. In a 2-pole machine θ and ω are equal to the instantaneous shaft angle θ_m and speed ω_m, respectively. In a P-pole machine they are related by Eqs. 3-1 and 3-3. The second and third terms on the right-hand sides of Eqs. 3-44 and 3-45 are $L \, di/dt$ induced voltages like those induced in stationary coupled circuits such as the windings of transformers. The fourth terms are caused by mechanical motion and are proportional to the instantaneous speed. They are the coupling terms relating the interchange of power between the electrical and mechanical systems.

The electromagnetic torque can be found from the coenergy in the magnetic field in the air gap. From Eq. 2-71

$$W'_{\text{fld}} = \tfrac{1}{2}L_{ss}i_s^2 + \tfrac{1}{2}L_{rr}i_r^2 + L_{sr}i_s i_r \cos \theta \tag{3-46}$$

and from Eq. 2-69

$$T = + \frac{\partial W'_{\text{fld}}(\theta_m, i_s i_r)}{\partial \theta_m} = + \frac{\partial W'_{\text{fld}}(\theta, i_s, i_r)}{\partial \theta} \frac{d\theta}{d\theta_m} \qquad (3\text{-}47)$$

where T is the electromagnetic torque acting in the positive direction of θ_m and the derivative must be taken with respect to actual *mechanical* angle θ_m because we are dealing here with mechanical variables. Differentiation of Eqs. 3-46 and 3-1 for a P-pole machine then gives

$$T = -\frac{P}{2} L_{\text{sr}} i_s i_r \sin \theta = -\frac{P}{2} L_{\text{sr}} i_s i_r \sin \frac{P}{2} \theta_m \qquad (3\text{-}48)$$

with T in newton-meters. The negative sign in Eq. 3-48 means that the electromagnetic torque acts in the direction to bring the magnetic fields of stator and rotor into alignment.

Equations 3-44, 3-45, and 3-48 are a set of three differential equations relating the electrical variables v_s, i_s, v_r, i_r and the mechanical variables T and θ_m. These equations, together with the constraints imposed on the electrical variables by the networks connected to its terminals (sources or loads and external impedances) and the constraints imposed on the mechanical variables (applied torques and inertial, frictional, and spring torques), determine the performance of the device as a coupling element. They are nonlinear differential equations and are difficult to solve except under special circumstances. We are not concerned with their solution here; rather, we are using them merely as steps in the development of the theory of rotating machines.

Now consider a uniform-air-gap machine with several stator and several rotor windings. The same general principles that apply to the elementary model of Fig. 3-29 also apply to the multiwinding machine. Each winding has its own self-inductances and mutual inductances with other windings. The self-inductances and mutual inductances between pairs of windings on the same side of the air gap are constant on the assumption of a uniform gap and negligible magnetic saturation. But the mutual inductances between pairs of stator and rotor windings vary as the cosine of the angle between the magnetic axes of the windings. The torque results from the tendency of the magnetic field of the rotor windings to line up with that of the stator windings. It can be expressed as the sum of terms like Eq. 3-48.

EXAMPLE 3-1

Consider the elementary rotating machine of Fig. 3-29. Its shaft
is coupled to a mechanical device which can be made to absorb
or deliver mechanical torque over a wide range of speeds. This
machine can be connected and operated in several ways. For example, suppose the rotor winding is excited with direct current I_r and
the stator winding is connected to an ac source which can either
absorb or deliver electrical power. Let the stator current be

$$i_s = I_s \cos \omega_s t$$

where $t = 0$ is arbitrarily chosen as the moment when the stator
current has its peak value.

a. Derive an expression for the magnetic torque developed by the
   machine as the speed is varied by control of the mechanical device
   connected to its shaft.
b. Find the speed at which average torque will be produced if the
   stator frequency is 60 Hz.
c. With the assumed current-source excitations what are the voltages
   induced in the stator and rotor windings at synchronous speed?

Solution

a. From Eq. 3-48 for a 2-pole machine

$$T = -L_{sr} i_s i_r \sin \theta_m$$

For the conditions of this problem

$$T = -L_{sr} I_s I_r \cos \omega_s t \sin (\omega_m t \pm \delta)$$

where ω_m is the clockwise angular velocity impressed on the rotor
by the mechanical drive and δ is the angular position of the rotor
at $t = 0$. From the trigonometric identity

$$\sin A \cos B = \tfrac{1}{2} \sin (A + B) + \tfrac{1}{2} \sin (A - B)$$
$$T = -\tfrac{1}{2} L_{sr} I_s I_r \{\sin [(\omega_m + \omega_s)t + \delta]$$
$$+ \sin [(\omega_m - \omega_s)t + \delta]\}$$

The torque consists of two sinusoidally time-varying terms of
frequencies $\omega_m + \omega_s$ and $\omega_m - \omega_s$.

b. Except when $\omega_m = \pm \omega_s$, the torque averaged over a sufficiently long time is zero. But if $\omega_m = +\omega_s$, the torque becomes

$$T = -\tfrac{1}{2}L_{\text{sr}}I_sI_r[\sin\,(2\omega_st + \delta) + \sin\,\delta]$$

The first sine term is a double-frequency component whose average value is zero. The second term is the average torque

$$T_{\text{av}} = -\tfrac{1}{2}L_{\text{sr}}I_sI_r\,\sin\,\delta$$

The other possibility is $\omega_m = -\omega_s$, which merely means rotation in the counterclockwise direction. The negative sign in the expression for T_{av} means that the magnetic torque tends to reduce δ. The machine is an idealized single-phase synchronous machine. (With polyphase synchronous machines the direction of rotation is determined by the phase sequence, as shown in Art. 3-4.)

With stator frequency of 60 Hz,

$$\omega_m = \omega_s = 2\pi(60) \qquad \text{rad/sec}$$

or 60 rev/sec, or 3,600 rpm.

c. From the second and fourth terms of Eq. 3-44 with $\omega_m = \omega_s$ the voltage induced in the stator is

$$e_s = -\omega_sL_{\text{ss}}I_s\,\sin\,\omega_st - \omega_sL_{\text{sr}}I_r\,\sin\,(\omega_st + \delta)$$

From the third and fourth terms of Eq. 3-45 the voltage induced in the rotor is

$$e_r = -\omega_sL_{\text{sr}}I_s[\sin\,\omega_st\,\cos\,(\omega_st + \delta) + \cos\,\omega_st\,\sin\,(\omega_st + \delta)]$$

$$= -\omega_sL_{\text{sr}}I_s\,\sin\,(2\omega_st + \delta)$$

The stator current induces a double-frequency voltage in the rotor.

---

b. The Magnetic-field Viewpoint

In the foregoing discussion the characteristics of the device as viewed from its electrical and mechanical terminals have been expressed in terms of its inductances. This viewpoint gives very little insight into the internal phenomena and gives no conception of the effects of physical

dimensions. An alternative formulation in terms of the interacting
magnetic fields in the air gap should serve to supply some of these
missing features.

Currents in the machine windings create magnetic flux in the air gap
between the stator and rotor, the flux paths being completed through
the stator and rotor iron. This condition corresponds to the appearance
of magnetic poles on both stator and rotor, centered on their respective
magnetic axes, as shown in Fig. 3-30a for a 2-pole machine with a smooth
air gap. Torque is produced by the efforts of the two component magnetic fields to line up their magnetic axes. The simple physical picture
is like that of two bar magnets pivoted at their centers on the same shaft.
The torque is proportional to the product of the amplitudes of the stator
and rotor mmf waves. It is also a function of the angle δ_{sr} between their
magnetic axes. We shall show that for a smooth air-gap machine the
torque is proportional to sin δ_{sr}.

Most of the flux produced by the stator and rotor windings (roughly
90 percent in typical machines) crosses the air gap and links both windings; this flux is termed the *mutual flux*. Small percentages of the flux,
however, do not cross the air gap but link only the rotor or stator winding; these are, respectively, the *rotor leakage flux* and the *stator leakage
flux*. They comprise slot and toothtip leakage, end-turn leakage, and
space harmonics in the air-gap field. It is only the mutual flux which is
of direct concern in torque production. The leakage fluxes do affect
machine performance, however, by virtue of the voltages they induce
in their own windings. Their effect on the electrical characteristics is

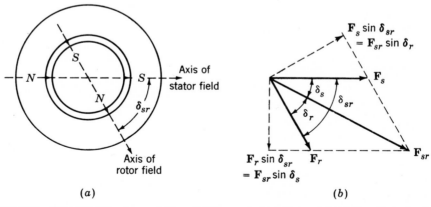

Fig. 3-30. Simplified 2-pole machine. (a) Elementary model and (b) vector diagram of
mmf waves.

accounted for by means of *leakage inductances* like those of a transformer. This effect, however, is an auxiliary one rather than a fundamental part of torque production.

Our analysis, then, will be in terms of the resultant mutual flux. We shall derive an expression for the magnetic coenergy stored in the air gap in terms of the stator and rotor mmfs and the angle δ_{sr} between their magnetic axes. The torque can then be found from the partial derivative of the coenergy with respect to angle δ_{sr}.

We shall now assume that the tangential component of the magnetic field in the air gap is negligible compared with the radial component—in other words, the mutual flux goes straight across the gap. We shall also assume that the radial length g of the gap (the clearance between rotor and stator) is small compared with the radius of the rotor or stator. On this assumption there is negligible difference between the flux density at the rotor surface, at the stator surface, or at any intermediate radial distance in the air gap, say, halfway between. The air-gap field then reduces to a radial field $\mathcal{H}$ or $\mathcal{B}$ whose intensity varies with angle around the periphery. The line integral of $\mathcal{H}$ across the gap then is simply $\mathcal{H}g$ and equals the resultant mmf $\mathcal{F}_{sr}$ of the stator and rotor windings; thus

$$\mathcal{H}g = \mathcal{F}_{sr} \tag{3-49}$$

where the script $\mathcal{F}$ denotes the mmf wave as a function of angle around the periphery.

The mmf waves of stator and rotor are spatial sine waves with δ_{sr} the phase angle between their magnetic axes in electrical degrees. They can be represented by the space vectors $\mathbf{F}_s$ and $\mathbf{F}_r$ drawn along the magnetic axes of the stator- and rotor-mmf waves, as in Fig. 3-30b. The resultant mmf $\mathbf{F}_{sr}$ acting across the air gap, also a sine wave, is the vector sum. From the trigonometric formula for the diagonal of a parallelogram, its peak value is found from

$$F_{sr}^2 = F_s^2 + F_r^2 + 2F_sF_r \cos \delta_{sr} \tag{3-50}$$

in which the F's are the peak values of the mmf waves. The resultant radial $\mathcal{H}$ field is a sinusoidal space wave whose peak value H_{peak} is, from Eq. 3-49

$$H_{\text{peak}} = \frac{F_{sr}}{g} \tag{3-51}$$

Now consider the magnetic-field coenergy stored in the air gap. The coenergy density at a point where the magnetic field intensity is $\mathfrak{IC}$ is

$$\frac{\mu_0}{2}\, \mathfrak{IC}^2$$

in rationalized mks units. The coenergy density averaged over the volume of the air gap is

$$\frac{\mu_0}{2} \quad (\text{average value of } \mathfrak{IC}^2)$$

The average value of the square of a sine wave is half its peak value. Hence,

$$\text{Average coenergy density} = \frac{\mu_0}{2}\frac{H_{\text{peak}}^2}{2} = \frac{\mu_0}{4}\left(\frac{F_{\text{sr}}}{g}\right)^2 \qquad (3\text{-}52)$$

The total coenergy is

$$W'_{\text{fld}} = (\text{average coenergy density})(\text{volume of air gap})$$

$$= \frac{\mu_0}{4}\left(\frac{F_{\text{sr}}}{g}\right)^2 \pi Dlg = \frac{\mu_0 \pi Dl}{4g} F_{\text{sr}}^2 \qquad (3\text{-}53)$$

where D is the average diameter at the air gap, l is its axial length, g is the air-gap clearance, and $\mu_0 = 4\pi \times 10^{-7}$. All dimensions are in meters. From Eq. 3-50, the coenergy stored in the air gap can now be expressed in terms of the peak amplitudes of the stator- and rotor-mmf waves and the space-phase angle between them; thus

$$W'_{\text{fld}} = \frac{\mu_0 \pi Dl}{4g}\, (F_s^2 + F_r^2 + 2F_s F_r \cos \delta_{\text{sr}}) \qquad (3\text{-}54)$$

An expression for the electromagnetic torque T can now be obtained in terms of the interacting magnetic fields by taking the partial derivative of the field coenergy with respect to angle. For a 2-pole machine

$$T = +\frac{\partial W'_{\text{fld}}}{\partial \delta_{\text{sr}}} = -\frac{\mu_0 \pi Dl}{2g}\, F_s F_r \sin \delta_{\text{sr}} \qquad (3\text{-}55)$$

For a P-pole machine Eq. 3-55 gives the torque per pair of poles. The torque for a P-pole machine then is

$$T = -\frac{P}{2}\frac{\mu_0}{2}\frac{\pi Dl}{g} F_s F_r \sin \delta_{sr} \qquad (3\text{-}56)$$

This important equation states that the torque is proportional to the peak values of the stator- and rotor-mmf waves F_s and F_r and the sine of the electrical space-phase angle δ_{sr} between them. The minus sign means that the fields tend to align themselves. Equal and opposite torques are exerted on stator and rotor. The torque on the stator is simply transmitted through the frame of the machine to the foundation.

On referring to Fig. 3-30b, it can be seen that $F_r \sin \delta_{sr}$ is the component of the F_r wave in electrical space quadrature with the F_s wave. Similarly $F_s \sin \delta_{sr}$ is the component of the F_s wave in quadrature with the F_r wave. Thus, the torque is proportional to the product of one magnetic field and the component of the other in quadrature with it, like the cross-product of vector analysis. Also note that in Fig. 3-30b

$$F_s \sin \delta_{sr} = F_{sr} \sin \delta_r \qquad (3\text{-}57)$$

and
$$F_r \sin \delta_{sr} = F_{sr} \sin \delta_s \qquad (3\text{-}58)$$

The torque can then be expressed in terms of the *resultant* mmf wave F_{sr} by substitution of either Eq. 3-57 or Eq. 3-58 in Eq. 3-56; thus

$$T = -\frac{P}{2}\frac{\pi}{2}\frac{\mu_0 Dl}{g} F_s F_{sr} \sin \delta_s \qquad (3\text{-}59)$$

$$T = -\frac{P}{2}\frac{\pi}{2}\frac{\mu_0 Dl}{g} F_r F_{sr} \sin \delta_r \qquad (3\text{-}60)$$

Comparison of Eqs. 3-56, 3-59, and 3-60 shows that the torque can be expressed in terms of the component magnetic fields due to *each* current acting alone as in Eq. 3-56, or in terms of the *resultant* field and *either* of the components as in Eqs. 3-59 and 3-60, *provided that we use the corresponding angle between the axes of the fields.* Ability to reason in any of these terms is a convenience in machine analysis.

In Eqs. 3-56, 3-59, and 3-60 the fields have been expressed in terms of the peak values of their mmf waves. When magnetic saturation is neglected, the fields may, of course, be expressed in terms of their flux-density waves or in terms of total flux per pole. Thus the peak value B

of the field due to a sinusoidally distributed mmf wave in a uniform-air-gap machine is $\mu_0 F/g$, where F is the peak value of the mmf wave. For example, the resultant mmf F_{sr} produces a resultant flux-density wave whose peak value is $\mu_0 F_{sr}/g$. Thus,

$$T = -\frac{P}{2}\frac{\pi Dl}{2} B_{sr}F_r \sin \delta_r \qquad (3\text{-}61)$$

One of the inherent limitations in the design of electromagnetic apparatus is the saturation flux density of magnetic materials. Because of saturation in the armature teeth the peak value B_{sr} of the resultant flux-density wave in the air gap is limited to about 1 weber/m² (64.5 kilolines/in.²). The maximum permissible value of the mmf wave is limited by temperature rise of the winding and other design requirements. Because the resultant flux density and mmf appear explicitly in Eq. 3-61, this equation is in a convenient form for design purposes.

Alternative forms arise when it is recognized that the resultant flux per pole is

$$\Phi = \text{(average value of } \mathcal{B} \text{ over a pole)(pole area)} \qquad (3\text{-}62)$$

and that the average value of a sinusoid over a half wavelength is $2/\pi$ times its peak value. Thus

$$\Phi = \frac{2}{\pi}B\frac{\pi Dl}{P} = \frac{2Dl}{P}B \qquad (3\text{-}63)$$

where B is the peak value of the corresponding flux-density wave. For example, substitution of Eq. 3-63 in Eq. 3-61 gives

$$T = -\frac{\pi}{2}\left(\frac{P}{2}\right)^2 \Phi_{sr}F_r \sin \delta_r \qquad (3\text{-}64)$$

where Φ_{sr} is the resultant flux produced by the combined effect of the stator and rotor mmfs.

To recapitulate, we now have several forms in which the torque of a uniform-air-gap machine may be expressed in terms of its magnetic fields. All of them are merely statements that the torque is proportional to the interacting fields and the sine of the electrical space angle between their magnetic axes. The negative sign indicates that the electromagnetic torque acts in a direction to decrease the displacement angle between the fields. In a preliminary discussion of machine types, Eq. 3-64 will be the preferred form.

One further remark may be made concerning the torque equations and the thought process leading up to them.   There is no restriction that the mmf wave or flux-density wave need remain stationary in space.   They may remain stationary, or they may be traveling waves as shown in Art. 3-4.   If the magnetic fields of stator and rotor are constant in amplitude and travel around the air gap at the same speed, a steady torque will be produced by the efforts of the stator and rotor fields to align themselves in accordance with the torque equations.

3-6
RÉSUMÉ

In this chapter we have presented a brief and elementary description of three basic types of rotating machines—synchronous, induction, and dc machines.   In all of them the basic principles are essentially the same. Voltages are generated by relative motion of a magnetic field with respect to a winding, and torques are produced by the interaction of the magnetic fields of the stator and rotor windings.   The characteristics of the various machine types are determined by the methods of connection and excitation of the windings, but the basic principles are still essentially similar.

The basic analytical tools for studying rotating machines are expressions for the generated voltages and for the electromagnetic torque. Taken together, they express the coupling between the electrical and mechanical systems.   In order to develop a reasonably quantitative theory without the confusion arising from too much detail, we have made several simplifying approximations.   In the study of ac machines we have assumed sinusoidal time variations of voltages and currents, and sinusoidal space waves of air-gap flux density and mmf.   Faraday's law then results in Eq. 3-16 for the rms voltage generated in an ac machine winding, or Eq. 3-19 for the average voltage generated between brushes in a dc machine.   On examination of the mmf of distributed ac windings, we found that the space-fundamental component is the most important, Eqs. 3-22 and 3-23, but that the mmf of dc machine armature windings is more nearly a triangular wave, Eq. 3-24.   For our preliminary study in this chapter, however, we have assumed sinusoidal mmf distributions for both ac and dc machines. We shall examine this assumption more thoroughly for dc machines in Chap. 5.

On examination of the mmf wave of a 3-phase winding, we found that balanced 3-phase currents produce a constant-amplitude magnetic field rotating in the air gap at synchronous speed, as shown in Fig. 3-28 and Eq. 3-36.   The importance of this fact cannot be overestimated, for

it means that the double-frequency time-varying torque inherently associated with the double-frequency component of the instantaneous power
in a single-phase system is eliminated in balanced 3-phase machines, as
shown in Appendix A. Imagine a multimegawatt 60-Hz generator or a
multihorsepower motor subjected to a multimegawatt instantaneous
power pulsation at 120 Hz! The discovery of rotating fields led to the
invention of the simple, rugged, reliable, self-starting polyphase induction
motor which will be described briefly in Art. 4-2 and analyzed in more
detail in Chap. 7. (A single-phase induction motor will not start itself—
it needs an auxiliary starting winding as shown in Chap. 11.) It also
made possible the construction of multimegawatt synchronous generators.
The most significant reason for the almost universal use of 3-phase
generation, transmission, and utilization (for loads above a few kilowatts)
is the rotating field in the generators and motors.

Having assumed sinusoidally distributed magnetic fields in the air
gap, we then derived expressions for the magnetic torque. The simple
physical picture is like that of two magnets, one on the stator and one
pivoted on the rotor, as shown schematically in Fig. 3-30a. The torque
acts in the direction to align the magnets. To get a reasonably close
quantitative analysis without being hindered by details, we assumed a
smooth air gap and neglected the reluctance of the magnetic paths in the
iron parts, with the mental note to look into these assumptions later in
Chaps. 5 and 6.

In Art. 3-5 we derived expressions for the magnetic torque from two
viewpoints both based on the fundamental principles of Chap. 2. The
first viewpoint regards the machine as a group of magnetically coupled
circuits with inductances which depend on the angular position of the
rotor, as in Art. 3-5a. The second regards the machine from the viewpoint of the magnetic fields in the air gap, as in Art. 3-5b. It is shown
that the torque can be expressed as the product of the stator field, the
rotor field, and the sine of the angle between their magnetic axes, as in
Eq. 3-56, or in any of the forms derived from Eq. 3-56. The two viewpoints are supplementary, and ability to reason in terms of both of them
is helpful in reaching an understanding of how machines work.

This chapter has been concerned with basic principles underlying
rotating-machine theory. By itself it is obviously incomplete. Many
questions remain unanswered. How do we apply these principles to the
determination of the characteristics of synchronous, induction, and
dc machines? What are some of the practical problems that arise from
the use of iron, copper, and insulation in physical machines? What
are some of the economic and engineering considerations affecting rotating-machine applications? What are the physical factors limiting the

conditions under which a machine may operate successfully? Chapter 4 discusses some of these problems. Taken together, Chaps. 3 and 4 serve as an introduction to the more detailed treatments of rotating machines in Chaps. 5 through 11.

PROBLEMS

3-1. The object of this problem is to illustrate how the armature windings of certain machines (i.e., dc machines) may be approximately represented by uniform current sheets, with the degree of correspondence growing better as the winding is distributed in a greater number of slots around the armature periphery. For this purpose, consider an armature with 8 slots uniformly distributed over 360 electrical degrees or one pair of poles. The air gap is of uniform width, the slot openings are very small, and the reluctance of the iron is negligible.

Lay out 360 electrical degrees of the armature with its slots in developed form in the manner of Fig. 3-24a, and number the slots 1 to 8 from left to right. The winding consists of 8 single-turn coils, each carrying a direct current of 10 amp. Coil sides which may be placed in slots 1 to 4 carry current directed into the paper; those which may be placed in slots 5 to 8 carry current out of the paper.

- *a.* Consider that all 8 coils are placed with one side in slot 1 and the other in slot 5. The remaining slots are empty. Draw the rectangular mmf wave produced by these coils.
- *b.* Next consider that 4 coils have one side in slot 1 and the other in slot 5, while the remaining 4 have one side in slot 3 and the other in slot 7. Draw the component rectangular mmf waves produced by each group of coils, and superimpose the components to give the resultant mmf wave.
- *c.* Now consider that 2 coils are placed in slots 1 and 5, 2 in 2 and 6, 2 in 3 and 7, and 2 in 4 and 8. Again superimpose the component rectangular waves to produce the resultant wave. Note that the task can be systematized and simplified by recognizing that the mmf wave is symmetrical about its axis and takes a step at each slot which is definitely related to the number of ampere-conductors in the slot.
- *d.* Let the armature now consist of 16 slots per 360 electrical degrees with one coil side in each slot. Draw the resultant mmf wave.
- *e.* Approximate each of the resultant waves of (*a*) to (*d*) by isosceles

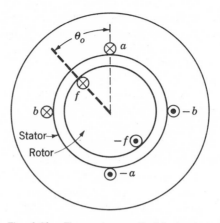

Fig. 3-31. Elementary cylindrical-rotor 2-phase synchronous machine, Prob. 3-2.

triangles, noting that the representation grows better as the winding is more finely distributed.

3-2. Figure 3-31 shows in cross section a machine having a rotor winding ff and 2 identical stator windings aa and bb. The self-inductance of each stator winding is L_{aa} henrys and of the rotor winding is L_{ff} henrys. The air gap is uniform. The stator windings are in quadrature. The mutual inductance between a stator winding and the rotor winding depends on the angular position θ_o of the rotor and may be assumed to be

$$M_{af} = M \cos \theta_o \qquad M_{bf} = M \sin \theta_o$$

where M is the maximum value of the mutual inductance. The resistance of each stator winding is R_a ohms.

a. Derive a general expression for the torque T in terms of the angle θ_o, the inductance constants, and the instantaneous currents i_a, i_b, and i_f. Does this expression apply at standstill? When the rotor is revolving?

b. Suppose the rotor is stationary and constant direct currents $I_a = 5$ amp, $I_b = 5$ amp, $I_f = 10$ amp are supplied to the windings in the directions indicated by the dots and crosses in Fig. 3-31. If the rotor is allowed to move, will it rotate continuously or will it tend to come to rest? If the latter, at what value is θ_o?

c. The rotor winding is now excited by a constant direct current I_f, and the stator windings carry balanced 2-phase currents

$$i_a = \sqrt{2} \, I_a \cos \omega t \qquad i_b = \sqrt{2} \, I_a \sin \omega t$$

The rotor is revolving at synchronous speed so that its instantaneous angular position θ_o is given by $\theta_o = \omega t - \delta$, where δ is a phase angle describing the position of the rotor at $t = 0$. The machine is an elementary 2-phase synchronous machine. Derive an expression for the torque under these conditions. Describe its nature.

d. Under the conditions of part c, derive an expression for the instantaneous terminal voltages of stator phases a and b.

3-3. Figure 3-32 shows in cross section a machine having 2 identical stator windings aa and bb arranged in quadrature on a laminated steel core. The salient-póle rotor is made of steel and carries a winding f connected to slip rings. The machine is an elementary 2-phase salient-pole synchronous machine.

Because of the nonuniform air gap, the self- and mutual inductances of the stator windings are functions of the angular position θ_o of the rotor, as follows:

$$L_{aa} = L_0 + L_2 \cos 2\theta_o$$
$$L_{bb} = L_0 - L_2 \cos 2\theta_o$$
$$M_{ab} = L_2 \sin 2\theta_o$$

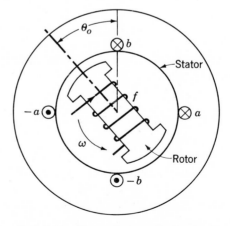

Fig. 3-32. Elementary salient-pole 2-phase synchronous machine, Prob. 3-3.

where L_0 and L_2 are positive constants.   The mutual inductances between the rotor and the stator windings are functions of θ_o as follows:

$$M_{af} = M \cos \theta_o \qquad M_{bf} = M \sin \theta_o$$

where M is a positive constant.   The self-inductance L_{ff} of the rotor winding is constant, independent of θ_o.

The rotor (or field) winding f is excited with direct current I_f, and the stator windings are connected to a balanced 2-phase voltage source. The currents in the stator windings are

$$i_a = \sqrt{2} \, I_a \cos \omega t \qquad i_b = \sqrt{2} \, I_a \sin \omega t$$

The rotor is revolving at synchronous speed so that its instantaneous angular position is given by
$$\theta_o = \omega t - \delta$$

where δ is a phase angle describing the position of the rotor at $t = 0$.

a. Derive an expression for the electromagnetic torque acting on the rotor.   Describe its nature.

b. Can the machine be operated as a motor?   As a generator? Explain.

c. Will the machine continue to run if the field current I_f is reduced to zero?   If so, give an expression for the torque, and a physical explanation.

3-4. Figure 3-33 shows a 2-pole rotor revolving inside a smooth stator which carries a coil of 100 turns.   The rotor produces a sinusoidal space distribution of flux at the stator surface, the peak value of the flux-density wave being 0.80 weber/m² when the current in the rotor is 10 amp.   The magnetic circuit is linear.   The inside diameter of the stator is 0.10 m, and its axial length is 0.10 m.   The rotor is driven at a speed of 60 rev/sec.

a. The rotor is excited by a direct current of 10 amp.   Taking zero time as the instant when the axis of the rotor is vertical, find the expression for the instantaneous voltage generated in the open-circuited stator coil.

b. The rotor is now excited by a 60-Hz sinusoidal alternating current whose rms value is 7.07 amp.   Consequently, the rotor current reverses every half revolution; it is timed to go through zero

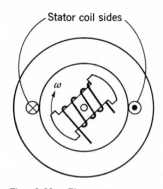

Fig. 3-33. Elementary generator, Prob. 3-4.

whenever the axis of the rotor is vertical. Taking zero time as
the instant when the axis of the rotor is vertical, find the expression for the instantaneous voltage generated in the open-circuited
stator coil. This scheme is sometimes suggested as a dc generator
without a commutator, the thought being that, if alternate half
cycles of the alternating voltage generated in part a are reversed
by reversal of the polarity of the field (rotor) winding, then a
pulsating direct voltage will be generated in the stator. Explain
whether this invention will work as described.

3-5. A small experimental 3-phase 4-pole alternator has the full-pitch
concentrated Y-connected armature winding shown diagrammatically in
Figs. 3-12b and 3-12c. Each coil (that represented by coil sides a and −a,
for example) has 2 turns, and all the turns in any one phase are connected
in series. The flux per pole is 0.25 weber and is sinusoidally distributed
in space. The rotor is driven at 1,800 rpm.

a. Determine the rms generated voltage to neutral.

b. Determine the rms generated voltage between lines.

c. Consider an abc phase order, and take zero time at the instant
   when the flux linkages with phase a are a maximum. Write a
   consistent set of time equations for the 3-phase voltages from
   terminals a, b, and c to neutral.

d. Under the conditions of c, write a consistent set of time equations for the three voltages between lines a and b, b and c, and
   c and a.

3-6. In a balanced 2-phase machine, the 2 windings are displaced 90 electrical degrees in space, and the currents in the 2 windings are phase-displaced 90 electrical degrees in time.   For such a machine, carry out the process leading up to an equation such as 3-36 for the rotating mmf wave.

3-7. The following statements are made in Art. 3-4 just after deriving and discussing Eq. 3-38: "In general it may be shown that a rotating field of constant amplitude will be produced by a q-phase winding excited by balanced q-phase currents when the respective phases are wound $2\pi/q$ electrical radians apart in space.   The constant amplitude will be $q/2$ times the maximum contribution of any one phase, and the speed will be $\omega = 2\pi f$ electrical radians per second."

Prove these statements.

4
rotating machines: engineering considerations

In Art. 3-1 we have taken a cursory look at the structural features and mode of operation of the common rotating-machine types. In the next four articles of this chapter, we shall investigate the basic principles underlying rotating-machine theory in the light of the voltage and torque equations of Arts. 3-2 and 3-5, and the rotating magnetic fields of Art. 3-4. At this point the treatment will not be complete or entirely rigorous; some details which must be clarified later will be glossed over. The purpose is primarily to lay a foundation on relatively simple physical reasoning for the more complete analyses which will come in Chaps. 5 to 11.

Our objective in this chapter is also to discuss some of the practical problems associated with machine application. Many of these problems are concerned with physical factors limiting the conditions under which a machine may operate successfully. Among them are saturation and the effects of losses on the rating and heating of the machine. These problems, common to all machine types, are discussed in this chapter.

INTRODUCTION TO POLYPHASE SYNCHRONOUS MACHINES

As indicated in Art. 3-1a, the synchronous machine is one in which alternating current flows in the armature winding and dc excitation is supplied to the field winding. The armature winding is almost invariably on the stator and is usually a 3-phase winding as described in Chap. 3. The field winding is on the rotor. The cylindrical-rotor construction shown in Figs. 3-11 and 3-12 is used for 2- and 4-pole turbine generators. The salient-pole construction shown in Fig. 3-9 is best adapted to multipolar slow-speed hydroelectric generators and most synchronous motors. The dc power required for excitation—approximately 1 to a few percent of the rating of the synchronous machine—usually is supplied through slip rings from a dc generator called an *exciter*, which is often mounted on the same shaft as the synchronous machine. Various excitation systems using ac exciters and solid-state rectifiers are used with large turbine generators. One of these systems is described in Art. 4-7.

A single synchronous generator supplying power to an impedance load acts as a voltage source whose frequency is determined by its prime-mover speed, as in Eq. 3-2. The current and power factor are then determined by the generator field excitation and the impedance of the generator and load.

Usually, however, the synchronous machine is connected to a power system containing many other synchronous machines. The voltage and frequency at its armature terminals are then substantially fixed by the system. A source of constant rms voltage and frequency is called an *infinite bus*. When carrying balanced polyphase currents the armature winding will produce a component magnetic field in the air gap rotating at synchronous speed (Eq. 3-37 or 3-38) as determined by the system frequency. But the field produced by the dc rotor winding revolves with the rotor. For the production of a steady unidirectional torque, the rotating fields of stator and rotor must be traveling at the same speed, and therefore the rotor must turn at precisely synchronous speed. A synchronous motor connected to a constant-frequency voltage source therefore operates at a constant steady-state speed regardless of load. A synchronous motor per se has no net starting torque and means must be provided for bringing it up to synchronous speed by induction-motor action, as described briefly at the end of Art. 4-2.

Behavior of a synchronous motor under running conditions can readily be visualized in terms of the torque equation 3-64; thus

$$T = \frac{\pi}{2}\left(\frac{P}{2}\right)^2 \Phi_{sr}F_r \sin \delta_r \qquad (4\text{-}1)$$

in which the minus sign of Eq. 3-64 has been omitted with the understanding that the electromagnetic torque acts in the direction to bring the interacting fields into alignment. Under normal conditions the armature-resistance voltage drop is negligible, and the armature leakage flux is small compared with the resultant air-gap flux Φ_{sr}. Consequently, the voltage generated by the air-gap flux wave must very nearly balance the terminal voltage V_t. From Eq. 3-16, then,

$$\Phi_{sr} = \frac{V_t}{4.44 f k_w N_{ph}} \qquad (4\text{-}2)$$

When the armature terminals are connected to a balanced polyphase "infinite bus," the resultant air-gap flux Φ_{sr} therefore is approximately constant, independent of shaft load. The rotor mmf F_r is determined by the dc field current and is also constant in normal operation. Variation in the torque requirements of the load must consequently be taken care of entirely by variation of the torque angle δ_r, as shown by the torque-angle curve in Fig. 4-1, in which positive values of T represent motor action and positive values of δ_r represent angles of lag of the rotor-mmf wave with respect to the resultant air-gap flux wave.

With a light shaft load, only a relatively small electromagnetic torque is required, and δ_r is small. When more shaft load is added, the rotor must drop back in space phase with respect to the rotating flux wave just enough so that δ_r assumes the value required to supply the necessary torque. The readjustment process is actually a dynamic one accompanied by a temporary decrease in the instantaneous mechanical speed of

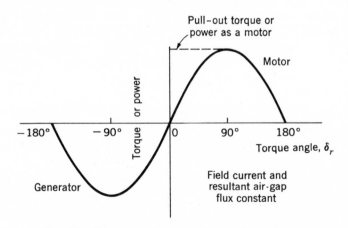

Fig. 4-1. Torque-angle curve of a synchronous machine.

the rotor and a damped mechanical oscillation, called *hunting*, of the rotor about its new space-phase position. In a practical machine, some changes in the amplitudes of the resultant flux-density and mmf waves may also occur because of factors (such as saturation and leakage-impedance drop) neglected in the present argument. The adjustment of the rotor to its new phase position following a load change may be observed experimentally in the laboratory by viewing the machine rotor with stroboscopic light having a flashing frequency which causes the rotor to appear stationary when it is turning at its normal synchronous speed.

When δ_r becomes 90°, the maximum possible torque or power, called *pull-out torque* or *pull-out power*, for a fixed terminal voltage and field current is reached. If the load requirements exceed this value, the motor slows down under the influence of the excess shaft torque and synchronous-motor action is lost because rotor and stator fields are no longer stationary with respect to each other. Under these conditions, the motor is usually disconnected from the line by the action of automatic circuit breakers. This phenomenon is known as *pulling out of step* or *losing synchronism*. Pull-out torque limits the short-time overload that may be placed on the motor. Increase of either field current or terminal voltage increases the pull-out torque.

The interrelation of generator and motor action and the associated influence on torque angle are illustrated in Fig. 4-1, where generator action is represented by merely extending the motor curve into the negative region. If the synchronous machine were connected to a constant-voltage constant-frequency ac system capable of either absorbing or supplying electrical power, it would supply power to that system when its rotor was driven mechanically so that the rotor-mmf wave was pushed ahead of the resultant air-gap flux wave. Synchronism is lost if the prime-mover applied torque exceeds the maximum generator pull-out torque. The speed will then increase rapidly, and quick-response governor action on the prime mover is required to prevent dangerous speeds.

4-2

INTRODUCTION TO POLYPHASE INDUCTION MACHINES

As indicated in Art. 3-1c, the induction motor is one in which alternating current is supplied to the stator directly and to the rotor by induction or transformer action from the stator. Like the synchronous machine, the stator winding is of the type discussed in Art. 3-4. When excited from a balanced polyphase source, it will produce a magnetic field in the air gap rotating at synchronous speed as determined by the number of poles and the applied stator frequency f (Eq. 3-38). The rotor may be one of two

Fig. 4-2. Cutaway view of a 3-phase induction motor with a wound rotor and slip rings connected to the 3-phase rotor winding. *(General Electric Company.)*

types.   A *wound rotor* carries a polyphase winding similar to and wound for the same number of poles as the stator.   The terminals of the rotor winding are connected to insulated slip rings mounted on the shaft. Carbon brushes bearing on these rings make the rotor terminals available external to the motor, as shown in the cutaway view in Fig. 4-2.   The motor in Fig. 3-18 has a *squirrel-cage rotor* with a winding consisting of conducting bars embedded in slots in the rotor iron and short-circuited at each end by conducting end rings.   The extreme simplicity and ruggedness of the squirrel-cage construction are outstanding advantages of the induction motor.

Now assume that the rotor is turning at the steady speed n rpm in the same direction as the rotating stator field.   Let the synchronous speed of the stator field be n_1 rpm as given by Eq. 3-38.   The rotor is then traveling at a speed $n_1 - n$ rpm in the backward direction with respect to the stator field, or the *slip* of the rotor is $n_1 - n$ rpm.   Slip is more usually expressed as a fraction of synchronous speed; i.e., the per-unit slip s is

$$s = \frac{n_1 - n}{n_1} \tag{4-3}$$

or

$$n = n_1(1 - s) \tag{4-4}$$

This relative motion of flux and rotor conductors induces voltages of frequency sf, called *slip frequency*, in the rotor.   Thus, the electrical behavior of an induction machine is similar to that of a transformer, but with the additional feature of frequency transformation.   A wound-rotor induction machine can be used as a frequency changer.

When used as an induction motor, the rotor terminals are short-circuited.   The rotor currents are then determined by the magnitudes of the induced voltages and the rotor impedance at slip frequency.   At starting, the rotor is stationary, the slip $s = 1$, and the rotor frequency equals the stator frequency f.   The field produced by the rotor currents therefore revolves at the same speed as the stator field, and a starting torque results, tending to turn the rotor in the direction of rotation of the stator-inducing field.   If this torque is sufficient to overcome the opposition to rotation created by the shaft load, the motor will come up to its operating speed.   The operating speed can never equal the synchronous speed n_1, however, for the rotor conductors would then be stationary with respect to the stator field and no voltage would be induced in them.

With the rotor revolving in the same direction of rotation as the stator field, the frequency of the rotor currents is sf, and the component rotor field set up by them will travel at sn_1 rpm *with respect to the rotor* in the forward direction.   But superimposed on this rotation is the mechanical rotation of the rotor at n rpm.   The speed of the rotor field in space is the sum of these two speeds and equals

$$sn_1 + n = sn_1 + n_1(1 - s) = n_1 \tag{4-5}$$

The stator and rotor fields are therefore stationary with respect to each other, a steady torque is produced, and rotation is maintained.   Such a torque which exists at any mechanical speed n other than synchronous speed is called an *asynchronous torque*.

Figure 4-3 shows a typical squirrel-cage induction-motor torque-speed characteristic.   The factors influencing the shape of this characteristic can be appreciated in terms of the torque equation 4-1.   In this equation recognize that the resultant air-gap flux Φ_{sr} is approximately constant when the stator-applied voltage and frequency are constant (Eq. 4-2).   Also recall that the rotor mmf F_r is proportional to the rotor current I_r.   Equation 4-1 then reduces to

$$T = KI_r \sin \delta_r \tag{4-6}$$

where K is a constant.   The rotor current is determined by the voltage induced in the rotor and its leakage impedance, both at slip frequency.

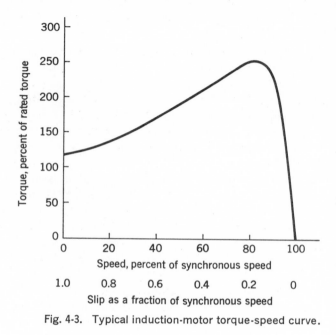

Fig. 4-3. Typical induction-motor torque-speed curve.

The rotor-induced voltage is proportional to slip. Under normal running conditions the slip is small—3 to 10 percent at full load in most squirrel-cage motors. The rotor frequency sf therefore is very low (of the order of 2 to 6 Hz in 60-Hz motors). Consequently, in this range the rotor impedance is largely resistive, and the rotor current is very nearly proportional to and in phase with the rotor voltage and is therefore very nearly proportional to slip. Furthermore, the rotor-mmf wave lags approximately 90 electrical degrees behind the resultant flux wave, and therefore $\sin \delta_r \approx 1$. (This point is discussed in Art. 7-1a.) Approximate linearity of torque as a function of slip is therefore to be expected in the range where the slip is small. As slip increases, the rotor impedance increases because of the increasing effect of rotor leakage inductance. Thus the rotor current is less than proportional to slip. Also the rotor current lags further behind the induced voltage, the mmf wave lags further behind the resultant flux wave, and $\sin \delta_r$ decreases. The result is that the torque increases with increasing slip up to a maximum value and then decreases, as shown in Fig. 4-3. The maximum torque, or *breakdown torque*, limits the short-time overload capability of the motor.

The squirrel-cage motor is substantially a constant-speed motor having a few percent drop in speed from no load to full load. Speed variation

Fig. 4-4.   Rotor of a 6-pole 1,200-rpm synchronous motor showing field coils, pole-face damper winding, and construction.   *(General Electric Company.)*

may be obtained by using a wound-rotor motor and inserting external resistance in the rotor circuit.   In the normal operating range, external resistance simply increases the rotor impedance, necessitating a higher slip for a desired rotor mmf and torque.

In Art. 4-1 it was mentioned that a synchronous motor per se has no starting torque.   To make a synchronous motor self-starting, a squirrel-cage winding, called an *amortisseur,* or *damper winding,* is inserted in the rotor pole faces, as shown in Fig. 4-4.   The rotor then comes up almost to synchronous speed by induction-motor action with the field winding unexcited.   If the load and inertia are not too great, the motor will then pull into synchronism when the field winding is energized from a dc source.

The dc machine differs in several respects from the ideal model of Art. 3-5. Although the basic concepts of Art. 3-5 are still valid a reexamination of the assumptions and a modification of the model are desirable. The crux of the matter is the effect of the commutator shown in the photographs of Figs. 3-1, 3-14, and 3-15.

Figure 4-5 shows diagrammatically the armature winding of Fig. 3-23 with the addition of the commutator, brushes, and connections of the coils to the commutator segments. The commutator is represented by the ring of segments in the center of the figure. The segments are insulated from each other and from the shaft. Two stationary brushes are shown by the black rectangles inside the commutator. Actually the brushes usually contact the outer surface, as shown in Figs. 3-14 and 3-15. The coil sides in the slots are shown in cross section by the small circles with dots and crosses in them, indicating currents toward and away from the reader, respectively, as in Fig. 3-23. The connections of the coils to the commutator segments are shown by the circular arcs. The end connections at the back of the armature are shown dotted for the 2 coils in slots 1 and 7, and the connections of these coils to adjacent commutator segments are shown by the heavy arcs. All coils are identical. The back end connections of the other coils have been omitted, to avoid complicating the figure, but they can easily be traced by remembering that each coil has one side in the top of a slot and the other side in the bottom of the diametrically opposite slot.

In Fig. 4-5a the brushes are in contact with commutator segments 1 and 7. Current entering the right-hand brush divides equally between two parallel paths through the winding. The first path leads to the inner coil side in slot 1 and finally ends at the brush on segment 7. The second path leads to the outer coil side in slot 6 and also finally ends at the brush on segment 7. The current directions in Fig. 4-5a can readily be verified by tracing these two paths. They are the same as in Fig. 3-23. The effect is identical to that of a coil wrapped around the armature with its magnetic axis vertical, and a clockwise magnetic torque is exerted on the armature, tending to align its magnetic field with that of the field winding.

Now suppose the machine is acting as a generator driven in the counterclockwise direction by an applied mechanical torque. Figure 4-5b shows the situation after the armature has rotated through the angle subtended by half a commutator segment. The right-hand brush is now in contact with both segments 1 and 2, and the left-hand brush is in contact

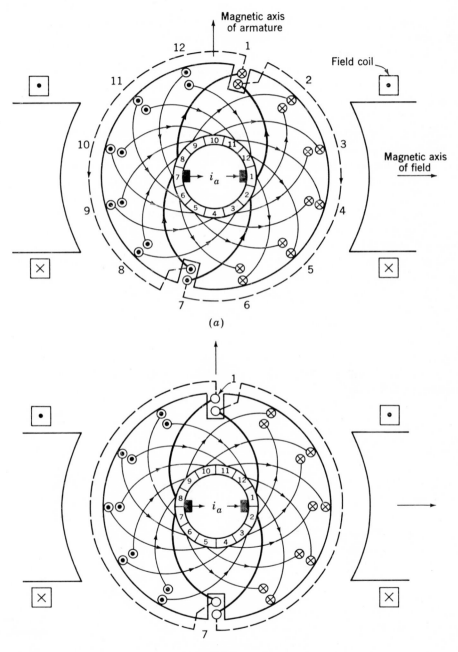

Fig. 4-5.   Direct-current-machine armature winding with commutator and brushes.
(a) and (b) Current directions for two positions of the armature.

with both segments 7 and 8.   The coils in slots 1 and 7 are now short-circuited by the brushes.   The currents in the other coils are shown by the dots and crosses, and they produce a magnetic field whose axis again is vertical.

After further rotation, the brushes will be in contact with segments 2 and 8, and slots 1 and 7 will have rotated into the positions which were previously occupied by slots 12 and 6 in Fig. 4-5a.   The current directions will be similar to those of Fig. 4-5a except that the currents in the coils in slots 1 and 7 will have reversed.   The magnetic axis of the armature is still vertical.

During the time when the brushes are simultaneously in contact with two adjacent commutator segments, the coils connected to these segments are temporarily removed from the main circuits through the winding, short-circuited by the brushes, and the currents in them are reversed. Ideally, the current in the coils being commutated should reverse linearly with time.   Serious departure from linear commutation will result in sparking at the brushes.   Means for obtaining sparkless commutation are discussed in Art. 5-2.   With linear commutation the waveform of the current in any coil as a function of time is trapezoidal, as shown in Fig. 4-6.

The winding of Fig. 4-5 is simpler than that used in most dc machines. Ordinarily more slots and commutator segments would be used, and except in small machines more than 2 poles are common.   Nevertheless, the simple winding of Fig. 4-5 includes the essential features of more complicated windings.

4-4

INTRODUCTION TO DC MACHINES

The essential features of dc machines are shown schematically in Fig. 4-7. The stator has salient poles excited by 1 or more field coils.   The air-gap

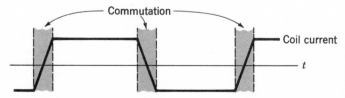

Fig. 4-6.   Waveform of current in an armature coil with linear commutation.

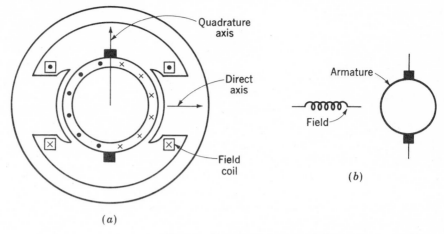

Fig. 4-7.   Schematic representations of a dc machine.

flux distribution created by the field windings is symmetrical about the
center line of the field poles.   This axis is called the *field axis*, or *direct
axis*.   The brushes are located so that commutation occurs when the coil
sides are in the neutral zone, midway between the field poles.   The axis
of the armature-mmf wave then is 90 electrical degrees from the axis of the
field poles, i.e., in the *quadrature axis*.   In the schematic representation of
Fig. 4-7a the brushes are shown in the quadrature axis, because this is the
position of the coils to which they are connected.   The armature-mmf
wave then is along the brush axis, as shown.   (The geometrical position
of the brushes in an actual machine is approximately 90 electrical degrees
from their position in the schematic diagram, because of the shape of the
end connections to the commutator.   For example, see Fig. 4-5.)   For
simplicity, the circuit representation usually will be drawn as in Fig. 4-7b.

Although the magnetic torque and the speed voltage appearing at the
brushes are independent of the spatial waveform of the flux distribution,
for convenience we shall continue to assume a sinusoidal flux-density
wave in the air gap.   The torque can then be found from the magnetic-
field viewpoint of Art. 3-5b.

The torque can be expressed in terms of the interaction of the direct-
axis air-gap flux per pole Φ_d and the space-fundamental component F_{a1} of
the armature-mmf wave, in a form similar to Eq. 3-64.   With the brushes
in the quadrature axis the angle between these fields is 90 electrical
degrees, and its sine equals unity.   Substitution in Eq. 3-64 then gives

for a P-pole machine

$$T = \frac{\pi}{2} \left(\frac{P}{2}\right)^2 \Phi_d F_{a1} \tag{4-7}$$

in which the minus sign has been dropped because the positive direction of the torque can be determined from physical reasoning. The peak value of the triangular armature-mmf wave is given by Eq. 3-24, and its space fundamental F_{a1} is $8/\pi^2$ times its peak. Substitution in Eq. 4-7 then gives

$$T = \frac{PZ_a}{2\pi a} \Phi_d i_a = K_a \Phi_d i_a \qquad \text{newton-meters} \tag{4-8}$$

where i_a is the current in the external armature circuit, Z_a is the total number of conductors in the armature winding, a is the number of parallel paths through the winding, and

$$K_a = \frac{PZ_a}{2\pi a} \tag{4-9}$$

is a constant fixed by the design of the winding.

The rectified voltage generated in the armature has already been found in Art. 3-2b for an elementary single-coil armature, and its waveform is shown in Fig. 3-20. The effect of distributing the winding in several slots is shown in Fig. 4-8, in which each of the rectified sine waves is the voltage generated in one of the coils, with commutation taking place at the moment when the coil sides are in the neutral zone. The generated voltage as observed from the brushes is the sum of the rectified voltages of all the coils in series between brushes and is shown by the rippling line labeled e_a in Fig. 4-8. With a dozen or so commutator segments per

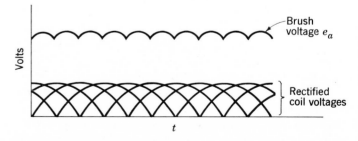

Fig. 4-8. Rectified coil voltages and resultant voltage between brushes in a dc machine.

pole, the ripple becomes very small and the average generated voltage observed from the brushes equals the sum of the average values of the rectified coil voltages. From Eq. 3-18 with changes in notation, the average coil voltage is

$$e_{a(\text{coil})} = \frac{PN_c}{\pi} \Phi_d \omega_m \qquad (4\text{-}10)$$

where N_c is the number of turns in one of the coils. For a distributed winding with C coils connected in a parallel paths between brushes, the brush voltage e_a is

$$e_a = \frac{C}{a} e_{a(\text{coil})} = \frac{PN_cC}{\pi a} \Phi_d \omega_m \qquad (4\text{-}11)$$

But $N_cC = Z_a/2$, where Z_a is the total number of active conductors in the winding. Hence

$$e_a = \frac{PZ_a}{2\pi a} \Phi_d \omega_m = K_a \Phi_d \omega_m \qquad (4\text{-}12)$$

where K_a is the design constant defined in Eq. 4-9. Compare with Eq. 3-19. The rectified voltage of a distributed winding has the same average value as that of a concentrated coil. The difference is that the ripple is greatly reduced.

From Eqs. 4-8 and 4-12, with all variables expressed in mks units,

$$e_a i_a = T \omega_m \qquad (4\text{-}13)$$

This equation simply says that the instantaneous electrical power associated with the speed voltage equals the instantaneous mechanical power associated with the magnetic torque, the direction of power flow being determined by whether the machine is acting as a motor or generator.

The direct-axis air-gap flux is produced by the combined mmf $\Sigma N_f i_f$ of the field windings, the flux-mmf characteristic being the *magnetization curve* for the particular iron geometry of the machine. A magnetization curve is shown in Fig. 4-9a, in which it is assumed that the armature mmf has no effect on the direct-axis flux because the axis of the armature-mmf wave is perpendicular to the field axis. It will be necessary to reexamine this assumption later in Chap. 5, where the effects of saturation are investigated more thoroughly. Because the armature emf is propor-

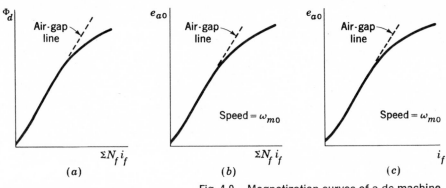

Fig. 4-9.   Magnetization curves of a dc machine.

tional to flux times speed, it is usually more convenient to express the magnetization curve in terms of the armature emf e_{a0} at a constant speed ω_{m0} as shown in Fig. 4-9b.   The emf e_a for a given flux at any other speed ω_m is proportional to the speed, i.e., from Eq. 4-12

$$\frac{e_a}{\omega_m} = K_a\Phi_d = \frac{e_{a0}}{\omega_{m0}} \tag{4-14}$$

or
$$e_a = \frac{\omega_m}{\omega_{m0}} e_{a0} \tag{4-15}$$

Figure 4-9c shows the magnetization curve with only 1 field winding excited.   This curve can easily be obtained by test methods, no knowledge of any design details being required.

   Over a fairly wide range of excitation the reluctance of the iron is negligible compared with that of the air gap.   In this region the flux is linearly proportional to the total mmf of the field windings, the constant of proportionality being the *direct-axis air-gap permeance* $\mathcal{P}_d$; thus

$$\Phi_d = \mathcal{P}_d\Sigma N_f i_f \tag{4-16}$$

The dotted straight line through the origin coinciding with the straight portion of the magnetization curves in Fig. 4-9 is called the *air-gap line*.

   The outstanding advantages of dc machines arise from the wide variety of operating characteristics which can be obtained by selection of the method of excitation of the field windings.   The field windings may be *separately excited* from an external dc source; or they may be *self-excited*, i.e., the machine may supply its own excitation.   Connection diagrams are shown in Fig. 4-10. The method of excitation profoundly influences not only the steady-state characteristics, which we shall describe briefly in this article, but also the dynamic behavior of the machine in control systems, discussed in Chap. 9.

The connection diagram of a separately excited generator is given in Fig. 4-10a. The required field current is a very small fraction of the rated armature current—of the order of 1 to 3 percent in the average generator. A small amount of power in the field circuit may control a relatively large amount of power in the armature circuit; i.e., the generator is a power amplifier. Its field winding is analogous to the grid of a vacuum-tube power amplifier, its armature is analogous to the plate circuit, and the mechanical power input from its mechanical drive is like the plate-circuit power supply. Separately excited generators are often used in feedback control systems when control of the armature voltage over a wide range is required.

The field windings of self-excited generators may be supplied in three different ways. The field may be connected in series with the armature (Fig. 4-10b), resulting in a *series generator*. The field may be connected in shunt with the armature (Fig. 4-10c), resulting in a *shunt generator*. Or the field may be in two sections (Fig. 4-10d), one of which is connected in series and the other in shunt with the armature, resulting in a *compound generator*. With self-excited generators residual magnetism must be present in the machine iron to get the self-excitation process started.

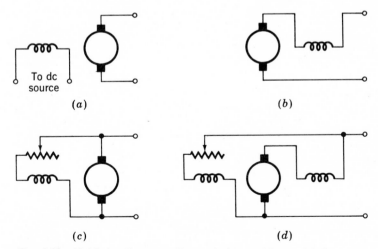

Fig. 4-10. Field-circuit connections of dc machines. (a) Separate excitation; (b) series; (c) shunt; (d) compound.

Typical steady-state volt-ampere characteristics are shown in Fig. 4-11, constant-speed prime movers being assumed. The relation between the steady-state generated emf E_a and the terminal voltage V_t is

$$V_t = E_a - I_a R_a \qquad\qquad (4\text{-}17)$$

where I_a is the armature current output and R_a is the armature circuit resistance. In a generator, E_a is larger than V_t, and the electromagnetic torque T is a countertorque opposing rotation.

The terminal voltage of a separately excited generator decreases slightly with increase in the load current, principally because of the voltage drop in the armature resistance. The field current of a series generator is the same as the load current, so that the air-gap flux and hence the voltage vary widely with load. As a consequence, series generators are not very often used. The voltage of shunt generators drops off somewhat with load, but not in a manner which is objectionable for many purposes. Compound generators are normally connected so that the mmf of the series winding aids that of the shunt winding. The advantage is that, through the action of the series winding, the flux per pole can increase with load, resulting in a voltage output which is nearly constant or which even rises somewhat as load increases. The shunt winding

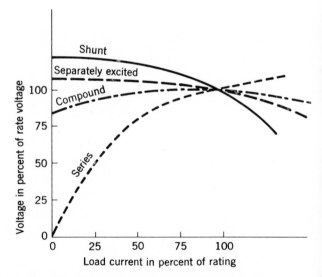

Fig. 4-11. Volt-ampere characteristics of dc generators.

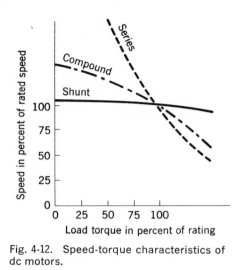

Fig. 4-12. Speed-torque characteristics of dc motors.

usually contains many turns of relatively small wire.   The series wind-ing, wound on the outside, consists of a few turns of comparatively heavy conductor because it must carry the full armature current of the machine. The voltage of both shunt and compound generators may be controlled over reasonable limits by means of rheostats in the shunt field.

Any of the methods of excitation used for generators may also be used for motors.   Typical steady-state speed-torque characteristics are shown in Fig. 4-12, in which it is assumed that the motor terminals are sup-plied from a constant-voltage source.   In a motor the relation between the emf E_a generated in the armature and the terminal voltage V_t is

$$V_t = E_a + I_a R_a \tag{4-18}$$

or

$$I_a = \frac{V_t - E_a}{R_a} \tag{4-19}$$

where I_a is now the armature current input.   The generated emf E_a is now smaller than the terminal voltage V_t, the armature current is in the opposite direction to that in a generator, and the electromagnetic torque is in the direction to sustain rotation of the armature.

In shunt and separately excited motors the field flux is nearly con-stant.   Consequently, increased torque must be accompanied by a very nearly proportional increase in armature current and hence by a small decrease in counter emf to allow this increased current through the small armature resistance. Since counter emf is determined by flux and speed (Eq. 4-12), the speed must drop slightly. Like the squirrel-cage induction motor, the shunt motor is substantially a constant-speed motor having about 5 percent drop in speed from no load to full load. A typical speed-load characteristic is shown by the solid curve in Fig. 4-12. Starting torque and maximum torque are limited by the armature current that can be commutated successfully.

An outstanding advantage of the shunt motor is ease of speed control. With a rheostat in the shunt-field circuit, the field current and flux per pole may be varied at will, and variation of flux causes the inverse variation of speed to maintain counter emf approximately equal to the impressed terminal voltage. A maximum speed range of about 4 or 5 to 1 may be obtained by this method, the limitation again being commutating conditions. By variation of the impressed armature voltage, very wide speed ranges may be obtained.

In the series motor, increase in load is accompanied by increases in the armature current and mmf and the stator field flux (provided the iron is not completely saturated). Because flux increases with load, speed must drop in order to maintain the balance between impressed voltage and counter emf; moreover, the increase in armature current caused by increased torque is smaller than in the shunt motor because of the increased flux. The series motor is therefore a varying-speed motor with a markedly drooping speed-load characteristic of the type shown dotted in Fig. 4-12. For applications requiring heavy torque overloads, this characteristic is particularly advantageous because the corresponding power overloads are held to more reasonable values by the associated speed drops. Very favorable starting characteristics also result from the increase in flux with increased armature current.

In the compound motor the series field may be connected either *cumulatively*, so that its mmf adds to that of the shunt field, or *differentially*, so that it opposes. The differential connection is very rarely used. As shown by the dashed curve in Fig. 4-12, a cumulatively compounded motor will have a speed-load characteristic intermediate between those of a shunt and a series motor, the drop of speed with load depending on the relative number of ampere-turns in the shunt and series fields. It does not have the disadvantage of very high light-load speed associated with a series motor, but it retains to a considerable degree the advantages of series excitation.

The application advantages of dc machines lie in the variety of performance characteristics offered by the possibilities of shunt, series, and compound excitation. Some of these characteristics have been touched upon briefly in this article. Still greater possibilities exist if additional sets of brushes are added so that other voltages may be obtained from the commutator.   Thus the versatility of dc-machine systems and their adaptability to control, both manual and automatic, are their outstanding features.   These characteristics will be discussed in Chaps. 5, 8, and 9 for both steady-state and dynamic operation.

4-5
THE NATURE OF MACHINERY PROBLEMS

As we begin to converge on the theory of rotating machines, we need to reflect momentarily on our objectives: what are the machine characteristics that we need to know in reasonably precise, quantitative form? The answer depends on what, specifically, the machines are intended to do for us.   The machine is one component in an electromechanical-energy-conversion system, and the machine characteristics often play the predominant part in the behavior of the complete system.

In many applications of electric motors, the motor is supplied with electric power from a constant-voltage source.   The motor drives a mechanical load whose torque requirements vary with the speed at which it is driven.   The steady-state operating speed is then fixed by the point at which the torque that the motor can furnish electromagnetically is equal to the torque that the load can absorb mechanically.   In Fig. 4-13, for instance, the solid curve is the speed-torque characteristic of an induction motor.   The dotted curve is a plot of the torque input required by a fan for various operating speeds.   When the fan and motor are coupled,

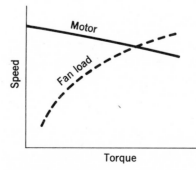

Fig. 4-13.   Superposition of motor and load characteristics.

the steady-state operating point of the combination is at the intersection of these two curves—where what the motor can give is the same as what the fan can take.

Motor power or torque requirements vary, of course, depending on conditions within the driven equipment. The requirements of some motor loads are satisfied by a speed which remains approximately constant as load varies; an ordinary hydraulic pump is an example. Others, like a phonograph turntable, require absolutely constant speed. Still others require a speed closely coordinated with another speed: the raising of both ends of a vertical-lift bridge, for instance. Some motor applications, such as cranes and many traction-type drives, inherently demand low speeds and heavy torques at one end of the range and relatively high speeds and light torques at the other—a varying-speed characteristic, in other words. Others may require an adjustable constant speed (e.g., some machine-tool drives in which the speed of operation may require adjustment over a wide range but must always be carefully predetermined) or an adjustable varying speed (a crane is again an example). In almost every application the torque which the motor is capable of supplying while starting, the maximum torque which it can furnish while running, and the current requirements are items of importance.

Many similar remarks can be made for generators. For example, the terminal voltage and power output of a generator are determined by the characteristics of both the generator and its load. Thus, the solid curve of Fig. 4-14 is a plot of the terminal voltage of a dc shunt generator as a function of its current output. The dotted curve is a plot of the volt-ampere characteristic of a load. When the load is connected to the generator terminals, the operating point of the combination is at the inter-

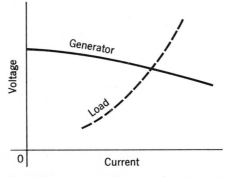

Fig. 4-14. Superposition of generator and load characteristics.

section of these two curves—where what the generator can give is the same as what the load can take. Often, as in the usual central station, the requirement is that terminal voltage shall remain substantially constant over a wide load range. Not infrequently, however, a motor is associated with its own individual generator in order to provide greater flexibility and more precise control. Then it may be desired that the terminal voltage vary with load in some particular fashion.

Among the features of outstanding importance, therefore, are the torque-speed characteristics of motors and the voltage-load characteristics of generators, together with knowledge of the limits between which these characteristics can be varied and ideas of how such variations may be obtained. Moreover, pertinent economic features are efficiency, power factor, comparative costs, and the effect of losses on the heating and rating of the machines. One of the objects of our analysis of machinery in Chaps. 5 to 8 is accordingly to study and compare these features for the various machine types. Of course there are many important, interesting, and complex engineering problems associated with the design, development, and manufacturing of machines for which these studies are but the introduction; most such problems are beyond the scope of the book.

A knowledge of the steady-state characteristics is insufficient, however, for an understanding of the role played by rotating machines in modern technology. In an increasingly important class of applications in the field of automatic control, the emphasis is rather on the dynamic behavior of the complete electromechanical system of which the machine is one component. For example, it may be desired to control the speed or position of a shaft driving a load in accordance with some specified function of time or of some other variable. A typical industrial application is the accurate control of tension in a process involving the winding of long strips of material, such as paper, on a reel. Dynamic controls of astounding accuracy and rapidity of response have been developed. In applications of this kind, the electromechanical transient behavior of the system as a whole is a major consideration; the system should respond accurately and rapidly to the control function, and oscillations should die out quickly. Not only the electrical characteristics but also the mechanical properties of the system, such as stiffness, inertia, and friction, must be considered and indeed may become the predominant factors.

In such studies, then, an adequate theory must be capable of treating the dynamic behavior of machines as system components. Since the analysis of a complete electromechanical system presents a complex problem, a theory of machines suitable for these purposes must be simplified as much as possible while still retaining the essential elements. The development of such a theory is the primary purpose of Chaps. 9 and 10.

4-6
MAGNETIC SATURATION

Both electromagnetic torque and generated voltage in all machines depend on rates of change of flux linkages with the windings of the machines. For specified mmfs in the windings, the fluxes depend on the reluctances of the iron portions of the magnetic circuits as well as of the air gaps. Saturation may therefore appreciably influence the characteristics of the machines. Another aspect of saturation, a more subtle one and one more difficult to evaluate without experimental and theoretical comparisons, concerns its influence on the basic premises from which the analytic approach to machinery is developed. Specifically, all relations for mmf are based on negligible reluctance in the iron. When these relations are applied to practical machines with varying degrees of saturation in the iron, the actual machine is, in effect, replaced for these considerations by an equivalent machine: one whose iron has negligible reluctance but whose air-gap length is increased by an amount sufficient to absorb the magnetic-potential drop in the iron of the actual machine. Incidentally, the effects of air-gap nonuniformities such as slots and ventilating ducts are also incorporated through the medium of an equivalent smooth gap, a replacement which, in contrast to that above, is made explicitly during magnetic-circuit computations for the machine structure. Thus, serious efforts are made to reproduce the magnetic conditions at the air gap correctly, and the computed performance of

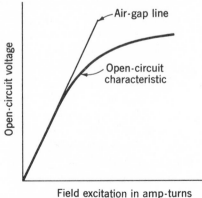

Fig. 4-15. Typical open-circuit characteristic and air-gap line.

machines is based largely on those conditions.   Final assurance of the legitimacy of the approach must, of course, be the pragmatic one given by close experimental checks.

Magnetic-circuit data essential to the handling of saturation are given by the *open-circuit characteristic, magnetization curve,* or *saturation curve.* An example is shown in Fig. 4-15.   Basically, this characteristic is the magnetization curve for the particular iron and air geometry of the machine under consideration.   Frequently, the abscissa is plotted in terms of field current or magnetizing current instead of mmf in ampere-turns.   Also, the generated voltage with zero armature current is directly proportional to the flux when the speed is constant.   For convenience

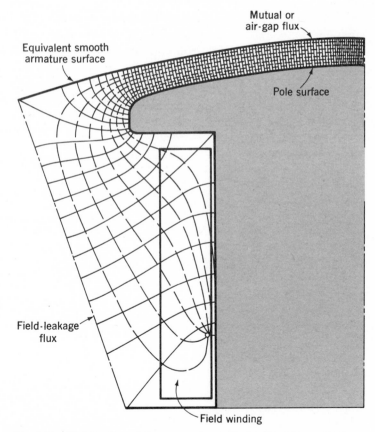

Fig. 4-16.   Distribution of flux around a salient pole.   The solid lines are flux lines; the dashed lines are loci of equal magnetic potential.

in use, then, open-circuit terminal voltage is plotted on the ordinate scale rather than air-gap flux per pole, and the entire curve is drawn for a stated fixed speed, usually rated speed. The straight line tangent to the lower portion of the curve is the *air-gap line* indicating very closely the mmf required to overcome the reluctance of the air gap. If it were not for the effects of saturation, the air-gap line and open-circuit character-istic would coincide, so that the departure of the curve from the air-gap line is an indication of the degree of saturation present. In a normal machine, the ratio at rated voltage of the total mmf to that required by the air gap alone usually is between 1.1 and 1.25.

The open-circuit characteristic may be calculated from design data by magnetic-circuit methods, often guided by flux mapping. A small sample map of the flux distribution around the pole of a salient-pole machine is given in Fig. 4-16. The distribution of air-gap flux found by means of this map, together with the fundamental- and third-harmonic components, are shown in Fig. 4-17. The map is drawn on the basis of infinite permeability in the iron and for a smooth armature surface with the air-gap width increased to compensate for the effect of armature slots on the flux per pole. Slot effects may be studied separately on either an analytical or a graphical basis. The influence of a slot on an otherwise uniform field is indicated graphically in Figs. 4-18 and 4-19. Note that

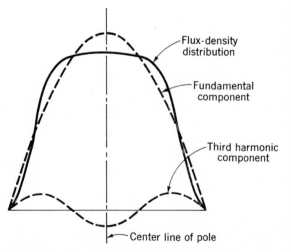

Fig. 4-17. Flux-density wave corresponding to Fig. 4-16, with its fundamental- and third-harmonic com-ponents.

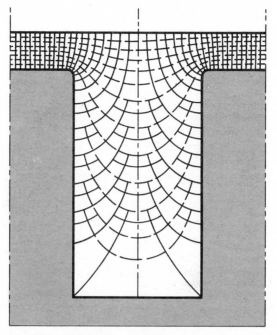

Fig. 4-18. Distribution of flux in a slot. The solid lines are flux lines; the dashed lines are loci of equal magnetic potential.

in Fig. 4-18 the scale to which the field is mapped increases at locations where a flux line is drawn only up to a point where it crosses an equipotential line. When this change is borne in mind, it is seen that the flux density in the slot is far lower than in the tooth. The general nature of the flux-density wave with slot effects superimposed is shown in Fig. 4-20. Slot effects are indicated in a pronounced form here because of the use of only a few relatively wide slots per pole. Flux maps of the type indicated in Figs. 4-16 and 4-18 yield precise, quantitative results, for they are graphical solutions of Laplace's equation for the assumed boundary conditions.

If the machine is an existing one, the magnetization curve is most commonly determined by operating it as an unloaded generator and reading the values of terminal voltage corresponding to a series of values of field current. For an induction motor, the machine is operated at or close to synchronous speed, and values of magnetizing current are obtained for a series of values of impressed stator voltage.

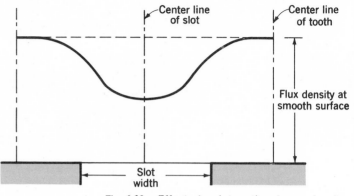

Fig. 4-19. Effect of a slot on the air-gap density.

4-7
EXCITATION SOURCES

The resultant flux in the magnetic circuit of a machine is established by the combined mmf of all the windings on the machine.   For the conventional dc machine, the bulk of the effective mmf is furnished by the field windings.   For the transformer, the net excitation may be furnished by either the primary or the secondary winding, or a portion may be furnished by each.   A similar situation exists in ac machines.   The furnishing of excitation to ac machines has associated with it two different operational aspects which are of economic importance in the application of the machines.

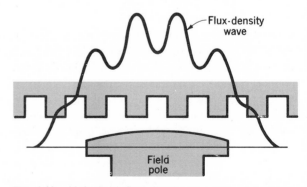

Fig. 4-20. Main-field flux distribution with slot effects superimposed. Slot effects are exaggerated because only a few wide slots per pole are shown.

a. Power Factor in AC Machines

The power factor at which ac machines operate is an economically important feature because of the cost of reactive kva. Low power factor adversely affects system operation in three principal ways. In the first place, generators, transformers, and transmission equipment are rated in terms of kva rather than kilowatts, because their losses and heating are very nearly determined by voltage and current regardless of power factor. The physical size and cost of ac apparatus is roughly proportional to its kva rating. The investment in generators, transformers, and transmission equipment for supplying a given useful amount of active power therefore is roughly inversely proportional to the power factor. In the second place, low power factor means more current and greater copper losses in the generating and transmitting equipment. A further disadvantage is poor voltage regulation.

Factors influencing reactive-kva requirements in motors can be visualized readily in terms of the relationship of these requirements to the establishment of magnetic flux. As in any electromagnetic device, the resultant flux necessary for motor operation must be established by a magnetizing component of current. It makes no difference either in the magnetic circuit or in the fundamental energy-conversion process whether this magnetizing current be carried by the rotor or stator winding, just as it makes no basic difference in a transformer which winding carries the exciting current. In some cases, part of it is supplied from each winding. If all or part of the magnetizing current is supplied to an ac winding, the input to that winding must include lagging reactive kva, because magnetizing current lags voltage drop by 90°. In effect, the lagging reactive kva sets up flux in the motor.

The only possible source of excitation in an induction motor is the stator input. The induction motor therefore must operate at a lagging power factor. This power factor is very low at no load and increases to about 85 to 90 percent at full load, the improvement being caused by the increased active-power requirements with increasing load.

With a synchronous motor, there are two possible sources of excitation: alternating current in the armature, or direct current in the field winding. If the field current is just sufficient to supply the necessary mmf, no magnetizing-current component or reactive kva is needed in the armature and the motor operates at unity power factor. If the field current is less (i.e., the motor is *underexcited*), the deficit in mmf must be made up by the armature and the motor operates at a lagging power factor. If the field current is greater (i.e., the motor is *overexcited*), the excess mmf must be counterbalanced in the armature and a leading component of current is present; the motor then operates at a leading power factor.

Because magnetizing current must be supplied to inductive loads, such as transformers and induction motors, the ability of overexcited synchronous motors to supply lagging current is a highly desirable feature which may have considerable economic importance. In effect, overexcited synchronous motors act as generators of lagging reactive kva and thereby relieve the power source of the necessity for supplying this component. They thus may perform the same function as a local capacitor installation. Sometimes unloaded synchronous machines are installed in power systems solely for power-factor correction or for control of reactive-kva flow. Such machines, called *synchronous condensers*, may be more economical in the larger sizes than static capacitors.

Both synchronous and induction machines may become self-excited when a sufficiently heavy capacitive load is present in their stator circuits. The capacitive current then furnishes the excitation and may cause serious overvoltage or excessive transient torques. Because of the inherent capacitance of transmission lines, the problem may arise when synchronous generators are energizing long unloaded or lightly loaded lines. The use of shunt reactors at the sending end of the line to compensate the capacitive current is sometimes necessary. For induction motors, it is normal practice to avoid self-excitation by limiting the size of any parallel capacitor when the motor and capacitor are switched as a unit.

b. Turbine-generator Excitation Systems

As the available ratings of turbine-generators have increased, the problems of supplying the dc field excitation (amounting to 4,000 amp or more in the larger units) have grown progressively more difficult. The conventional excitation source is a dc generator whose output is supplied to the alternator field through brushes and slip rings. Cooling and maintenance problems are inevitably associated with slip rings, commutators, and brushes. Many modern excitation systems have minimized these problems by minimizing the use of sliding contacts and brushes.

A schematic diagram of one modern system is given in Fig. 4-21. At the heart of the system are the silicon diode rectifiers which are mounted on the same shaft as the generator field and which furnish dc excitation directly to the field. An ac exciter with a rotating armature feeds power along the shaft to the revolving rectifiers. The stationary field of the ac exciter is fed through a magnetic amplifier which controls and regulates the output voltage of the main generator. To make the system self-contained and free of sliding contacts, the excitation power

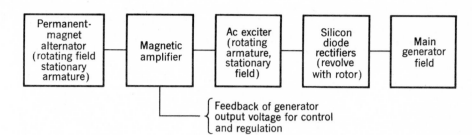

Fig. 4-21. Diagrammatic representation of brushless excitation system.

for the magnetic amplifier is obtained from the stationary armature of a small permanent-magnet alternator also driven from the main shaft. The voltage and frequency of the ac exciter are chosen so as to optimize the performance and design of the overall system. Time delays in the response to a controlling signal are all short compared with the time constant of the main generator field. The system may have the additional advantage of doing away with need for spare exciters, generator-field circuit breakers, and field rheostats.

Another excitation system uses a shaft-driven alternator of conventional design as the main exciter. This alternator has a stationary armature and a rotating field winding. Its frequency may be 180 or 240 Hz. Its output is fed to a stationary solid-state rectifier which in turn supplies the turbine-generator field through slip rings.

4-8
LOSSES

Consideration of machine losses is important for three reasons: losses determine the efficiency of the machine and appreciably influence its operating cost; losses determine the heating of the machine and hence the rating or power output that may be obtained without undue deterioration of the insulation; and the voltage drops or current components associated with supplying the losses must be properly accounted for in a machine representation. Machine efficiency, like that of transformers or of any energy-transforming device, is given by

$$\text{Efficiency} = \frac{\text{output}}{\text{input}} \qquad (4\text{-}20)$$

which can also be expressed as

$$\text{Efficiency} = \frac{\text{input} - \text{losses}}{\text{input}} = 1 - \frac{\text{losses}}{\text{input}} \qquad (4\text{-}21)$$

$$\text{Efficiency} = \frac{\text{output}}{\text{output} + \text{losses}} \qquad (4\text{-}22)$$

Rotating machines in general operate efficiently except at light loads. The full-load efficiency of average motors, for example, is in the neighborhood of 74 percent for 1-hp size, 89 percent for 50-hp, 93 percent for 500-hp, and 97 percent for 5,000 hp. The efficiency of slow-speed motors is usually lower than that of high-speed motors, the total spread being 3 or 4 percent.

The forms given by Eqs. 4-21 and 4-22 are often used for electric machines, for their efficiency is most commonly determined by measurement of losses instead of by directly measuring the input and output under load. Loss measurements have the advantage of convenience and economy and of yielding more accurate and precise values of efficiency because a given percentage error in measuring losses causes only about one-tenth of that percentage error in the efficiency. Efficiencies determined from loss measurements can be used in comparing competing machines provided that exactly the same methods of measurement and computation are used in each case. For this reason, the various losses and the conditions for their measurement are precisely defined by the American National Standards Association (ANSI). The following discussion of individual losses incorporates many of these provisions as given in American National Standard C50, although no attempt is made to present all the details.

1 *Copper losses*, or I^2R losses, are, of course, found in all the windings of the machine. By convention, these losses are computed on the basis of the dc resistances of the winding at 75°C. Actually the I^2R loss depends on the effective resistance of the winding under the operating frequency and flux conditions. The increment in loss represented by the difference between dc and effective resistances is included with stray load losses, discussed below. In the field circuits of synchronous and dc machines, only the losses in the field winding are charged against the machine; the I^2R loss in the rheostat controlling the field current and all losses in external sources supplying the excitation are charged against the plant of which the machine is a part. Closely associated with I^2R loss is the *brush-contact loss* at slip rings and commutators. By conven-

tion, this loss is normally neglected for induction and synchronous machines, and for industrial-type dc machines the voltage drop at the brushes is regarded as constant at 2 volts total when carbon and graphite brushes with shunts (pigtails) are used.

2 *Mechanical losses* consist of brush and bearing friction, windage, and the power required to circulate the air through the machine and ventilating system, if one is provided, whether by self-contained or external fans (except for the power required to force air through long or restricted ducts external to the machine). Friction and windage losses may be measured by determining the input to the machine running at the proper speed but unloaded and unexcited. Frequently they are lumped with core loss and determined at the same time.

3 *Open-circuit*, or *no-load, core loss* consists of the hysteresis and eddy-current losses arising from changing flux densities in the iron of the machine with only the main exciting winding energized. In dc and synchronous machines, these losses are confined largely to the armature iron, although the flux pulsations arising from slot openings will cause losses in the field iron as well, particularly in the pole shoes or surfaces of the field iron. In induction machines, the losses are confined largely to the stator iron. Open-circuit core loss may be found by measuring the input to the machine when it is operating unloaded at rated speed or frequency and under the appropriate flux or voltage conditions, and deducting the friction and windage loss and, if the machine is self-driven during the test, the no-load armature copper loss (no-load stator copper loss for an induction motor). Usually, data are taken for a curve of core loss as a function of armature voltage in the neighborhood of rated voltage. The core loss under load is then considered to be the value at a voltage equal to rated voltage corrected for armature ohmic-resistance drop under load (a phasor correction for an ac machine). For induction motors, however, this correction is dispensed with, and the core loss at rated voltage is used. For efficiency determination alone, there is no need to segregate open-circuit core loss and friction and windage loss; the sum of these two losses is termed the *no-load rotational loss*.

Eddy-current loss is dependent on the squares of the flux density, frequency, and thickness of laminations. Under normal machine conditions, it may be expressed to a sufficiently close approximation as

$$P_e = K_e(B_{\max}f\tau)^2 \qquad\qquad (4\text{-}23)$$

where τ is the lamination thickness, $B_{\max}$ the maximum flux density, f the frequency, and K_e a proportionality constant whose value depends on the units used, the volume of iron, and the resistivity of the iron. Variation of hysteresis loss can be expressed in equation form only on an empirical basis. The most commonly used relation is

$$P_h = K_h f B_{\max}^n \qquad (4\text{-}24)$$

where K_h is a proportionality constant dependent on the characteristics and volume of iron and the units used and the exponent n ranges from 1.5 to 2.5 with a value of 2.0 often used for estimating purposes in machines. In both Eqs. 4-23 and 4-24, frequency may be replaced by speed and flux density by the appropriate voltage when the proportionality constants are changed accordingly. Such replacements are implied when the core-loss tests are made at rated speed and the appropriate voltage.

When the machine is loaded, the space distribution of flux density is significantly changed by the mmf of the load currents. The actual core losses increase noticeably. For example, mmf harmonics cause appreciable losses in the iron near the air-gap surfaces. The total increment in core loss is classified as part of the stray load loss.

4 *Stray load loss* consists of the losses arising from nonuniform current distribution in the copper and the additional core losses produced in the iron by distortion of the magnetic flux by the load current. It is a difficult loss to determine accurately. By convention it is taken as 1.0 percent of the output for dc machines. For synchronous and induction machines it may be found by test.

Study of the foregoing classification of the losses in a machine shows it to have a few features which, from a fundamental viewpoint, are somewhat artificial. Illustrations are offered by the division of iron losses into no-load core loss and an increment which appears under load, the division of copper losses into ohmic I^2R losses and an increment created by nonuniform current distribution, and the lumping of these two increments in the scavengerlike stray-load-loss category. These features are dictated by ease of testing. They are justified by the fact that the principal motivation is the determination of the total losses and efficiency suitable for economic comparison of machines and at the same time as nearly equal to the actual values as possible. Because of this seeming dominance of efficiency aspects, it may be appropriate to emphasize once more that losses play more than a bookkeeping type of part in machine operation.

In a generator, for example, components of mechanical input torque to the shaft are obviously required to supply copper and iron losses as well as friction and windage losses and the generator output. These losses may therefore be appreciable factors in the damping of electrical and mechanical transients in the machine. Components of the stray load loss, although they may be individually only a fraction of a percent of the output, may be of first importance in the design of the machine. Thus rotor heating is usually a limiting factor in the design of large high-speed alternators, and the components of stray loss on the surface of the rotor structure are of great importance because they directly affect the dimensions of an alternator of given output. Of more direct concern in theoretical aspects is the influence of hysteresis and eddy currents in causing flux to lag behind mmf. There is a small angle of lag between the rotating mmf waves in a machine and the corresponding component flux-density waves. Associated with this influence is a torque on magnetic material in a rotating field, a torque proportional to the hysteresis and eddy-current losses in the material. Although the torque accompanying these losses is relatively small in normal machines, direct use of it is made in one type of small motor, the hysteresis motor.

4-9
RATING AND HEATING

One of the most common and important questions in the application of machines, transformers, and other electrical equipment is "What maximum output may be obtained?" The answer, of course, depends on various factors, for the machine, while providing this output, must in general meet definite performance standards. A universal requirement is that the life of the machine shall not be unduly shortened by overheating. The temperature rise resulting from the losses considered in the previous article is therefore a major factor in the rating of a machine.

The operating temperature of a machine is closely associated with its life expectancy because deterioration of the insulation is a function of both time and temperature. Such deterioration is a chemical phenomenon involving slow oxidation and brittle hardening and leading to loss of mechanical durability and dielectric strength. In many cases the deterioration rate is such that the life of the insulation is an exponential

$$\text{Life} = A\epsilon^{B/T} \tag{4-25}$$

where A and B are constants and T is the absolute temperature. Thus, according to Eq. 4-25, when life is plotted to a logarithmic scale against the reciprocal of absolute temperature on a uniform scale, a straight line should result. Such plots form valuable guides in the thermal evaluation of insulating materials and systems. A very rough idea of the life-temperature relation can be obtained from the old and more or less obsolete rule of thumb that the time to failure of organic insulation is halved for each 8 to 10°C rise.

The evaluation of insulating materials and complete systems of insulation (which may include widely different materials and techniques in combination) is to a large extent a functional one based on accelerated life tests. Both normal life expectancy and service conditions will vary widely for different classes of electrical equipment. Life expectancy, for example, may be a matter of minutes in some military and missile applications, may be 500 to 1,000 hr in certain aircraft and electronic equipment, and ranges from 10 to 30 years or more in large industrial equipment. The test procedures will accordingly vary with the type of equipment. Accelerated life tests on models, called *motorettes*, are commonly used in insulation evaluation. Such tests, however, cannot be easily applied to all equipment, especially the insulation systems of large machines.

Life tests generally attempt to simulate service conditions. They usually include the following elements:

1 Thermal shock resulting from heating to the test temperature
2 Sustained heating at that temperature
3 Thermal shock resulting from cooling to room temperature or below
4 Vibration and mechanical stress such as may be encountered in actual service
5 Exposure to moisture
6 Dielectric testing to determine the condition of the insulation

A sufficiently large number of samples must be tested so that statistical methods may be applied in analyzing the results. The life-temperature relations obtained from these tests lead to the classification of the insulation or insulating system in the appropriate temperature class.

For the allowable temperature limits of insulating systems used commercially, the latest standards of the ANSI, IEEE, and National Electrical Manufacturers Association (NEMA) should be consulted. The three NEMA insulation-system classes of chief interest for industrial machines are Class B, Class F, and Class H.   Class B insulation includes mica, glass fiber, asbestos, and similar materials with suitable bonding substances.   Class F insulation also includes mica, glass fiber, and synthetic substances similar to those in Class B, but the system must be capable of withstanding higher temperatures.   Class H insulation, intended for still higher temperatures, may consist of materials such as silicone elastomer and combinations including mica, glass fiber, asbestos, etc., with bonding substances such as appropriate silicone resins.   Experience and tests showing the material or system to be capable of operation at the recommended temperature form the important classifying criteria.

When the temperature class of the insulation is established, the permissible observable temperature rises for the various parts of industrial-type machines may be found by consulting the appropriate Standards.   Reasonably detailed distinctions are made with respect to type of machine, method of temperature measurement, machine part involved, whether the machine is enclosed or not, and the type of cooling (air-cooled, fan-cooled, hydrogen-cooled, etc.).   Distinctions are also made between general-purpose machines and definite- or special-purpose machines. The term *general-purpose motor* refers to one of standard rating "up to 200 hp with standard operating characteristics and mechanical construction for use under usual service conditions without restriction to a particular application or type of application."   In contrast a *special-purpose motor* is "designed with either operating characteristics or mechanical construction, or both, for a particular application."   For the same class of insulation, the permissible rise of temperature is lower for a general-purpose motor than for a special-purpose motor, largely to allow a greater factor of safety where service conditions are unknown.   Partially compensating the lower rise, however, is the fact that general-purpose motors are allowed a service factor of 1.15 when operated at rated voltage; *service factor* is a multiplier which, applied to the rated output, indicates a permissible loading which may be carried continuously under the conditions specified for that service factor.

Examples of allowable temperature rises may be seen from the table below, excerpted from NEMA standards.   The table applies to integral-horsepower induction motors, is based on 40°C ambient temperature, and assumes measurement of temperature rise by determining the increase of winding resistances.

The most common machine rating is the *continuous rating* defining the output (in kilowatts for dc generators, kilovolt-amperes at a specified power factor for ac generators, and horsepower for motors) which can be carried indefinitely without exceeding established limitations.   For intermittent, periodic, or varying duty a machine may be given a *short-*

	ALLOWABLE TEMPERATURE RISE, °C		
	CLASS B	CLASS F	CLASS H
Motors with 1.15 service factor	90	115	
Motors, 1.00 service factor, encapsulated windings	85	110	
Totally enclosed, fan-cooled motors	80	105	125
Totally enclosed, nonventilated motors	85	110	135

time rating defining the load which can be carried for a specified time. Standard periods for short-time ratings are 5, 15, 30, and 60 min. Speeds, voltages, and frequencies are also specified in machine ratings, and provision is made for possible variations in voltage and frequency. Motors, for example, must operate successfully at voltages 10 percent above and below rated voltage and, for ac motors, at frequencies 5 percent above and below rated frequency; the combined variation of voltage and frequency may not exceed 10 percent. Other performance conditions are so established that reasonable short-time overloads may be carried. Thus, the user of a motor may expect to be able to apply for a short time an overload of, say, 25 percent at 90 percent of normal voltage with an ample margin of safety.

The converse problem to the rating of machinery, that of choosing the size of machine for a particular application, is a relatively simple one when the load requirements remain substantially constant. For many motor applications, however, the load requirements vary more or less cyclically and over a wide range. The duty cycle of a typical crane or hoist motor may readily be visualized as an example. From the thermal viewpoint, the average heating of the motor must be found by detailed study of the motor losses during the various parts of the cycle. Account must be taken of changes in ventilation with motor speed for open and semienclosed motors. Judicious selection is based on a large number of experimental data and considerable experience with the motors involved. For estimating the required size of motors operating at substantially constant speeds, it is sometimes assumed that the heating of the insulation varies as the square of the horsepower load, an assumption which obviously overemphasizes the role of armature I^2R loss at the expense of the core loss. The rms ordinate of the horsepower-time curve representing the duty cycle is obtained by the same technique used to find the rms value of periodically varying currents, and a motor rating is chosen on the basis of the result; i.e.,

$$\text{rms hp} = \sqrt{\frac{\Sigma \ (\text{hp})^2 \times \text{time}}{\text{running time} + (\text{standstill time}/k)}} \qquad (4\text{-}26)$$

where the constant k accounts for the poorer ventilation at standstill and equals approximately 4 for an open motor.   The time for a complete cycle must be short compared with the time for the motor to reach a steady temperature.

Although crude, the rms-horsepower method is used fairly often. The necessity for rounding off the result to a commercially available motor size obviates the need for precise computations; if the rms horsepower were 87, for example, a 100-hp motor would be chosen.   Special consideration must be given to motors that are frequently started or reversed, for such operations are thermally equivalent to heavy overloads. Consideration must also be given to duty cycles having such high torque peaks that motors with continuous ratings chosen on purely thermal bases would be unable to furnish the torques required.   It is to such duty cycles that special-purpose motors with short-time ratings are often applied.   Short-time-rated motors in general have better torque-producing ability than motors rated to produce the same power output continuously, although, of course, they have a lower thermal capacity. Both these properties follow from the fact that a short-time-rated motor is designed for high flux densities in the iron and high current densities in the copper.   In general, the ratio of torque capacity to thermal capacity increases as the period of the short-time rating decreases. Higher temperature rises are allowed than for general-purpose motors. A motor with a 150-hp 1-hr 50°C rating, for example, may have the torque ability of a 200-hp continuously rated motor; it will be able to carry only about 0.8 of its rated output, or 120 hp, continuously, however. In many cases it will be the economical solution for a drive requiring a continuous thermal capacity of 120 hp but having torque peaks which require the ability of a 200-hp continuously rated motor.

4-10
COOLING MEANS FOR ELECTRIC MACHINES

The cooling problem in electric apparatus in general increases in difficulty with increasing size.   The surface area from which the heat must be carried away increases roughly as the square of the dimensions, whereas the heat developed by the losses is roughly proportional to the volume and therefore increases approximately as the cube of the dimensions. This problem is a particularly serious one in large turbine generators, where economy, mechanical requirements, shipping, and erection all demand compactness, especially for the rotor forging.   Even in moderate sizes of machines (e.g., above a few thousand kva for generators), a closed ventilating system is commonly used.   Rather elaborate systems of cooling ducts must be provided to ensure that the cooling medium effectively removes the heat arising from the losses.

For turbine generators, hydrogen is commonly used as the cooling medium in the totally enclosed ventilating system.   Hydrogen has the following properties which make it well suited to the purpose:

1 Its density is only about 0.07 that of air at the same temperature and pressure, and therefore windage and ventilating losses are much less.

2 Its specific heat on an equal-weight basis is about 14.5 times that of air.   This means that, for the same temperature and pressure, hydrogen and air are about equally effective in their heat-storing capacity per unit volume.   But the heat transfer by forced convection between the hot parts of the machine and the cooling gas is considerably greater with hydrogen than with air.

3 The life of the insulation is increased, and maintenance expenses are decreased, because of the absence of dirt, moisture, and oxygen.

4 The fire hazard is minimized.   A hydrogen-air mixture will not explode if the hydrogen content is above about 70 percent.

The result of the first two properties is that for the same operating conditions the heat which must be dissipated is reduced and at the same time the ease with which it can be carried off is improved.

The machine and its water-cooled heat exchanger for cooling the hydrogen must be sealed in a gastight envelope.   The crux of the problem is in sealing the bearings.   The system is maintained at a slight pressure (at least 0.5 psi) above atmospheric so that gas leakage is outward and an explosive mixture cannot accumulate within the machine.   At this pressure, the rating of the machine can be increased by about 30 percent above its air-cooled rating and the full-load efficiency increased by about 0.5 percent.   The trend is toward the use of higher pressures (15 to 60 psi).   Increasing the hydrogen pressure from 0.5 to 15 psi increases the output for the same temperature rise by about 15 percent; a further increase to 30 psi provides about an additional 10 percent.

An important step which has made it possible almost to double the output of a hydrogen-cooled turbine generator of given physical size is the development of *conductor cooling*, also called *inner cooling*.   Here the coolant (liquid or gas) is forced through hollows or ducts inside the conductor or conductor strands.   Examples of such conductors may be

(a)   (b)   (c)

Fig. 4-22.   Cross sections of bars for two-layer stator winding of turbine generators.   Insulation system consists of synthetic resin with vacuum impregnation.
Type A: indirectly cooled bar with tubular strands.   Type B: water-cooled bars,
two-wire-wide mixed strands.   Type C: water-cooled bars, four-wire-wide mixed
strands.   *(Brown Boveri Corporation.)*

seen in Fig. 4-22.   Thus, the thermal barrier presented by the electrical
insulation is largely circumvented, and the conductor losses may be
absorbed directly by the coolant.   Hydrogen is usually the cooling
medium for the rotor conductors.   Either gas or liquid cooling may be
used for the stator conductors.   Hydrogen is the coolant in the former
case, and transil oil or water is commonly used in the latter.   A sectional
view of a conductor-cooled turbine generator is given in Fig. 4-23.   A
large hydroelectric generator in which both stator and rotor are water-
cooled is shown in Figs. 3-2 and 3-9.

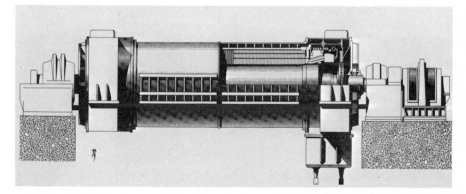

Fig. 4-23.   Cutaway view of 4-pole 1,800-rpm turbine generator rated 1,333 Mva, 0.85
power factor, 26 kv, 60 Hz, 60 psig H_2 pressure.   Stator winding is water-cooled, rotor
winding is hydrogen-cooled.   *(Brown Boveri Corporation.)*

4-11
RESUMÉ

As we go deeper into the study of rotating machines, detailed differences among the various machine types begin to appear. For example, commutation is a problem unique to dc machines. Magnetic saturation and salient poles may have significant effects on dc and synchronous machines. Speed control of induction motors presents problems. An understanding of motor speed-control systems using solid-state rectifiers requires study of the rectifier and motor together. Important problems concerning transient and dynamic characteristics arise in many machine applications. Fractional-horsepower motors have their unique characteristics, not the least of which is a highly competitive market. Chapters 5 to 11 are concerned with some of these problems.

It should be recognized that the performance limitation of a machine is basically determined by the properties of the materials of which it is composed. Thus, much of the great progress in electric machinery over the years has stemmed from improvements in the quality and characteristics of steel and insulating material and in the cooling of the machines.

A group of interrelated problems common to all machine types is created by the losses in the machine and the necessity of dissipating the associated heat. The rating of the machine is closely connected with its ability to operate at temperatures compatible with reasonable life of the insulation and of the machine as a whole. Matters of rating, allowable temperature rise, and determination of losses are all subjects of standardization by professional organizations such as the IEEE, NEMA, and ANSI. Such matters are obviously of great importance in the economics and engineering of a project involving machinery.

PROBLEMS

4-1. A synchronous motor with its stator connected to a balanced polyphase source is operating at 1.00 power factor and constant load torque equal to one-half of its full-load value. State which way (in the direction of rotation or counter to the direction of rotation) the rotor will move relative to the resultant air-gap flux-density wave as the field current is increased.

4-2. A synchronous motor fed from constant-voltage mains is supplying a constant-torque load. The effects of losses and of the leakage reactance of the armature may be ignored. The field current is initially adjusted so that the motor is operating at unity power factor. Describe with reasons the effect of decreasing the field current on the following quantities:

a. The magnitude of the resultant flux wave

b. The component of armature current in phase with the voltage

c. The space-phase angle δ between the armature mmf and the resultant flux wave

4-3. Electrical power is to be supplied to a 3-phase 25-Hz system from a 3-phase 60-Hz system through a motor-generator set consisting of two directly coupled synchronous machines.

a. What is the minimum number of poles which the motor may have?

b. What is the minimum number of poles which the generator may have?

c. At what speed in rpm will the set specified in a and b operate?

4-4. A 3-phase induction motor runs at almost 1,200 rpm at no load and 1,140 rpm at full load when supplied with power from a 60-Hz 3-phase line.

a. How many poles has the motor?

b. What is the percent slip at full load?

c. What is the corresponding frequency of the rotor voltages?

d. What is the corresponding speed (1) of the rotor field with respect to the rotor?   (2) Of the rotor field with respect to the stator? (3) Of the rotor field with respect to the stator field?

e. What speed would the rotor have at a slip of 10 percent?

f. What is the rotor frequency at this speed?

g. Repeat part d for a slip of 10 percent.

4-5. A linear motor based on the induction-motor principle consists of a car riding on a track.   The track is a developed squirrel-cage winding, and the car, which is 12 ft long, 3½ ft wide, and only 5½ in. high, has a developed 3-phase 8-pole winding.   The center-line distance between adjacent poles is $1\frac{2}{8} = 1\frac{1}{2}$ ft.   Power at 60 Hz is fed to the car from arms extending through slots to rails below ground level.

a. What is the synchronous speed in miles per hour?

b. Will the car reach this speed?   Explain your answer.

c. To what slip frequency does a car speed of 75 mph correspond?

Linear induction motors have been proposed for a variety of applications including high-speed ground transportation.

4-6. Describe the effect on the normal torque-speed characteristic of an induction motor produced by:

a. Halving the applied voltage with normal frequency
b. Halving both the applied voltage and frequency

Sketch the associated torque-speed characteristics in their approximate relative positions with respect to the normal one. Neglect the effects of stator resistance and leakage reactance.

4-7. The stator of an unloaded 3-phase 6-pole wound-rotor induction motor is connected to a 60-Hz source; the rotor is connected to a 25-Hz source.

a. Is a starting torque produced?
b. At what speed will steady-state motor action result? There are two possible answers, depending on circumstances in a particular case.
c. What determines at which of the two speeds in *b* the motor will operate in a specific case?
d. Suppose now that the rotor supply frequency is varied over the range 0 to 25 Hz. Sketch curves showing motor speed in rpm as a function of rotor frequency, interpreting zero frequency as direct current.
e. What changes are made in the foregoing answers if the motor is fully loaded instead of unloaded?

4-8. Figure 4-24 shows a 3-phase wound-rotor induction machine whose shaft is rigidly coupled to the shaft of a 3-phase synchronous motor. The terminals of the 3-phase rotor winding of the induction machine are brought out to slip rings as shown. The induction machine is driven by the synchronous motor at the proper speed and in the proper direction of rotation so that 3-phase 120-Hz voltages are available at the slip rings. The induction machine has a 6-pole stator winding.

a. How many poles must the rotor winding of the induction machine have?

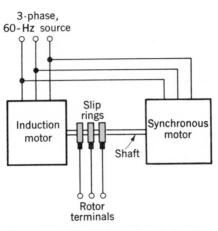

Fig. 4-24. Interconnected induction and synchronous machines, Probs. 4-8 and 4-9.

b. If the stator field in the induction machine rotates in a clockwise direction, what must be the direction of rotation of its rotor?

c. What must be the speed in rpm?

d. How many poles must the synchronous motor have?

4-9. The system shown in Fig. 4-24 is used to convert balanced 60-Hz voltages to other frequencies. The synchronous motor has 2 poles and drives the interconnecting shaft in the clockwise direction. The induction machine has 12 poles, and its stator windings are connected to the lines to produce a counterclockwise rotating field (in the opposite direction to the synchronous motor). The machine has a wound rotor whose terminals are brought out through slip rings.

a. At what speed does the motor run?

b. What is the frequency of the rotor voltages in the induction machine?

4-10. a. Compare the effect on the speed of a dc shunt motor of varying the line voltage with that of varying only the armature terminal voltage, so that the field current remains fixed.

b. Compare both these effects with that of varying only the shunt-field current, the armature terminal voltage remaining fixed.

4-11. State approximately how the armature current and speed of a dc shunt motor would be affected by each of the following changes in the operating conditions:

a. Halving the armature terminal voltage, the field current and load torque remaining constant

b. Halving the armature terminal voltage, the field current and horsepower output remaining constant

c. Doubling the field flux, the armature terminal voltage and load torque remaining constant

d. Halving both the field flux and armature terminal voltage, the horsepower output remaining constant

e. Halving the armature terminal voltage, the field flux remaining constant and the load torque varying as the square of the speed

In each case, only brief quantitative statements of the order of magnitude of the changes are desired, e.g., "speed approximately doubled."

4-12. A 25-kw 250-volt dc machine has an armature resistance of 0.10 ohm. Its magnetization curve at a constant speed of 1,200 rpm is shown in Fig. 4-25.

Its field is separately excited, and it is driven by a synchronous motor at a constant speed of 1,200 rpm.

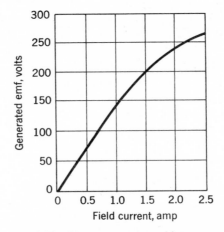

Fig. 4-25. Direct-current-machine magnetization curve at 1,200 rpm, Probs. 4-12 and 4-13.

Plot a family of curves of armature terminal voltage vs. armature current for constant field currents of 2.5, 2.0, 1.5, and 1.0 amp.

4-13. The dc machine of Prob. 4-12 is operated as a separately excited motor.

a. For a constant field current of 2.0 amp, plot a family of curves of speed in rpm versus torque in newton-meters for applied armature terminal voltages of 250, 200, 150, and 100 volts.

b. For a constant applied armature terminal voltage of 200 volts, plot a family of curves of speed vs. torque for field currents of 2.5, 2.0, 1.5, and 1.0 amp.

4-14. An eddy-current brake of the type used for load tests on motors requires a torque of 100 lb-ft to drive it at a speed of 1,000 rpm when the current in its magnetizing coils is 10.0 amp.

This brake is driven by a dc series motor rated to deliver 20 hp at 1,000 rpm with an applied voltage of 230 volts.

a. Plot a family of five torque-speed curves for the brake at coil currents of 6, 8, 10, 12, and 14 amp, respectively. For the purposes of this problem, assume that the flux is linearly proportional to the coil current and that the magnetic effect of the eddy currents generated in the brake disk is negligible. Also, neglect windage and friction torques.

b. On the same curve sheet plot a family of four torque-speed curves for the series motor at applied voltages of 120, 100, 80, and 60 percent of rated voltage. Assume that all losses are negligible, that the motor flux is linearly proportional to the motor field current, and that the resistive voltage drops in the field and armature windings are negligible.

c. Plot curves of torque and speed against brake coil current when the motor is supplied with rated voltage.

d. Plot curves of torque and speed against motor applied voltage when the brake coil current is constant at 10 amp.

4-15. General-purpose 3-phase 60-Hz induction motors are available in 2-, 4-, 6-, and 8-pole designs and in the following horsepower ratings: 2, 3, 5, 7.5, 10, 15, 20, etc. These motors develop rated output at a slip of about 5 percent and develop a maximum torque of 200 percent of rated torque at a slip of about 15 percent.

Select the appropriate motor for an application requiring a torque of 50 lb-ft at a speed of about 1,500 rpm for a period of 30 sec, followed by 4 min running at no load, followed by repetitions of the same load cycle. Specify the horsepower and synchronous-speed ratings.

4-16. A dc compound motor is to be selected for the operation of a lift. The motor is to drive continuously a steel cable which runs over pulleys at the bottom and the top of the lift. When the load is descending, the motor becomes a generator and pumps power back into the line, the resulting torque supplying a braking action.

The operating cycle is as follows and is repeated continuously throughout the day:

Load going up (1 min) = 75 hp
Loading period at top (2 min) = 5 hp
Load going down (1 min) = -60 hp
Loading period at bottom (3 min) = 5 hp

On the basis of heating, select the smallest motor suitable for this application. Motors are available in the following sizes: 25, 30, 40, 50, 60, 75, and 100 hp.

What other factors, besides heating, should be considered?

4-17. In the design of a grab-bucket hoist for unloading coal from a barge into a bunker, a study is made of the mechanical requirements to determine the motor duty cycle. The results are given in the following table for an average cycle:

Part of cycle	Elapsed time, sec	Required output, hp
Close bucket	6	40
Hoist	10	80
Open bucket	3	30
Lower bucket	10	45
Rest	16	0

Because of the conditions of service, a dustproof enclosed motor without forced ventilation is to be used, and the constant k associated with the standstill time may be taken as unity.

a. Using the rms method, specify the continuous horsepower rating of the motor.   Choose a commercially available motor size.

b. Proposals to furnish this motor are submitted by two manufacturers.   These proposals contain the following efficiency guarantees and prices:

Efficiencies, percent

Motor	¼ load	½ load	¾ load	1.0 load	1¼ load	1½ load	Net prices,* dollars
A	83.4	90.5	90.3	88.0	86.8	85.0	2,500
B	81.3	88.6	90.3	90.6	90.3	89.6	3,000

*Both prices f.o.b. factory, freight allowed.

The average net cost of energy at this plant is 1.5 cents per kilowatt-hour.   Total fixed charges on invested capital are 25 percent.   The hoist will be in operation an average of 2,000 hr/year.

Which of these two motors would you recommend?

5
dc machines

In order to place the dc machine on a fully realistic basis, we first reexamine the flux and mmf conditions in the machine. This reexamination, together with consideration of the switching action at the commutator, brings out conditions limiting the capability of the machine as well as means for combating the conditions. It also leads to methods for including saturation and armature-mmf effects in the analysis of the machine. The results of the analysis illustrate the versatility of the dc machine by itself and in combination with other dc machines.

5-1
EFFECT OF ARMATURE MMF

Armature mmf has definite effects on both the space distribution of the air-gap flux and the magnitude of the net flux per pole. The effect on flux distribution is important because the limits of successful commutation are directly influenced; the effect on flux magnitude is important

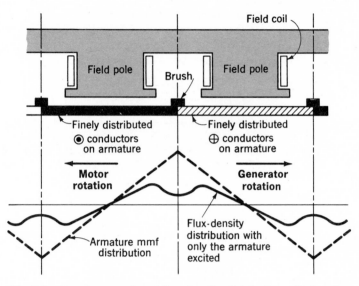

Fig. 5-1. Armature mmf and flux-density distribution with brushes on neutral and only the armature excited.

because both the generated voltage and torque per unit of armature current are influenced thereby. These effects and the problems arising from them are described in this article.

It is shown in Art. 3-3b and by Fig. 3-24 that the armature-mmf wave may be closely approximated by a triangle, corresponding to the wave produced by a finely distributed armature winding or current sheet. For a machine with brushes in the neutral position, the idealized mmf wave is again shown by the dotted triangle in Fig. 5-1, in which a positive mmf ordinate denotes flux lines leaving the armature surface. Current directions in all windings other than the main field are indicated by black and crosshatched bands. Because of the salient-pole field structure found in almost all dc machines, the associated space distribution of flux will not be triangular. The distribution of air-gap flux density with only the armature excited is given by the solid curve of Fig. 5-1. As may readily be seen, it is appreciably decreased by the long air path in the interpolar space.

The axis of the armature mmf is fixed at 90 electrical degrees from the main-field axis by the brush position. The corresponding flux follows the paths shown in Fig. 5-2. The effect of the armature mmf is seen to be that of creating flux sweeping across the pole faces; thus its path in the pole shoes crosses the path of the main-field flux. For this reason, arma-

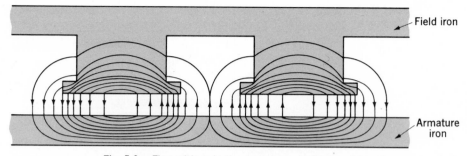

Fig. 5-2.   Flux with only the armature excited and brushes on neutral.

ture reaction of this type is called *cross-magnetizing armature reaction*. It evidently causes a decrease in the resultant air-gap flux density under one half of the pole and an increase under the other half.

When the armature and field windings are both excited, the resultant air-gap flux-density distribution is of the form given by the solid curve of Fig. 5-3. Superimposed on this figure are the flux distributions with only the armature excited (dashed curve) and only the field excited (dotted curve). The effect of cross-magnetizing armature reaction in decreasing the flux under one pole tip and increasing it under the other may be seen by comparing the solid and dotted curves. In general, the solid curve is not the algebraic sum of the dotted and dashed curves because of the non-linearity of the iron magnetic circuit. Because of saturation of the iron, the flux density is decreased by a greater amount under one pole tip than it is increased under the other. Accordingly, the resultant flux per pole is lower than would be produced by the field winding alone, a consequence known as the *demagnetizing effect of cross-magnetizing armature reaction*. Since it is caused by saturation, its magnitude is a nonlinear function of both the field current and the armature current. For normal machine operation at the flux densities used commercially, the effect is usually significant, especially at heavy loads, and must often be taken into account in analyses of performance.

The distortion of the flux distribution caused by cross-magnetizing armature reaction may be a detrimental influence on ability to commutate the current, especially if the distortion becomes excessive. In fact, this distortion is usually an important factor limiting the short-time overload of a dc machine. Tendency toward distortion of flux distribution is most pronounced in a machine, such as a shunt motor, where the field excitation remains substantially constant while the armature mmf may reach very significant proportions at heavy loads. The tendency is least

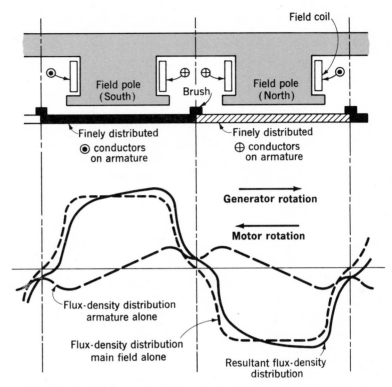

Fig. 5-3. Armature, main-field, and resultant flux-density distributions with brushes on neutral.

pronounced in a series-excited machine, such as the series motor, for both the field and armature mmfs increase with load.

The effect of cross-magnetizing armature reaction may be limited in the design and construction of the machine.   The mmf of the main field should exert predominating control on the air-gap flux, so that the condition of weak-field mmf and strong armature mmf may be avoided.   The reluctance of the cross-flux path—essentially, the armature teeth, pole shoes, and the air gap, especially at the pole tips—may be increased by increasing the degree of saturation in the teeth and pole faces, by avoiding too small an air gap, and by using a chamfered or eccentric pole face, which increases the air gap at the pole tips.   These expedients affect the path of the main flux as well, but the influence on the cross flux is much greater.   The best but also the most expensive curative measure is to compensate the armature mmf by means of a winding embedded in the pole faces, a measure which is discussed in Art. 5-3.

If the brushes are not in the neutral position the axis of the arma-ture-mmf wave is not 90 degrees from the main-field axis. The armature mmf then produces not only cross-magnetization but also a direct demag-netizing or magnetizing effect depending on the direction of brush shift. Shifting of the brushes from the neutral is usually inadvertent due to incorrect positioning of the brushes or a poor brush fit. Prior to the invention of interpoles, however, shifting of the brushes was a common method of securing satisfactory commutation, the direction of the shift being such that demagnetizing action was produced. It can be shown that brush shift in the direction of rotation in a generator or against rotation in a motor produces a direct demagnetizing mmf which may result in unstable operation of a motor or excessive drop in voltage of a gen-erator. Incorrectly placed brushes can be detected by means of a load test. If the brushes are on neutral, the terminal voltage of a generator or the speed of a motor should be the same for identical conditions of field excitation and armature current when the direction of rotation is reversed.

5-2

COMMUTATION AND INTERPOLES

One of the most important limiting factors on satisfactory operation of a dc machine is the ability to transfer the necessary armature current through the brush contact at the commutator without sparking and with-out excessive local losses and heating of the brushes and commutator. Sparking causes destructive blackening, pitting, and wear of both com-mutator and brushes, conditions which rapidly become worse and lead to burning away of the copper and carbon. It may be caused by faulty mechanical conditions, such as chattering of the brushes or a rough, unevenly worn commutator, or, as in any switching problem, by elec-trical conditions. The latter conditions are seriously influenced by the armature mmf and the resultant flux wave.

As indicated in Art. 4-3, a coil undergoing commutation is in transi-tion between two groups of armature coils: at the end of the commutation period, the coil current must be equal but opposite to that at the begin-ning. Figure 4-5b shows the armature in an intermediate position during which the coils formed by inductors are being commutated. The commutated coils are short-circuited by the brushes. During this period, the brushes must continue to conduct the armature current I_a from the armature winding to the external circuit. The short-circuited coil con-stitutes an inductive circuit with time-varying resistances at the brush contact, with, in general, rotational voltages induced in the coil, and with both conductive and inductive coupling to the rest of the armature winding.

The attainment of good commutation is more an empirical art than a quantitative science. The principal obstacle to quantitative analysis lies in the electrical behavior of the carbon-copper contact film. Its resistance is nonlinear and is a function of current density, current direction, temperature, brush material, moisture, and atmospheric pressure. Its behavior in some respects is like that of an ionized gas. The most significant fact is that an unduly high current density in a portion of the brush surface (and hence an unduly high energy density in that part of the contact film) results in sparking and a breakdown of the film at that point. The boundary film also plays an important part in the mechanical behavior of the rubbing surfaces. At high altitudes, definite steps must be taken to preserve it, or extremely rapid brush wear takes place.

The empirical basis of securing sparkless commutation, then, is to avoid excessive current densities at any point in the copper-carbon contact. This basis, combined with the principle of utilizing all material to the fullest extent, indicates that optimum conditions are obtained when the current density is uniform over the brush surface during the entire commutation period. A linear change of current with time in the commutated coil, corresponding to linear commutation as shown in Fig. 4-6, brings about this condition and is accordingly the optimum.

The principal factors tending to produce linear commutation are changes in brush-contact resistance resulting from the linear decrease in area at the trailing brush edge and linear increase in area at the leading edge. Several electrical factors militate against linearity. Resistance in the commutated coil is one example. Usually, however, the voltage drop at the brush contacts is sufficiently large (of the order of 1.0 volt) in comparison with the resistance drop in a single armature coil so that the latter may be ignored. Coil inductance is a much more serious factor. Both the voltage of self-induction in the commutated coil and the voltage of mutual induction from other coils (particularly those in the same slot) undergoing commutation at the same time oppose changes in current in the commutated coil. The sum of these two voltages is often referred to as the *reactance voltage*. Its result is that current values in the short-circuited coil lag in time the values dictated by linear commutation. This condition is known as *undercommutation*, or *delayed commutation*.

Armature inductance thus tends to produce high losses and sparking at the trailing brush tip. For best commutation, inductance must be held to a minimum by using the fewest possible number of turns per armature coil and by using a multipolar design with a short armature. The effect of a given reactance voltage in delaying commutation is minimized when the resistive brush-contact voltage drop is significant compared with it. This fact is one of the main reasons for the use of carbon brushes with their appreciable contact drop. When good commutation is secured by virtue of resistance drops, the process is referred to as *resistance commutation*. It is used today as the exclusive means only in fractional-horsepower machines.

Another important factor in the commutation process is the rotational voltage induced in the short-circuited coil. Depending on its sign, this voltage may hinder or aid commutation. In Fig. 5-3, for example, cross-magnetizing armature reaction creates a definite flux in the interpolar region. The direction of the corresponding rotational voltage in the commutated coil is the same as the current under the immediately preceding pole face. This voltage then encourages the continuance of current in the old direction and, like the reactance voltage, opposes its reversal. To aid commutation, the rotational voltage must oppose the reactance voltage. The general principle of producing in the coil undergoing commutation a rotational voltage which approximately compensates for the reactance voltage, a principle called *voltage commutation*, is used in almost all modern commutating machines. The appropriate flux density is introduced in the commutating zone by means of small, narrow poles located between the main poles. These auxiliary poles are called *interpoles*, or *commutating poles*.

The general appearance of interpoles and an approximate map of the flux produced when they alone are excited may be seen in Fig. 5-4. (The

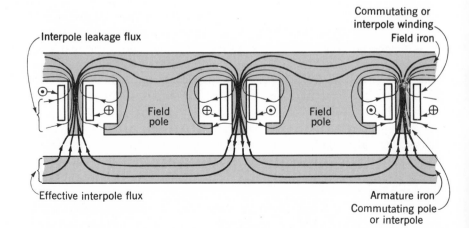

Fig. 5-4. Interpoles and their associated component flux.

interpoles are the smaller poles between the larger main poles in Fig. 5-6.) The polarity of a commutating pole must be that of the main pole just ahead of it (i.e., in the direction of rotation) for a generator and just behind it for a motor.   The interpole mmf must be sufficient to neutralize the cross-magnetizing armature mmf in the interpolar region and enough more to furnish the flux density required for the rotational voltage in the short-circuited armature coil to cancel the reactance voltage.   Since both the armature mmf and the reactance voltage are proportional to the armature current, the commutating winding must be connected in series with the armature.   To preserve the desired linearity, the commutating pole should operate at low saturations.   By the use of commutating fields, then, sparkless commutation is secured over a wide range in modern machines.   In accordance with the performance standards of the NEMA, general-purpose dc machines must be capable of carrying with successful commutation for 1 min loads of 150 percent of the current corresponding to the continuous rating with the field rheostat set for rated-load excitation.

5-3
COMPENSATING WINDINGS

For machines subjected to heavy overloads, rapidly changing loads, or operation with a weak main field, there is the possibility of trouble other than simply sparking at the brushes.   At the instant when an armature coil is located at the peak of a badly distorted flux wave, the coil voltage may be high enough to break down the air between the adjacent segments to which the coil is connected and result in *flashover*, or arcing, between segments.   The breakdown voltage here is not high, because the air near the commutator is in a condition favorable to breakdown.   The maximum allowable voltage between segments is of the order of 30 to 40 volts, a fact which limits the average voltage between segments to lower values and thus determines the minimum number of segments which may be used in a proposed design.   Under transient conditions, high voltages between segments may result from the induced voltages associated with growth and decay of armature flux.   Inspection of Fig. 5-2, for instance, may enable one to visualize very appreciable voltages of this nature being induced in a coil under the pole centers by the growth or decay of the armature flux shown in the sketch.   Consideration of the sign of this induced voltage will show that it adds to the normal rotational emf when load is dropped from a generator or added

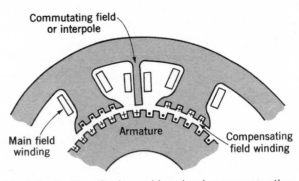

Fig. 5-5. Section of a dc machine showing compensating winding.

to a motor. Flashing between segments may quickly spread around the entire commutator and, in addition to its possibly destructive effects on the commutator, constitutes a direct shortcircuit on the line. Even with interpoles present, therefore, armature reaction under the poles definitely limits the conditions under which a machine may operate.

These limitations may be considerably extended by compensating or

Fig. 5-6. Section of a dc motor stator or field showing shunt and series coils, interpoles, and pole-face compensating winding. *(Westinghouse Electric Corporation.)*

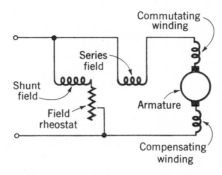

Fig. 5-7. Schematic connection diagram of a dc machine.

neutralizing the armature mmf under the pole faces.   Such compensation can be achieved by means of a *compensating*, or *pole-face*, *winding* (Fig. 5-5) embedded in slots in the pole face and having a polarity opposite to that of the adjoining armature winding.   The physical appearance of such a winding may be seen in the stator of Fig. 5-6.   Since the axis of the compensating winding is the same as that of the armature, it will almost completely neutralize the armature reaction of the armature inductors under the pole faces when it is given the proper number of turns.   It must be connected in series with the armature in order that it may carry a proportional current.   The net effect of the main field, armature, commutating winding, and compensating winding on the air-gap flux is that, except for the commutation zone, the resultant flux-density distribution is substantially the same as that produced by the main field alone (Fig. 5-3).   Furthermore, the addition of a compensating winding improves the speed of response, because it reduces the armature-circuit time constant.

The main disadvantage of pole-face windings is their expense.   They are used in machines designed for heavy overloads or rapidly changing loads (steel-mill motors are a good example of machines subjected to severe duty cycles) or in motors intended to operate over wide speed ranges by shunt-field control.   By way of a schematic summary, Fig. 5-7 shows the circuit diagram of a compound machine with a compensating winding.   The relative position of the coils in this diagram indicates that the commutating and compensating fields act along the armature axis and the shunt and series fields act along the axis of the main poles. Rather complete control of air-gap flux around the entire armature periphery is thus achieved.

ANALYTICAL FUNDAMENTALS: ELECTRIC–CIRCUIT ASPECTS

From Eqs. 4-8 and 4-12, the electromagnetic torque and generated
voltage of a dc machine are, respectively,

$$T = K_a \Phi_d I_a \tag{5-1}$$

$$E_a = K_a \Phi_d \omega_m \tag{5-2}$$

where
$$K_a = \frac{PZ_a}{2\pi a} \tag{5-3}$$

Here the capital-letter symbols E_a for generated voltage and I_a for armature current are used to emphasize that we are primarily concerned with
steady-state considerations in this chapter.   The remaining symbols are
as defined in Art. 4-4.   These are basic equations for analysis of the
machine.   The quantity $E_a I_a$ is frequently referred to as the *electromagnetic power;* from Eqs. 5-1 and 5-2, it is related to electromagnetic
torque by the relation

$$T = \frac{E_a I_a}{\omega_m} \tag{5-4}$$

Figures 5-8 and 5-9 present in graphical form power balances for dc
generators and motors, respectively, with both shunt and series fields.
The connection diagram is given in Fig. 5-10.   When either the shunt or

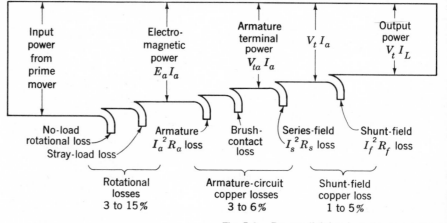

Fig. 5-8.   Power division in a dc generator.

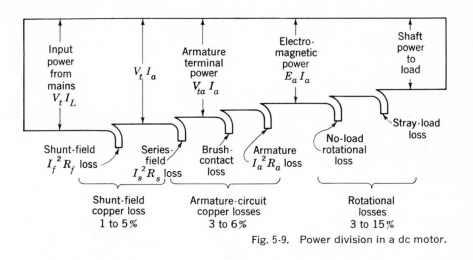

Fig. 5-9.   Power division in a dc motor.

the series field is not present in the machine, the associated entry is omitted from Figs. 5-8 to 5-10.   In these diagrams, V_t is the machine terminal voltage, V_{ta} the armature terminal voltage, I_L the line current, I_s the series-field current (equal to I_a for the connections shown in Fig. 5-10), I_f the shunt-field current, R_a the armature resistance, R_f the shunt-field resistance, and R_s the series-field resistance.   Included in R_a is the resistance of any commutating and compensating winding.   The armature-circuit copper losses, field-circuit copper losses, and rotational losses are those originally considered in Art. 4-8; typical full-load orders of magnitude of these losses, expressed in percent of the machine input, are quoted in Figs. 5-8 and 5-9 for general-purpose generators and motors in

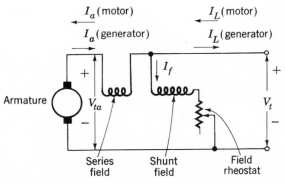

Fig. 5-10.   Motor or generator connection diagram with current directions.

the 1- to 100-kw or 1- to 100-hp range, with the smaller percentages applying to the larger ratings.

The electromagnetic power differs from the mechanical power at the machine shaft by the rotational losses and differs from the electrical power at the machine terminals by the copper losses. The electromagnetic power is that measured at the points across which E_a exists; numerical addition of the rotational losses for generators and subtraction for motors yield the mechanical power at the shaft.

The interrelations between voltage and current are immediately evident from the connection diagram. Thus,

$$V_{ta} = E_a \pm I_a R_a \tag{5-5}$$

$$V_t = E_a \pm I_a(R_a + R_s) \tag{5-6}$$

and
$$I_L = I_a \pm I_f \tag{5-7}$$

where the plus sign is used for a motor and the minus sign for a generator. Some of the terms in Eqs. 5-5 to 5-7 may be omitted when the machine connections are simpler than those shown in Fig. 5-10. The resistance R_a is to be interpreted as that of the armature plus brushes unless specifically stated otherwise. Sometimes, R_a is taken as the resistance of the armature winding alone, and the brush-contact drop is accounted for as a separate item, usually assumed to be 2 volts.

For compound machines, another variation may occur. Figure 5-10 shows a so-called *long-shunt connection* in that the shunt field is connected directly across the line terminals with the series field between it and the armature. An alternative possibility is the *short-shunt connection*, illustrated in Fig. 5-11 for a compound generator, with the shunt field directly across the armature and the series field between it and the line terminals. The series-field current is then I_L instead of I_a, and the voltage equations

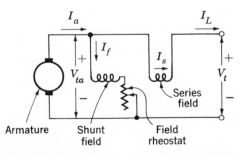

Fig. 5-11. Short-shunt compound generator connections.

are modified accordingly.   There is so little practical difference between these two connections that the distinction may usually be ignored; unless otherwise stated, compound machines will be treated as though they were long-shunt-connected.

Although the difference between terminal voltage V_t and armature generated voltage E_a is comparatively small for normal operation, it has a definite bearing on performance characteristics.   In effect, this difference, acting in conjunction with the circuit resistances and energy-conversion requirements, affects the value of armature current I_a and hence the rotor-field strength.   Complete determination of machine behavior requires a similar investigation of factors influencing the stator-field strength or, more particularly, the net flux per pole Φ_d.

5-5

ANALYTICAL FUNDAMENTALS: MAGNETIC–CIRCUIT ASPECTS

The flux per pole is that resulting from the combined armature and field mmfs.   The interdependence of armature generated voltage E_a and magnetic-circuit conditions in the machine is accordingly a function of the sum of all the mmfs on the polar- or direct-axis flux path.   We shall first consider the mmf purposely placed on the stator main poles to create the working flux—i.e., the *main-field mmf*—and then include armature-mmf effects.

a. Armature Reaction Neglected

With no load on the machine or with armature-reaction effects ignored, the resultant mmf is the algebraic sum of the mmfs on the main- or direct-axis poles.   For the usual compound generator or motor having N_f shunt-field turns per pole and N_s series-field turns per pole,

$$\text{Main-field mmf} = N_f I_f \pm N_s I_s \tag{5-8}$$

Additional terms will appear in this equation when there are additional field windings on the main poles (and, unlike the compensating windings of Art. 5-3, wound concentric with the normal field windings) to permit specialized control.   In Eq. 5-8, the plus sign is used when the two mmfs are aiding, or when the two fields are cumulatively connected; the minus sign is used when the series field opposes the shunt field, or for a differential connection.   When either the series or the shunt field is absent, the corresponding term in Eq. 5-8 naturally is omitted.

Equation 5-8 thus sums up in ampere-turns per pole the gross mmf of the main-field windings acting on the main magnetic circuit. The magnetization curve for a dc machine is generally given in terms of current in only the principal field winding, which is almost invariably the shunt-field winding when one is present. The mmf units of such a magnetization curve and of Eq. 5-8 may be made the same by one of two rather obvious steps. The field current on the magnetization curve may be multiplied by the turns per pole in that winding, giving a curve in terms of ampere-turns per pole; or both sides of Eq. 5-8 may be divided by N_f, converting the units to the equivalent current in the N_f coil alone which produces the same mmf. Thus

$$\text{Gross mmf} = I_f \pm \frac{N_s}{N_f} I_s \qquad \text{equivalent shunt-field amp} \qquad (5\text{-}9)$$

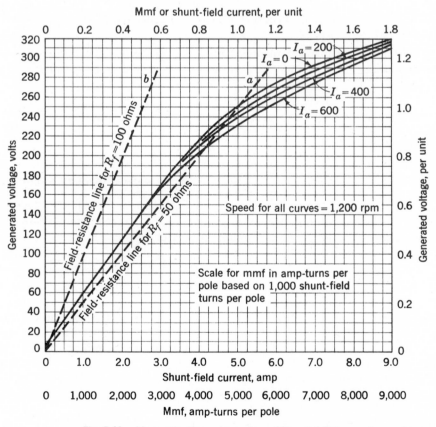

Fig. 5-12.   Magnetization curves for a 250-volt 1,200-rpm dc machine.

The latter procedure is often the more convenient and the one more commonly adopted.

An example of a no-load magnetization characteristic is given by the curve for $I_a = 0$ in Fig. 5-12. The numerical scales on the left-hand and lower axes give representative values for a 100-kw 250-volt 1,200-rpm generator; the mmf scale is given in both shunt-field current and ampere-turns per pole, the latter being derived from the former on the basis of a 1,000-turn-per-pole shunt field. The characteristic may also be presented in *normalized,* or *per-unit, form,* as shown by the upper mmf and right-hand voltage scale. On these scales, 1.0-*per-unit field current* or mmf is that required to produce rated voltage at rated speed when the machine is unloaded; similarly, 1.0-*per-unit voltage* equals rated voltage.

Use of the magnetization curve with generated voltage rather than flux plotted on the vertical axis may be somewhat complicated by the fact that the speed of a dc machine need not remain constant, and speed enters into the relation between flux and generated voltage. Hence generated-voltage ordinates correspond to a unique machine speed. The generated voltage E_a at any speed ω_m is, in accordance with Eq. 5-2, given by

$$E_a = E_{a0} \frac{\omega_m}{\omega_{m0}} \tag{5-10}$$

where ω_{m0} is the magnetization-curve speed and E_{a0} the corresponding armature emf.

EXAMPLE 5-1

A 100-kw 250-volt 400-amp long-shunt compound generator has armature resistance (including brushes) of 0.025 ohm, a series-field resistance of 0.005 ohm, and the magnetization curve of Fig. 5-12. There are 1,000 shunt-field turns per pole and 3 series-field turns per pole.

Compute the terminal voltage at rated current output when the shunt-field current is 4.7 amp and the speed is 1,150 rpm. Neglect armature reaction.

Solution

$I_s = I_a = I_L + I_f = 400 + 4.7 = 405$ amp. From Eq. 5-9, the main-field gross mmf is

$$4.7 + \frac{3}{1,000} \times 405 = 5.9 \text{ equivalent shunt-field amp}$$

By entering the $I_a = 0$ curve of Fig. 5-12 with this current, one reads 274 volts. Accordingly, the actual emf is

$$E_a = 274 \times \frac{1,150}{1,200} = 262 \text{ volts}$$

Then $\qquad V_t = E_a - I_a(R_a + R_s)$

$$= 262 - 405(0.025 + 0.005) = 250 \text{ volts}$$

---

b. Effect of Armature MMF

As described in Art. 5-1, excitation of the armature winding gives rise to a demagnetizing effect caused by cross-magnetizing armature reaction. Analytic inclusion of this effect is not a straightforward task, because of the nonlinearities involved. One common approach is to base the work on experimentally determined data for the machine involved or for one of similar design and frame size. Data are taken with both the field and armature excited, and the tests are conducted so that the effects on generated emf of varying both main-field excitation and the armature mmf may be noted.

One form of summarizing and correlating the results is illustrated in Fig. 5-12. Curves are plotted not only for the no-load characteristic $(I_a = 0)$ but for a family of values of I_a. In the analysis of machine performance, then, the inclusion of armature reaction becomes simply a matter of using the magnetization curve corresponding to the armature current involved. Note that the ordinates of all these curves give values of armature generated voltage E_a, not terminal voltage under load. Note also that all the curves tend to merge with the air-gap line as saturation of the iron decreases.

The load saturation curves are displaced to the right of the no-load curve by an amount which is a function of I_a. The effect of armature reaction then is approximately the same as a demagnetizing mmf AR acting on the main-field axis. The *net* direct-axis mmf is then assumed to be

$$\text{Net mmf} = \text{gross mmf} - AR$$

$$= N_f I_f \pm N_s I_s - AR \qquad (5\text{-}11)$$

The no-load magnetization curve can then be used as the relation between generated emf and net excitation under load with the armature reaction accounted for as a demagnetizing mmf. Over the normal operating range (about 240 to about 300 volts for the machine of Fig. 5-12) the demagnetizing effect of armature reaction may be assumed to be approximately proportional to the armature current.

The amount of armature reaction present in Fig. 5-12 is chosen so that some of its disadvantageous effects will appear in a pronounced form in subsequent numerical examples and problems illustrating generator and motor performance features. It is definitely more than one would expect to find in a normal, well-designed machine operating at normal currents.

5-6
ANALYSIS OF STEADY–STATE PERFORMANCE

Although identically the same principles apply to analysis of a dc machine acting as a generator as to one acting as a motor, the general nature of the problems ordinarily encountered is somewhat different for the two methods of operation. For a generator, the speed is usually fixed by the prime mover, and problems often met are to determine the terminal voltage corresponding to a specified load and excitation or to find the excitation required for a specified load and terminal voltage. For a motor, on the other hand, problems frequently encountered are to determine the speed corresponding to a specified load and excitation or to find the excitation required for specified load and speed conditions; terminal voltage is often fixed at the value of the available supply mains. The routine techniques of applying the common basic principles therefore differ to the extent that the problems differ.

a. Generator Analysis

Since the main-field current is independent of the generator voltage, separately excited generators are the simplest to analyze. For a given load, the main-field excitation is given by Eq. 5-9, and the associated armature generated voltage E_a is determined by the appropriate magnetization curve. This voltage, together with Eq. 5-5 or 5-6, fixes the terminal voltage.

In self-excited generators, the shunt-field excitation depends on the terminal voltage and the series-field excitation on the armature current. Dependence of shunt-field current on terminal voltage may be incorporated graphically in an analysis by drawing the *field-resistance line*, the line 00a in Fig. 5-12, on the magnetization curve. The field-resistance line 00a is simply a graphical representation of Ohm's law for the shunt field. It is the locus of the terminal-voltage vs. shunt-field-current operating point. Thus, the line $0a$ is drawn for $R_f = 50$ ohms and hence passes through the origin and the point (1.0 amp, 50 volts).

One instance of the interdependence of magnetic- and electric-circuit conditions may be seen by examining the *build-up of voltage* for an unloaded shunt generator. When the field circuit is closed, the small voltage from residual magnetism (the 6-volt intercept of the magnetization curve, Fig. 5-12) causes a small field current. If the flux produced by the resulting ampere-turns adds to the residual flux, progressively greater voltages and field currents are obtained. If the field ampere-turns oppose the residual magnetism, the shunt-field terminals must be reversed to obtain build-up. Build-up continues until the volt-ampere relations represented by the magnetization curve and the field-resistance line are simultaneously satisfied (i.e., at their intersection, 250 volts for line $00a$ in Fig. 5-12). This statement ignores the extremely small voltage drop caused by the shunt-field current in the armature-circuit resistance. Notice that if the field resistance is too high, as shown by line $00b$ for $R_f = 100$ ohms, the intersection is at very low voltage and build-up is not obtained. Notice also that if the field-resistance line is essentially tangent to the lower part of the magnetization curve, corresponding to 57 ohms in Fig. 5-12, the intersection may be anywhere from about 60 to 170 volts, resulting in very unstable conditions. The corresponding resistance is the *critical field resistance*, beyond which build-up will not be obtained. The same build-up process and the same conclusions apply to compound generators; in a long-shunt compound generator, the series-field mmf created by the shunt-field current is entirely negligible.

This build-up of voltage is evidently a transient process in which, at any particular point, the vertical difference between the field-resistance line and the magnetization curve is the voltage serving to increase the current through the shunt-field inductance. The transient process is discussed in Art. 9-6b.

For a shunt generator, the magnetization curve for the appropriate value of I_a is the locus of E_a versus I_f. The field-resistance line is the locus V_t versus I_f. With steady-state operation and at any value of I_f, therefore, the vertical distance between the line and the curve must be the $I_a R_a$ drop at the load corresponding to that condition. Determination of the terminal voltage for a specified armature current is then simply a matter of finding where the line and curve are separated vertically by the proper amount; the ordinate of the field-resistance line at that field current is then the terminal voltage. For a compound generator, however, the series-field mmf causes corresponding points on the line and curve to be displaced horizontally as well as vertically. The horizontal displacement equals the series-field mmf measured in equivalent shunt-field amperes, and the vertical displacement is still the I_aR_a drop.

Great precision is evidently not obtained from the foregoing computational process.   The uncertainties caused by magnetic hysteresis in dc machines make high precision unattainable in any event.   In general, the magnetization curve on which the machine operates on any given occasion may range from the rising to the falling part of the rather fat hysteresis loop for the magnetic circuit of the machine, depending essentially on the magnetic history of the iron just prior to that occasion. The curve used for analysis is usually the mean magnetization curve, and thus the results obtained are substantially correct on the average.   Significant departures from the average may be encountered in the performance of any dc machine at a particular time, however.

EXAMPLE 5-2

A 100-kw 250-volt 400-amp 1,200-rpm dc shunt generator has the magnetization curves (including armature-reaction effects) of Fig. 5-12.   The armature-circuit resistance, including brushes, is 0.025 ohm.   The generator is driven at a constant speed of 1,200 rpm, and the excitation is adjusted to give rated voltage at no load.

a. Determine the terminal voltage at an armature current of 400 amp.
b. A series field of 4 turns per pole having a resistance of 0.005 ohm is to be added.   There are 1,000 turns per pole in the shunt field. The generator is to be flat-compounded so that the full-load voltage is 250 volts when the shunt-field rheostat is adjusted to give a no load of 250 volts.   Show how a resistance across the series field (a so-called *series-field diverter*) may be adjusted to produce the desired performance.

Solution

a. The field-resistance line $00a$ (Fig. 5-12) passes through the 250-volt 5.0-amp point of the no-load magnetization curve.   At $I_a = 400$ amp,

$$I_aR_a = 400 \times 0.025 = 10 \text{ volts}$$

A vertical distance of 10 volts exists between the magnetization curve for $I_a = 400$ amp and the field-resistance line at a field current of 4.1 amp, corresponding to $V_t = 205$ volts.   The associated line current is

$$I_L = I_a - I_f = 400 - 4 = 396 \text{ amp}$$

Note that a vertical distance of 10 volts also exists at a field current of 1.2 amp, corresponding to $V_t = 60$ volts. The voltage-load curve is accordingly double-valued in this region. The point for which $V_t = 205$ is the normal operating point.

b. For the no-load voltage to be 250 volts, the shunt-field resistance must be 50 ohms, and the field-resistance line is $00a$ (Fig. 5-12). At full load, $I_f = 5.0$ amp because $V_t = 250$ volts. Then,

$$I_a = 400 + 5.0 = 405 \text{ amp}$$

and $\qquad E_a = 250 + 405(0.025 + 0.005) = 262 \text{ volts}$

In the last equation, the effect of the diverter in reducing the series-field circuit resistance is ignored, a neglect which is permissible in view of the degree of precision warranted. From the 400-amp magnetization curve, an E_a of 262 volts requires a main-field excitation of 5.95 equivalent shunt-field amperes. (Strictly speaking, of course, a curve for $I_a = 405$ amp should be used, but such a small distinction is obviously meaningless.) From Eq. 5-9,

$$5.95 = 5.0 + \frac{4}{1,000} I_s$$

$$I_s = 238 \text{ amp}$$

Hence, only 238 of the total 405 amp of armature current must pass through the series field, a process requiring that the series field be shunted by a resistor of

$$\frac{238 \times 0.005}{405 - 238} = 0.0071 \text{ ohm}$$

b. Motor Analysis

Since the terminal voltage of motors is usually substantially constant at a specified value, there is no dependence of shunt-field excitation on a varying voltage as in shunt and compound generators. Hence, motor analysis most nearly resembles that for separately excited generators, although speed is now an important variable and often the one whose value is to be found. Analytical essentials include Eqs. 5-5 and 5-6 relating terminal voltage and generated or counter emf, Eq. 5-9 for main-field excitation, the magnetization curve for the appropriate armature current as the graphical relation between counter emf and excitation, Eq. 5-1 showing the dependence of electromagnetic torque on flux and armature current, and Eq. 5-2 relating counter emf with flux and speed. The last two relations are particularly significant in motor analysis. The former is pertinent because the interdependence of torque and the stator and rotor field strengths must often be examined.   The latter is the usual medium for determining motor speed from other specified operating conditions.

Motor speed corresponding to a given armature current I_a may be found by first computing the actual generated voltage E_a from Eq. 5-5 or 5-6.   Next obtain the main-field excitation from Eq. 5-9.   Since the magnetization curve will be plotted for a constant speed ω_{m0} which in general will be different from the actual motor speed ω_m, the generated voltage read from the magnetization curve at the foregoing main-field excitation will correspond to the correct flux conditions but to the speed ω_{m0}.   Substitution in Eq. 5-10 then yields the actual motor speed.

It will be noted that knowledge of the armature current is postulated at the start of this process.   When, as is frequently the case, the speed at a stated shaft power or torque output is to be found, successive trials based on assumed values of I_a usually form the simplest procedure. Plotting of the successive trials permits speedy determination of the correct armature current and speed at the desired output.

EXAMPLE 5-3

A 100-hp 250-volt dc shunt motor has the magnetization curves (including armature-reaction effects) of Fig. 5-12.   The armature circuit resistance, including brushes, is 0.025 ohm.   No-load rotational losses are 2,000 watts, and stray load losses equal 1.0 percent of the output.   The field rheostat is adjusted for a no-load speed of 1,100 rpm.

a. As an example of computing points on the speed-load characteristic, determine the speed in rpm and output in horsepower corresponding to an armature current of 400 amp.
b. Because the speed-load characteristic referred to in (a) is considered undesirable, a *stabilizing winding* consisting of $1\frac{1}{2}$ cumulative series turns per pole is to be added.   The resistance of this winding is negligible.   There are 1,000 turns per pole in the shunt field.   Compute the speed corresponding to an armature current of 400 amp.

Solution

a. At no load, $E_a = 250$ volts. The corresponding point on the 1,200-rpm no-load saturation curve is

$$E_{a0} = 250 \times \frac{1,200}{1,100} = 273 \text{ volts}$$

for which $I_f = 5.90$ amp. The field current remains constant at this value.

At $I_a = 400$ amp, the actual counter emf is

$$E_a = 250 - 400 \times 0.025 = 240 \text{ volts}$$

From Fig. 5-12 with $I_a = 400$ and $I_f = 5.90$, the value of E_a would be 261 volts if the speed were 1,200 rpm. The actual speed is then

$$n = \frac{240}{261} \times 1,200 = 1,100 \text{ rpm}$$

The electromagnetic power is

$$E_a I_a = 240 \times 400 = 96,000 \text{ watts}$$

Deduction of the rotational losses leaves 94,000 watts. With stray load losses accounted for, the power output P_o is given by

$$94,000 - 0.01 P_o = P_o$$

or $$P_o = 93.1 \text{ kw} = 124.7 \text{ hp}$$

Note that the speed at this load is the same as at no load, indicating that armature-reaction effects have caused an essentially flat speed-load curve.

b. With $I_f = 5.90$ amp and $I_s = I_a = 400$ amp, the main-field mmf in equivalent shunt-field amperes is

$$5.90 + \frac{1.5}{1,000} \times 400 = 6.50$$

From Fig. 5-12, the corresponding value of E_a at 1,200 rpm would be 273 volts.   Accordingly, the speed is now

$$n = \frac{240}{273} \times 1,200 = 1,055 \text{ rpm}$$

The power output is the same as in (a).   The speed-load curve is now drooping.

EXAMPLE 5-4

To limit the starting current to the value which the motor can commutate successfully, all except very small dc motors are started with external resistance in series with their armatures.   This resistance is cut out either manually or automatically as the motor comes up to speed.   In Fig. 5-13, for example, the contactors $1A$, $2A$, and $3A$ cut out successive steps R_1, R_2, and R_3 of the starting resistor.

Consider that a motor is to be started with normal field flux. Armature reaction and armature inductance are to be ignored.   During starting, the armature current and hence the electromagnetic torque are not to exceed twice the rated values, and a step of the starting resistor is to be cut out whenever the armature current drops to its rated value.   Except in part f, computations are to be made in the per-unit system with magnitudes expressed as fractions of base values.   (Base voltage equals rated line voltage, base armature current equals full-load armature current, and base resistance equals the ratio of base voltage to base current.)

$a.$ What is the minimum per-unit value of armature resistance which will permit these conditions to be met by a three-step starting resistor?

$b.$ Above what per-unit value of armature resistance will a two-step resistor suffice?

$c.$ For the armature resistance of (a), what are the per-unit resistance values R_1, R_2, and R_3 of the starting resistor?

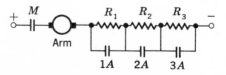

Fig. 5-13.   Starting resistors and accelerating contactors for a dc motor.

d. For a motor with the armature resistance of (a), the contactors are to be closed by voltage-sensitive relays connected across the armature (called the *counter-emf method*). At what fractions of rated line voltage should the contactors close?

e. For a motor with the armature resistance of (a), sketch approximate curves of armature current, electromagnetic torque, and speed during the starting process, and label the ordinates with the appropriate per-unit values at significant instants of time.

f. For a 10-hp 230-volt 500-rpm dc shunt motor having a full-load armature current of 37 amp and fulfilling the conditions of (a), list numerical values in their usual units for armature resistance, the results of (c) and (d), and the ordinate labelings of (e).

Solution

a. In order that the armature current not exceed 2.00 per unit at the instant main contactor M closes,

$$R_1 + R_2 + R_3 + R_a = \frac{V_t}{I_a} = \frac{1.00}{2.00} = 0.50$$

When the current has dropped to 1.00 per unit,

$$E_{a1} = V_t - I_a(R_1 + R_2 + R_3 + R_a) = 1.00 - 1.00 \times 0.50$$
$$= 0.50$$

At the instant that accelerating contactor $1A$ closes, short-circuiting R_1, the counter emf has attained this numerical value. Then, in order that the allowable armature current shall not be exceeded,

$$R_2 + R_3 + R_a = \frac{V_t - E_{a1}}{I_a} = \frac{1.00 - 0.50}{2.00} = 0.25$$

When the current has again dropped to 1.00 per unit,

$$E_{a2} = V_t - I_a(R_2 + R_3 + R_a) = 1.00 - 1.00 \times 0.25 = 0.75$$

Repetition of this procedure for the closing of accelerating contactors $2A$ and $3A$ yields the following results:

$$R_3 + R_a = 0.125$$
$$E_{a3} = 0.875$$
$$R_a = 0.0625$$

and   Final E_a at full load $= 0.938$

The desired minimum per-unit value of R_a is therefore 0.0625, because a lower value will allow the armature current to exceed twice the rated value when contactor $3A$ is closed.

b. If a two-step resistor is to suffice, R_3 must be zero.   Since, from a,

$$R_3 + R_a = 0.125$$

it follows that a three-step resistor is not required when R_a is equal to or greater than 0.125.

Under the specified starting conditions, a three-step resistor is appropriate for motors whose armature-circuit resistances are between 0.0625 and 0.125 per unit.   For general-purpose continuously rated shunt motors, these values correspond to the lower integral-horsepower sizes.   On the average, motor sizes up to about 10 hp will conform to these requirements, although the size limit will be lower for high-speed motors and higher for slow-speed motors.   For larger motors, either additional steps must be provided, or the limit on current and torque peaks must be relaxed. The results of this analysis are conservative because the armature resistance under transient conditions is higher than the static value.

c. From the relations in part a, the per-unit starting resistances are

$$R_3 = 0.125 - 0.0625 = 0.0625$$

$$R_2 = 0.25 - 0.0625 - 0.0625 = 0.125$$

and $$R_1 = 0.50 - 0.0625 - 0.0625 - 0.125 = 0.25$$

d. Just before contactor $1A$ closes,

$$V_{ta1} = E_{a1} + I_a R_a = 0.50 + 1.00 \times 0.0625 = 0.563$$

In like manner,

$$V_{ta2} = 0.75 + 1.00 \times 0.625 = 0.813$$

and $$V_{ta3} = 0.875 + 1.00 \times 0.0625 = 0.938$$

Acceleration contactors $1A$, $2A$, and $3A$, respectively, should pick up at these fractions of rated line voltage.

e. Consider that main contactor M closes at $t = 0$ and that accelerating contactors $1A$, $2A$, and $3A$ close, respectively, at times t_1, t_2, and t_3. These values of time are not known (when armature and load inertias and torque-speed curve of the load are given, values of time can be computed by the methods of Chap. 9), so that only the general shapes of the current, electromagnetic torque, and speed curves can be given. They are indicated in Fig. 5-14.

The labeling of the speed curve follows from the fact that a counter emf $E_a = 0.938$ corresponds to rated speed at rated load and hence to unity speed. Other speeds are in proportion to E_a;

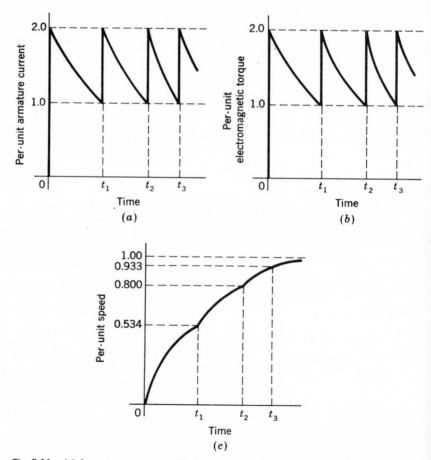

Fig. 5-14. (a) Armature current; (b) electromagnetic torque; and (c) speed during starting of a dc motor.

thus, at t_1, t_2, and t_3, respectively,

$$n_1 = \frac{0.50}{0.938} \times 1.00 = 0.534$$

$$n_2 = \frac{0.75}{0.938} \times 1.00 = 0.800$$

and
$$n_3 = \frac{0.875}{0.938} \times 1.00 = 0.933$$

f. Base quantities for this motor are as follows:

Base voltage = 230 volts

Base armature current = 37 amp

Base armature-circuit resistance = $^{230}\!/_{37}$ = 6.22 ohms

Base speed = 500 rpm

Base electromagnetic torque

$$= \frac{60}{2\pi n} E_a I_a$$

$$= \frac{60}{2\pi \times 500} (230 - 37 \times 0.0625 \times 6.22) \times 37$$

$$= 152 \text{ newton-meters}$$

Note that rated electromagnetic torque will be greater than rated shaft torque because of rotational and stray load losses.

The motor armature resistance is

$$R_a = 0.0625 \times 6.22 = 0.389 \text{ ohm}$$

Values for the other quantities desired are listed in Table 5-1.

TABLE 5-1
ABSOLUTE VALUES FOR EXAMPLE 5-4*f*

Part *c*	Part *d*	Part *e*, scales of Fig. 5-14
$R_1 = 1.56$ ohms	Relay $1A$: 129 volts	1.0 armature current = 37 amp
$R_2 = 0.778$ ohm	Relay $2A$: 187 volts	1.0 electromagnetic torque = 152 newton-m
$R_3 = 0.389$ ohm	Relay $3A$: 216 volts	1.0 speed = 500 rpm

MOTOR SPEED CONTROL

Direct-current machines are in general much more adaptable to adjustable speed service than are the ac machines associated with a constant-
speed rotating field.   Indeed, the ready susceptibility of dc motors to
adjustment of their operating speed over wide ranges and by a variety
of methods is one of the important reasons for the strong competitive
position of dc machinery in modern industrial applications.

The three most common speed-control methods are adjustment of the
flux, usually by means of a shunt-field rheostat, adjustment of the resistance associated with the armature circuit, and adjustment of the armature terminal voltage.

Shunt-field-rheostat control is the most commonly used of the three
methods and forms one of the outstanding advantages of shunt motors.
The method is, of course, also applicable to compound motors.   Adjustment of field current and hence the flux and speed by adjustment of the
shunt-field circuit resistance is accomplished simply, inexpensively, and
without much change in motor losses.

The lowest speed obtainable is that corresponding to full field or zero
resistance in the field rheostat; the highest speed is limited electrically
by the effects of armature reaction under weak-field conditions in causing
motor instability or poor commutation.   Addition of a stabilizing winding increases the speed range appreciably, and the alternative addition of
a compensating winding still further increases the range.   With a compensating winding, the overall range may be as high as 8 to 1 for a small
integral-horsepower motor.   Economic factors limit the feasible range
for very large motors to about 2 to 1, however, with 4 to 1 often regarded
as the limit for the average-sized motor.

To examine approximately the limitations on the allowable continuous motor output as the speed is changed, neglect the influence of changing ventilation and changing rotational losses on the allowable output.
The maximum armature current I_a is then fixed at the nameplate value in
order that the motor shall not overheat, and the counter emf E_a remains
constant because the effect of a speed change is compensated by the
change of flux causing it.   The $E_a I_a$ product and hence the allowable
motor output then remain substantially constant over the speed range.
The dc motor with shunt-field-rheostat speed control is accordingly
referred to as a *constant-horsepower drive*.   Torque, on the other hand,
varies directly with flux and therefore has its highest allowable value at
the lowest speed.   Field-rheostat control is thus best suited to drives

requiring increased torque at low speeds.   When a motor so controlled is used with a load requiring constant torque over the speed range, the rating and size of the machine are determined by the product of the torque and the highest speed.   Such a drive is inherently oversize at the lower speeds, which is the principal economic factor limiting the practical speed range of large motors.

Armature-circuit-resistance control consists in obtaining reduced speeds by the insertion of external series resistance in the armature circuit. It may be used with series, shunt, and compound motors; for the last two types, the series resistor must be connected between the shunt field and the armature, not between the line and the motor.   It is the common method of speed control for series motors and is generally analogous in action to wound-rotor induction-motor control by series rotor resistance.

For a fixed value of series armature resistance, the speed will vary widely with load, since the speed depends on the voltage drop in this resistance and hence on the armature current demanded by the load. For example, a 1,200-rpm shunt motor whose speed under load is reduced to 750 rpm by series armature resistance will return to almost 1,200-rpm operation when the load is thrown off, because the effect of the no-load current in the series resistance is insignificant.   The disadvantage of poor speed regulation may not be important in a series motor, which is used only where varying speed service is required or satisfactory anyway.

Also, the power loss in the external resistor is large, especially when the speed is greatly reduced.   In fact, for a constant-torque load, the power input to the motor plus resistor remains constant, while the power output to the load decreases in proportion to the speed.   Operating costs are therefore comparatively high for long-time running at reduced speeds. Because of its low initial cost, however, the series-resistance method (or the variation of it discussed in the next paragraph) will often be attractive economically for short-time or intermittent slowdowns.   Unlike shunt-field control, armature-resistance control offers a *constant-torque drive* because both flux and, to a first approximation, allowable armature current remain constant as speed changes.

A variation of this control scheme is given by the *shunted-armature method*, which may be applied to a series motor as in Fig. 5-15a or a shunt motor as in Fig. 5-15b.   In effect, resistors R_1 and R_2 act as a voltage divider applying a reduced voltage to the armature.   Greater flexibility is possible because two resistors may now be adjusted to provide the desired performance.   For series motors, the no-load speed may be adjusted to a finite, reasonable value, and the scheme is therefore applicable to the production of slow speeds at light loads.   For shunt motors,

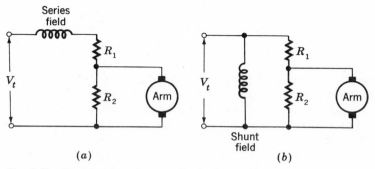

Fig. 5-15. Shunted-armature method of speed control applied to (a) series motor and (b) shunt motor.

the speed regulation in the low-speed range is appreciably improved because the no-load speed is definitely lower than the value with no controlling resistors.

Armature-terminal-voltage control utilizes the fact that a change in the armature terminal voltage of a shunt motor is accompanied in the steady state by a substantially equal change in the counter emf and, with constant motor flux, a consequent proportional change in motor speed. Usually, the power available is constant-voltage alternating current, so that auxiliary equipment in the form of a rectifier or a motor-generator set is required to provide the controlled armature voltage for the motor. The development of solid-state controlled rectifiers capable of handling many kilowatts has opened up a whole new field of applications where precise control of motor speed is required. These applications are of such importance that they are given special treatment in Chap. 8.

The conventional scheme, also called the *Ward Leonard system*, shown schematically in Fig. 5-16, requires an individual motor-generator set to

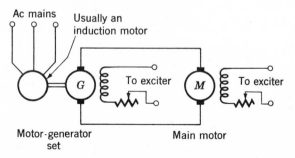

Fig. 5-16. Adjustable-armature-voltage, or Ward Leonard, method of speed control.

supply power to the armature of the motor whose speed is to be controlled.   Control of the armature voltage of the main motor M is obtained
by field-rheostat adjustment in the separately excited generator G, permitting close control of speed over a wide range.   An obvious disadvantage is the initial investment in three full-size machines in contrast to
that in a single motor.   The speed-control equipment is located in low-
power field circuits, however, rather than in the main power circuits.
The smoothness and versatility of control are such that the method or
one of its variants is often applied.

Frequently the control of generator voltage is combined with motor-
field control, as indicated by the rheostat in the field of motor M in
Fig. 5-16, in order to achieve the widest possible speed range.   With such
dual control, *base speed* may be defined as the normal-armature-voltage
full-field speed of the motor.   Speeds above base speed are obtained by
motor-field control; speeds below base speed are obtained by armature-
voltage control.   As discussed in connection with shunt-field-rheostat
control, the range above base speed is that of a constant-horsepower drive.
The range below base speed is that of a constant-torque drive because,
as in armature-resistance control, the flux and the allowable armature
current remain approximately constant.   The overall output limitations
are therefore as shown in Fig. 5-17a for approximate allowable torque and
Fig. 5-17b for approximate allowable horsepower.   The constant-torque
characteristic is well suited to many applications in the machine-tool
industry, where many loads consist largely in overcoming the friction of
moving parts and hence have essentially constant-torque requirements.

The speed regulation and the limitations on the speed range above
base speed are those already presented with reference to shunt-field-

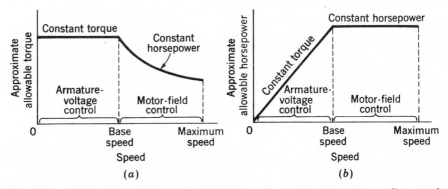

Fig. 5-17. (*a*) Torque and (*b*) power limitations of combined armature-voltage and
field-rheostat methods of speed control.

rheostat control; the maximum speed thus does not ordinarily exceed four times base speed, and preferably not twice base speed. In the region of armature-voltage control, the principal limitation in the basic system is residual magnetism in the generator, although considerations of speed regulation may also be determining. For conventional machines, the lower limit for reliable and stable operation is about 0.1 of base speed, corresponding to a total maximum-to-minimum range not exceeding 40 to 1. With armature reaction ignored, the decrease in speed from no-load to full-load torque is caused entirely by the full-load armature-resistance drop in the dc generator and motor. This full-load armature-resistance drop is constant over the voltage-control range, since full-load torque and hence full-load current are usually regarded as constant in that range. When measured in rpm, therefore, the speed decrease from no-load to full-load torque is a constant, independent of the no-load speed. The torque-speed curves accordingly are closely approximated by a series of parallel straight lines for the various generator-field adjustments. Now a speed decrease of, say, 40 rpm from a no-load speed of 1,200 rpm is often of little importance; a decrease of 40 rpm from a no-load speed of 120 rpm, however, may at times be of critical importance and require corrective steps in the layout of the system.

Many detailed variations of the basic Ward Leonard system have been devised to overcome these limitations when precise speed control over a wide range is required and also to utilize fully the versatility of dc machines in providing a variety of inherent performance characteristics. Flat compounding of the dc generator, for example, is one simple method of improving the natural speed regulation. Control-type generators such as the amplidyne, discussed in Chap. 9, may be used either as the main generator G (Fig. 5-16) in low-power systems or to supply and control the excitation for this generator in heavy-power systems. Greater flexibility is thereby provided, the residual-magnetism limitation may be alleviated, and excellent speed regulation over a range as high as 120 to 1 may be obtained. As illustrated in Chap. 9, the basic method is frequently used in closed-cycle control systems.

When a motor armature is supplied from its own individual dc generator, the shape of the speed-torque curve of the system may also be controlled by incorporating special features in the generator. An example is the use of a *three-field generator* to supply the armature of a shunt motor. This special-purpose generator has a separately excited winding, a shunt-connected winding, and a differentially connected series winding, a combination which permits the placing of an adjustable limit on the torque output of the system and in particular on the torque and current when the motor is stalled. In other systems, the torque and current while accelerating may be limited by the appropriate controls. In effect, the degrees of freedom are such that, within rather wide limits, a tailor-made drive may be devised.

5-8

RÉSUMÉ. DC MACHINE APPLICATIONS

Discussion of dc machine applications involves recapitulation of the highlights of the machine's performance features, together with economic and technical evaluation of the machine's position with respect to competing energy-conversion devices. For dc machines in general, the outstanding advantage lies in their flexibility and versatility. The principal disadvantage is likely to be the initial investment concerned. Yet the advantages of dc motors are such that they retain a strong competitive position for industrial applications.

Direct-current generators are the obvious answer to the problem of converting mechanical energy to electrical energy in dc form. When the consumer of electrical energy is geographically removed from the site of energy conversion by any appreciable distance, however, the advantages of ac generation, voltage transformation, and transmission are such that energy conversion and transmission in ac form are almost always adopted with ac-to-dc transformation taking place at or near the consumer. For ac-to-dc transformation, the dc generator as part of an ac-to-dc motor-generator set must compete with mercury-arc rectifiers, ignitrons, and semiconductor rectifiers. When large-power rectification from ac to constant-voltage dc form is involved, the electronic methods usually possess determining economic advantages. The principal applications of dc generators, therefore, are to cases where ability freely to control the output voltage in a prescribed manner is necessary or where the primary energy conversion occurs very near the point of consumption.

Among dc generators themselves, separately excited and cumulatively compounded self-excited machines are the most common. Separately excited generators have the advantage of permitting a wide range of output voltages, whereas self-excited machines may produce unstable voltages in the lower ranges where the field-resistance line becomes essentially tangent to the magnetization curve. Cumulatively compounded generators may produce a substantially flat voltage characteristic or one which rises with load, whereas shunt or separately excited generators (assuming no series field in the latter, which, of course, is not at all a practical restriction) produce a drooping voltage characteristic unless external regulating means are added. So far as the control potentialities of dc generators are concerned, the control-type generators (amplidynes and similar machines) discussed in Chap. 9 represent the results of a fuller exploration of the inherent possibilities.

Among dc motors, the outstanding characteristics of each type are as follows: The series motor operates with a decidedly drooping speed as load is added, the no-load speed usually being prohibitively high; the torque is proportional to almost the square of the current at low saturations and to some power between 1 and 2 as saturation increases. The shunt motor at constant field current operates at a slightly drooping but almost constant speed as load is added, the torque being almost proportional to armature current; equally important, however, is the fact that its speed may be controlled over wide ranges by shunt-field control or armature-voltage control, or a combination of both. Depending on the relative strengths of shunt and series field, the cumulatively compounded motor is intermediate between the other two and may be given essentially the advantages of one or the other.

By virtue of its ability to handle heavy torque overloads while cushioning the associated power overload with a speed drop, and by virtue of its ability to withstand severe starting duties, the series motor is best adapted to hoist, crane, and traction-type loads. Its ability is almost unrivaled in this respect. Speed changes are usually achieved by armature-resistance control. In some instances, the wound-rotor induction motor with rotor-resistance control competes with the series motor, but the principal argument concerns the availability and economics of a dc power supply rather than inherent motor characteristics.

Compound motors with a heavy series field have performance features approaching those of series motors except that the shunt field limits the no-load speed to safe values; the general remarks for series motors therefore apply. Compound motors with lighter series windings not infrequently find competition from squirrel-cage induction motors with high-resistance rotors—so-called *high-slip* motors (referred to in Chap. 7 as Class D induction motors). Both motors provide a definitely drooping speed-load characteristic such as is desirable, for example, when flywheels are used as load equalizers to smooth out intermittent load peaks. Complete economic comparison of the two competing types must reflect both the usually higher initial cost of a compound-motor installation and the usually higher cost of losses in the high-slip induction motor.

Because of the comparative simplicity, cheapness, and ruggedness of the squirrel-cage induction motor, the shunt motor is not in a favorable competitive position for constant-speed service except at low speeds, where it becomes difficult and expensive to build high-performance induction motors with the requisite number of poles. The comparison

at these low speeds is often likely to be between synchronous and dc
motors.   The outstanding feature of the shunt motor is its adaptability
to adjustable-speed service by means of armature-resistance control for
speeds below the full-field speed, field-rheostat control for speeds above
the full-field speed, and armature-voltage, or Ward Leonard, control for
speeds below (and, at times, somewhat above) the normal-voltage full-
field speed.   The combination of armature-voltage control and shunt-
field control, together with the possibility of additional field windings in
either the motor or the associated generator to provide desirable inherent
characteristics, gives the dc drives an enviable degree of flexibility.   The
solid-state controlled rectifiers of Chap. 8 and the control-type dc generators of Chap. 9 definitely reinforce the competitive position of dc
machines where complete control of operation is important.

It should be emphasized that the choice of equipment for a significant
engineering application to adjustable-speed drives is rarely a cut-and-
dried matter or one to be decided from a mere verbal list of advantages
and disadvantages.   In general, specific, quantitative, economic, and
technical comparison of all possibilities should be undertaken.   Consideration must be given to the transient- and dynamic-response details
of Chap. 9.   Local conditions and the characteristics of the driven equipment (e.g., constant-horsepower, constant-torque, and variable-horse-
power variable-torque requirements) invariably play an important role.
One should also remember that comparative studies of motor cost and
characteristics are based on the combination of motor and control equipment, for the latter plays an important part in determining motor performance under specific conditions and represents a by no means negligible portion of the total initial cost.   Control equipment coupled with
susceptibility to control makes dc machines the versatile energy-conversion devices that they are.

PROBLEMS

5-1. A cumulatively compounded generator with interpoles and with
its brushes on neutral is to be used as a compound motor.

If no changes are made in the internal connections, will the motor be
cumulatively or differentially compounded?   Will the polarity of the
interpoles be correct?   Will the direction of rotation be the same as or
opposite to the direction in which it was driven as a generator?

5-2. A self-excited dc machine with interpoles is adjusted for proper
operation as an overcompounded generator.   The machine is shut down,

the connections to the shunt field are reversed, and the machine is then started with the direction of rotation reversed. The machine builds up normal terminal voltage. Answer the following questions, and give a brief explanation:

Is the terminal-voltage polarity the same as before? Is the machine still cumulatively compounded? Do the interpoles have the proper polarity for good commutation?

5-3. A 10-kw 230-volt 1,150-rpm shunt generator is driven by a prime mover whose speed is 1,195 rpm when the generator delivers no load. The speed falls to 1,150 rpm when the generator delivers 10 kw and may be assumed to decrease in proportion to the generator output. The generator is to be changed into a short-shunt compound generator by equipping it with a series field which will cause its voltage to rise from 230 volts at no load to 250 volts for a load of 43.5 amp. It is estimated that the series field will have a resistance of 0.09 ohm. The armature resistance (including brushes) is 0.26 ohm. The shunt-field winding has 1,800 turns per pole.

In order to determine the necessary series-field turns, the machine is run as a separately excited generator and the following load data obtained: armature terminal voltage, 254 volts; armature current, 44.7 amp; field current, 1.95 amp; speed, 1,145 rpm.

The magnetization curve at 1,195 rpm is as follows:

E_a, volts	230	240	250	260	270
I_f, amp	1.05	1.13	1.26	1.46	1.67

a. Determine the necessary number of series-field turns per pole.

b. Determine the armature reaction in equivalent demagnetizing ampere-turns per pole for $I_a = 44.7$ amp.

5-4. A small, lightweight dc shunt generator for use in aircraft has a rating of 9 kw, 30 volts, 300 amp. It is driven by one of the main engines of the airplane through an auxiliary power shaft. The generator speed is proportional to the main-engine speed and may have any value from 4,500 rpm to 8,000 rpm. The terminal voltage of the generator is held constant at 30 volts for all speeds and loads by means of a voltage regulator which automatically adjusts a carbon-pile field rheostat whose minimum resistance is 0.75 ohm. The resistance of the shunt-field, commutating, and armature (including brushes) windings are, respectively, 2.50, 0.0040, and 0.0120 ohms.

Data for the magnetization curve at 4,550 rpm are:

I_f, amp	0	2.0	4.0	5.0	6.0	8.0	11.7
E_a, volts	1.0	18.0	30.5	33.6	35.5	38.0	40.5

In a load test at 4,550 rpm, the field current required to maintain rated terminal voltage at rated load is 7.00 amp.

Determine the following characteristics of this generator:

a. Maximum resistance required in the field rheostat

b. Maximum power dissipated in the field rheostat

c. Demagnetizing effect of armature reaction at rated load and 4,550 rpm, expressed in terms of equivalent shunt-field current

5-5. Assume that the demagnetizing effect of armature reaction in the aircraft generator of Prob. 5-4 is equivalent to a demagnetizing mmf proportional to the armature current.

a. Plot the curve of terminal voltage as a function of line current for minimum field-rheostat resistance and 4,550 rpm.

b. When the airplane is on the ground and the engines are idling, the generator speed may be below 4,500 rpm. Plot a curve of terminal voltage as a function of speed with minimum field-rheostat resistance and a constant line current of 300 amp covering the subnormal speed range from 4,550 to 3,500 rpm. Estimate the minimum speed at which the generator is capable of delivering 300 amp.

5-6. A dc series motor operates at 750 rpm with a line current of 80 amp from the 230-volt mains. Its armature-circuit resistance is 0.14 ohm, and its field resistance is 0.11 ohm.

Assuming that the flux corresponding to a current of 20 amp is 40 percent of that corresponding to a current of 80 amp, find the motor speed at a line current of 20 amp at 230 volts.

5-7. A certain series motor is so designed that flux densities in the iron part of the magnetic circuit are low enough to result in a linear relationship between field flux and field current throughout the normal range of operation. The rating of this motor is 50 hp, 190 amp, 220 volts, 600 rpm. Losses at full load in percentage of motor input are:

Armature copper loss (including brush loss) = 3.7 percent

Field copper loss = 3.2 percent

Rotational loss = 2.8 percent

Rotational loss may be assumed constant; armature reaction and stray load loss may be neglected.

When this motor is operating from a 220-volt supply with a current of half the rated value, what will be:

a. The speed in rpm?

b. The shaft power output in horsepower?

5-8. A 150-hp 600-volt 600-rpm dc series-wound railway motor has a combined field and armature resistance (including brushes) of 0.155 ohm. The full-load current at rated voltage and speed is 206 amp. The magnetization curve at 400 rpm is as follows:

Induced volts	375	400	425	450	475
Field amp	188	216	250	290	333

Determine the internal starting torque when the starting current is limited to 350 amp. Assume armature reaction to be equivalent to a demagnetizing mmf which varies as the square of the current.

5-9. Following are the nameplate data of a certain dc motor: 230 volts, 75.7 amp, 20 hp, 900 rpm full-load, 50°C 1-hr rating, series-wound. The field winding has 33 turns per pole and a hot resistance of 0.06 ohm; the hot armature-circuit resistance is 0.09 ohm (including brushes). Points on the magnetization curve at 900 rpm are as follows:

Amp-turns per pole	500	1,000	1,500	2,000	2,500	3,000
Generated voltage	95	150	188	212	229	243

To determine the fitness of this motor for driving a skip hoist, points on the motor speed-load curve are to be computed.

a. For currents equal to $\frac{1}{3}$, $\frac{2}{3}$, 1, and $\frac{4}{3}$ of the nameplate value, compute the speed of the motor. Neglect armature reaction.

b. For the same currents, compute the shaft-horsepower outputs. For this purpose, consider rotational losses to remain constant at the value determined by nameplate conditions.

c. Compute the pulley torques in pound-feet corresponding to the values in b. Arrange the results of parts a, b, and c in tabular form for convenience in checking.

d. The maximum safe speed for the motor is 250 percent of full-load speed. What is the motor power input at this point?

e. What value of resistance connected in series with the motor will enable the production of full-load electromagnetic torque at a speed of 500 rpm?

5-10. A 10-hp 230-volt shunt motor has an armature-circuit resistance of 0.30 ohm and a field resistance of 170 ohms. At no load and rated voltage, the speed is 1,200 rpm, and the armature current is 2.7 amp. At full load and rated voltage, the line current is 38.4 amp, and, because of armature reaction, the flux is 4 percent less than its no-load value.

What is the full-load speed?

5-11. A 36-in. axial-flow disk pressure fan is rated to deliver 27,120 ft³ of air per minute against a static pressure of $\frac{1}{2}$ in. of water when rotating at a speed of 1,165 rpm. This fan has the following speed-load characteristics:

Speed, rpm	700	800	900	1,000	1,100	1,200
Input, hp	2.9	3.9	5.2	6.7	8.6	11.1

It is proposed to drive the fan by a 10-hp 230-volt 37.5 amp 4-pole dc shunt motor. The motor has an armature winding with two parallel paths and $Z = 666$ active inductors. Armature-circuit resistance is 0.267 ohm. The armature flux per pole is $\Phi = 10^6$ lines; armature reaction is negligible. No-load rotational losses (considered constant) are estimated at 600 watts, a typical value for such a motor.

Determine the shaft-horsepower output and the operating speed of the motor when it is connected to the fan load.

5-12. A 100-hp 250-volt dc shunt motor has the magnetization curve of Fig. 5-12 and an armature resistance (including brushes) of 0.025 ohm. There are 1,000 turns per pole on the shunt field.

When the shunt-field rheostat is set for a motor speed of 1,200 rpm at no load, the armature current is 8.0 amp. How many series-field turns per pole must be added if the speed is to be 950 rpm for a load requiring an armature current of 250 amp? Neglect the added resistance of the series field.

5-13. A shunt motor operating from a 230-volt line draws a full-load armature current of 38.5 amp and runs at a speed of 1,200 rpm at both no load and full load. The following data are available on this motor:

Armature-circuit resistance (including brushes) = 0.21 ohm

Shunt-field turns per pole = 2,000 turns

Magnetization curve taken as a generator at no load and 1,200 rpm

E_a, volts	180	200	220	240	250
I_f, amp	0.74	0.86	1.10	1.45	1.70

a. Determine the shunt-field current of this motor at no load and 1,200 rpm when connected to a 230-volt line. Assume negligible armature-circuit resistance drop and armature reaction at no load.

b. Determine the effective armature reaction at full load in ampere-turns per pole.

c. How many series-field turns should be added to make this machine into a long-shunt cumulatively compounded motor whose speed will be 1,090 rpm when the armature current is 38.5 amp and the applied voltage is 230 volts? The series field will have a resistance of 0.052 ohm.

d. If a series-field winding having 25 turns per pole and a resistance of 0.052 ohm is installed, determine the speed when the armature current is 38.5 amp and the applied voltage is 230 volts.

5-14. A 10-hp 230-volt shunt motor has 2,000 shunt-field turns per pole, an armature resistance (including brushes) of 0.20 ohm, and a commutating-field resistance of 0.041 ohm. The shunt-field resistance (exclusive of rheostat) is 235 ohms. When the motor is operated at no load with rated terminal voltage and varying field resistance, the following data are taken:

Speed, rpm	1,110	1,130	1,160	1,200	1,240
I_f, amp	0.932	0.880	0.830	0.770	0.725

The no-load armature current is negligible. When the motor is operated at full load and rated terminal voltage, the armature current is 37.5 amp, the field current is 0.770 amp, and the speed is 1,180 rpm.

a. Calculate the full-load armature reaction in equivalent demagnetizing ampere-turns per pole.

b. Calculate the full-load electromagnetic torque.

c. What starting torque will the motor exert with maximum field current if the starting armature current is limited to 75 amp? The armature reaction under these conditions is 160 amp-turns per pole.

d. Design a series field to give a full-load speed of 1,100 rpm when the no-load speed is 1,200 rpm.

5-15. When operated at rated voltage, a 230-volt shunt motor runs at 1,600 rpm at full load and also at no load. The full-load armature current is 50.0 amp. The shunt-field winding has 1,000 turns per pole. The resistance of the armature circuit (including brushes and interpoles) is 0.20 ohm. The magnetization curve at 1,600 rpm is:

E_a, volts	200	210	220	230	240	250
I_f, amp	0.80	0.88	0.97	1.10	1.22	1.43

a. Compute the demagnetizing effect of armature reaction at full load, in ampere-turns per pole.

b. A long-shunt cumulative series-field winding having 5 turns per pole and a resistance of 0.05 ohm is added to the machine. Compute the speed at full-load current and rated voltage, with the same shunt-field circuit resistance as in (a).

c. With the series-field winding of (b) installed, compute the internal starting torque in newton-meters if the starting armature current is limited to 100 amp and the shunt-field current has its normal value. Assume that the corresponding demagnetizing effect of armature reaction is 260 amp-turns per pole.

5-16. A weak shunt-field winding is to be added to a 50-hp 230-volt 600-rpm series hoist motor for the purpose of preventing excessive speeds at very light loads. Its resistance will be 230 ohms. The combined resistance of the interpole and armature winding (including brushes) is 0.055 ohm. The series-field winding has 24 turns per pole with a total resistance of 0.021 ohm.

In order to determine its design, the following test data were obtained before the shunt field was installed:

Load test as a series motor (output not measured):

$$V_t = 230 \text{ volts} \qquad I_a = 184 \text{ amp} \qquad n = 600 \text{ rpm}$$

No-load test with series field separately excited:

Voltage applied to armature, volts	Speed, rpm	Armature current, amp	Series-field current, amp
230	1,500	10.0	60
230	1,200	9.2	74
230	900	8.0	103
215	700	7.7	135
215	600	7.5	175
215	550	7.2	201
215	525	7.1	225
215	500	7.0	264

  a. Determine the number of shunt-field turns per pole if the no-load speed at rated voltage is to be 1,500 rpm. The armature, series-field, and interpole winding resistance drops are negligible at no load.

  b. Determine the speed after installation of the shunt field when the motor is operated at rated voltage with a load which results in a line current of 185 amp. Assume that the demagnetizing mmf of armature reaction is unchanged by addition of the shunt field.

5-17. *a.* A 230-volt dc shunt-wound motor is used as an adjustable-speed drive over the range from 0 to 1,000 rpm. Speeds from 0 to 500 rpm are obtained by adjusting the armature terminal voltage from 0 to 230 volts with the field current kept constant. Speeds from 500 to 1,000 rpm are obtained by decreasing the field current with the armature terminal voltage maintained at 230 volts. Over the entire speed range, the torque required by the load remains constant.

Show the general form of the curve of armature current vs. speed over the entire range. Ignore machine losses and armature-reaction effects.

b. Suppose that, instead of keeping the load torque constant, the armature current is not to exceed a specified value. Show the general form of the curve of allowable load torque vs. speed. Conditions otherwise are as in a.

5-18. a. Two adjustable-speed dc shunt motors have maximum speeds of 1,650 rpm and minimum speeds of 450 rpm. Speed adjustment is obtained by field-rheostat control. Motor A drives a load requiring constant horsepower over the speed range; motor B drives one requiring constant torque. All losses and armature reaction may be neglected.

1 If the horsepower outputs are equal at 1,650 rpm and the armature currents are each 100 amp, what will be the armature currents at 450 rpm?

2 If the horsepower outputs are equal at 450 rpm and the armature currents are each 100 amp, what will be the armature currents at 1,650 rpm?

b. Answer part a for speed adjustment by armature-voltage control with conditions otherwise the same.

5-19. A 230-volt dc shunt motor has an armature-circuit resistance of 0.1 ohm. This motor operates on the 230-volt mains and takes an armature current of 100 amp. An external resistance of 1.0 ohm is now inserted in series with the armature, and the electromagnetic torque and field-rheostat setting are unchanged.

a. Give the percentage change in the total current taken by the motor from the mains.

b. Give the percentage change in the speed of the motor, and state whether this will be an increase or a decrease.

5-20. A punch press is found to operate satisfactorily when driven by a 10-hp 230-volt compound motor having a no-load speed of 1,800 rpm and a full-load speed of 1,200 rpm when the torque is 43.8 lb-ft. The motor is temporarily out of service, and the only available replacement is a compound motor with the following characteristics:

Rating = 230 volts, 12.5 hp
No-load current = 4 amp
No-load speed = 1,820 rpm
Full-load speed = 1,600 rpm
Full-load current = 57.0 amp
Full-load torque = 43.8 lb-ft
Armature-circuit resistance = 0.2 ohm
Shunt-field current = 1.6 amp

It is desired to use this motor as an emergency drive for the press without making any change in its field windings.

a. How can it be made to have the desired speed regulation?
b. Draw the pertinent circuit diagram, and give complete specifications of the necessary apparatus.

5-21. Consider a dc shunt motor connected to constant-voltage mains and driving a load requiring constant electromagnetic torque. Show that, if $E_a > 0.5V_t$ (the normal situation), increasing the resultant air-gap flux decreases the speed, whereas, if $E_a < 0.5V_t$ (as might be brought about by inserting a relatively high resistance in series with the armature), increasing the resultant air-gap flux increases the speed.

5-22. Two identical 5-hp 230-volt 17-amp dc shunt machines are to be used as the generator and motor, respectively, in a Ward Leonard system. The generator is driven by a synchronous motor whose speed is constant at 1,200 rpm. The armature-circuit resistance of each machine is 0.47 ohm (including brushes). Armature reaction is negligible. Data for the magnetization curve of each machine at 1,200 rpm are as follows:

I_f, amp	0.2	0.4	0.6	0.8	1.0	1.2
E_a, volts	108	183	230	254	267	276

a. Compute the maximum and minimum values of generator-field current needed to give the motor a speed range from 300 to 1,500 rpm at full-load armature current (17.0 amp), with the motor-field current held constant at 0.50 amp.

b. Compute the speed regulation of the motor for the conditions of maximum speed and minimum speed found in part *a*.

c. Compute the maximum motor speed obtainable at full-load armature current if the motor-field current is reduced to 0.20 amp and the generator-field current is not allowed to exceed 1.10 amp.

5-23. One of the commonest industrial applications of dc series motors is for crane and hoist drives. This problem relates to the computation of selected motor performance characteristics for such a drive. The specific motor concerned is a series-wound 230-volt totally enclosed motor having a ½-hr crane rating of 65 hp with a 75°C temperature rise. The performance characteristics of the motor alone on 230 volts as taken from the manufacturer's catalog are listed in Table 5-2.

TABLE 5-2

Line current, amp	Shaft torque, lb-ft	Speed, rpm
50	80	940
100	210	630
150	380	530
200	545	475
250	730	438
300	910	407
350	1,105	385
400	1,265	370

The resistance of the armature (including brushes) plus commutating field is 0.090 ohm, and that of the series-field winding is 0.040 ohm. Armature reaction should be ignored. The handling of rotational and stray load losses is to be discussed in part *i* of the problem.

The motor is to be connected as in Fig. 5-18a for hoisting and Fig. 5-18b for lowering. The former connection is simply one for series-resistance control. The latter connection is one for lowering by dynamic

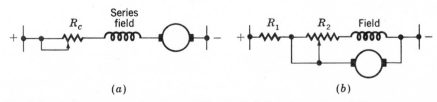

Fig. 5-18. Series crane motor, Prob. 5-23. (a) Hoisting connection; (b) lowering connection.

braking with the field reconnected in shunt and having an adjustable resistance in series with it.

A few samples of the torque-speed curves determining the suitability of the motor and control for its particular application are to be plotted. Plot all these curves on the same sheet, torque horizontally and speed vertically, covering about the torque-magnitude range embraced in Table 5-2. Provide for both positive and negative values of speed, corresponding, respectively, to hoisting and lowering; provide also for both positive and negative values of torque, corresponding, respectively, to torque in the direction of raising the load and torque in the direction of lowering the load; thus, use all four quadrants of the conventional rectangular coordinate system.

a. For the hoisting connection, plot torque-speed curves for the control resistor R_c set at 0, 0.65, and 1.30 ohms. If any of these curves extend into the fourth quadrant within the range of torques covered, plot them in that region, and interpret physically what operation there means.

b. Discuss the suitability of these characteristics for the hoisting operation.

c. For the lowering connection, plot a torque-speed curve for $R_1 = 0.65$ ohm and R_2 set at 0.65 ohm. The most important portion of this curve is in the fourth quadrant, but if it extends into the third quadrant, that region should also be plotted and interpreted physically.

d. In (c) what is the lowering speed corresponding to rated torque?

e. How is the speed in (d) affected by decreasing R_2? Why?

f. How is the speed in (d) affected by decreasing R_1? Why?

g. How would the speed of (d) be affected by adding resistance in series with the motor armature? Why?

h. Discuss the suitability of these characteristics for the lowering operation.

i. What assumptions, if any, concerning rotational and stray load losses have you found it necessary to make because of having limited data on the motor? Discuss this point.

5-24. An automatic starter is to be designed for a 15-hp 230-volt shunt motor. The resistance of the armature circuit is 0.162 ohm. When operated at rated voltage and loaded until its armature current is 32 amp, the motor runs at a speed of 1,100 rpm with a field-circuit resistance of 115 ohms. When the motor is delivering rated output, the armature current is 56 amp.

The motor is to be started with a load which requires a torque proportional to speed and which under running conditions requires 15 hp. The field winding is connected across the 230-volt mains, and the resistance in series with the armature is to be adjusted automatically so that during the starting period the armature current does not exceed 200 percent of rated value or fall below rated value. That is, the machine is to start with 200 percent of rated armature current, and as soon as the current falls to rated value, sufficient series resistance is to be cut out to restore current to 200 percent. This process is repeated until all the series resistance has been cut out.

a. What should be the total resistance of the starter?

b. How much resistance should be cut out at each step in the starting operation?

5-25. Figure 5-19 shows schematically the connections of the three-field Ward Leonard system mentioned at the end of Art. 5-7. The separately excited motor M drives the forward motion of the scoop on a very large power shovel used for open-pit strip mining of coal. The motor has a commutating winding C and a separately excited field winding F_m. Its armature is connected to the armature of a generator G having a commutating winding C and 3 field windings: a separately excited control field F_1, a self-excited shunt field F_2, and a differential series field S. The purpose of field S is to limit the armature current if the motor should be stalled. The exciter E supplies a constant voltage of 250 volts to the motor field F_m and to the generator control field F_1.

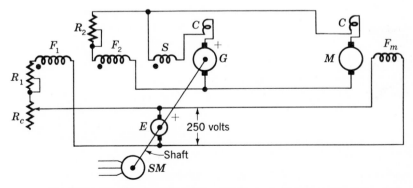

Fig. 5-19. Ward Leonard system with three-field generator, Prob. 5-25.

The generator G and exciter E are driven by a 2,300-volt 3-phase synchronous motor SM.

Main generator G (rating 500 kw, 500 volts):

S_1: 1 turn per pole, total resistance = 0.001 ohm

F_1: 200 turns per pole, resistance = 25 ohms

F_2: 100 turns per pole, resistance = 12 ohms

R_1 and R_2 are fixed resistors

R_c is the controller

Resistance of armature plus commutating winding = 0.009 ohm

Armature reaction negligible

Data for the generator magnetization curve are as follows:

Field excitation, amp-turns per pole	500	1,000	1,500	2,000	2,500	3,000
Generated volts	250	450	540	585	615	640

Motor M (rating 500 volts, 1,000 amp):

F_m: Excited at 250 volts

With 500 volts applied to motor armature, no-load speed = 600 rpm

Resistance of armature plus commutating winding plus cables = 0.015 ohm

Armature reaction negligible

The following results are required:

a. Find the resistance of R_2 which makes the excitation line of F_2 coincide with the magnetization-curve air-gap line.

b. Find the resistance of R_1 which limits the stalled torque of the motor to 1.5 per unit with $R_c = 0$.

c. With the above settings of R_1 and R_2, plot the generator volt-ampere characteristic and the motor speed-torque characteristic with $R_c = 0$. Use per-unit values of the variables, with 600 rpm, 500 volts, 1,000 amp, and motor torque at 1,000 amp as base values.

5-26. Dc electric drives are used in some ship propulsion systems where maneuverability is of prime importance. This problem concerns a dc electric drive for a twin-screw ice breaker.

Specifications

1 The propulsion prime movers will comprise four 3,000-bhp, 810-rpm diesel engines and two 3,000-bhp, geared-down output 810-rpm gas turbines divided equally between the two shafts.

2 Each prime mover will drive a double-armature dc generator. Each armature will be rated 1,060 kw, 900 volts, 1,180 amp. On each shaft, three prime movers driving six generators connected in parallel will supply power to the propulsion motors.

3 Each propulsion motor is to be double armature. The armatures will be connected in parallel. Each armature will be rated 3,150-kw input at 890 volts at a base speed (maximum motor-field excitation) of 150 rpm with the vessel stalled in the ice. Under these conditions the losses in cables are estimated to be 60 kw. Taking into account the motor losses, the maximum input of 6,300 kw from three engine-generator sets will give a motor output of approximately 8,000 shp per shaft, or 16,000 shp for the ship.

4 Control of the speed of the propulsion motors below base speed will be obtained by controlling the generator voltage. Generator voltage will be controlled by a combination of varying the generator field current and prime-mover speed. Motor direction of rotation will be determined by controlling the generator polarity.

5 With all three prime movers per shaft providing power, the motor can transmit 100 percent shaft power at 100 percent base speed (150 rpm) with the vessel stalled in the ice. As the vessel breaks free the propeller speed can be increased still maintaining 100 percent shaft power by motor-field weakening until 146 percent speed (220 rpm) is reached when free running in open water.

Results required

1 Plot curves of per-unit shaft power vs. per-unit base shaft speed, (*a*) with vessel stalled and (*b*) running free. Assume power is proportional to the cube of the speed.

2 On the same sheet plot per-unit propeller shaft power for (*a*) three, (*b*) two, (*c*) one engine-generator sets running. Assume that the generator emfs are adjusted by controlling the speed and flux so as to maintain rated generator armature currents. Make assumptions regarding the power consumed in losses.

3 With two prime movers per shaft, what will be the maximum shaft horsepower and corresponding shaft speed (*a*) with the vessel stalled, and (*b*) running free.   Repeat with one prime mover.

4 What sort of excitation system would you recommend for (*a*) the motors, (*b*) the generators?   That is, should they be self-excited shunt machines? separately excited from constant-speed individual exciters? from a constant-voltage exciter bus?   what voltage?   should they have series fields?   any other scheme?   As a rough approximation the excitation requirements for machines of this size are about 2 percent of their rating.

5 Would you suggest some other arrangement of the electrical equipment?

6
synchronous machines, steady state

A synchronous machine is an ac machine whose speed under steady-state conditions is proportional to the frequency of the current in its armature. At synchronous speed, the rotating magnetic field created by the armature currents travels at the same speed as the field created by the field current, and a steady torque results. An elementary picture of how a synchronous machine works has already been given in Art. 4-1 with emphasis on torque production in terms of the interactions among its magnetic fields.

Analytical methods of examining the steady-state performance of polyphase synchronous machines will be developed in this chapter. Initial consideration will be confined to cylindrical-rotor machines, with the effects of salient poles taken up in Arts. 6-6 and 6-7.

6-1
FLUX AND MMF WAVES IN SYNCHRONOUS MACHINES

Developed sketches of the armature and field windings of a cylindrical-rotor generator are given in Figs. 6-1 and 6-2. As far as the armature winding is concerned, these are the same type of winding used in discussing rotating magnetic fields in Art. 3-4.   The results as well as the underlying assumptions of that article apply to the two cases.

In both figures, the space-fundamental mmf produced by the field winding is shown by the sinusoid F.   As indicated by the alternative designation B_f, this wave may also represent the corresponding component flux-density wave.   Both Figs. 6-1a and 6-2a show the F wave at the specific instant when the excitation emf of phase a has its maximum value.   The axis of the field is then 90° ahead of the axis of phase a in order that the time rate of change of flux linkages with phase a shall be a maximum.   The excitation emf is represented by the rotating time phasor E_f in Figs. 6-1b and 6-2b.   The projection of this phasor on the reference axis for phase a is proportional to the instantaneous emf in the arrow direction defined by the dots and crosses (representing the heads and tails of arrows) in the phase-a conductors.

The mmf wave created by the armature current, commonly called the *armature-reaction mmf*, can now be superimposed through use of the principles in Art. 3-4.   Recall that balanced polyphase currents in a symmetrical polyphase winding create an mmf wave whose space-fundamental component rotates at synchronous speed.   Recall also that the mmf wave is directly opposite phase a at the instant when the phase a current has its maximum value.   Figure 6-1a is drawn with I_a and E_f in phase; hence the armature reaction wave A is drawn opposite phase a because at this instant both I_a and E_f have their maximum values. Figure 6-2a is drawn with I_a lagging E_f by the time-phase angle ϕ_{lag}; hence A is drawn behind its position in Fig. 6-1a by the space angle ϕ_{lag} because I_a has not yet reached its maximum value.   In both figures the armature-reaction wave bears the alternative designation B_{ar} to indicate that, in the absence of saturation, the armature-reaction flux-density wave is proportional to the A wave.

The resultant magnetic field in the machine is the sum of the two components produced by the field current and armature reaction.   The resultant mmf waves R (also labeled B_r to denote that the resultant flux-density wave can be similarly represented) in Figs. 6-1a and 6-2a are obtained by graphically adding the F and A waves.   Because sinusoids can conveniently be added by phasor methods, the same addition can be performed by means of the phasor diagrams of Figs. 6-1c and 6-2b.   In these diagrams phasors are also drawn to represent the fundamental flux per pole Φ_f, Φ_{ar}, and Φ_r produced, respectively, by the mmfs F, A, and R and proportional to these mmfs with a uniform air gap and no saturation.

The air-gap flux and mmf conditions in a synchronous machine can therefore be represented by phasor diagrams like those of Figs. 6-1c and

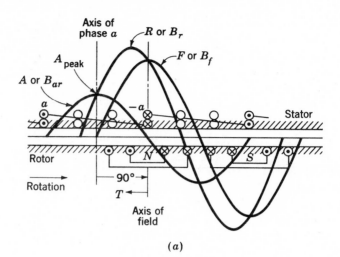

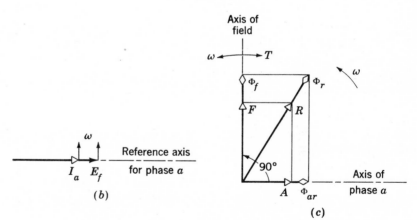

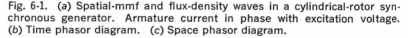

Fig. 6-1. (a) Spatial-mmf and flux-density waves in a cylindrical-rotor syn-chronous generator. Armature current in phase with excitation voltage. (b) Time phasor diagram. (c) Space phasor diagram.

6-2b without troubling to draw the wave diagrams. For example, the corresponding phasor diagrams for motor action are given in Fig. 6-3 for unity power factor with respect to excitation voltage and in Fig. 6-4 for lagging power factor with respect to that voltage. To maintain the same conventions as in Figs. 6-1 and 6-2, it is recognized that the phasor $-I_a$ rather than I_a should be in phase with or lag E_f in Figs. 6-3 and 6-4.

These phasor diagrams show how the space-phase position of the armature mmf wave with respect to the field poles depends on the time-phase angle between armature current and excitation voltage. They are

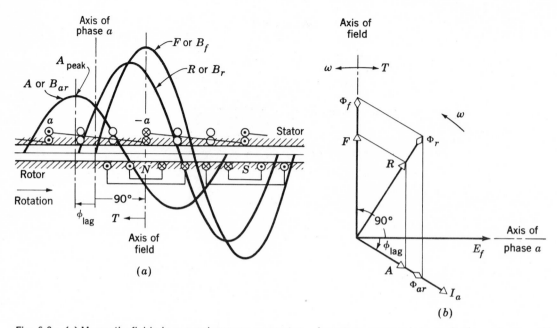

Fig. 6-2. (a) Magnetic fields in a synchronous generator. Armature current lags excitation voltage. (b) Combined space and time phasor diagram.

also helpful in correlating the simple, physical picture of torque production with the way in which the armature current adjusts itself to the operating conditions.

The electromagnetic torque on the rotor acts in a direction to urge the field poles into alignment with the resultant air-gap flux and armature-

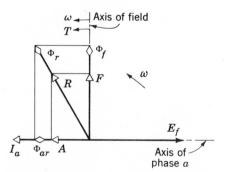

Fig. 6-3. Phasor diagram of a synchronous motor. Unity power factor with respect to excitation voltage.

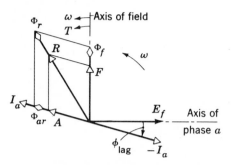

Fig. 6-4. Phasor diagram of a synchronous motor. Lagging power factor with
respect to excitation voltage.

reaction flux waves, as shown by the arrows labeled T attached to the
field axes in Figs. 6-1 to 6-4.   If the field poles lead the resultant air-gap
flux wave, as in Figs. 6-1 and 6-2, the electromagnetic torque on the
rotor acts in opposition to the rotation—in other words, the machine
must be acting as a generator.   On the other hand, if the field poles lag
the resultant air-gap flux wave, as in Figs. 6-3 and 6-4, the electromagnetic torque acts in the direction of rotation—i.e., the machine must be
acting as a motor.   An alternative statement is that for generator action
the field poles must be driven ahead of the resultant air-gap flux wave by
the forward torque of a prime mover, while for motor action the field
poles must be dragged behind the resultant air-gap flux by the retarding
torque of a shaft load.

The magnitude of the torque can be expressed in terms of the resultant fundamental air-gap flux per pole Φ_r and the peak value F of the
space-fundamental field-mmf wave.   Corresponding to Eq. 4-1,

$$ T = \frac{\pi}{2}\left(\frac{\text{poles}}{2}\right)^2 \Phi_r F \sin \delta_{\text{RF}} \tag{6-1}$$

where δ_{RF} is the space-phase angle in electrical degrees between the
resultant-flux and field-mmf waves.   When F and Φ_r are constant, the
machine adjusts itself to changing torque requirements by adjusting the
torque angle δ_{RF}.

EXAMPLE 6-1

Consider a synchronous machine with negligible armature resistance
and leakage reactance and negligible losses to be connected to an

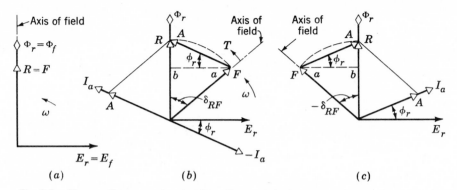

Fig. 6-5.   Phasor diagrams showing the effects of shaft torque.   (a) No load; (b) motor action; (c) generator action.

infinite bus (i.e., to a system so large that its voltage and frequency remain constant regardless of the power delivered or absorbed). The field current is kept constant at the value which causes the armature current to be zero at no load.

   With the aid of phasor diagrams, describe how the machine readjusts itself to varying torque requirements.   Include both motor and generator action.

Solution

The resultant air-gap flux Φ_r generates the voltage E_r in each armature phase.   It is usually called the air-gap voltage.   In the absence of resistance and leakage reactance, E_r must remain constant at the value of the infinite-bus voltage.   At no load, the torque and δ_{RF} are zero.   With I_a also zero, A is zero, and the phasor diagram is that of Fig. 6-5a.

   When shaft load is now added, causing the machine to become a motor, the rotor momentarily slows down slightly under the influence of the retarding torque and the field poles slide back in space phase with respect to the resultant air-gap flux wave; that is, δ_{RF} increases, and the machine develops motor torque.   After a transient period, steady-state operation at synchronous speed is resumed when δ_{RF} has assumed the value required to supply the load torque, as shown by point m in the torque-angle characteristic of Fig. 6-6.   The phasor diagram is now as shown in Fig. 6-5b.   The field mmf is no longer in phase with the resultant-flux wave, and the discrepancy in mmf must be made up by the armature reaction, thus giving rise to the armature current needed to supply the electrical power input

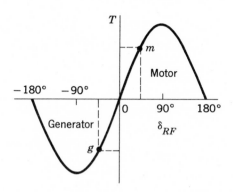

Fig. 6-6. Torque-angle characteristic.

corresponding to the mechanical power output.   Note that

$$F \sin \delta_{\mathrm{RF}} = A \cos \phi_r$$

as indicated by the dashed line ab, where ϕ_r is the power-factor angle of the armature current with respect to the air-gap voltage E_r.   But $A \cos \phi_r$ is proportional to the active-power component $I_a \cos \phi_r$ of the armature current, and, from Eq. 6-1, $F \sin \delta_{\mathrm{RF}}$ is proportional to the torque.   That is, the electrical active-power input is proportional to the mechanical torque output as, of course, it must be.

If, instead of being loaded as a motor, the shaft is driven forward by the torque of a prime mover, the field poles advance in phase ahead of the resultant-flux wave to an angle $-\delta_{\mathrm{RF}}$ where the counter torque $-T$ developed by the machine equals the driving torque of the prime mover, as shown by point g in Fig. 6-6.   The effects on the armature reaction and armature current are shown in the phasor diagram of Fig. 6-5c.   The machine has now become a generator.

In Fig. 6-5b and c, note that, for the components of F and A in phase with R,

$$F \cos \delta_{\mathrm{RF}} + A \sin \phi_r = R$$

That is, not only must the active-power component $I_a \cos \phi_r$ of the armature current adjust itself to supply the torque, but also the reactive component $I_a \sin \phi_r$ must adjust itself so that the corresponding component $A \sin \phi_r$ of the armature-reaction mmf combines with the component $F \cos \delta_{\mathrm{RF}}$ of the field mmf to produce the required resultant mmf R.   The reactive kva can therefore be controlled by adjusting the field excitation.

6-2
THE SYNCHRONOUS MACHINE AS AN IMPEDANCE

A very useful and refreshingly simple equivalent circuit representing the steady-state behavior of a cylindrical-rotor synchronous machine under balanced, polyphase conditions is obtained if the effect of the armature-reaction flux is represented by an inductive reactance. For the purpose of this preliminary discussion, consider an unsaturated cylindrical-rotor machine. Although neglect of magnetic saturation may appear to be a rather drastic simplification, it will be shown that the results which we are about to obtain can be modified so as to take saturation into account.

The resultant air-gap flux in the machine can be considered as the phasor sum of the component fluxes created by the field and armature-reaction mmfs, respectively, as shown by phasors Φ_f, Φ_{ar}, and Φ_r in Fig. 6-7. From the viewpoint of the armature windings, these fluxes manifest themselves as generated emfs. The resultant air-gap voltage E_r can then be considered as the phasor sum of the excitation voltage E_f generated by the field flux and the voltage E_{ar} generated by the armature-reaction flux. The component emfs E_f and E_{ar} are proportional to the field and armature currents, respectively, and each lags the flux which generates it by 90°. The armature-reaction flux Φ_{ar} is in phase with the armature current I_a, and consequently the armature-reaction emf E_{ar} lags the armature current by 90°. Thus,

$$E_f - jI_a x_\varphi = E_r \tag{6-2}$$

where x_φ is the constant of proportionality relating the rms values of E_{ar} and I_a. Equation 6-2 also applies to that portion of the circuit of Fig. 6-8a to the left of E_r. The effect of armature reaction therefore is

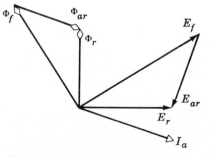

Fig. 6-7. Phasor diagram of component fluxes and corresponding voltages.

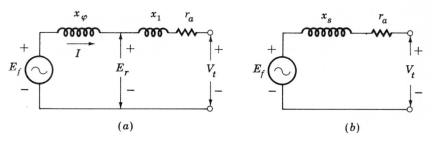

(a)   (b)

Fig. 6-8.   Equivalent circuits

simply that of an inductive reactance x_φ accounting for the component voltage generated by the space-fundamental flux created by armature reaction. This reactance is commonly called the *magnetizing reactance*, or *reactance of armature reaction*.

The air-gap voltage E_r differs from the terminal voltage by the armature-resistance and leakage-reactance voltage drops, as shown to the right of E_r in Fig. 6-8a, wherein r_a is the armature resistance, x_l is the armature leakage reactance, and V_t is the terminal voltage. All quantities are per phase (line to neutral in a Y-connected machine). The armature leakage reactance accounts for the voltages induced by the component fluxes which are not included in the air-gap voltage E_r. These fluxes include not only leakage across the armature slots and around the coil ends but also those associated with the space-harmonic fields created by the departure from a sinusoid necessarily present in the actual armature-mmf wave.

Finally, the equivalent circuit for an unsaturated cylindrical-rotor machine under balanced polyphase conditions reduces to the form shown in Fig. 6-8b in which the machine is represented on a per-phase basis by its excitation voltage E_f in series with a simple impedance. This impedance is called the *synchronous impedance*. Its reactance x_s is called the *synchronous reactance*. In terms of the magnetizing and leakage reactances,

$$x_s = x_\varphi + x_l \tag{6-3}$$

The synchronous reactance x_s takes into account all the flux produced by balanced polyphase armature currents, while the excitation voltage takes into account the flux produced by the field current. In an unsaturated cylindrical-rotor machine at constant frequency, the synchronous reactance is a constant. Furthermore, the excitation voltage is proportional to the field current and equals the voltage which would appear at the terminals if the armature were open-circuited, the speed and field current being held constant.

It is helpful to have a rough idea as to the order of magnitude of the impedance components. For machines with ratings above a few hundred kva, the armature-resistance voltage drop at rated current usually is less than 0.01 of rated voltage; i.e., the armature resistance usually is less than 0.01 per unit on the machine rating as a base. (The per-unit system is described in Chap. 1, Art. 1-10.) The armature leakage reactance usually is in the range from 0.1 to 0.2 per unit, and the synchronous reactance is in the vicinity of 1.0 per unit. In general, the per-unit armature resistance increases and the per-unit synchronous reactance decreases with decreasing size of the machine. In small machines, such as those in educational laboratories, the armature resistance may be in the vicinity of 0.05 per unit and the synchronous reactance in the vicinity of 0.5 per unit. In all but small machines, the armature resistance usually is neglected except insofar as its effect on losses and heating is concerned.

6-3

OPEN-CIRCUIT AND SHORT-CIRCUIT CHARACTERISTICS

Two basic sets of characteristic curves for a synchronous machine are involved in the inclusion of saturation effects and in the determination of the appropriate machine constants. These sets are discussed here. Except for a few remarks on the degree of validity of certain assumptions, the discussions apply to both cylindrical-rotor and salient-pole machines.

a. Open-circuit Characteristic and No-load Rotational Losses

Like the magnetization curve for a dc machine, the open-circuit characteristic of a synchronous machine is a curve of the armature terminal voltage on open circuit as a function of the field excitation when the machine is running at synchronous speed, as shown by the curve *occ* in Fig. 6-9a. The curve often is plotted in per-unit terms as in Fig. 6-9b, where unity voltage is the rated voltage and unity field current is the excitation corresponding to rated voltage on the air-gap line. Essentially, the open-circuit characteristic represents the relation between the space-fundamental component of the air-gap flux and the mmf on the magnetic circuit when the field winding constitutes the only mmf source. When the machine is an existing one, the open-circuit characteristic is usually determined experimentally by driving it mechanically at synchronous speed with its armature terminals on open circuit and reading the terminal voltage corresponding to a series of values of field current. If the mechanical power required to drive the synchronous machine dur-

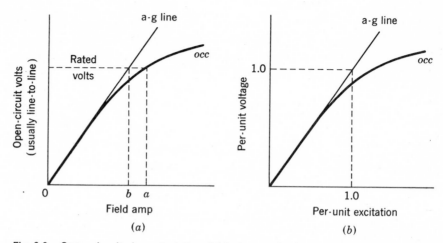

Fig. 6-9.   Open-circuit characteristic.   (a) In terms of volts and field amperes; (b) in per unit.

ing the open-circuit test is measured, the no-load rotational losses can be obtained.   These losses comprise friction, windage, and core loss corresponding to the flux in the machine at no load.   The friction and windage losses at synchronous speed are constant, while the open-circuit core loss is a function of the flux, which in turn is proportional to the open-circuit voltage.

The mechanical power required to drive the machine at synchronous speed and unexcited is its friction and windage loss.   When the field is excited, the mechanical power equals the sum of the friction, windage, and open-circuit core loss.   The open-circuit core loss therefore can be found from the difference between these two values of mechanical power. A curve of open-circuit core loss as a function of open-circuit voltage is shown in Fig. 6-10.

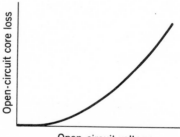

Fig.   6-10.   Open-circuit   core-loss curve.

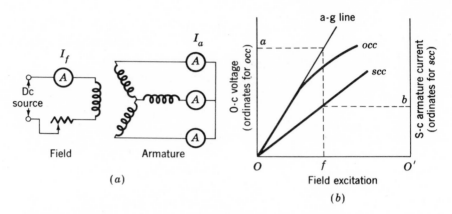

Fig. 6-11. (a) Connections for short-circuit test; (b) open-circuit and short-circuit characteristics.

b. Short-circuit Characteristic and Short-circuit Load Loss

If the armature terminals of a synchronous machine which is being driven as a generator at synchronous speed are short-circuited through suitable ammeters, as shown in Fig. 6-11a, and the field current is gradually increased until the armature current has reached a maximum safe value (perhaps twice rated current), data can be obtained from which the short-circuit armature current can be plotted against the field current. This relation is known as the *short-circuit characteristic*.   An open-circuit characteristic *occ* and a short-circuit characteristic *scc* are shown in Fig. 6-11b.

The phasor relation between the excitation voltage E_f and the steady-state armature current I_a under polyphase short-circuit conditions is

$$E_f = I_a(r_a + jx_s) \tag{6-4}$$

The phasor diagram is shown in Fig. 6-12.   Because the resistance is much smaller than the synchronous reactance, the armature current lags the excitation voltage by very nearly 90°.   Consequently the armature-reaction-mmf wave is very nearly in line with the axis of the field poles and in opposition to the field mmf, as shown by the phasors A and F representing the space waves of armature-reaction and field mmf, respectively.

The resultant mmf creates the resultant air-gap flux wave which generates the air-gap voltage E_r equal to the voltage consumed in armature resistance r_a and leakage reactance x_l; as an equation,

$$E_r = I_a(r_a + jx_l) \tag{6-5}$$

In most synchronous machines the armature resistance is negligible, and the leakage reactance is between 0.10 and 0.20 per unit—a representative value is about 0.15 per unit.   That is, at rated armature current the leakage-reactance voltage drop is about 0.15 per unit.   From Eq. 6-5, therefore, the air-gap voltage at rated armature current on short circuit is about 0.15 per unit; that is to say, the resultant air-gap flux is only about 0.15 of its normal-voltage value.   Consequently, the machine is operating in an unsaturated condition.   The short-circuit armature current therefore is directly proportional to the field current over the range from zero to well above rated armature current.

The unsaturated synchronous reactance can be found from the open-circuit and short-circuit data.   At any convenient field excitation, such as Of in Fig. 6-11b, the armature current on short circuit is $O'b$, and the excitation voltage for the same field current corresponds to Oa read from the air-gap line.   Note that the voltage on the air-gap line should be used, because the machine is operating on short circuit in an unsaturated condition.   If the voltage per phase corresponding to Oa is $E_{f(ag)}$ and the armature current per phase corresponding to $O'b$ is $I_{a(sc)}$, then from Eq. 6-4, with armature resistance neglected, the unsaturated value $x_{s(ag)}$

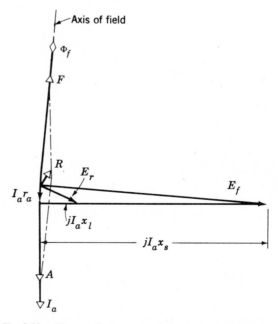

Fig. 6-12.   Phasor diagram for short-circuit conditions.

of the synchronous reactance is

$$x_{s(\text{ag})} = \frac{E_{f(\text{ag})}}{I_{a(\text{sc})}} \tag{6-6}$$

where the subscripts (ag) indicate air-gap-line conditions. If $E_{f(\text{ag})}$ and $I_{a(\text{sc})}$ are expressed in per unit, the synchronous reactance will be in per unit. If $E_{f(\text{ag})}$ and $I_{a(\text{sc})}$ are expressed in volts per phase and amperes per phase, respectively, the synchronous reactance will be in ohms per phase.

For operation at or near rated terminal voltage, it is sometimes assumed that the machine is equivalent to an unsaturated one whose magnetization curve is a straight line through the origin and the rated-voltage point on the open-circuit characteristic, as shown by the dashed line Op in Fig. 6-13. According to this approximation, the saturated value of the synchronous reactance at rated voltage V_t is

$$x_s = \frac{V_t}{I'_{a(\text{sc})}} \tag{6-7}$$

where $I'_{a(\text{sc})}$ is the armature current $O'c$ read from the short-circuit characteristic at the field current Of' corresponding to V_t on the open-circuit characteristic, as shown in Fig. 6-13. This method of handling the effects of saturation usually gives satisfactory results when great accuracy is not required.

The *short-circuit ratio* is defined as the ratio of the field current

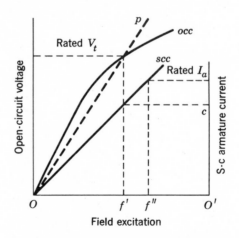

Fig. 6-13. Open-circuit and short-circuit characteristics.

required for rated voltage on open circuit to the field current required for rated armature current on short circuit.   That is, in Fig. 6-13 the short-circuit ratio SCR is

$$\text{SCR} = \frac{Of'}{Of''} \tag{6-8}$$

It can be shown that the short-circuit ratio is the reciprocal of the per-unit value of the saturated synchronous reactance given by Eq. 6-7.

EXAMPLE 6-2

The following data are taken from the open-circuit and short-circuit characteristics of a 45-kva 3-phase Y-connected 220-volt (line to line) 6-pole 60-Hz synchronous machine:

From open-circuit characteristic:

> Line-to-line voltage = 220 volts
>
> Field current = 2.84 amp

From short-circuit characteristic:

Armature current, amp	118	152
Field current, amp	2.20	2.84

From air-gap line:

> Field current = 2.20 amp
>
> Line-to-line voltage = 202 volts

Compute the unsaturated value of the synchronous reactance, its saturated value at rated voltage in accordance with Eq. 6-7, and the short-circuit ratio.   Express the synchronous reactance in ohms per phase and also in per unit on the machine rating as a base.

Solution

At a field current of 2.20 amp, the voltage to neutral on the air-gap line is

$$E_{f(\text{ag})} = \frac{202}{\sqrt{3}} = 116.7 \text{ volts}$$

and, for the same field current, the armature current on short circuit is

$$I_{a(sc)} = 118 \text{ amp}$$

From Eq. 6-6,

$$x_{s(ag)} = \frac{116.7}{118} = 0.987 \text{ ohm per phase}$$

Note that rated armature current is $45,000/\sqrt{3}\,(220) = 118$ amp. Therefore, $I_{a(sc)} = 1.00$ per unit. The corresponding air-gap-line voltage is

$$E_{f(ag)} = \frac{202}{220} = 0.92 \text{ per unit}$$

From Eq. 6-6 in per unit,

$$x_{s(ag)} = \frac{0.92}{1.00} = 0.92 \text{ per unit}$$

From the open-circuit and short-circuit characteristics and Eq. 6-7,

$$x_s = \frac{220}{\sqrt{3}\,(152)} = 0.836 \text{ ohm per phase}$$

In per unit, $I'_{a(sc)} = {}^{152}\!/_{118} = 1.29$, and from Eq. 6-7

$$x_s = \frac{1.00}{1.29} = 0.775 \text{ per unit}$$

From the open-circuit and short-circuit characteristics and Eq. 6-8,

$$\text{SCR} = \frac{2.84}{2.20} = 1.29$$

If the mechanical power required to drive the machine is measured while the short-circuit test is being made, information can be obtained regarding the losses caused by the armature current. The mechanical power required to drive the synchronous machine during the short-circuit

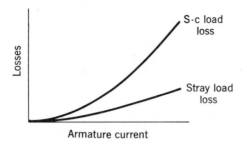

Fig. 6-14.   Short-circuit load-loss and stray
load-loss curves.

test equals the sum of friction and windage plus losses caused by the
armature current.   The losses caused by the armature current can then
be found by subtracting friction and windage from the driving power.
The losses caused by the short-circuit armature current are known collec-
tively as the *short-circuit load loss*.   A curve of short-circuit load loss
plotted against armature current is shown in Fig. 6-14.   It is approximately parabolic.

The short-circuit load loss comprises copper loss in the armature
winding, local core losses caused by the armature leakage flux, and a very
small core loss caused by the resultant flux.   The dc resistance loss can
be computed if the dc resistance is measured and corrected, when necessary, for the temperature of the windings during the short-circuit test.
For copper conductors

$$\frac{r_T}{r_t} = \frac{234.5 + T}{234.5 + t} \qquad (6\text{-}9)$$

where r_T and r_t are the resistances at centigrade temperatures T and t,
respectively.   If this dc resistance loss is subtracted from the short-
circuit load loss, the difference will be the loss due to skin effect and eddy
currents in the armature conductors plus the local core losses caused by
the armature leakage flux.   (The core loss caused by the resultant flux on
short circuit is customarily neglected.)   This difference between the
short-circuit load loss and the dc resistance loss is the additional loss
caused by the alternating current in the armature.   It is the stray load
loss described in Art. 4-8 and is commonly considered to have the same
value under normal load conditions as on short circuit.   It is a function
of the armature current, as shown by the curve in Fig. 6-14.

As with any ac device, the effective resistance of the armature is the power loss attributable to the armature current divided by the square of the current. On the assumption that the stray load loss is a function of only the armature current, the effective resistance $r_{a(eff)}$ of the armature can be determined from the short-circuit load loss; thus,

$$r_{a(eff)} = \frac{\text{short-circuit load loss}}{(\text{short-circuit armature current})^2} \qquad (6\text{-}10)$$

If the short-circuit load loss and armature current are in per unit, the effective resistance will be in per unit. If they are in watts per phase and amperes per phase, respectively, the effective resistance will be in ohms per phase. Usually it is sufficiently accurate to find the value of $r_{a(eff)}$ at rated current and then to assume it to be constant.

EXAMPLE 6-3

For the 45-kva 3-phase Y-connected synchronous machine of Example 6-2, at rated armature current (118 amp) the short-circuit load loss (total for 3 phases) is 1.80 kw at a temperature of 25°C. The dc resistance of the armature at this temperature is 0.0335 ohm per phase.

Compute the armature effective resistance, in per unit and in ohms per phase at 25°C.

Solution

In per unit, the short-circuit load loss is

$$\frac{1.80}{45} = 0.040$$

at $I_a = 1.00$ per unit. Therefore,

$$r_{a(eff)} = \frac{0.040}{(1.00)^2} = 0.040 \text{ per unit}$$

On a per-phase basis the short-circuit load loss is

$$\frac{1,800}{3} \text{ watts per phase}$$

and consequently the effective resistance is

$$r_{a(eff)} = \frac{1,800}{(3)(118)^2} = 0.043 \text{ ohm per phase}$$

The ratio of ac to dc resistance is

$$\frac{r_{a(eff)}}{r_{a(dc)}} = \frac{0.043}{0.0335} = 1.28$$

Because this is a small machine, its per-unit resistance is relatively high. The armature resistance of machines with ratings above a few hundred kva usually is less than 0.01 per unit.

STEADY-STATE OPERATING CHARACTERISTICS

The principal steady-state operating characteristics are the interrelations among terminal voltage, field current, armature current, and power factor, and the efficiency. A selection of performance curves which are of importance in practical application of the machines is presented here. All of them can be computed for application studies by the methods presented in this chapter.

Consider a synchronous generator delivering power at constant frequency to a load whose power factor is constant. The curve showing the field current required to maintain rated terminal voltage as the constant-power-factor load is varied is known as a *compounding curve*. Three compounding curves at various constant power factors are shown in Fig. 6-15.

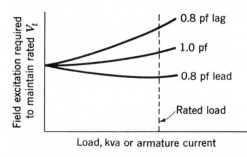

Fig. 6-15. Generator compounding curves.

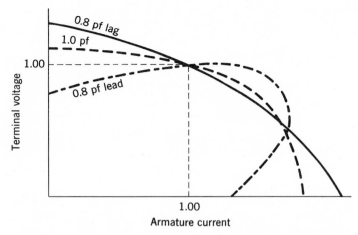

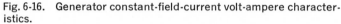

Fig. 6-16.   Generator constant-field-current volt-ampere characteristics.

If the field current is held constant while the load varies, the terminal
voltage will vary.   Characteristic curves of terminal voltage plotted
against armature current for three constant power factors are shown in
Fig. 6-16.   Each curve is drawn for a different value of constant field
current.   In each case, the field current equals the value required to give
rated terminal voltage at rated armature current and corresponds to the
rated-armature-current value read from the compounding curves (Fig.
6-15).

Synchronous generators are usually rated in terms of the maximum
kva load at a specific voltage and power factor (often 80, 85, or 90 percent lagging) which they can carry continuously without overheating.
The active power output of the generator is usually limited to a value
within the kva rating by the capability of its prime mover.   By virtue
of its voltage-regulating system, the machine normally operates at a
constant voltage whose value is within ±5 percent of rated voltage.
When the active-power loading and voltage are fixed, the allowable
reactive-power loading is limited by either armature or field heating.   A
typical set of reactive-power capability curves for a large turbine-generator are shown in Fig. 6-17.   They give the maximum reactive-power
loadings corresponding to various power loadings with operation at rated
voltage.   Armature heating is the limiting factor in the region from unity
to rated power factor (0.85 in Fig. 6-17).   For lower power factors, field
heating is limiting.   Such a set of curves forms a valuable guide in
planning and operating the system of which the generator is a part.

Also shown in Fig. 6-17 is the effect of increased hydrogen pressure on allowable machine loadings.

The power factor at which a synchronous motor operates, and hence its armature current, can be controlled by adjusting its field excitation. The curve showing the relation between armature current and field current at a constant terminal voltage and with a constant shaft load is known as a V *curve* because of its characteristic shape. A family of V curves is shown in Fig. 6-18. For constant power output, armature current is, of course, a minimum at unity power factor and increases as power factor decreases. The dashed lines are loci of constant power factor. They are the synchronous-motor compounding curves showing how the field current must be varied as load is changed in order to maintain constant power factor. Points to the right of the unity-power-factor compounding curve correspond to overexcitation and leading current input; points to the left correspond to underexcitation and lagging current input. The synchronous-motor compounding curves are very similar to the generator compounding curves of Fig. 6-15. (Note the inter-

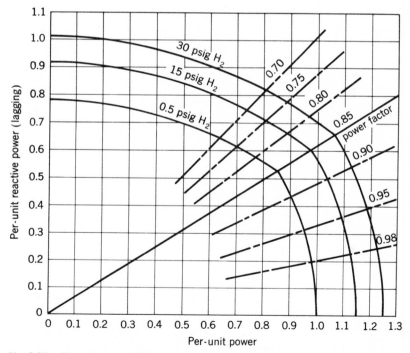

Fig. 6-17.   Reactive-capability curves of hydrogen-cooled turbine generator, 0-85 power factor, 0.80 short-circuit ratio. Base kva is rated kva at 0.5 lb hydrogen.

change of armature-current and field-current axes when comparing Figs. 6-15 and 6-18.) In fact, if it were not for the small effects of armature resistance, the motor and generator compounding curves would be identical except that the lagging- and leading-power-factor curves would be interchanged.

As in all electromagnetic machines, the losses in synchronous machines comprise I^2R losses in the windings, core losses, and mechanical losses. The conventional efficiency is computed in accordance with a set of rules agreed upon by the ANSI. The general principles upon which these rules are based are described in Art. 4-8. The purpose of the following example is to show how these rules are applied specifically to synchronous machines.

EXAMPLE 6-4

Data are given in Fig. 6-19 with respect to the losses of the 45-kva synchronous machine of Examples 6-2 and 6-3. Compute its efficiency when running as a synchronous motor at a terminal voltage of 230 volts and with a power input to its armature of 45 kw at 0.80 power factor, leading current. The field current measured in a load test taken under these conditions is I_f (test) $= 5.50$ amp.

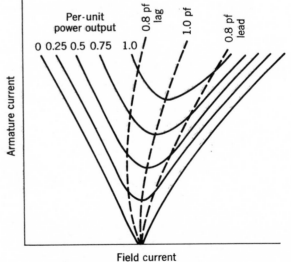

Fig. 6-18. Synchronous-motor V curves.

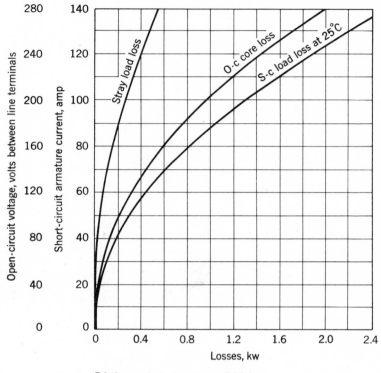

Friction and windage loss = 0.91 kw

Armature dc resistance at 25°C = 0.0335 ohm per phase

Field-winding resistance at 25°C = 29.8 ohms

Fig. 6-19.   Losses in 3-phase 45-kva Y-connected 220-volt 60-Hz 6-pole synchronous machine, Example 6-4.

Solution

For the specified operating conditions, the armature current is

$$I_a = \frac{45{,}000}{\sqrt{3}\,(230)(0.80)} = 141 \text{ amp}$$

The copper losses are to be computed on the basis of the dc resistances of the windings at 75°C.   Correcting the winding resistances by means of Eq. 6-9 gives

Field-winding resistance r_f at 75°C = 35.5 ohms

Armature dc resistance r_a at 75°C = 0.0399 ohm per phase

The field copper loss is

$$I_f^2 r_f = (5.50)^2(35.5) = 1,070 \text{ watts or } 1.07 \text{ kw}$$

According to the ANSI Standards, field-rheostat and exciter losses are not charged against the machine. The armature copper loss is

$$3I_a^2 r_a = (3)(141)^2(0.0399) = 2,380 \text{ watts, or } 2.38 \text{ kw}$$

and from Fig. 6-19, at $I_a = 141$ amp, stray load loss = 0.56 kw. According to the ANSI Standards, no temperature correction is to be applied to the stray load loss.

The core loss is read from the open-circuit core-loss curve at a voltage equal to the internal voltage behind the resistance of the machine. The stray load loss is considered to account for the losses caused by the armature leakage flux. For motor action this internal voltage is, as a phasor,

$$V_t - I_a r_a = \frac{230}{\sqrt{3}} - 141(0.80 + j0.60)(0.0399)$$

$$= 128.4 - j3.4$$

The magnitude is 128.4 volts per phase, or 222 volts between line terminals. From Fig. 6-19, open-circuit core loss = 1.20 kw. Also, friction and windage loss = 0.91 kw. All losses have now been found.

$$\text{Total losses} = 1.07 + 2.38 + 0.56 + 1.20 + 0.91 = 6.12 \text{ kw}$$

The power input is the sum of the ac input to the armature and the dc input to the field, or

$$\text{Input} = 46.07 \text{ kw}$$

Therefore $\text{Efficiency} = 1 - \dfrac{\text{losses}}{\text{input}} = 1 - \dfrac{6.12}{46.1} = 0.867$

6-5
STEADY-STATE POWER-ANGLE CHARACTERISTICS

The maximum short-time overload which a synchronous machine can deliver is determined by the maximum torque which can be applied with-

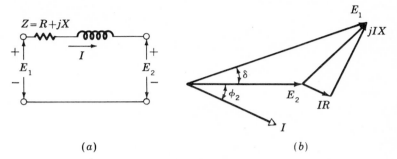

(a)   (b)

Fig. 6-20.   (a) Impedance interconnecting two voltages; (b) phasor diagram.

out loss of synchronism.   The purpose of this article is to derive expressions for the steady-state power limits of simple systems with gradually applied loads.   The heretofore neglected effects of external impedance will also be included.

Since the machine can be represented by a simple impedance, the study of power limits becomes merely a special case of the more general problem of the limitations on power flow through an inductive impedance. The impedance can include that of a line and transformer bank as well as the synchronous impedance of the machine.

Consider the simple circuit of Fig. 6-20a comprising two ac voltages E_1 and E_2 connected by an impedance Z through which the current is I. The phasor diagram is shown in Fig. 6-20b.   The power P_2 delivered through the impedance to the load end E_2 is

$$P_2 = E_2 I \cos \phi_2 \tag{6-11}$$

where ϕ_2 is the phase angle of I with respect to E_2.   The phasor current is

$$I = \frac{E_1 - E_2}{Z} \tag{6-12}$$

If the phasor voltages and the impedance are expressed in polar form,

$$I = \frac{E_1\underline{/\delta} - E_2\underline{/0°}}{Z\underline{/\phi_Z}} = \frac{E_1}{Z}\underline{/\delta - \phi_Z} - \frac{E_2}{Z}\underline{/-\phi_Z} \tag{6-13}$$

wherein E_1 and E_2 are the magnitudes of the voltages, δ is the phase angle by which E_1 leads E_2, Z is the magnitude of the impedance, and ϕ_Z is its angle in polar form.   The real part of the phasor equation 6-13 is the component of I in phase in E_2, whence

$$I \cos \phi_2 = \frac{E_1}{Z} \cos (\delta - \phi_z) - \frac{E_2}{Z} \cos (-\phi_z) \qquad (6\text{-}14)$$

Substitution of Eq. 6-14 in Eq. 6-11, it being noted that

$$\cos (-\phi_z) = \cos \phi_z = R/Z$$

gives

$$P_2 = \frac{E_1 E_2}{Z} \cos (\delta - \phi_z) - \frac{E_2^2 R}{Z^2} \qquad (6\text{-}15)$$

$$P_2 = \frac{E_1 E_2}{Z} \sin (\delta + \alpha_z) - \frac{E_2^2 R}{Z^2} \qquad (6\text{-}16)$$

where

$$\alpha_z = 90° - \phi_z = \tan^{-1} \frac{R}{X} \qquad (6\text{-}17)$$

and usually is a small angle.

Similarly the power P_1 at source end E_1 of the impedance can be expressed as

$$P_1 = \frac{E_1 E_2}{Z} \sin (\delta - \alpha_z) + \frac{E_1^2 R}{Z^2} \qquad (6\text{-}18)$$

If, as is frequently the case, the resistance is negligible,

$$P_1 = P_2 = \frac{E_1 E_2}{X} \sin \delta \qquad (6\text{-}19)$$

If the resistance is negligible and the voltages are constant, the maximum power is

$$P_{1\,max} = P_{2\,max} = \frac{E_1 E_2}{X} \qquad (6\text{-}20)$$

and occurs when $\delta = 90°$.

When Eq. 6-19 is compared with Eq. 6-1 for torque in terms of interacting flux and mmf waves, it is seen that they are of the same form. This is no coincidence. First remember that torque and power are linearly proportional when, as here, speed is constant. Then what we are really saying is that Eq. 6-1, when applied specifically to the idealized cylindrical-rotor machine and translated to circuit terms, becomes Eq. 6-19. A quick mental review of the background of each relation should show that they stem from the same fundamental considerations.

EXAMPLE 6-5

A 2,000-hp 1.0-power-factor 3-phase Y-connected 2,300-volt 30-pole 60-Hz synchronous motor has a synchronous reactance of 1.95 ohms per phase.   For the purposes of this problem all losses may be neglected.

a. Compute the maximum torque in pound-feet which this motor can deliver if it is supplied with power from a constant-voltage constant-frequency source, commonly called an *infinite bus,* and if its field excitation is constant at the value which would result in 1.00 power factor at rated load.

b. Instead of the infinite bus of part *a,* suppose that the motor were supplied with power from a 3-phase Y-connected 2,300-volt 1,750-kva 2-pole 3,600-rpm turbine generator whose synchronous reactance is 2.65 ohms per phase.   The generator is driven at rated speed, and the field excitations of generator and motor are adjusted so that the motor runs at 1.00 power factor and rated terminal voltage at full load.   The field excitations of both machines are then held constant, and the mechanical load on the synchronous motor is gradually increased.   Compute the maximum motor torque under these conditions.   Also compute the terminal voltage when the motor is delivering its maximum torque.

Solution

Although this machine is undoubtedly of the salient-pole type, we shall solve the problem by simple cylindrical-rotor theory.   The solution accordingly neglects reluctance torque.   The machine actually would develop a maximum torque somewhat greater than our computed value.

a. The equivalent circuit is shown in Fig. 6-21a and the phasor diagram at full load in Fig. 6-21b, wherein E_{fm} is the excitation voltage of the motor and x_{sm} is its synchronous reactance.   From the motor rating with losses neglected,

Rated kva $= 2,000 \times 0.746$
$$= 1,492 \text{ kva, 3-phase} = 497 \text{ kva per phase}$$

Rated voltage $= \dfrac{2,300}{\sqrt{3}} = 1,330$ volts to neutral

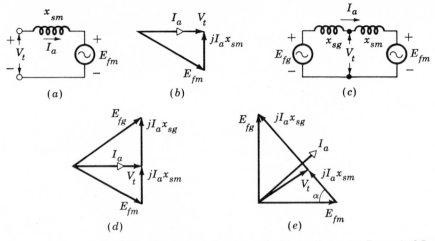

Fig. 6-21. Equivalent circuits and phasor diagrams, Example 6-5.

$$\text{Rated current} = \frac{497,000}{1,330} = 374 \text{ amp per phase Y}$$

$$I_a x_{sm} = 374 \times 1.95 = 730 \text{ volts per phase}$$

From the phasor diagram at full load

$$E_{fm} = \sqrt{V_t^2 + (I_a x_{sm})^2} = 1,515 \text{ volts}$$

When the power source is an infinite bus and the field excitation is constant, V_t and E_{fm} are constant. Substitution of V_t for E_1, E_{fm} for E_2, and x_{sm} for X in Eq. 6-20 then gives

$$P_{max} = \frac{V_t E_{fm}}{x_{sm}}$$

$$= \frac{1,330 \times 1,515}{1.95} = 1,030 \times 10^3 \text{ watts per phase}$$

$$= 3,090 \text{ kw for 3 phases}$$

(In per unit, $P_{max} = 3,090/1,492 = 2.07$.) With 30 poles at 60 Hz, synchronous speed = 4 rev/sec.

$$T_{max} = \frac{P_{max}}{\omega_s} = \frac{3,090 \times 10^3}{2\pi \times 4} = 123 \times 10^3 \text{ newton-meters}$$

$$= 0.738(123 \times 10^3) = 90,600 \text{ lb-ft}$$

b. When the power source is the turbine generator, the equivalent circuit becomes that shown in Fig. 6-21c, wherein E_{fg} is the excitation voltage of the generator and x_{sg} is its synchronous reactance. The phasor diagram at full motor load, 1.00 power factor, is shown in Fig. 6-21d. As before,

$$V_t = 1{,}330 \text{ volts at full load}$$

$$E_{fm} = 1{,}515 \text{ volts}$$

The synchronous-reactance drop in the generator is

$$I_a x_{sg} = 374 \times 2.65 = 991 \text{ volts}$$

and from the phasor diagram

$$E_{fg} = \sqrt{V_t^2 + (I_a x_{sg})^2} = 1{,}655 \text{ volts}$$

Since the field excitations and speeds of both machines are constant, E_{fg} and E_{fm} are constant. Substitution of E_{fg} for E_1, E_{fm} for E_2, and $x_{sg} + x_{sm}$ for X in Eq. 6-20 then gives

$$P_{max} = \frac{E_{fg}E_{fm}}{x_{sg} + x_{sm}}$$

$$= \frac{1{,}655 \times 1{,}515}{4.60} = 545 \times 10^3 \text{ watts per phase}$$

$$= 1{,}635 \text{ kw for 3 phases}$$

(In per unit, $P_{max} = 1{,}635/1{,}492 = 1.095$.)

$$T_{max} = \frac{P_{max}}{\omega_s} = \frac{1{,}635 \times 10^3}{2\pi \times 4} = 65 \times 10^3 \text{ newton-meters}$$

$$= 48{,}000 \text{ lb-ft}$$

Synchronism would be lost if a load torque greater than this value were applied to the motor shaft. The motor would stall, the generator would tend to overspeed, and the circuit would be opened by circuit-breaker action.

With fixed excitations, maximum power occurs when E_{fg}

leads E_{fm} by 90°, as shown in Fig. 6-21e. From this phasor diagram

$$I_a(x_{\text{sg}} + x_{\text{sm}}) = \sqrt{E_{\text{fg}}^2 + E_{\text{fm}}^2} = 2{,}240 \text{ volts}$$

$$I_a = \frac{2{,}240}{4.60} = 488 \text{ amp}$$

$$I_a x_{\text{sm}} = 488 \times 1.95 = 951 \text{ volts}$$

$$\cos \alpha = \frac{E_{\text{fm}}}{I_a(x_{\text{sg}} + x_{\text{sm}})} = \frac{1{,}515}{2{,}240} = 0.676$$

$$\sin \alpha = \frac{E_{\text{fg}}}{I_a(x_{\text{sg}} + x_{\text{sm}})} = \frac{1{,}655}{2{,}240} = 0.739$$

The phasor equation for the terminal voltage is

$$V_t = E_{\text{fm}} + jI_a x_{\text{sm}} = E_{\text{fm}} - I_a x_{\text{sm}} \cos \alpha + jI_a x_{\text{sm}} \sin \alpha$$
$$= 1{,}515 - 643 + j703 = 872 + j703$$

The magnitude of V_t is

$$V_t = 1{,}120 \text{ volts to neutral}$$
$$= 1{,}940 \text{ volts, line to line}$$

When the source is the turbine generator, as in part b, the effect of its impedance causes the terminal voltage to decrease with increasing load, thereby reducing the maximum power from 3,090 kw in part a to 1,635 kw in part b.

6-6

EFFECTS OF SALIENT POLES.
INTRODUCTION TO TWO-REACTANCE THEORY

a. Flux and MMF Waves

The flux produced by an mmf wave in the uniform-air-gap machine is independent of the spatial alignment of the wave with respect to the field poles. The salient-pole machine, on the other hand, has a preferred direction of magnetization determined by the protruding field poles. The

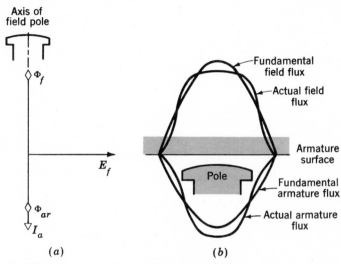

Fig. 6-22.   Direct-axis air-gap fluxes in a salient-pole synchronous machine.

permeance along the polar, or direct, axis is appreciably greater than that along the interpolar, or quadrature, axis.

We have seen that the armature-reaction flux wave lags the field flux wave by a space angle of $90° + \phi_{\text{lag}}$, where ϕ_{lag} is the time-phase angle by which the armature current in the direction of the excitation emf lags the excitation emf.   If the armature current I_a lags the excitation emf E_f by $90°$, the armature-reaction flux wave Φ_{ar} is directly opposite the field poles and in the opposite direction to the field flux Φ_f, as shown in the phasor diagram of Fig. 6-22a.   The corresponding component flux-density waves at the armature surface produced by the field current and by the synchronously rotating space-fundamental component of armature-reaction mmf are shown in Fig. 6-22b, in which the effects of slots are neglected. The waves consist of a space fundamental and a family of odd-harmonic components.   The harmonic effects usually are small (see Art. 3-3a). Accordingly only the space-fundamental components will be considered. It is the fundamental components which are represented by the flux per pole phasors Φ_f and Φ_{ar} in Fig. 6-22a.

Conditions are quite different when the armature current is in phase with the excitation emf, as shown in the phasor diagram of Fig. 6-23a. The axis of the armature-reaction wave then is opposite an interpolar space, as shown in Fig. 6-23b.   The armature-reaction flux wave is badly distorted, comprising principally a fundamental and a prominent third space harmonic.   The third-harmonic flux wave generates third-harmonic

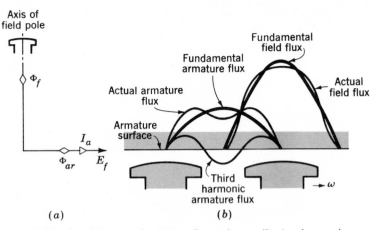

Fig. 6-23. Quadrature-axis air-gap fluxes in a salient-pole synchronous machine.

emfs in the armature phases, but these voltages do not appear between the line terminals.

Because of the high reluctance of the air gap between poles, the space-fundamental armature-reaction flux when the armature reaction is in quadrature with the field poles (Fig. 6-23) is less than the space-fundamental armature-reaction flux which would be created by the same armature current if the armature flux wave were directly opposite the field poles (Fig. 6-22). Hence, the magnetizing reactance is less when the armature current is in time phase with the excitation emf (Fig. 6-23) than when it is in time quadrature with respect to the excitation emf (Fig. 6-22a).

The effects of salient poles can be taken into account by resolving the armature current I_a into two components, one in time quadrature with and the other in time phase with the excitation voltage E_f, as shown in the phasor diagram of Fig. 6-24. This diagram is drawn for an unsaturated salient-pole generator operating at a lagging power factor. The component I_d of the armature current, in time quadrature with the excitation voltage, produces a component fundamental armature-reaction flux Φ_{ad} along the axes of the field poles, as in Fig. 6-22. The component I_q in phase with the excitation voltage, produces a component fundamental armature-reaction flux Φ_{aq} in space quadrature with the field poles, as in Fig. 6-23. The subscripts d and q refer to the space phase of the armature-reaction fluxes, and not to the time phase of the component currents producing them. Thus a *direct-axis* quantity is one whose magnetic effect is centered on the axes of the field poles. Direct-axis mmfs

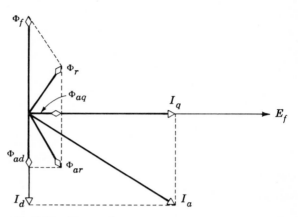

Fig. 6-24.   Phasor diagram of a salient-pole synchronous generator.

act on the main magnetic circuit. A *quadrature-axis* quantity is one whose magnetic effect is centered on the interpolar space. For an unsaturated machine, the armature-reaction flux Φ_{ar} is the sum of the components Φ_{ad} and Φ_{aq}. As in Fig. 6-5, the resultant flux Φ_r is the sum of Φ_{ar} and the main-field flux Φ_f.

b. Equivalent-circuit Aspects

With each of the component currents I_d and I_q there is associated a component synchronous-reactance voltage drop, jI_dx_d and jI_qx_q, respectively. The reactances x_d and x_q are, respectively, the direct- and quadrature-axis synchronous reactances. The synchronous reactances account for the inductive effects of all the fundamental-frequency-generating fluxes created by the armature currents, including both armature-leakage and armature-reaction fluxes. Thus, the inductive effects of the direct- and quadrature-axis armature-reaction flux waves can be accounted for by *direct-* and *quadrature-axis magnetizing reactances* $x_{\varphi d}$ and $x_{\varphi q}$, respectively, similar to the magnetizing reactance x_φ of cylindrical-rotor theory. The direct- and quadrature-axis synchronous reactances then are

$$x_d = x_l + x_{\varphi d} \tag{6-21}$$

$$x_q = x_l + x_{\varphi q} \tag{6-22}$$

where x_l is the armature leakage reactance and is assumed to be the same for direct- and quadrature-axis currents. Compare with Eq. 6-3. As shown in the generator phasor diagram (Fig. 6-25), the excitation voltage

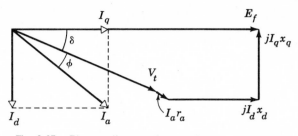

Fig. 6-25. Phasor diagram for synchronous generator.

E_f equals the phasor sum of the terminal voltage V_t plus the armature-resistance drop $I_a r_a$ and the component synchronous-reactance drops $jI_d x_d + jI_q x_q$.

The reactance x_q is less than the reactance x_d because of the greater reluctance of the air gap in the quadrature axis.   Usually, x_q is between 0.6 and 0.7 of x_d.   Typical values are given in Table 6-1.   Note that a small salient-pole effect is present in turboalternators, even though they are cylindrical-rotor machines, because of the effect of the rotor slots on the quadrature-axis reluctance.

TABLE 6-1
TYPICAL PER-UNIT VALUES OF MACHINE REACTANCES
(Machine KVA Rating as Base)

	Synchronous motors		Synchronous condensers	Water-wheel generators	Turbine generators
	High-speed	Low-speed			
x_d	0.65 (min)	0.80	. . . .	0.60	
	0.80 (av)	1.10	1.60	1.00	1.15
	0.90 (max)	1.50	. . . .	1.25	
x_q	0.50 (min)	0.60	. . . .	0.40	
	0.65 (av)	0.80	1.00	0.65	1.00
	0.70 (max)	1.10	. . . .	0.80	

In using the phasor diagram of Fig. 6-25, the armature current must be resolved into its d- and q-axis components.   This resolution assumes that the phase angle $\phi + \delta$ of the armature current with respect to the excitation voltage is known.   Often, however, the power-factor angle ϕ at the machine terminals is explicitly known, rather than the internal power-factor angle $\phi + \delta$.   The phasor diagram of Fig. 6-25 is repeated by the solid-line phasors in Fig. 6-26.   Study of this phasor diagram shows that the dashed phasor $o'a'$, perpendicular to I_a, equals $jI_a x_q$.   This result follows geometrically from the fact that triangles $o'a'b'$ and oab are

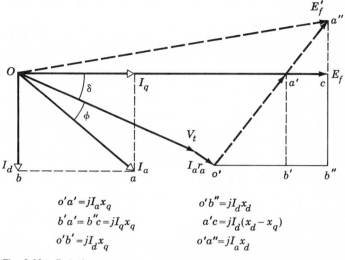

$$o'a' = jI_a x_q \qquad\qquad o'b'' = jI_d x_d$$
$$b'a' = b''c = jI_q x_q \qquad a'c = jI_d(x_d - x_q)$$
$$o'b' = jI_d x_q \qquad\qquad o'a'' = jI_a x_d$$

Fig. 6-26.   Relations among component voltages in phasor diagram.

similar, because their corresponding sides are perpendicular.   Thus

$$\frac{o'a'}{oa} = \frac{b'a'}{ba} \tag{6-23}$$

or
$$o'a' = \frac{b'a'}{ba}\, oa = \frac{jI_q x_q}{I_q}\, I_a = jI_a x_q \tag{6-24}$$

The phasor sum $V_t + I_a r_a + jI_a x_q$ then locates the angular position of the excitation voltage E_f and therefore the d and q axes.   Physically this must be so, because all the field excitation in a normal machine is in the direct axis.   One use of these relations in determining the excitation requirements for specified operating conditions at the terminals of a salient-pole machine is illustrated in Example 6-6.

EXAMPLE 6-6

The reactances x_d and x_q of a salient-pole synchronous generator are 1.00 and 0.60 per unit, respectively.   The armature resistance is negligible.

Compute the excitation voltage when the generator delivers rated kva at 0.80 power factor, lagging current, and rated terminal voltage.

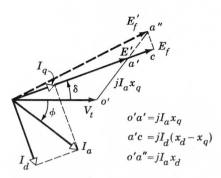

Fig. 6-27. Generator phasor diagram,
Example 6-6.

Solution

First, the phase of E_f must be found so that I_a can be resolved into
its d- and q-axis components. The phasor diagram is shown in
Fig. 6-27.

$$I_a = 0.80 - j0.60 = 1.00\underline{/-36.9°}$$

$$jI_a x_q = j(0.80 - j0.60)(0.60) = 0.36 + j0.48$$
$$V_t = \text{reference phasor} = 1.00 + j0$$
$$\text{Phasor sum} = E' = \overline{1.36 + j0.48} = 1.44\underline{/19.4°}$$

The angle $\delta = 19.4°$, and the phase angle between E_f and I_a is
$36.9° + 19.4° = 56.3°$.

The armature current can now be resolved into its d- and q-axis
components. Their magnitudes are

$$I_d = 1.00 \sin 56.3° = 0.832$$

$$I_q = 1.00 \cos 56.3° = 0.555$$

As phasors, $I_d = 0.832\underline{/-90° + 19.4°} = 0.832\underline{/-70.6°}$

$$I_q = 0.555\underline{/19.4°}$$

We can now find E_f by adding numerically the length

$$a'c = I_d(x_d - x_q)$$

to the magnitude of E'; thus, the magnitude of the excitation voltage is the algebraic sum

$$E_f = E' + I_d(x_d - x_q)$$

$$= 1.44 + (0.832)(0.40) = 1.77 \text{ per unit}$$

As a phasor, $E_f = 1.77\underline{/19.4°}$

In the simplified theory of Art. 6-2, the synchronous machine is assumed to be representable by a single reactance, the synchronous reactance of Eq. 6-3. The question naturally arises as to how serious an approximation is involved if a salient-pole machine is treated in this simple fashion. Suppose the salient-pole machine of Figs. 6-26 and 6-27 were treated by cylindrical-rotor theory as if it had a single synchronous reactance equal to its direct-axis value x_d. For the same conditions at its terminals, the synchronous-reactance drop jI_ax_d would be the phasor $o'a''$, and the equivalent excitation voltage would be E'_f as shown in these figures. Because ca'' is perpendicular to E_f, there is little difference in magnitude between the correct value E_f and the approximate value E'_f for a normally excited machine. Recomputation of the excitation voltage on this basis for Example 6-6 gives a value of $1.79\underline{/26.6°}$.

Insofar as the interrelations among terminal voltage, armature current, power, and excitation over the normal operating range are concerned, the effects of salient poles usually are of minor importance, and such characteristics of a salient-pole machine usually can be computed with satisfactory accuracy by the simple cylindrical-rotor theory. Only at small excitations will the differences between cylindrical-rotor and salient-pole theory become important.

There is, however, considerable difference in the phase angles of E_f and E'_f in Figs. 6-26 and 6-27. This difference is caused by the reluctance torque in a salient-pole machine. Its effect is investigated in the following article.

6-7
POWER-ANGLE CHARACTERISTICS OF SALIENT-POLE MACHINES

We shall limit the discussion to the simple system shown in the schematic diagram of Fig. 6-28a comprising a salient-pole synchronous machine SM connected to an infinite bus of voltage E_e through a series impedance of reactance x_e per phase. Resistance will be neglected because it usually is small. Consider that the synchronous machine is acting as a generator.

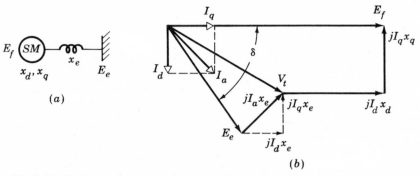

Fig. 6-28. Salient-pole synchronous machine and series impedance. (a) Single-line diagram; (b) phasor diagram.

The phasor diagram is shown by the solid-line phasors in Fig. 6-28b. The dashed phasors show the external reactance drop resolved into components due to I_d and I_q.   The effect of the external impedance is merely to add its reactance to the reactances of the machine; i.e., the total values of reactance interposed between the excitation voltage E_f and the bus voltage E_e are

$$X_d = x_d + x_e \tag{6-25}$$

$$X_q = x_q + x_e \tag{6-26}$$

If the bus voltage E_e is resolved into components $E_e \sin \delta$ and $E_e \cos \delta$ in phase with I_d and I_q, respectively, the power P delivered to the bus per phase is

$$P = I_d E_e \sin \delta + I_q E_e \cos \delta \tag{6-27}$$

Also, from Fig. 6-28b,

$$I_d = \frac{E_f - E_e \cos \delta}{X_d} \tag{6-28}$$

$$I_q = \frac{E_e \sin \delta}{X_q} \tag{6-29}$$

Substitution of Eqs. 6-28 and 6-29 in Eq. 6-27 gives

$$P = \frac{E_f E_e}{X_d} \sin \delta + E_e^2 \frac{X_d - X_q}{2X_d X_q} \sin 2\delta \tag{6-30}$$

This power-angle characteristic is shown in Fig. 6-29.   The first term is the same as the expression obtained for a cylindrical-rotor machine. This term is merely an extension of the basic concepts of Chap. 3 to include the effects of series reactance.   The second term introduces the effect of salient poles.   It represents the fact that the air-gap flux wave creates torque tending to align the field poles in the position of minimum reluctance.   This term is the power corresponding to the *reluctance torque* and is of the same general nature as the reluctance torque discussed in Art. 2-6.   Note that the reluctance torque is independent of field excitation.   Note, also, that if $X_d = X_q$, as in a uniform-air-gap machine, there is no preferential direction of magnetization, the reluctance torque is zero, and Eq. 6-30 reduces to the power-angle equation for a cylindrical-rotor machine whose synchronous reactance is X_d.

Figure 6-30 shows a family of power-angle characteristics at various values of excitation and constant terminal voltage.   Only positive values of δ are shown.   The curves for negative values of δ are the same except for a reversal in the sign of P.   That is, the generator and motor regions are alike if the effects of resistance are negligible.   For generator action E_f leads E_e; for motor action E_f lags E_e.   Steady-state operation is stable over the range where the slope of the power-angle characteristic is positive.   Because of the reluctance torque, a salient-pole machine is stiffer than one with a cylindrical rotor—i.e., for equal voltages and equal values of X_d, a salient-pole machine develops a given torque at a smaller value

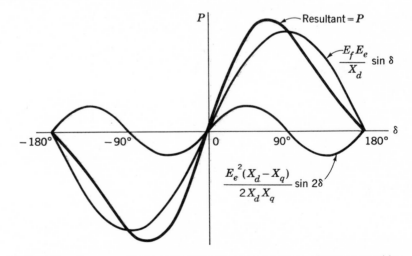

Fig. 6-29.   Power-angle characteristic of a salient-pole synchronous machine showing fundamental component due to field excitation and second-harmonic component due to reluctance torque.

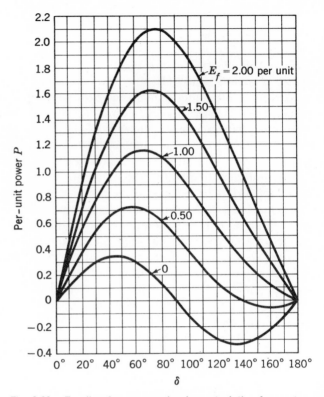

Fig. 6-30. Family of power-angle characteristics for system of Fig. 6-28a with $E_e = 1.00$, $R = 0$, $X_d = 1.00$, $X_q = 0.60$.

of δ, and the maximum torque which can be developed is somewhat greater.

Equation 6-30 contains six quantities—the two variables P and δ, and the four parameters E_f, E_e, X_d, and X_q. To simplify the notation, let the maximum power due to the field excitation be designated by $P_{f\,max}$ and the maximum power due to the reluctance torque be designated by $P_{r\,max}$. Then Eq. 6-30 can be expressed as

$$P = P_{f\,max} \sin \delta + P_{r\,max} \sin 2\delta \qquad (6\text{-}31)$$

A further reduction of the number of parameters can be obtained if Eq. 6-31 is divided by $P_{f\,max}$; thus,

$$\frac{P}{P_{f\,max}} = \sin \delta + \frac{P_{r\,max}}{P_{f\,max}} \sin 2\delta \qquad (6\text{-}32)$$

Equation 6-32 is in normalized form.   It applies to all possible combinations of a synchronous machine and an external system, as in Fig. 6-28a, so long as the resistance is negligible.   A family of curves can be plotted from Eq. 6-32, as shown in Fig. 6-31.   The maximum value $P_{max}/P_{f\,max}$ of the power ratio and the angle $\delta_{max\ P}$ at which maximum power occurs are shown as functions of the reluctance-power ratio $P_{r\,max}/P_{f\,max}$ in Fig. 6-32.   These curves correspond to the dashed locus of the maximum points on the curves of Fig. 6-31.   Use of these curves for computing steady-state power limits is illustrated in Example 6-7.

EXAMPLE 6-7

The 2,000-hp 1.0-power-factor 3-phase Y-connected 2,300-volt synchronous motor of Example 6-5 has reactances of $x_d = 1.95$ and $x_q = 1.40$ ohms per phase.   All losses may be neglected.

Compute the maximum mechanical power in kilowatts which this motor can deliver if it is supplied with electrical power from an

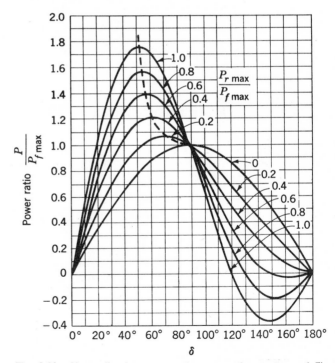

Fig. 6-31.   Normalized power-angle curves for system of Fig. 6-28a with $R = 0$.

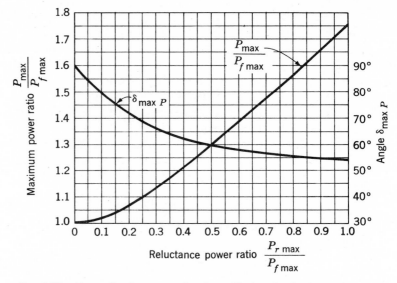

Fig. 6-32. Normalized curves showing effects of reluctance torque on steady-state power limits.

infinite bus (Fig. 6-33a) at rated voltage and frequency, and if its field excitation is constant at the value which would result in 1.00 power factor at rated load. The shaft load is assumed to be increased gradually so that transient swings are negligible and the steady-state power limit applies. Include the effects of salient poles.

Solution

The first step is to compute the synchronous-motor excitation at rated voltage, full load, 1.0 power factor. As in Example 6-5, the full-load terminal voltage and current are 1,330 volts to neutral and

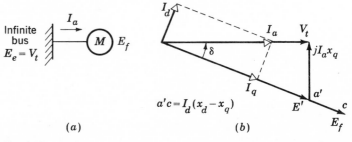

Fig. 6-33. (a) Single-line diagram and (b) phasor diagram for motor of Example 6-7.

374 amp per phase Y. The phasor diagram for the specified full-load conditions is shown in Fig. 6-33b. The only essential difference between this phasor diagram and the generator phasor diagram of Fig. 6-27 is that I_a in Fig. 6-33 represents motor *input* current. The phasor voltage equation then becomes

$$E_f = V_t - jI_dx_d - jI_qx_q$$

In Fig. 6-33b,

$$E' = V_t - jI_ax_q$$

$$= 1{,}330 + j0 - j(374)(1.40) = 1{,}429/\underline{-21.5°}$$

That is, the angle δ is 21.5°, with E_f lagging V_t. The magnitude of I_d is

$$I_d = I_a \sin \delta = (374)(0.367) = 137 \text{ amp}$$

The magnitude of E_f can now be found by adding numerically the length $a'c = I_d(x_d - x_q)$ to the magnitude of E'; thus

$$E_f = E' + I_d(x_d - x_q)$$

$$= 1{,}429 + (137)(0.55) = 1{,}504 \text{ volts to neutral}$$

The maximum values of the field-excitation and reluctance-torque components of the power can now be computed. From Eqs. 6-31 and 6-32,

$$P_{f\,\text{max}} = \frac{(1{,}504)(1{,}330)}{1.95} = 1{,}025 \times 10^3 \text{ watts per phase}$$

$$P_{r\,\text{max}} = \frac{(1{,}330)^2(0.55)}{2(1.95)(1.40)} = 178 \times 10^3 \text{ watts per phase}$$

whence

$$\frac{P_{r\,\text{max}}}{P_{f\,\text{max}}} = 0.174$$

From Fig. 6-32, the corresponding value of the maximum-power ratio is

$$\frac{P_{\text{max}}}{P_{f\,\text{max}}} = 1.05$$

whence the maximum power is

$$P_{max} = 1.05 P_{f\,max} = (1.05)(1{,}025 \times 10^3) = 1{,}080 \text{ kw per phase}$$
$$= 3{,}240 \text{ kw for 3 phases}$$

Compare with $P_{max} = 3{,}090$ kw found in Example 6-5, where the effects of salient poles were neglected. The error caused by neglecting saliency is slightly less than 5 percent.

The effect of salient poles on the power limits increases as the reluctance-power ratio $P_{r\,max}/P_{f\,max}$ increases, as shown in Fig. 6-32. For a normally excited machine the effect of salient poles usually amounts to a few percent at most. Only at small excitations does the reluctance torque become important. Except at small excitations or when exceptionally accurate results are required, a salient-pole machine usually can be treated by simple cylindrical-rotor theory.

6-8

INTERCONNECTED SYNCHRONOUS GENERATORS

Synchronous generators can readily be operated in parallel, and, in fact, the electricity supply systems of industrialized countries may have scores or even hundreds of alternators operating in parallel, interconnected by hundreds of miles of transmission lines, and supplying electrical energy to loads scattered over areas of hundreds of thousands of square miles. These huge systems have grown in spite of the necessity for designing the system so that synchronism will be maintained following disturbances, and the problems, both technical and administrative, which must be solved to coordinate the operation of such a complex system of machines and personnel. The principal reasons for these interconnected systems are continuity of service and economies in plant investment and operating costs.

To illustrate the basic features of parallel operation on a simple scale, consider an elementary system comprising two identical 3-phase generators G_1 and G_2 with their prime movers PM_1 and PM_2 supplying power to a load L, as shown in the single-line diagram of Fig. 6-34. Suppose generator G_1 is supplying the load at rated voltage and frequency, with generator G_2 disconnected. Generator G_2 can be paralleled with G_1 by driving it at synchronous speed and adjusting its field rheostat so that its voltage equals that of the bus. If the frequency of the incoming machine

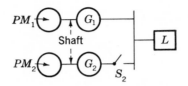

Fig. 6-34. Parallel operation of two synchronous generators.

is not exactly equal to that of the bus, the phase relation between its voltage and that of the bus will vary at a frequency equal to the difference between the frequencies of the two voltages—perhaps a fraction of a cycle per second.   The switch S_2 should be closed when the two voltages are momentarily in phase and the voltage across the switch is zero.   A device for indicating the appropriate moment is known as a *synchroscope*.   After G_2 has been synchronized in this manner, each machine can be made to take its share of the active- and reactive-power load by appropriate adjustments of the prime-mover throttles and field rheostats.

In contrast with dc generators, paralleled synchronous generators must run at exactly the same steady-state speed (for the same number of poles).   Consequently, the way in which the active power divides between them depends almost wholly on the speed-power characteristics of their prime movers.   In Fig. 6-35 the sloping solid lines PM_1 and PM_2 represent the speed-power characteristics of the two prime movers for constant throttle openings.   All practical prime movers have drooping speed-power characteristics.   The total load P_L is shown by the solid horizontal line AB, and the generator power outputs are P_1 and P_2 (losses being neglected).   Now suppose the throttle opening of PM_2 is increased, translating its speed-power curve upward to the dotted line PM_2'.   The dotted line $A'B'$ now represents the load power.   Note that

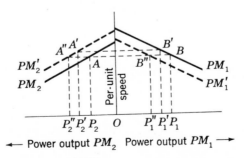

Fig. 6-35. Prime-mover speed-power characteristics.

the power output of generator 2 has now increased from P_2 to P'_2 while
that of generator 1 has decreased from P_1 to P'_1. At the same time, the
system frequency has increased. The frequency can be restored to normal with a further load shift from generator 1 to generator 2 by closing
the throttle on generator 1, lowering its speed-power curve to the dotted
line PM'_1. The load power is now represented by $A''B''$, and the power
outputs of the generators are P''_1 and P''_2. Thus, the system frequency
and the division of active power between the generators can be controlled
by means of the prime-mover throttles.

Changes in excitation affect the terminal voltage and reactive-kva
distribution. For example, let the two identical generators of Fig. 6-34
be adjusted to share the active and reactive loads equally. The phasor
diagram is shown by the solid lines in Fig. 6-36, wherein V_t is the terminal
voltage, I_L is the load current, I_a is the armature current in each generator, and E_f is the excitation voltage. The synchronous-reactance
drop in each generator is $jI_a x_s$, and the resistance drops are neglected.
Now suppose the excitation of generator 1 is increased. The bus voltage
V_t will increase. It can then be restored to normal by decreasing the
excitation of generator 2. The final condition is shown by the dotted
phasors in Fig. 6-36. The terminal voltage, load current, and load
power factor have been unchanged. Since the prime-mover throttles
have not been touched, the power output and inphase components of
the generator armature currents have not been changed. The excitation
voltages E_{f1} and E_{f2} have shifted in phase so that $E_f \sin \delta$ remains
constant. The generator with the increased excitation has now taken
on more of the lagging reactive-kva load. For the condition shown by
the dotted phasors in Fig. 6-36, generator 1 is supplying all the reactive

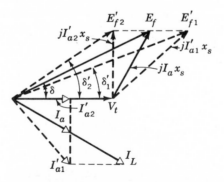

Fig. 6-36. Effects of changing excitations in two paralleled synchronous
generators.

kva, and generator 2 is operating at unity power factor.   Thus the terminal voltage and reactive-kva distribution between the generators can be controlled by means of the field rheostats.

Usually the prime-mover throttles are controlled by governors and automatic frequency regulators so that system frequency is maintained very nearly constant and power is divided properly among the generators. Voltage and reactive-kva flow often are automatically regulated by voltage regulators acting on the field circuits of the generators and by transformers with automatic tap-changing devices.

6-9
RÉSUMÉ

The physical picture of the internal workings of a synchronous machine in terms of rotating magnetic fields is rather simple.   It is that of Art. 3-5: interaction of component fields of rotor and stator when the two are stationary with respect to each other.   For both round-rotor and salient-pole machines, the component fields and mmfs, together with the associated voltages and currents, can be represented on phasor diagrams like those of Figs. 6-2b and 6-24.   The phasor diagrams in turn lead to the concept of the synchronous reactances x_s, x_d, and x_q.   These reactances are derived by replacing the effect of the rotating armature-reaction wave by magnetizing reactions x_φ or $x_{\varphi d}$ and $x_{\varphi q}$.

The unsaturated synchronous reactance x_s or x_d can be found from the results of an open-circuit and a short-circuit test.   These test methods are a variation of a testing technique applicable not only to synchronous machines but also to anything whose behavior can be approximated by a linear equivalent circuit to which Thévenin's theorem applies. From the Thévenin-theorem viewpoint, an open-circuit test gives the internal emf, and a short-circuit test gives information regarding the internal impedance.   From the more specific viewpoint of electromagnetic machinery, an open-circuit test gives information regarding excitation requirements, core losses, and (for rotating machines) friction and windage losses; a short-circuit test gives information regarding the magnetic reactions of the load current, leakage impedances, and losses associated with the load current such as copper and stray load losses.   The only real complication arises from the effects of magnetic nonlinearity, effects which can be taken into account approximately by considering the machine to be equivalent to an unsaturated one whose magnetization curve is the straight line Op of Fig. 6-13 and whose synchronous reactance is empirically adjusted for saturation as in Eq. 6-7.

Prediction of the steady-state synchronous-machine characteristics then becomes merely a study of power flow through a simple impedance with constant or easily determinable voltages at its ends. Study of the maximum-power limits for short-time overloads is simply a special case of the limitations on power flow through an inductive impedance. The power flow through such an impedance can be expressed conveniently in terms of the voltages at its sending and receiving ends and the phase angles associated with these voltages, as in Eq. 6-19 for a cylindrical-rotor machine and 6-30 for a salient-pole machine. These analyses show that saliency has relatively little effect on the interrelations among field excitation, terminal voltage, armature current, and power; but the power-angle characteristics are affected by the presence of a reluctance-torque component. Because of the reluctance torque, a salient-pole machine is stiffer than one with a cylindrical rotor.

PROBLEMS

6-1. *a.* Draw wave and phasor diagrams like those of Figs. 6-1 and 6-2 but for a synchronous generator whose armature current leads the excitation voltage.

b. Draw a phasor diagram like those of Figs. 6-3 and 6-4 but for a synchronous motor having leading power factor with respect to the excitation voltage.

6-2. A synchronous generator is supplying power to a large system with its field current adjusted so that the armature current lags the terminal voltage. Armature resistance and leakage reactance may be neglected.

The field current is now increased 10 percent without changing the driving torque of the prime mover. Qualitatively, what changes occur in power output, in magnitude and phase of the armature current, and in magnitude of the torque angle δ_{RF}? Explain by means of phasor diagrams representing the flux and mmf waves.

If, instead of changing the field current, the driving torque of the prime mover is increased 10 percent, what changes will occur?

6-3. A synchronous motor is operating at half load. An increase in its field excitation causes a decrease in armature current. Before the increase, was the motor delivering or absorbing lagging reactive kva?

6-4. The full-load torque angle δ_{RF} of a synchronous motor at rated voltage and frequency is 30 electrical degrees. Neglect the effects of armature resistance and leakage reactance. If the field current is constant, how would the torque angle be affected by the following changes in operating conditions?

a. Frequency reduced 10 percent, load torque constant

b. Frequency reduced 10 percent, load power constant

c. Both frequency and applied voltage reduced 10 percent, load torque constant

d. Both frequency and applied voltage reduced 10 percent, load power constant

6-5. A dc shunt motor is mechanically coupled to a 3-phase cylindrical-rotor synchronous generator. The dc motor is connected to a 230-volt constant-potential dc supply, and the ac generator is connected to a 230-volt (line-to-line) constant-potential constant-frequency 3-phase supply. The 4-pole Y-connected synchronous machine is rated 25 kva, 230 volts, and has a synchronous reactance of 1.60 ohms per phase. The 4-pole dc machine is rated 25 kw, 230 volts. All losses are to be neglected.

a. If the two machines act as a motor-generator set receiving power from the dc mains and delivering power to the ac mains, what is the excitation voltage of the ac machine in volts per phase (line to neutral) when it delivers rated kva at 1.00 power factor?

b. Leaving the field current of the ac machine as in part a, what adjustment can be made to reduce the power transfer (between ac and dc) to zero? Under this condition of zero transfer, what is the armature current of the dc machine? What is the armature current of the ac machine?

c. Leaving the field current of the ac machine as in parts a and b, what adjustment can be made to cause 25 kw to be taken from the ac mains and delivered to the dc mains? Under these conditions, what is the armature current of the dc machine? What are the magnitude and phase of the current of the ac machine?

6-6. The following readings are taken from the results of an open-circuit and a short-circuit test on a 9,375-kva 3-phase Y-connected 13,800-volt (line-to-line) 2-pole 60-Hz turbine generator driven at synchronous speed:

Field current	169	192
Armature current, short-circuit test	392	446
Line voltage, open-circuit characteristic	13,000	13,800
Line voltage, air-gap line	15,400	17,500

The armature resistance is 0.064 ohm per phase. The armature leakage reactance is 0.10 per unit on the generator rating as a base.

    a. Find the unsaturated value of the synchronous reactance in ohms per phase and also in per unit.

    b. Find the short-circuit ratio.

6-7. *a.* Compute the field current required in the generator of Prob. 6-6 at rated voltage, rated kva load, 0.80-power-factor lagging. Account for saturation under load by the method described in the paragraph relating to Eq. 6-7.

    b. In addition to the data given in Prob. 6-6, more points on the open-circuit characteristic are given below:

Field current	200	250	300	350
Line voltage	14,100	15,200	16,000	16,600

Find the voltage regulation for the load of part *a.* *Voltage regulation* is defined as the rise in voltage when load is removed, the speed and field excitation being held constant. It is usually expressed as a percentage of the voltage under load.

6-8. Loss data for the generator of Prob. 6-6 are as follows:

Open-circuit core loss at 13,800 volts = 68 kw
Short-circuit load loss at 392 amp, 75°C = 50 kw
Friction and windage = 87 kw
Field-winding resistance at 75°C = 0.285 ohm

Compute the efficiency at rated load, 0.80 power factor lagging.

6-9. A 3-phase synchronous generator is rated 12,000 kva, 13,800 volts, 0.80 power factor, 60 Hz. What should be its kva and voltage rating at 0.80 power factor and 50 Hz if the field and armature copper losses are to be the same as at 60 Hz? If its voltage regulation at rated load and 60 Hz is 18 percent, what will be the value of the voltage regulation at its rated load for 50-Hz operation? The effect of armature-resistance voltage drop on regulation may be neglected.

6-10. A 150-hp 0.8-power-factor 2,300-volt 38.0-amp 60-Hz 3-phase synchronous motor has a direct-connected exciter to supply its field current. For the purposes of this problem, the efficiency of the exciter may be assumed constant at a value of 80 percent. The synchronous motor is run at no load from a 2,300-volt 60-Hz circuit, with its field current supplied by its exciter, and the following readings taken:

Armature voltage between terminals = 2,300 volts
Armature current = 38.0 amp per terminal
Three-phase power input = 13.7 kw
Field current = 20.0 amp
Voltage applied to field = 300 volts

When the synchronous motor is loaded so that its input is 38.0 amp at 0.80 power factor and 2,300 volts between terminals, its field current is found to be 17.3 amp. Under these conditions, what is the efficiency of the synchronous motor exclusive of the losses in its exciter? What is the useful mechanical power output in horsepower?

6-11. From the phasor diagram of a synchronous machine with constant synchronous reactance x_s operating at constant terminal voltage V_t and constant excitation voltage E_f, show that the locus of the tip of the armature-current phasor is a circle. On a phasor diagram with terminal voltage chosen as the reference phasor indicate the position of the center of this circle and its radius. Express the coordinates of the center and the radius of the circle in terms of V_t, E_f, and x_s.

6-12. A synchronous generator is connected to an infinite bus through two parallel 3-phase transmission circuits each having a reactance of 0.60 per unit including step-up and step-down transformers at each end. The synchronous reactance of the generator (which may be handled on a cylindrical-rotor basis) is 0.90 per unit. All resistances are negligible, and reactances are expressed on the generator rating as a base. The infinite-bus voltage is 1.00 per unit.

a. The power output and excitation of the generator are adjusted so that it delivers rated current at 1.0 power factor at its terminals in the steady state. Compute the generator terminal and excitation voltages, the power output, and the reactive power delivered to the infinite bus.

b. The throttle of the prime mover is now adjusted so that there is no power transfer between the generator and the infinite bus. The field current of the generator is adjusted until 0.50-per-unit lagging reactive kva is delivered to the infinite bus. Under these conditions, compute the terminal and excitation voltages of the generator.

c. The system is then returned to the operating conditions of part a. One of the two parallel transmission circuits is disconnected by tripping the circuit breakers at its ends. The generator excitation is kept constant. Will the generator remain in synchronism? After comparing the desired power transfer with the maximum under these conditions, give an opinion regarding the adequacy of the transmission system.

6-13. A 1,000-hp 2,300-volt Y-connected 3-phase 60-Hz 20-pole synchronous motor has a synchronous reactance of 4.00 ohms per phase. In this problem cylindrical-rotor theory may be used, and all losses may be neglected.

a. This motor is operated from an infinite bus supplying rated voltage at rated frequency, and its field excitation is adjusted so that the power factor is unity when the shaft load is such as to require an input of 800 kw. If the shaft load is slowly increased, with the field excitation held constant, determine the maximum torque (in pound-feet) that the motor can deliver.

b. Instead of the infinite bus of part a, suppose that the power supply is a 1,000-kva 2,300-volt Y-connected synchronous generator whose synchronous reactance is also 4.00 ohms per phase. The frequency is held constant by a governor, and the field excitations of motor and generator are held constant at the values which result in rated terminal voltage when the motor absorbs 800 kw at unity power factor. If the shaft load on the synchronous motor is slowly increased, determine the maximum torque (in pound-feet). Also determine the armature current, terminal voltage, and power factor at the terminals corresponding to this maximum load.

c. Determine the maximum motor torque if, instead of remaining constant as in part b, the field currents of the generator and motor are slowly increased so as always to maintain rated terminal voltage and unity power factor while the shaft load is increased.

6-14. Draw the steady-state dq phasor diagram for an overexcited synchronous motor (i.e., one whose field current is sufficiently high so that lagging reactive kva is delivered to the supply system). From this phasor diagram show that the torque angle δ between the excitation and terminal voltage phasors is given by

$$\tan \delta = \frac{I_a x_q \cos \phi + I_a r_a \sin \phi}{V_t + I_a x_q \sin \phi - I_a r_a \cos \phi}$$

6-15. What percent of its rated output will a salient-pole synchronous motor deliver without loss of synchronism when the applied voltage is normal and the field excitation is zero, if $x_d = 0.80$ per unit and $x_q = 0.50$ per unit? Compute the per-unit armature current at maximum power.

6-16. A salient-pole synchronous motor has $x_d = 0.80$ and $x_q = 0.50$ per unit. It is running from an infinite bus of $V_t = 1.00$ per unit. Neglect all losses. What is the minimum per-unit excitation for which the machine will stay in synchronism with full-load torque?

7

induction motors, steady state

In the induction motor, alternating current is supplied to the stator winding directly and to the rotor winding by induction from the stator. Balanced polyphase stator and rotor currents create stator- and rotor-component-mmf waves of constant amplitude rotating in the air gap at synchronous speed and therefore stationary with respect to each other regardless of the mechanical speed of the rotor. The resultant of these mmfs creates the resultant air-gap flux-density wave. Interaction of the flux wave and the rotor-mmf wave gives rise to torque. All of the conditions are fulfilled for the production of a steady value of torque at all speeds other than synchronous speed.

The objects of this chapter are to develop equivalent circuits for the polyphase induction motor from which both the effects of the motor on its supply circuit and the characteristics of the motor itself can be determined, and to study these effects and characteristics. The general form of equivalent circuit is suggested by the similarity of an induction machine to a transformer.

7-1

FLUX AND MMF WAVES IN INDUCTION MACHINES

When the stator winding of a polyphase induction machine is excited by balanced polyphase voltages, a rotating magnetic field is produced in the air gap in the manner described in Art. 3-4.   The rotating field is traveling at synchronous speed as given by Eq. 3-38.   To examine the air-gap flux and mmf waves, let us consider conditions existing when the rotor is turning at a speed n corresponding to a per-unit slip s.   The space-fundamental component of the resultant air-gap flux wave then travels past the rotor at slip speed and induces slip-frequency emfs in the rotor circuits. These emfs give rise to slip-frequency currents in the short-circuited rotor phases or bars.   With a cage rotor or with a coil-wound rotor wound for the same number of poles as the stator, the slip-frequency rotor currents create an mmf whose space fundamental also travels at slip speed with respect to the rotor.   The mmf and flux-density waves are thus stationary relative to each other, and a steady torque is produced by their interaction.   The torque magnitude is dependent on the space angle between the two waves (see Eq. 3-64).

a. Reactions of the Rotor

For a coil-wound rotor, the flux-mmf situation may be seen with the aid of Fig. 7-1.   This sketch shows a development of a simple 2-pole 3-phase rotor winding in a 2-pole field.   It therefore conforms with the restriction that a wound rotor must have the same number of poles as the stator (although the number of phases need not be the same).   The flux-density wave is moving to the right at slip speed with respect to the winding.   It is shown in Fig. 7-1 in the position of maximum instantaneous voltage in phase a.

If rotor leakage reactance is very small compared with rotor resistance (which is very nearly the case at the small slips corresponding to normal operation), the phase a current will also be a maximum.   As shown in Art. 3-4, the rotor-mmf wave will then be centered on phase a. It is so shown in Fig. 7-1a.   The displacement angle, or torque angle, δ under these conditions is at its optimum value of 90°.

If the rotor leakage reactance is appreciable, however, the phase a current lags the induced voltage by the power-factor angle ϕ_2 of the leakage impedance.   The phase a current will not be a maximum until a correspondingly later time.   The rotor-mmf wave will then not be centered on phase a until the flux wave has traveled ϕ_2 degrees farther down the gap, as shown in Fig. 7-1b.   The angle δ is now 90° + ϕ_2.   In

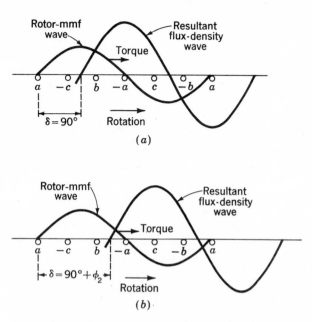

Fig. 7-1. Developed rotor winding of induction motor
with flux-density and mmf waves in their relative positions
for (a) zero and (b) nonzero rotor leakage reactance.

general, therefore, the torque angle of an induction motor is

$$\delta = 90° + \phi_2 \qquad\qquad (7\text{-}1)$$

It departs from the optimum value by the power-factor angle of the rotor
leakage impedance at slip frequency.   The electromagnetic rotor torque
is directed toward the right in Fig. 7-1, or in the direction of the rotating
flux wave.

The comparable picture for a squirrel-cage rotor is given in Fig. 7-2.
A 16-bar rotor, placed in a 2-pole field, is shown in developed form.   For
simplicity of drafting, only a relatively small number of rotor bars is
chosen, and the number is an integral multiple of the number of poles, a
choice normally avoided in order to prevent harmful harmonic effects.
In Fig. 7-2a, the sinusoidal flux-density wave induces a voltage in each
bar which has an instantaneous value indicated by the solid vertical lines.
At a somewhat later instant of time, the bar currents assume the instantaneous values indicated by the solid vertical lines in Fig. 7-2b, the time
lag being the rotor power-factor angle ϕ_2.   In this time interval, the flux-
density wave has traveled in its direction of rotation with respect to the

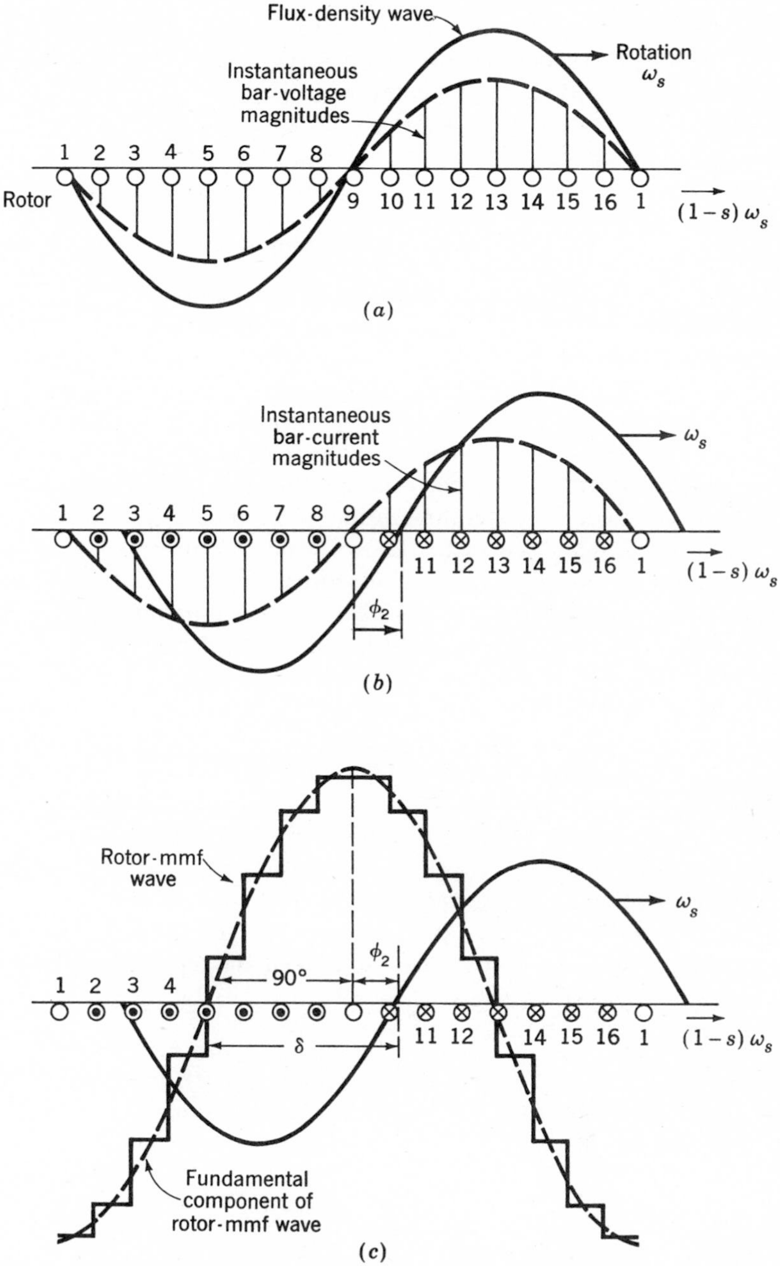

Fig. 7-2. Reactions of a squirrel-cage rotor in a 2-pole field.

rotor through a space angle ϕ_2 and is then in the position shown in Fig. 7-2b. The corresponding rotor-mmf wave is shown by the step wave of Fig. 7-2c. The fundamental component is shown by the dashed sinusoid and the flux-density wave by the solid sinusoid. Study of these figures confirms the general principle that the number of rotor poles in a squirrel-cage rotor is determined by the inducing flux wave.

b. Referring Rotor Quantities to the Stator

Thus we see that, insofar as fundamental components are concerned, both squirrel-cage and wound rotors react by producing an mmf wave having the same number of poles as the inducing flux wave, traveling at the same speed as the flux wave, and with a torque angle 90° greater than the rotor power-factor angle. The reaction of the rotor-mmf wave on the stator calls for a compensating load component of stator current and thereby enables the stator to absorb from the line the power needed to sustain the torque created by the interaction of the flux and mmf waves. The only way in which the stator knows what is happening is through the medium of the air-gap flux and rotor-mmf waves. Consequently, if the rotor were replaced by one having the same mmf and power factor at the same speed, the stator would be unable to detect the change. Such replacement leads to the idea of referring rotor quantities to the stator, an idea which is of great value in translating flux-mmf considerations into an equivalent circuit for the motor.

Consider, for example, a coil-wound rotor, wound for the same number of poles and phases as the stator. The number of effective turns per phase in the stator winding is a times the number in the rotor winding. Compare the magnetic effect of this rotor with that of a magnetically equivalent rotor having the same number of turns as the stator. For the same flux and speed, the relation between the voltage E_{rotor} induced in the actual rotor and the voltage E_{2s} induced in the equivalent rotor is

$$E_{2s} = aE_{\text{rotor}} \qquad (7\text{-}2)$$

If the rotors are to be magnetically equivalent, their ampere-turns must be equal, and the relation between the actual rotor current I_{rotor} and the current I_{2s} in the equivalent rotor must be

$$I_{2s} = \frac{I_{\text{rotor}}}{a} \qquad (7\text{-}3)$$

Consequently the relation between the slip-frequency leakage impedance Z_{2s} of the equivalent rotor and the slip-frequency leakage impedance

Z_{rotor} of the actual rotor must be

$$Z_{2s} = \frac{E_{2s}}{I_{2s}} = \frac{a^2 E_{\text{rotor}}}{I_{\text{rotor}}} = a^2 Z_{\text{rotor}} \qquad (7\text{-}4)$$

The voltages, currents, and impedances in the equivalent rotor are defined as their values *referred to the stator*. The thought process is essentially like that involved in referring secondary quantities to the primary in static-transformer theory (see Arts. 1-6 and 1-7). The referring factors are ratios of effective turns and are the same in essence as in transformer theory.

The referring factors must, of course, be known when one is concerned specifically with what is happening in the actual rotor circuits. From the viewpoint of the stator, however, the reflected effects of the rotor show up in terms of the referred quantities, and the theory of both coil-wound and cage rotors can be formulated in terms of the referred rotor. We shall assume, therefore, that the referred rotor constants are known.

Since the rotor is short-circuited, the phasor relation between the slip-frequency emf E_{2s} generated in the reference phase of the referred rotor and the current I_{2s} in this phase is

$$\frac{E_{2s}}{I_{2s}} = Z_{2s} = r_2 + jsx_2 \qquad (7\text{-}5)$$

where Z_{2s} is the slip-frequency rotor leakage impedance per phase referred to the stator, r_2 the referred effective resistance, and sx_2 the referred leakage reactance at slip frequency. The reactance is expressed in this way because it is proportional to rotor frequency and therefore to slip. Thus x_2 is defined as the value the referred rotor leakage reactance would have at stator frequency. The slip-frequency equivalent circuit for 1 phase of the referred rotor is shown in Fig. 7-3.

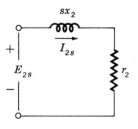

Fig. 7-3.   Rotor equivalent circuit for a polyphase induction motor.

THE EQUIVALENT CIRCUIT

The foregoing considerations of flux and mmf waves can readily be translated into the steady-state equivalent circuit for the machine. Only machines with symmetrical polyphase windings excited by balanced polyphase voltages are considered. As in many other discussions of polyphase devices, it may be helpful to think of 3-phase machines as Y-connected, so that currents are always line values and voltages always line-to-neutral values.

First consider conditions in the stator. The synchronously rotating air-gap flux wave generates balanced polyphase counter emfs in the phases of the stator. The stator terminal voltage differs from the counter emf by the voltage drop in the stator leakage impedance, the phasor relation for the phase under consideration being

$$V_1 = E_1 + I_1(r_1 + jx_1) \qquad (7\text{-}6)$$

where V_1 is the stator terminal voltage, E_1 is the counter emf generated by the resultant air-gap flux, I_1 is the stator current, r_1 is the stator effective resistance, and x_1 is the stator leakage reactance. The positive directions are shown in the equivalent circuit of Fig. 7-4.

The resultant air-gap flux is created by the combined mmfs of the stator and rotor currents. Just as in the transformer analog, the stator current can be resolved into two components, a load component and an exciting component. The load component I_2 produces an mmf which exactly counteracts the mmf of the rotor current. The exciting component I_φ is the additional stator current required to create the resultant air-gap flux and is a function of the emf E_1. The exciting current can be resolved into a core-loss component I_c in phase with E_1 and a magnetizing component I_m lagging E_1 by 90°. In the equivalent circuit, the exciting

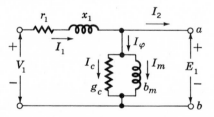

Fig. 7-4. Stator equivalent circuit for a polyphase induction motor.

current can be accounted for by means of a shunt branch, formed by
core-loss conductance g_c and magnetizing susceptance b_m in parallel, connected across E_1, as in Fig. 7-4.   Both g_c and b_m are usually determined at
rated stator frequency and for a value of E_1 close to the expected operating value; they are then assumed to remain constant for the small departures from that value associated with normal operation of the motor.

So far, the equivalent circuit representing stator phenomena is
exactly like that for the primary of a transformer.   To complete the
circuit, the effects of the rotor must be incorporated.   This is done by
considering stator and rotor voltages and currents in terms of referred
rotor quantities.

The stator sees a flux wave and an mmf wave rotating at synchronous
speed.   The flux wave induces the slip-frequency rotor voltage E_{2s} and
the stator counter emf E_1.   If it were not for the effect of speed, the
referred rotor voltage would equal the stator voltage, since the referred
rotor winding is identical with the stator winding.   Because the relative
speed of the flux wave with respect to the rotor is s times its speed with
respect to the stator, the relation between the effective values of stator
and rotor emfs is

$$E_{2s} = sE_1 \tag{7-7}$$

The rotor-mmf wave is opposed by the mmf of the load component I_2 of
stator current, and therefore, for effective values,

$$I_{2s} = I_2 \tag{7-8}$$

Division of Eq. 7-7 by Eq. 7-8 then gives

$$\frac{E_{2s}}{I_{2s}} = \frac{sE_1}{I_2} \tag{7-9}$$

Furthermore, the mmf wave created by the stator load current I_2
must be space-displaced from the resultant flux wave by the same space
angle as that between the rotor-mmf wave and the resultant flux wave,
viz., the torque angle δ.   The time-phase angle between the stator voltage
E_1 and the stator load current I_2 therefore must equal the corresponding
time angle for the rotor, viz., the rotor power-factor angle ϕ_2.   The fact
that the rotor and stator mmfs are in opposition is accounted for, since the
rotor current I_{2s} is created by the rotor emf E_{2s}, whereas the stator current I_2 is flowing against the stator counter emf E_1.   Therefore Eq. 7-9
is true, not only for effective values, but also in a phasor sense.   Through

substitution of Eq. 7-5 in the phasor equivalent of Eq. 7-9,

$$\frac{sE_1}{I_2} = \frac{E_{2s}}{I_{2s}} = r_2 + jsx_2 \qquad (7\text{-}10)$$

Division by s then gives

$$\frac{E_1}{I_2} = \frac{r_2}{s} + jx_2 \qquad (7\text{-}11)$$

That is, the stator sees magnetic conditions in the air gap which result in stator counter emf E_1 and stator load current I_2, and by Eq. 7-11 these conditions are identical with the result of connecting an impedance $(r_2/s) + jx_2$ across E_1. Consequently, the effect of the rotor can be incorporated in the equivalent circuit of Fig. 7-4 by this impedance connected across the terminals ab. The final result is shown in Fig. 7-5. The combined effect of shaft load and rotor resistance appears as a reflected resistance r_2/s, a function of slip and therefore of the mechanical load. The current in the reflected rotor impedance equals the load component I_2 of stator current; the voltage across this impedance equals the stator emf E_1. It should be noted that, when rotor currents and voltages are reflected into the stator, their frequency is also changed to stator frequency. All rotor electrical phenomena, when viewed from the stator, become stator-frequency phenomena, because the stator winding simply sees mmf and flux waves traveling at synchronous speed.

7-3
ANALYSIS OF THE EQUIVALENT CIRCUIT

Among the important performance aspects in the steady state are the variations of current, speed, and losses as the load-torque requirements

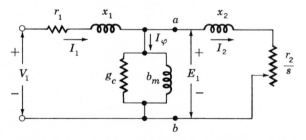

Fig. 7-5. Equivalent circuit for a polyphase induction motor.

change, the starting torque, and the maximum torque.   All these characteristics can be determined from the equivalent circuit.

The equivalent circuit shows that the total power P_{g1} transferred
across the air gap from the stator is

$$P_{g1} = q_1 I_2^2 \frac{r_2}{s} \tag{7-12}$$

where q_1 is the number of stator phases.   The total rotor copper loss is
evidently

$$\text{Rotor copper loss} = q_1 I_2^2 r_2 \tag{7-13}$$

The internal mechanical power P developed by the motor is therefore

$$P = P_{g1} - \text{rotor copper loss} = q_1 I_2^2 \frac{r_2}{s} - q_1 I_2^2 r_2 \tag{7-14}$$

$$= q_1 I_2^2 r_2 \frac{1-s}{s} \tag{7-15}$$

$$= (1 - s) P_{g1} \tag{7-16}$$

We see, then, that of the total power delivered to the rotor the
fraction $1 - s$ is converted to mechanical power and the fraction s is
dissipated as rotor-circuit copper loss.   From this it is evident that an
induction motor operating at high slip is an inefficient device.   When
power aspects are to be emphasized, the equivalent circuit is frequently
redrawn in the manner of Fig. 7-6.   The internal mechanical power per
stator phase is equal to the power absorbed by the resistance $r_2(1 - s)/s$.

The internal electromagnetic torque T corresponding to the internal
power P can be obtained by recalling that mechanical power equals torque
times angular velocity.   Thus, when ω_s is the synchronous angular velocity of the rotor in mechanical radians per second,

$$P = (1 - s) \omega_s T \tag{7-17}$$

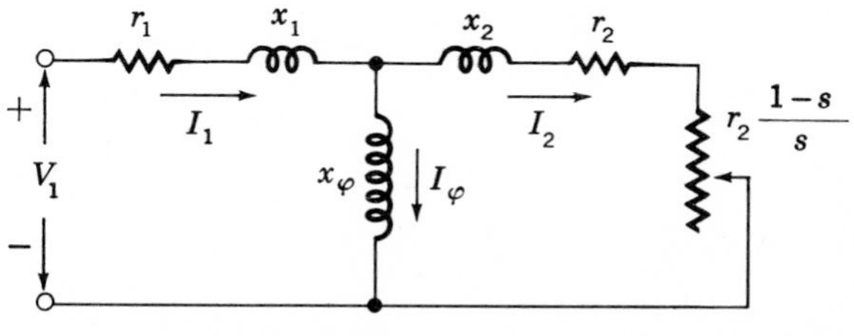

Fig. 7-6. Alternative form of equivalent circuit.

with T in newton-meters.   By use of Eq. 7-15,

$$T = \frac{1}{\omega_s} q_1 I_2^2 \frac{r_2}{s}$$

(7-18)

with the synchronous angular velocity ω_s given by

$$\omega_s = \frac{4\pi f}{\text{poles}}$$

(7-19)

The torque T and power P are not the output values available at the shaft because friction, windage, and stray load losses remain to be accounted for.   It is obviously correct to subtract friction and windage effects from T or P, and it is generally assumed that stray load effects may be subtracted in the same manner.   The final remainder is available in mechanical form at the shaft for useful work.

In static-transformer theory, analysis of the equivalent circuit is often simplified by either neglecting the exciting branch entirely or adopting the approximation of moving it out directly to the primary terminals.   Such approximations are not permissible for the induction motor under normal running conditions because the presence of the air gap makes necessary a much higher exciting current—30 to 50 percent of full-load current—and because the leakage reactances are also necessarily higher.   Some simplification of the induction-motor equivalent circuit results if the shunt conductance g_c is omitted and the associated core-loss effect deducted from T or P at the same time that friction, windage, and stray load effects are subtracted.   The equivalent circuit then becomes that of Fig. 7-7a or b, and the error introduced is negligible.   Such a procedure also has an advantage during motor testing, for no-load core loss need not then be separated from friction and windage.   These last circuits will be used in subsequent discussions.

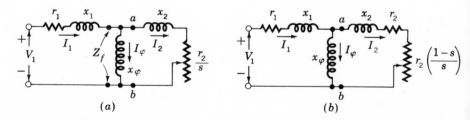

Fig. 7-7.   Equivalent circuits.

EXAMPLE 7-1

A 3-phase Y-connected 220-volt (line-to-line) 10-hp 60-Hz 6-pole induction motor has the following constants in ohms per phase referred to the stator:

$$r_1 = 0.294 \qquad r_2 = 0.144$$

$$x_1 = 0.503 \qquad x_2 = 0.209 \qquad x_\varphi = 13.25$$

The total friction, windage, and core losses may be assumed to be constant at 403 watts, independent of load.

For a slip of 2.00 percent, compute the speed, output torque and power, stator current, power factor, and efficiency when the motor is operated at rated voltage and frequency.

Solution

The impedance Z_f (Fig. 7-7a) represents physically the per-phase impedance presented to the stator by the air-gap field, both the reflected effect of the rotor and the effect of the exciting current being included therein. From Fig. 7-7a,

$$Z_f = R_f + jX_f = \frac{r_2}{s} + jx_2 \qquad \text{in parallel with } jx_\varphi$$

Substitution of numerical values gives, for $s = 0.0200$,

$$
\begin{aligned}
R_f + jX_f &= 5.41 + j3.11 \\
r_1 + jx_1 &= 0.29 + j0.50 \\
\text{Sum} &= \overline{5.70 + j3.61} = 6.75 \underline{/32.4^\circ} \text{ ohms}
\end{aligned}
$$

$$\text{Applied voltage to neutral} = \frac{220}{\sqrt{3}} = 127 \text{ volts}$$

$$\text{Stator current } I_1 = \frac{127}{6.75} = 18.8 \text{ amp}$$

$$\text{Power factor} = \cos 32.4^\circ = 0.844$$

$$\text{Synchronous speed} = \frac{2f}{p} = \frac{120}{6} = 20 \text{ rev/sec, or } 1{,}200 \text{ rpm}$$

$$\omega_s = 2\pi(20) = 125.6 \text{ rad/sec}$$

$$
\begin{aligned}
\text{Rotor speed} &= (1 - s) \times (\text{synchronous speed}) \\
&= (0.98)(1{,}200) = 1{,}176 \text{ rpm}
\end{aligned}
$$

From Eq. 7-12,

$$P_{g1} = q_1 I_2^2 \frac{r_2}{s} = q_1 I_1^2 R_f$$

$$= (3)(18.8)^2(5.41) = 5,740 \text{ watts}$$

From Eqs. 7-12 and 7-15, the internal mechanical power is

$$P = (0.98)(5,740) = 5,630 \text{ watts}$$

Deducting losses of 403 watts gives

Output power $= 5,630 - 403 = 5,230$ watts, or 7.00 hp

$$\text{Output torque} = \frac{\text{output power}}{\omega_{\text{rotor}}} = \frac{5,230}{(0.98)(125.6)}$$

$$= 42.5 \text{ newton-meters, or } 31.4 \text{ lb-ft}$$

The efficiency is calculated from the losses.

Total stator copper loss $= (3)(18.8)^2(0.294)$	$=$	312 watts
Rotor copper loss (from Eq. 7-13) $= (0.0200)(5,740)$	$=$	115
Friction, windage, and core losses	$=$	403
Total losses	$=$	830 watts
Output		$= 5,230$
Input		$= 6,060$ watts

$$\frac{\text{Losses}}{\text{Input}} = \frac{830}{6,060} = 0.137 \qquad \text{Efficiency} = 1.000 - 0.137 = 0.863$$

   The complete performance characteristics of the motor can be determined by repeating these calculations for other assumed values of slip.

7-4

TORQUE AND POWER BY USE OF THÉVENIN'S THEOREM

When torque and power relations are to be emphasized, considerable simplification results from application of Thévenin's network theorem to the induction-motor equivalent circuit.

   In its general form, Thévenin's theorem permits the replacement of

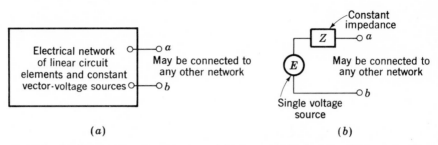

(a)   (b)

Fig. 7-8. (a) General linear network and (b) its equivalent at terminals ab by Thé-venin's theorem.

any network of linear circuit elements and constant phasor voltage sources, as viewed from two terminals a and b in Fig. 7-8a, by a single phasor voltage source E in series with a single impedance Z (Fig. 7-8b). The voltage E is that appearing across terminals a and b of the original network when these terminals are open-circuited; the impedance Z is that viewed from the same terminals when all voltage sources within the network are short-circuited.   For application to the induction-motor equivalent circuit, points a and b are taken as those so designated in Fig. 7-7a and b.   The equivalent circuit then assumes the forms given in Fig. 7-9. So far as phenomena to the right of points a and b are concerned, the circuits of Figs. 7-7 and 7-9 are identical when the voltage V_{1a} and the impedance $R_1 + jX_1$ have the proper values.   According to Thévenin's theorem, the equivalent source voltage V_{1a} is the voltage that would appear across terminals a and b of Fig. 7-7 with the rotor circuits open and is

$$V_{1a} = V_1 - I_0(r_1 + jx_1) = V_1 \frac{jx_\varphi}{r_1 + jx_{11}} \qquad (7\text{-}20)$$

where I_0 is the zero-load exciting current and

$$x_{11} = x_1 + x_\varphi \qquad (7\text{-}21)$$

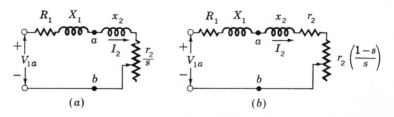

(a)   (b)

Fig. 7-9.   Induction-motor equivalent circuits simplified by Thévenin's theorem.

is the self-reactance of the stator per phase and very nearly equals the reactive component of the zero-load motor impedance.   For most induction motors, negligible error results from neglecting the stator resistance in Eq. 7-20.   The Thévenin equivalent stator impedance $R_1 + jX_1$ is the impedance between terminals a and b of Fig. 7-7, viewed toward the source with the source voltage short-circuited, and therefore is

$$R_1 + jX_1 = r_1 + jx_1 \qquad \text{in parallel with } jx_\varphi \qquad (7\text{-}22)$$

From the Thévenin equivalent circuit (Fig. 7-9) and the torque expression (Eq. 7-18) it can be seen that

$$T = \frac{1}{\omega_s} \frac{q_1 V_{1a}^2 (r_2/s)}{(R_1 + r_2/s)^2 + (X_1 + x_2)^2} \qquad (7\text{-}23)$$

The general shape of the torque-speed or torque-slip curve is that shown in Fig. 7-10.   Both the motor region $(s > 0)$ and the generator region $(s < 0)$ are shown for completeness.

Curves of stator load-component current I_2, internal torque T, and

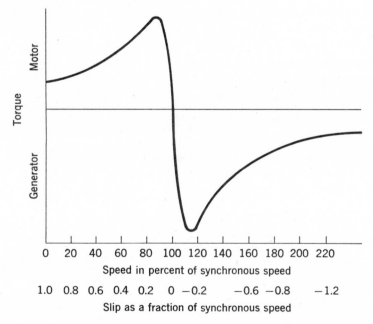

Fig. 7-10.   Induction-machine torque-slip curve in both motor and generator region.

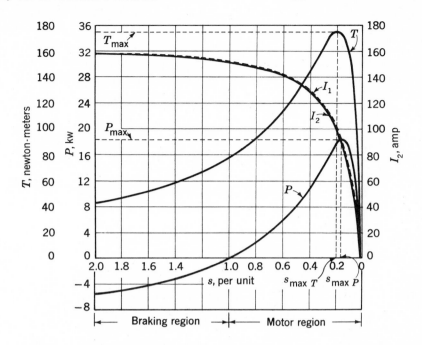

Fig. 7-11.   Computed torque, power, and current curves for 10-hp induction motor, Examples 7-1 and 7-2.

internal power P as functions of slip s are shown in Fig. 7-11.   Data for these curves are computed in Example 7-2.   Starting conditions are those for $s = 1$.   In order physically to obtain operation in the region of s greater than 1, the motor must be driven backward, against the direction of rotation of its magnetic field, by a source of mechanical power capable of counteracting the internal torque T.   The chief practical usefulness of this region is in bringing motors to a quick stop by a method called *plugging*.   By interchange of two stator leads in a 3-phase motor, the phase sequence, and hence the direction of rotation of the magnetic field, is reversed suddenly; the motor comes to a stop under the influence of the torque T and is disconnected from the line before it can start in the other direction.   Accordingly, the region from $s = 1.0$ to $s = 2.0$ is labeled *Braking region* in Fig. 7-11.

The maximum internal, or breakdown, torque T_{max} and the maximum internal power P_{max}, indicated in Fig. 7-11, can be obtained readily from circuit considerations.   Note that maximum torque and maximum power do not occur at the same speed.   Internal torque is a maximum when the power delivered to r_2/s in Fig. 7-9a is a maximum.   Now, by the familiar impedance-matching principle in circuit theory, this power will be greatest when the impedance of r_2/s equals the magnitude of the impedance between it and the constant voltage V_{1a}, or at a value $s_{\max T}$ of slip for which

$$\frac{r_2}{s_{\max T}} = \sqrt{R_1^2 + (X_1 + x_2)^2} \qquad (7\text{-}24)$$

The slip $s_{\max T}$ at maximum torque is therefore

$$s_{\max T} = \frac{r_2}{\sqrt{R_1^2 + (X_1 + x_2)^2}} \qquad (7\text{-}25)$$

and the corresponding torque is, from Eq. 7-23,

$$T_{\max} = \frac{1}{\omega_s} \frac{0.5q_1 V_{1a}^2}{R_1 + \sqrt{R_1^2 + (X_1 + x_2)^2}} \qquad (7\text{-}26)$$

EXAMPLE 7-2

For the motor of Example 7-1, determine:

a. The load component I_2 of the stator current, the internal torque T, and the internal power P for a slip $s = 0.03$
b. The maximum internal torque and the corresponding speed
c. The internal starting torque and the corresponding stator-load current I_2

Solution

First reduce the circuit to its Thévenin-theorem form. From Eq. 7-20, $V_{1a} = 122.3$; and from Eq. 7-22, $R_1 + jX_1 = 0.273 + j0.490$.

a. At $s = 0.03$, $r_2/s = 4.80$. Then, from Fig. 7-9a,

$$I_2 = \frac{122.3}{\sqrt{(5.07)^2 + (0.699)^2}} = 23.9 \text{ amp}$$

From Eq. 7-18,

$$T = \frac{1}{125.6} (3)(23.9)^2(4.80) = 65.5 \text{ newton-meters}$$

From Eq. 7-15,

$$P = (3)(23.0)^2(4.80)(0.97) = 7,970 \text{ watts}$$

Data for the curves of Fig. 7-11 were computed by repeating these calculations for a number of assumed values of s.

b. At the maximum-torque point, from Eq. 7-25,

$$s_{\max T} = \frac{0.144}{\sqrt{(0.273)^2 + (0.699)^2}} = \frac{0.144}{0.750} = 0.192$$

Speed at $T_{\max} = (1 - 0.192)(1,200) = 970$ rpm

From Eq. 7-26,

$$T_{\max} = \frac{1}{125.6} \frac{(0.5)(3)(122.3)^2}{0.273 + 0.750} = 175 \text{ newton-meters}$$

c. At starting, $s = 1$, and r_2 will be assumed constant. Therefore,

$$\frac{r_2}{s} = r_2 = 0.144 \qquad R_1 + \frac{r_2}{s} = 0.417$$

$$I_{2\,\text{start}} = \frac{122.3}{\sqrt{(0.417)^2 + (0.699)^2}} = 150.5 \text{ amp}$$

From Eq. 7-18,

$$T_{\text{start}} = \frac{1}{125.6} (3)(150.5)^2(0.144)$$

$$= 78.0 \text{ newton-meters}$$

It is thus seen that the conventional induction motor with a squirrel-cage rotor is substantially a constant-speed motor having about 5 percent drop in speed from no load to full load. Speed variation may be obtained by using a wound-rotor motor and inserting external resistance in the rotor circuit. In the normal operating range, the external resistance simply increases the rotor impedance, necessitating a higher slip for a desired

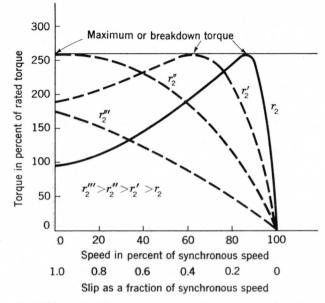

Fig. 7-12. Induction-motor torque-slip curves showing effect of changing rotor-circuit resistance.

rotor mmf and torque.   The influence of increased rotor resistance on the torque-speed characteristic is shown by the dashed curves in Fig. 7-12. Variation of starting torque with rotor resistance may be seen from these curves by noting the variation of the zero-speed ordinates.

Notice from Eqs. 7-25 and 7-26 that the slip at maximum torque is directly proportional to rotor resistance r_2 but the value of the maximum torque is independent of r_2.   When r_2 is increased by inserting external resistance in the rotor of a wound-rotor motor, the maximum internal torque is therefore unaffected but the speed at which it occurs may be directly controlled.

In applying the induction-motor equivalent circuit, the idealizations on which it is based should be kept in mind.   This is particularly necessary when investigations are carried out over a wide speed range, as in motor-starting problems.   Saturation under the heavy inrush currents associated with starting has a significant effect on the motor reactances. Moreover, the rotor currents are at slip frequency, which, of course, varies from stator frequency at zero speed to a low value at full-load speed. The current distribution in the rotor conductors and hence the rotor resistance may vary very significantly over this range.   Errors from these causes may be kept to a minimum by using equivalent-circuit constants determined by simulating the proposed operating conditions as closely as possible.[1]

7-5
NORMALIZED TORQUE-SLIP CURVES

A feature common to all branches of engineering is that equations expressing the performance of devices may present somewhat complicated arrays involving a multitude of different quantities.   Equation 7-23, for example, contains nine quantities, including the dependent and independent variables T and s.   It is frequently of value to simplify such equations by writing them in dimensionless form as relations between ratios rather than between absolute magnitudes.   For the induction motor, such a result is obtained by expressing the torque-slip relation as one between the ratios $T/T_{\max}$ and $s/s_{\max T}$.   From Eqs. 7-23 and 7-26,

$$\frac{T}{T_{\max}} = \frac{2[R_1 + \sqrt{R_1^2 + (X_1 + x_2)^2}]\dfrac{r_2}{s}}{\left(R_1 + \dfrac{r_2}{s}\right)^2 + (X_1 + x_2)^2} \tag{7-27}$$

Since the final result is to be a function of $s/s_{\max T}$ instead of simply s, r_2 in Eq. 7-27 must now be replaced by its value in terms of $s_{\max T}$ from Eq. 7-25.   After algebraic reduction, this processs yields

$$\frac{T}{T_{\max}} = \frac{1 + \sqrt{Q^2 + 1}}{1 + \dfrac{1}{2}\sqrt{Q^2 + 1}\left(\dfrac{s}{s_{\max T}} + \dfrac{s_{\max T}}{s}\right)} \tag{7-28}$$

where
$$Q = \frac{X_1 + x_2}{R_1} \tag{7-29}$$

The symbol Q is used because of the similarity of this ratio to the quality factor Q or reactance-to-resistance ratio in circuit theory.

In a similar manner, the ratio of stator load current I_2 to that at

---

[1] See, for instance, R. F. Horrell and W. E. Wood, A Method of Determining Induction Motor Speed-Torque-Current Curves from Reduced Voltage Tests, *Trans. AIEE*, **73**(III):670–674 (1954).

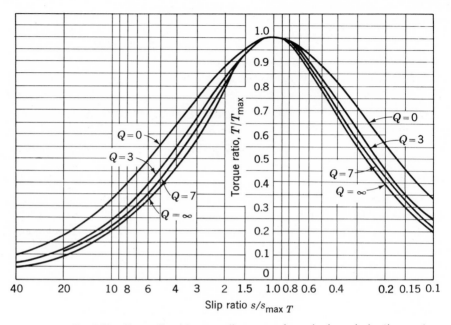

Fig. 7-13.   Normalized torque-slip curves for polyphase induction motors.

maximum torque $I_{2\,\text{max}\,T}$ can be shown to be

$$\frac{I_2}{I_{2\,\text{max}\,T}} = \sqrt{\frac{(1 + \sqrt{1 + Q^2})^2 + Q^2}{\left(1 + \dfrac{s_{\text{max}\,T}}{s}\sqrt{1 + Q^2}\right)^2 + Q^2}} \qquad (7\text{-}30)$$

Curves of T/T_{max} are plotted as functions of the appropriate slip ratio for several values of the Q ratio in Fig. 7-13, and curves of the current ratio $I_2/I_{2\,\text{max}\,T}$ in Fig. 7-14.   Most induction motors will fall in the region between $Q = 3$ and $Q = 7$, and the average will lie about midway between the curves for these two values.   Notice the rather small influence which variation of Q has on these curves; bear in mind, however, that the curves are plots of ratios, not of absolute magnitudes.   As a result of the small influence of Q, a simple approximate expression for the torque-slip relation can be obtained by substituting $Q = \infty$ in Eq. 7-28. Such substitution is equivalent to saying that the stator resistance R_1 has only a negligible influence.   The result is

$$\frac{T}{T_{\text{max}}} = \frac{2}{s/s_{\text{max}\,T} + s_{\text{max}\,T}/s} \qquad (7\text{-}31)$$

One characteristic feature of simple induction motors is shown by the very fact that the torque-slip curves can be normalized in the manner of Fig. 7-13: except for the relatively small effect of the Q ratio, if the maximum torque and the slip at which it occurs are specified, the speed-torque characteristic is approximately fixed throughout the entire speed range. This statement is, of course, subject to the limitation that the parameters of the motor are constant and therefore does not apply to motors with variable rotor resistance.

EXAMPLE 7-3

An induction motor with constant rotor resistance develops a maximum torque of 2.5 times its full-load torque at a slip of 0.20.   Estimate its slip at full load and its starting torque at rated voltage.

Solution

At full load, $T/T_{\max} = 0.40$.   From Fig. 7-13, the corresponding value of $s/s_{\max\,T}$ lies between 0.17 and 0.19 for values of Q between 3

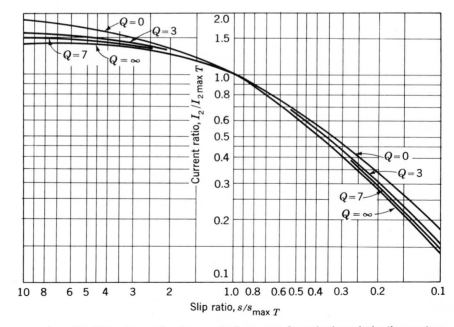

Fig. 7-14. Normalized current-slip curves for polyphase induction motors.

and 7, the range for normal motors.   Consequently, the full-load slip lies between $(0.17)(0.20) = 0.034$ and $(0.19)(0.20) = 0.038$.

   At starting, $s/s_{\max T} = 1/0.20 = 5.0$.   From Fig. 7-13 the corresponding value of $T/T_{\max}$ lies between 0.42 and 0.45 for values of Q between 7 and 3.   The starting torque therefore lies between $(0.42)(2.5) = 1.05$ and $(0.45)(2.5) = 1.13$ times full-load torque.

EXAMPLE 7-4

When operated at rated voltage and frequency with its rotor windings short-circuited, a 500-hp wound-rotor induction motor develops its rated full-load output at a slip of 1.5 percent.   The maximum torque which this motor can develop is 200 percent of full-load torque.   The Q of its Thévenin equivalent circuit is 7.0.   For the purposes of this example, rotational and stray load losses may be neglected.   Determine:

a. The rotor I^2R loss at full load, in kilowatts
b. The slip at maximum torque
c. The rotor current at maximum torque
d. The torque at a slip of 20 percent
e. The rotor current at a slip of 20 percent

Express the torque and rotor currents in per unit based on their full-load values.

Solution

   a. *Rotor I^2R at Full Load.*   The power P_{g1} absorbed from the stator divides between mechanical power P and rotor I^2R in the ratio $(1 - s)/s$.   Consequently, at full load (neglecting rotational and stray load losses)

$$P_{g1} = \frac{P}{1 - s} = \frac{(500)(0.746)}{0.985} = 379 \text{ kw}$$

$$\text{Rotor } I^2R = sP_{g1} = (0.015)(379) = 5.69 \text{ kw}$$

   Parts *b* to *e* can readily be solved by means of the normalized curves (Figs. 7-13 and 7-14).

b. *Slip at Maximum Torque.*   From the data, $T_{fl}/T_{max} = 0.50$, where the subscripts *fl* indicate full load.   From Fig. 7-13 at $Q = 7.0$ and $T/T_{max} = 0.50$,

$$\frac{s}{s_{max\,T}} = \frac{s_{fl}}{s_{max\,T}} = 0.25$$

whence $\qquad s_{max\,T} = \frac{s_{fl}}{0.25} = \frac{0.015}{0.25} = 0.060$

c. *Rotor Current at Maximum Torque.*   From Fig. 7-14 at $Q = 7.0$ and a slip ratio $s/s_{max\,T} = 0.25$ at full load, the corresponding current ratio is

$$\frac{I_2}{I_{2\,max\,T}} = \frac{I_{2fl}}{I_{2\,max\,T}} = 0.355$$

whence $\qquad I_{2\,max\,T} = \frac{I_{2fl}}{0.355} = 2.82 I_{2fl}$

d and e.   *Torque and Rotor Current at $s = 0.20$.*   The slip ratio is

$$\frac{s}{s_{max\,T}} = \frac{0.20}{0.060} = 3.33$$

The corresponding torque and current ratios can be read from the curves of Figs. 7-13 and 7-14 at $Q = 7.0$ and $s/s_{max\,T} = 3.33$.   From Fig. 7-13

$$\frac{T}{T_{max}} = 0.60 \qquad or \qquad T = 0.60 T_{max} = 1.20 T_{fl}$$

From Fig. 7-14

$$\frac{I_2}{I_{2\,max\,T}} = 1.40 \qquad or \qquad I_2 = 1.40 I_{2\,max\,T}$$

and from c

$$I_2 = (1.40)(2.82 I_{2fl}) = 3.95 I_{2fl}$$

EFFECTS OF ROTOR RESISTANCE.
DOUBLE-SQUIRREL-CAGE ROTORS

A basic limitation of induction motors with constant rotor resistance is that the rotor design has to be a compromise. High efficiency under normal running conditions requires a low rotor resistance; but a low rotor resistance results in a low starting torque and high starting current at a low starting power factor.

a. Wound-rotor Motors

The use of a wound rotor is one effective way of avoiding the necessity for compromise. The terminals of the rotor winding are connected to slip rings in contact with brushes. For starting, resistors may be connected in series with the rotor windings, the result being increased starting torque and reduced starting current at an improved power factor. The general nature of the effects on the torque-speed characteristics caused by varying rotor resistance is shown in Fig. 7-12. By use of the appropriate value of rotor resistance, the maximum torque can be made to occur at standstill if high starting torque is needed. As the rotor speeds up, the external resistances can be decreased, making maximum torque available throughout the accelerating range. Since most of the rotor I^2R loss is dissipated in the external resistors, the rotor temperature rise during starting is lower than it would be if the resistance were incorporated in the rotor winding. For normal running, the rotor winding can be short-circuited directly at the brushes. The rotor winding is designed to have low resistance so that running efficiency is high and full-load slip is low. Besides their use when starting requirements are severe, wound-rotor induction motors may be used for adjustable-speed drives. Their chief disadvantage is greater cost as compared with squirrel-cage motors.

The principal effects of varying rotor resistance on the starting and running characteristics of induction motors can be shown quantitatively by means of the following example.

EXAMPLE 7-5

The rotor winding of the motor of Example 7-4 is 3-phase, Y-connected, and has a resistance of r_{rotor}.

If the rotor-circuit resistance is increased to $5r_{\text{rotor}}$ by connecting noninductive resistances in series with each rotor slip ring, determine:

   a. The slip at which the motor will develop the same full-load torque as in Example 7-4

   b. The total rotor-circuit I^2R loss at full-load torque

   c. The horsepower output at full-load torque

   d. The slip at maximum torque

   e. The rotor current at maximum torque

   f. The starting torque

   g. The rotor current at starting

Express the torques and rotor currents in per unit based on the full-load-torque values.

Solution

The solution involves recognition of the fact that the only way in which the stator is cognizant of the happenings in the rotor is through the effect of the resistance r_2/s.   Examination of the equivalent circuit shows that for specified applied voltage and frequency everything concerning the stator performance is fixed by the value of r_2/s, the other impedance elements being constant.   For example, if r_2 is doubled and s is simultaneously doubled, the stator is unaware that any change has been made.   The stator current and power factor, the power delivered to the air gap, and the torque are constant so long as the ratio r_2/s is the same.

   Added physical significance can be given to the argument by examining the effects of simultaneously doubling r_2 and s from the viewpoint of the rotor.   An observer on the rotor then sees the resultant air-gap flux wave traveling past him at twice the original slip speed, generating twice the original rotor voltage at twice the original slip frequency.   The rotor reactance therefore is doubled, and since the original premise is that the rotor resistance also is doubled, the rotor impedance is doubled but the rotor power factor is unchanged.   Since rotor voltage and impedance are both doubled, the effective value of the rotor current remains the same; only its frequency is changed.   The air gap still has the same synchronously rotating flux and mmf waves with the same torque angle.   The observer on the rotor therefore agrees with his counterpart on the stator that the torque is unchanged when both rotor resistance and slip are changed proportionally.

   The observer on the rotor, however, is aware of two changes not apparent in the stator: (1) the rotor I^2R loss has doubled, and (2) the rotor is turning more slowly and therefore developing less mechanical power with the same torque. In other words, more of the power absorbed from the stator goes into I^2R heat in the rotor, and less is available for mechanical power.

The preceding thought processes now can readily be applied to the solution of Example 7-5.

a. *Slip at Full-load Torque.* If the rotor resistance is increased 5 times, the slip must increase 5 times for the same value of r_2/s and therefore for the same torque. But the original slip at full load, as given in Example 7-4, is 0.015. The new slip at full-load torque therefore is $(5)(0.015) = 0.075$.

b. *Rotor I^2R at Full-load Torque.* The effective value of the rotor current is the same as its full-load value in Example 7-4, and therefore the rotor I^2R loss is 5 times the full-load value of 5.69 kw found in part a of Example 7-4; or

$$\text{Rotor } I^2R = (5)(5.69) = 28.45 \text{ kw}$$

c. *Power Output at Full-load Torque.* The increased slip has caused the per-unit speed at full-load torque to drop from $1 - s = 0.985$ in Example 7-4 down to $1 - s = 0.925$ with added rotor resistance. The torque is the same. The power output therefore has dropped proportionally, or

$$P = \frac{0.925}{0.985} (500) = 469.5 \text{ hp}$$

The decrease in output equals the increase in rotor I^2R loss.

d. *Slip at Maximum Torque.* If rotor resistance is increased 5 times, the slip at maximum torque simply increases 5 times. But the original slip at maximum torque is 0.060, as found in part b of Example 7-4. The new slip at maximum torque with the added rotor resistance therefore is

$$s_{\text{max } T} = (5)(0.060) = 0.30$$

e. *Rotor Current at Maximum Torque.* The effective value of the rotor current at maximum torque is independent of rotor resistance; only its frequency is changed when rotor resistance is varied. Therefore, from part c of Example 7-4,

$$I_{2 \text{ max } T} = 2.82 I_{2\text{fl}}$$

f. Starting Torque.   With the rotor resistance increased 5 times, the starting torque will be the same as the original running torque at a slip of 0.20 and therefore equals the running torque in part *d* of Example 7-4, viz.,

$$T_{\text{start}} = 1.20 T_{\text{fl}}$$

g. Rotor Current at Starting.   The rotor current at starting with the added rotor resistances will be the same as the rotor current when running at a slip of 0.20 with the slip rings short-circuited as in part *e* of Example 7-4, viz.,

$$I_{2\,\text{start}} = 3.95 I_{2\text{fl}}$$

---

b. Deep-bar and Double-squirrel-cage Rotors

An ingenious and simple way for obtaining a rotor resistance which will automatically vary with speed makes use of the fact that, at standstill, the rotor frequency equals the stator frequency; as the motor accelerates, the rotor frequency decreases to a very low value—perhaps 2 or 3 Hz at full load in a 60-Hz motor.   By use of suitable shapes and arrangements for rotor bars, squirrel-cage rotors can be designed so that their effective resistance at 60 Hz is several times their resistance at 2 or 3 Hz. The various schemes all make use of the inductive effect of the slot-leakage flux on the current distribution in the rotor bars.   The phenomena are basically the same as the skin and proximity effect in any system of conductors with alternating current in them.

Consider first a squirrel-cage rotor having deep, narrow bars like that shown in cross section in Fig. 7-15.   The general character of the slot-leakage field produced by the current in the bar within this slot is

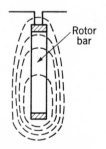

Fig.   7-15.   Deep rotor bar and slot-leakage flux.

shown in the figure. If the rotor iron had infinite permeability, all the leakage-flux lines would close in paths below the slot, as shown. Now, imagine the bar to consist of an infinite number of layers of differential depth; one at the bottom and one at the top are indicated crosshatched in Fig. 7-15. The leakage inductance of the bottom layer is greater than that of the top layer, because the bottom layer is linked by more leakage flux. But all the layers are electrically in parallel. Consequently, with alternating current, the current in the low-reactance upper layers will be greater than that in the high-reactance lower layers; the current will be forced toward the top of the slot, and the current in the upper layers will lead the current in the lower ones. The nonuniform current distribution results in an increase in the effective resistance and a smaller decrease in the effective leakage inductance of the bar. Since the distortion in current distribution depends on an inductive effect, the effective resistance is a function of the frequency. It is also a function of the depth of the bar and of the permeability and resistivity of the bar material. Figure 7-16 shows a curve of the ratio of ac effective resistance to dc resistance as a function of frequency computed for a copper bar 1.00 in. deep. A squirrel-cage rotor with deep bars can readily be designed to have an effective resistance at stator frequency (standstill) several times greater than its dc resistance. As the motor accelerates, the rotor frequency decreases and therefore the rotor effective resistance decreases, approaching its dc value at small slips.

An alternative way of attaining similar results is the double-cage arrangement shown in Fig. 7-17. The squirrel-cage winding consists of two layers of bars short-circuited by end rings. The upper bars are of smaller cross-sectional area than the lower bars and consequently have

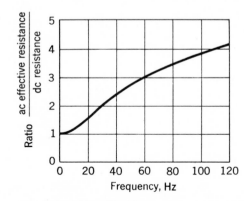

Fig. 7-16. Skin effect in a copper rotor bar 1.00 in. deep.

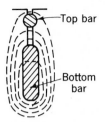

Fig. 7-17. Double-squirrel-cage rotor bars and slot-leakage flux.

higher resistance.   The general nature of the slot-leakage field is shown in Fig. 7-17, from which it can be seen that the inductance of the lower bars is greater than that of the upper ones, because of the flux crossing the slot between the two layers.   The difference in inductance can be made quite large by properly proportioning the constriction in the slot between the two bars.   At standstill, when rotor frequency equals stator frequency, there is relatively little current in the lower bars because of their high reactance; the effective resistance of the rotor at standstill then approximates that of the high-resistance upper layer. At the low rotor frequencies corresponding to small slips, however, reactance becomes unimportant, and the rotor resistance then approaches that of the two layers in parallel.

Note that, since the effective resistance and leakage inductance of double-cage and deep-bar rotors vary with frequency, the parameters r_2 and x_2 representing the referred effects of rotor resistance and leakage inductance as viewed from the stator are not constant.   The normalizing processes of Art. 7-5 are therefore no longer strictly applicable, and their use in such cases is more or less of an approximation.   A more complicated form of equivalent circuit is required if the reactions of the rotor are to be represented by the effects of slip together with constant resistance and reactance elements.

The simple equivalent circuit derived in Art. 7-2 still correctly represents the motor, however, but now r_2 and x_2 are functions of slip.   All the basic relations still apply to the motor if the values of r_2 and x_2 are properly adjusted with changes in slip.   For example, in computing the starting performance, r_2 and x_2 should be taken as their effective values at stator frequency; in computing the running performance at small slips, however, r_2 should be taken as its effective value at a low frequency, and x_2 should be taken as the stator-frequency value of the reactance corresponding to a low-frequency effective value of the rotor leakage inductance.   Over the normal running range of slips, the rotor resistance and leakage inductance usually can be considered constant at substantially their dc values.

c. Motor-application Considerations

By use of double-cage and deep-bar rotors, squirrel-cage motors can be designed to have the good starting characteristics resulting from high rotor resistance and at the same time the good running characteristics resulting from low rotor resistance.   The design is necessarily somewhat of a compromise, however, and the motor lacks the flexibility of the wound-rotor machine with external rotor resistance.   The wound-rotor motor should be used when starting requirements are very severe.

To meet the usual needs of industry, integral-horsepower 3-phase squirrel-cage motors are available from manufacturers' stock in a range of standard ratings up to 200 hp at various standard frequencies, voltages, and speeds.   (Larger motors are generally regarded as special-purpose rather than general-purpose motors.)   According to the terminology established by the NEMA, several standard designs are available to meet various starting and running requirements.   Representative torque-speed characteristics of the four commonest designs are shown in Fig. 7-18. These curves are fairly typical of 1,800-rpm (synchronous-speed) motors in ratings from 7.5 to 200 hp, although it should be understood that individual motors may differ appreciably from these average curves.   Briefly, the characteristic features of these designs are as follows:

DESIGN CLASS A.   *Normal starting torque, normal starting current, low slip.*   This design usually has a low-resistance single-cage rotor.   It emphasizes good running performance at the expense of starting.   The full-load slip is low and the full-load efficiency high.   The maximum torque usually is well over 200 percent of full-load torque and occurs at a small slip (less than 20 percent).   The starting torque at full voltage varies from about 200 percent of full-load torque in small motors to about 100 percent in large motors.   The high starting current (500 to 800 percent of full-load current when started at rated voltage) is the principal disadvantage of this design.   In sizes below about 7.5 hp, these starting currents usually are within the limits on inrush current which the distribution system supplying the motor can withstand, and across-the-line starting at full voltage then can be used; otherwise, reduced-voltage starting must be used.   Reduced-voltage starting results in a decrease in starting torque, because the starting torque is proportional to the volt-ampere input to the motor, which in turn is proportional to the square of the voltage applied to the motor terminals.   The reduced

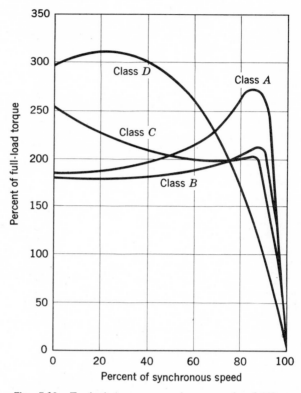

Fig. 7-18. Typical torque-speed curves for 1,800-rpm general-purpose induction motors.

voltage for starting is usually obtained from an autotransformer, called a *starting compensator*, which may be manually operated or automatically operated by relays which cause full voltage to be applied after the motor is up to speed.   A circuit diagram of one type of compensator is shown in Fig. 7-19.   If a smoother start is necessary, series resistance or reactance in the stator may be used.

The Class A motor is the basic standard design in sizes below about 7.5 and above about 200 hp.   It is also used in intermediate ratings wherein design considerations may make it difficult to meet the starting-current limitations of the Class B design.   Its field of application is about the same as that of the Class B design described below.

DESIGN CLASS B.   *Normal starting torque, low starting current, low slip*.   This design has approximately the same starting torque as the Class A design with but 75 percent of the starting current.   Full-voltage starting therefore may be used with larger sizes than with Class A.   The

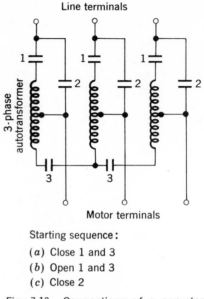

Starting sequence:

(*a*) Close 1 and 3
(*b*) Open 1 and 3
(*c*) Close 2

Fig. 7-19. Connections of a one-step starting autotransformer.

starting current is reduced by designing for relatively high leakage reactance, and the starting torque is maintained by use of a double-cage or deep-bar rotor. The full-load slip and efficiency are good—about the same as for the Class A design. However, the use of high reactance slightly decreases the power factor and decidedly lowers the maximum torque (usually only slightly over 200 percent of full-load torque being obtainable).

This design is the commonest in the 7.5- to 200-hp range of sizes. It is used for substantially constant-speed drives where starting-torque requirements are not severe, such as in driving fans, blowers, pumps, and machine tools.

DESIGN CLASS C. *High starting torque, low starting current.* This design uses a double-cage rotor with higher rotor resistance than the Class B design. The result is higher starting torque with low starting current but somewhat lower running efficiency and higher slip than the Class A and Class B designs. Typical applications are in driving compressors and conveyors.

DESIGN CLASS D. *High starting torque, high slip.* This design usually has a single-cage high-resistance rotor (frequently brass bars). It produces very high starting torque at low starting current, high maximum torque at 50 to 100 percent slip, but runs at a high slip at full load (7 to 11 percent) and consequently has low running efficiency. Its principal uses are for driving intermittent loads involving high accelerating duty and for driving high-impact loads such as punch presses and shears. When driving high-impact loads, the motor is generally aided by a fly-wheel which helps supply the impact and reduces the pulsations in power drawn from the supply system. A motor whose speed falls appreciably with increase in torque is required in order that the flywheel may slow down and deliver some of its kinetic energy to the impact.

7-7
SPEED CONTROL OF INDUCTION MOTORS

The simple induction motor fulfills admirably the requirements of sub-stantially constant-speed drives. Many motor applications, however, require several speeds, or even a continuously adjustable range of speeds. From the earliest days of ac power systems engineers have been interested in the development of adjustable-speed ac motors.

The synchronous speed of an induction motor can be changed by (*a*) changing the number of poles or (*b*) varying the line frequency. The slip can be changed by (*c*) varying the line voltage, (*d*) varying the rotor resistance, or (*e*) inserting voltages of the appropriate frequency in the rotor circuits. The salient features of speed-control methods based on these five possibilities are discussed in the following five sections of this article. Control methods involving solid-state devices will be mentioned only briefly because they are more fully treated in the next chapter.

a. Pole-changing Motors

The stator winding can be designed so that by simple changes in coil connections the number of poles can be changed in the ratio 2 to 1. Either of two synchronous speeds can be selected. The rotor is almost always of the squirrel-cage type. A cage winding always reacts by producing a rotor field having the same number of poles as the inducing stator field. If a wound rotor is used, additional complications are introduced because the rotor winding also must be rearranged for pole changing. With two independent sets of stator windings, each arranged for pole changing, as many as four synchronous speeds can be obtained in a squirrel-cage motor—for example, 600, 900, 1,200, and 1,800 rpm.

The basic principles of the pole-changing winding are shown in Fig. 7-20, in which *aa* and *a'a'* are 2 coils comprising part of the phase *a* stator

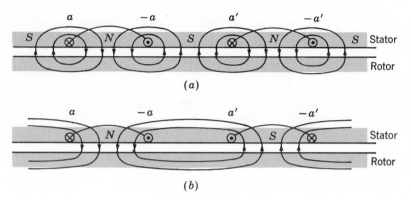

Fig. 7-20.   Principles of pole-changing winding.

winding.   An actual winding would, of course, consist of several coils in each group.   The windings for the other stator phases (not shown in the figure) would be similarly arranged.   In Fig. 7-20*a* the coils are con-nected to produce a 4-pole field; in Fig. 7-20*b* the current in the $a'a'$ coil has been reversed by means of a controller, the result being a 2-pole field.   At the same time that the controller reverses the $a'a'$ coils, the connections of the two groups of coils may be changed from series to parallel and the connections among the phases from Y to Δ, or vice versa. By these means the air-gap flux density can be adjusted to produce the desired torque-speed characteristics on the two connections.   Figure 7-21 shows three possibilities and their corresponding torque-speed char-acteristics for three motors having identical characteristics on the high-speed connection.   Figure 7-21*a* results in approximately the same maximum torque on both speeds and is applicable to drives requiring approximately the same torque on both speeds (loads in which friction predominates, for example).   Figure 7-21*b* results in approximately twice the maximum torque on the low speed and is applicable to drives requir-ing approximately constant power (such as machine tools and winches). Figure 7-21*c* results in considerably less maximum torque on the low speed and is applicable to drives requiring less torque on the low speed (fans and centrifugal pumps, for example).   The constant-horsepower type is the most expensive because it is physically the largest.

b. Line-frequency Control

The synchronous speed of an induction motor can be controlled by vary-ing the line frequency.   In order to maintain approximately constant flux density, the line voltage should also be varied directly with the fre-quency.   The maximum torque then remains very nearly constant.   An

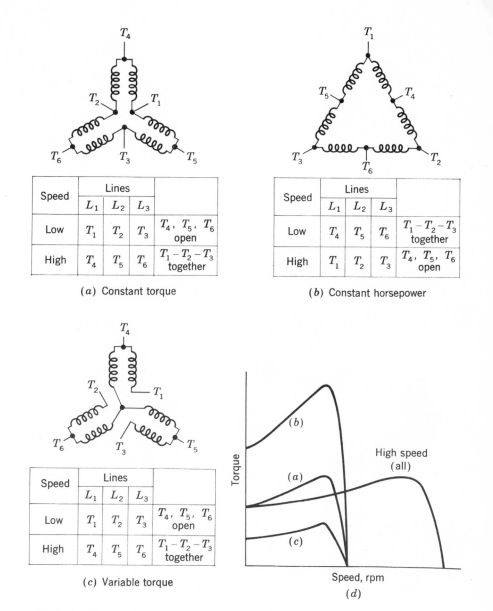

(a) Constant torque

Speed	Lines			
	L_1	L_2	L_3	
Low	T_1	T_2	T_3	T_4, T_5, T_6 open
High	T_4	T_5	T_6	$T_1 - T_2 - T_3$ together

(b) Constant horsepower

Speed	Lines			
	L_1	L_2	L_3	
Low	T_4	T_5	T_6	$T_1 - T_2 - T_3$ together
High	T_1	T_2	T_3	T_4, T_5, T_6 open

(c) Variable torque

Speed	Lines			
	L_1	L_2	L_3	
Low	T_1	T_2	T_3	T_4, T_5, T_6 open
High	T_4	T_5	T_6	$T_1 - T_2 - T_3$ together

(d)

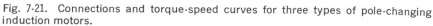

Fig. 7-21. Connections and torque-speed curves for three types of pole-changing induction motors.

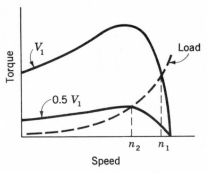

Fig. 7-22. Speed control by means of line voltage.

induction motor used in this way has characteristics similar to those of a separately excited dc motor with constant flux and variable armature voltage.

The major problem is to determine the most effective and economical source of adjustable frequency. One method is to use a wound-rotor induction machine as a frequency changer. Another, considered in Chap. 8, is to use solid-state frequency converters.

c. Line-voltage Control

The internal torque developed by an induction motor is proportional to the square of the voltage applied to its primary terminals, as shown by the two torque-speed characteristics in Fig. 7-22. If the load has the torque-speed characteristic shown by the dashed line, the speed will be reduced from n_1 to n_2. This method of speed control is commonly used with small squirrel-cage motors driving fans.

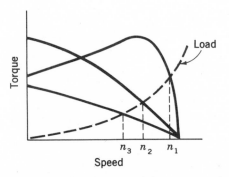

Fig. 7-23. Speed control by means of rotor resistance.

d. Rotor-resistance Control

The possibility of speed control of a wound-rotor motor by changing its rotor-circuit resistance has already been pointed out in Art. 7-6a. The torque-speed characteristics for three different values of rotor resistance are shown in Fig. 7-23. If the load has the torque-speed characteristic shown by the dashed line, the speeds corresponding to each of the values of rotor resistance are n_1, n_2, and n_3. This method of speed control has characteristics similar to those of dc shunt-motor speed control by means of resistance in series with the armature.

The principal disadvantages of both line-voltage and rotor-resistance control are low efficiency at reduced speeds and poor speed regulation with respect to change in load.

e. Control of Slip by Auxiliary Devices

In considering schemes for speed control by varying the slip, the fundamental laws relating the flow of power in induction machines should be borne in mind. The fraction s of the power absorbed from the stator is transformed by electromagnetic induction to electric power in the rotor circuits. If the rotor circuits are short-circuited, this power is wasted as rotor copper loss and operation at reduced speeds is inherently inefficient.

Numerous schemes have been invented for recovering this slip-frequency electric power. Although some of them are rather complicated in their details, they all comprise a means for introducing adjustable voltages of slip frequency into the rotor circuits of a wound-rotor induction motor. Broadly, they can be classified in two types, as shown in Fig. 7-24, where IM represents a 3-phase wound-rotor induction motor whose speed is to be regulated. In Fig. 7-24a the rotor circuits of IM are connected to auxiliary frequency-changing apparatus, represented by

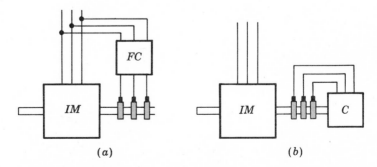

(a)   (b)

Fig. 7-24. Two basic schemes for induction-motor speed control by auxiliary machines.

the box FC, in which the slip-frequency electric power generated in the rotor of the main motor is converted to electric power at line frequency and returned to the line.   In Fig. 7-24b the rotor circuits of IM are connected to auxiliary apparatus, represented by the box C, in which the slip-frequency electric power is converted to mechanical power and added to the shaft power developed by the main motor.   In both these schemes the speed and power factor of the main motor can be adjusted by controlling the magnitude and phase of the slip-frequency emfs of the auxiliary machines.   The auxiliary apparatus may be a fairly complicated system of rotating machines and adjustable-ratio transformers or, in the case of Fig. 7-24a, may consist of solid-state frequency conversion.

7-8
RÉSUMÉ

Examination of the flux-mmf interactions in a polyphase induction motor shows that, electrically, the machine is a generalized transformer.   The synchronously rotating air-gap flux wave in the induction machine is the counterpart of the mutual core flux in the transformer.   The rotating field induces emfs of stator frequency in the stator windings and of slip frequency in the rotor windings for all rotor speeds other than synchronous speed.   Thus, the induction machine transforms voltages and at the same time changes frequency.   When viewed from the stator, all rotor electric and magnetic phenomena are transformed to stator frequency.   The rotor mmf reacts on the stator windings in the same manner that the mmf of the secondary current in a transformer reacts on the primary.

Pursuit of this line of reasoning leads to the equivalent circuit for the machines.   The effects of saturation on the equivalent circuit are less serious than in the corresponding steady-state circuit for synchronous machines.   This is largely because, as in the transformer, the performance is determined to a considerably greater extent by the leakage impedances than by the magnetizing impedance.   Care must be used in both testing and analysis, however, to reflect the effects of saturation on leakage reactances as well as of nonuniformity of current distribution on rotor resistance.

One of the salient facts affecting induction-motor applications is that the slip at which maximum torque occurs can be controlled by varying the rotor resistance.   A high rotor resistance gives optimum starting conditions but poor running performance.   A low rotor resistance, on the other hand, may result in unsatisfactory starting conditions.   The design of a squirrel-cage motor is therefore quite likely to be a compromise. Marked improvement in the starting performance with relatively little sacrifice in running performance can be built into a squirrel-cage motor by using a deep-bar or double-cage rotor whose effective resistance increases with slip. A wound-rotor motor can be used for very severe starting conditions or when speed control by rotor resistance is required. A wound-rotor motor is more expensive than a squirrel-cage motor.

For applications requiring a substantially constant speed without excessively severe starting conditions, the squirrel-cage motor usually is unrivaled, because of its ruggedness, simplicity, and relatively low cost. Its only disadvantage is its relatively low power factor (about 0.85 to 0.90 at full load for 4-pole 60-Hz motors—considerably lower at light loads and for lower-speed motors). The low power factor is a consequence of the fact that all the excitation must be supplied by lagging reactive kva taken from the ac mains. At speeds below about 500 rpm and ratings above about 50 hp or at medium speeds (500 to 900 rpm) and ratings above about 500 hp, a synchronous motor may cost less than an induction motor.

The induction motor is at a disadvantage for adjustable-speed drives. A machine dependent on a constant-speed rotating magnetic field prefers to be a constant-speed machine. Pole changing is a good solution when only two, or perhaps four, speeds are required. Speed control by varying the slip is inherently inefficient unless one of the more or less elaborate schemes described in Art. 7-7e is used for recovering the slip energy. The economic comparison for adjustable-speed drives often is between the cost of a dc motor plus ac-to-dc conversion equipment and controls, on the one hand, and the relatively elaborate schemes of induction-motor speed control described in Art. 7-7, on the other. The possibilities of a variable-speed mechanical transmission interposed between a constant-speed motor and the driven load must also be considered. In very large sizes, or at high speeds, or when space is at a premium, the simplicity and compactness of the induction motor are a big advantage.

PROBLEMS

7-1. When an induction machine is driven above synchronous speed, it is capable of acting as a generator. Sketch the rotor-mmf wave and the resultant flux-density wave in the manner of Fig. 7-1a and b, but for generator action. Show that the torque angle is $-(90° + \phi_2)$.

7-2. Redraw Fig. 7-2 for the same rotor placed in a sinusoidal 4-pole field.

7-3. A 3-phase 8-pole wound-rotor induction machine driven by an adjustable speed dc motor is to be used as an adjustable-frequency source. The stator of the machine is fed from a 60-Hz source, and the output is taken from the rotor slip rings. The output-frequency range is to be from 150 to 360 Hz. The output power is to be 80 kw at 0.80 lagging power factor and is independent of the output frequency.

For these purposes neglect all machine losses, exciting currents, and leakage reactances. Compute:

a. The required speed range of the dc motor
b. The stator kva rating of the induction machine
c. The maximum torque on the shaft of the dc motor

7-4. A 100-hp 3-phase Y-connected 440-volt 60-Hz 8-pole squirrel-cage induction motor has the following equivalent-circuit constants in ohms per phase referred to the stator:

$$r_1 = 0.085 \qquad r_2 = 0.067$$

$$x_1 = 0.196 \qquad x_2 = 0.161 \qquad x_\varphi = 6.65$$

No-load rotational loss = 2.7 kw. Stray load loss = 0.5 kw. The rotational and stray load losses may be considered constant.

a. Compute the horsepower output, stator current, power factor, and efficiency at rated voltage and frequency for a slip of 3.00 percent.
b. Compute the starting current and the internal starting torque in pound-feet at rated voltage and frequency.

7-5. A 10-hp 3-phase 60-Hz 6-pole induction motor runs at a slip of 3.0 percent at full load. Rotational and stray load losses at full load are 4.0 percent of the output power. Compute:

a. The rotor copper loss at full load
b. The electromagnetic torque at full load, in newton-meters
c. The power delivered by the stator to the air gap at full load

7-6. A 10-hp 230-volt 3-phase Y-connected 60-Hz 4-pole squirrel-cage induction motor develops full-load internal torque at a slip of 0.04 when operated at rated voltage and frequency. For the purposes of this problem rotational and core losses can be neglected. Impedance data on the motor are as follows:

Stator resistance $r_1 = 0.36$ ohm per phase

Leakage reactances $x_1 = x_2 = 0.47$ ohm per phase

Magnetizing reactance $x_\varphi = 15.5$ ohms per phase

Determine the maximum internal torque at rated voltage and frequency, the slip at maximum torque, and the internal starting torque at rated voltage and frequency. Express the torques in newton-meters.

7-7. Suppose the induction motor of Prob. 7-6 is supplied from a 240-volt constant-voltage 60-Hz source through a feeder whose impedance is $0.50 + j0.30$ ohm per phase. Determine the maximum internal torque that the motor can deliver and the corresponding values of stator current and terminal voltage.

7-8. A 3-phase induction motor, at rated voltage and frequency, has a starting torque of 160 percent and a maximum torque of 200 percent of full-load torque. Neglect stator resistance and rotational losses, and assume constant rotor resistance. Determine:

a. The slip at full load
b. The slip at maximum torque
c. The rotor current at starting, in per unit of full-load rotor current

7-9. When operated at rated voltage and frequency, a 3-phase squirrel-cage induction motor (of the design classification known as a high-slip motor) delivers full load at a slip of 8.5 percent and develops a maximum torque of 250 percent of full-load torque at a slip of 50 percent. Neglect core and rotational losses, and assume that the resistances and inductances of the motor are constant.

Determine the torque and rotor current at starting with rated voltage and frequency. Express the torque and rotor current in per unit based on their full-load values.

7-10. For a 25-hp 230-volt 3-phase 60-Hz squirrel-cage motor operated at rated voltage and frequency, the rotor copper loss at maximum torque is 9.0 times that at full-load torque, and the slip at full-load torque is 0.030. Stator resistance and rotational losses may be neglected and the reactances and rotor resistance assumed to remain constant. Find:

   a. The slip at maximum torque

   b. The maximum torque

   c. The starting torque

Express the torques in per unit of full-load torque.

7-11. A squirrel-cage induction motor runs at a slip of 5.0 percent at full load. The rotor current at starting is 5.0 times the rotor current at full load. The rotor resistance is independent of rotor frequency, and rotational losses, stray load losses, and stator resistance may be neglected.

   a. Compute the starting torque.

   b. Compute the maximum torque and the slip at which maximum torque occurs.

Express the torques in per unit of full-load torque.

7-12. The maximum internal power P_{max} of an induction motor occurs at the slip $s_{max\ P}$. Show that the normalized curves of Fig. 7-13 also give the relations between the power ratio P/P_{max} and the slip ratio $s(1 - s_{max\ P})/s_{max\ P}(1 - s)$ with the parameter $Q = (X_1 + x_2)/(R_1 + r_2)$.

7-13. A 50-hp 440-volt 3-phase 4-pole 60-Hz wound-rotor induction motor develops a maximum internal torque of 250 percent at a slip of 16 percent when operating at rated voltage and frequency with its rotor short-circuited directly at the slip rings. Stator resistance and rotational losses may be neglected, and the rotor resistance may be assumed to be constant, independent of rotor frequency. Determine:

   a. The slip at full load, in percent

   b. The rotor copper loss at full load, in watts

   c. The starting torque at rated voltage and frequency, in newton-meters

If the rotor resistance is now doubled (by inserting external series resistances), determine:

   d. The torque in newton-meters when the stator current has its full-load value

   e. The corresponding slip

7-14. A 50-hp 3-phase 440-volt 4-pole wound-rotor induction motor develops its rated full-load output at a speed of 1,746 rpm when operated at rated voltage and frequency with its slip rings short-circuited. The maximum torque it can develop at rated voltage and frequency is 200 percent of full-load torque. The resistance of the rotor winding is 0.10 ohm per phase Y. Rotational and stray load losses and stator resistance may be neglected.

    a. Compute the rotor copper loss at full load.

    b. Compute the speed at maximum torque.

    c. How much resistance must be inserted in series with the motor to produce maximum starting torque?

The motor is now run from a 50-Hz supply with the applied voltage adjusted so that the air-gap flux wave has the same amplitude at the same torque as on 60 Hz.

    d. Compute the 50-Hz applied voltage.

    e. Compute the speed at which the motor will develop a torque equal to its 60-Hz full-load value with its slip rings short-circuited.

7-15. A 220-volt 3-phase 4-pole 60-Hz wound-rotor induction motor develops an internal torque of 150 percent with a line current of 155 percent at a slip of 5.0 percent when running at rated voltage and frequency with its rotor terminals short-circuited. (Torque and current are expressed as percentages of their full-load values.) The rotor resistance is 0.100 ohm between each pair of rotor terminals and may be assumed to be constant. What should be the resistance of each of three balanced Y-connected resistors inserted in series with each rotor terminal if the starting current at rated voltage and frequency is to be limited to 155 percent? What internal starting torque will be developed?

7-16. A 220-volt 3-phase 4-pole 60-Hz squirrel-cage induction motor develops a maximum internal torque of 250 percent at a slip of 16 percent when operating at rated voltage and frequency. If the effect of stator resistance is neglected, determine the maximum internal torque that this motor would develop if it were operated at 200 volts and 50 Hz. Under these conditions, at what speed in rpm would maximum torque be developed?

7-17. A frequency-changer set is to be designed for supplying variable-frequency power to induction motors driving the propellers on scale-model airplanes for wind-tunnel testing, as described in Art. 7-7b. The frequency changer is a wound-rotor induction machine driven by a dc motor whose speed can be controlled. The 3-phase stator winding of the induction machine is excited from a 60-Hz source, and variable-frequency 3-phase power is taken from its rotor winding. The set must meet the following specifications:

Output frequency range = 120 to 450 Hz

Maximum speed not to exceed 3,000 rpm

Maximum power output = 80 kw at 0.80 power factor and 450 Hz

The power required by the induction-motor load drops off rapidly with decreasing frequency, so that the maximum-speed condition determines the sizes of the machines.

On the basis of negligible exciting current, losses, and voltage drops in the induction machine, find:

a. The minimum number of poles for the induction machine
b. The corresponding maximum and minimum speeds
c. The kva rating of the stator winding of the induction machine
d. The horsepower rating of the dc machine

7-18. The resistance measured between each pair of slip rings of a 3-phase 60-Hz 300-hp 16-pole induction motor is 0.035 ohm. With the slip rings short-circuited, the full-load slip is 0.025, and it may be assumed that the slip-torque curve is a straight line from no load to full load. This motor drives a fan which requires 300 hp at the full-load speed of the motor. The torque required to drive the fan varies as the square of the speed. What resistances should be connected in series with each slip ring so that the fan will run at 300 rpm?

7-19. An adjustable-speed drive is to be furnished for a large fan in an industrial plant. The fan is to be driven by two wound-rotor induction motors coupled mechanically to the fan shaft and arranged so that lower speeds will be carried on the smaller motor and higher speeds on the larger motor. The speed control is to be arranged in steps so that the control will be uninterrupted from minimum to maximum speed and so that there will not be a sudden change in speed during the transfer from one motor to the other.

The larger motor is to be a 2,300-volt 3-phase 500-hp 60-Hz 6-pole motor; the smaller motor is to be a 200-hp 60-Hz 8-pole motor. The following motor data are furnished by the manufacturer:

Constants	200-hp motor	500-hp motor
Stator resistance	0.57 ohm	0.14 ohm
Rotor resistance	0.93	0.24
Stator plus rotor leakage reactance	2.6	0.98

These values are per-phase values (Y connection) referred to the stator. Motor rotational losses and exciting requirements may be ignored.

The minimum operating speed of the fan is to be 450 rpm; the maximum speed is to be approximately 1,170 rpm. The fan requires 450 hp at 1,170 rpm, and the power required at other speeds varies nearly as the cube of the speed.

The proposed control scheme is based on the stators of both motors being connected to the line at all times when the drive is in operation. In the lower speed range, the rotor of the 500-hp motor is open-circuited. This lower range is obtained by adjustment of external resistance in the rotor of the 200-hp motor. Above this lower speed range, the rotor of the 200-hp motor is open-circuited; speed adjustment in the upper range is obtained by adjustment of external resistance in the rotor of the 500-hp motor.

The transition between motors from the lower to the upper speed range is to be handled as follows:

1 All external rotor resistance is cut out of the 200-hp motor.
2 By means of a closed-before-open type of contactor, the rotor circuit of the 500-hp motor is closed, and then the rotor circuit of the 200-hp motor is opened. The external rotor resistance in the 500-hp motor for this step has such a value that the speed will be the same as in the first step.
3 Higher speeds are obtained by cutting out rotor resistance in the 500-hp motor.

This procedure is essentially reversed in going from the higher to the lower speed range.

As plant engineer, you are asked to give consideration to some features of this proposal. In particular, you are asked to:

a. Determine the range of external rotor resistance (referred to the stator) which must be available for insertion in the 200-hp motor.
b. Determine the range of external rotor resistance which must be available for insertion in the 500-hp motor.
c. Discuss any features of the scheme which you may not like, and suggest alternatives.

8
solid-state motor control

In most applications, motors are operated directly from a supply line under their own inherent torque-speed characteristics and at operating conditions determined by the mechanical load. However, in many other applications, the motors are provided with control equipment by which their characteristics can be adjusted and their operating conditions with respect to the mechanical load varied to suit the particular requirements. The most common control adjustment is motor speed, but torque and acceleration can be adjusted as well. The control equipment consists of relays, contactors, magnetic components, and solid-state devices such as diodes, thyristors, and transistors. The combination of the motor and the control equipment is called a drive system.

8-1
APPLICATION OF SOLID-STATE CONTROLS

Each type of motor can be controlled to provide adjustment of speed and torque with respect to the mechanical load. The difference among the motors lies in the amount and cost of the equipment required to achieve the control. Direct-current motors are easily controllable and have dominated the adjustable-speed drive field. Alternating-current motors are more expensive to control and are used in drive systems when special features of the ac motors, such as absence of commutators and brushes, must be utilized.

a. Terminology

In comparing one drive system with another, and in assigning numerical values to performance, one must use a terminology that is understood in the industry. We will define key terms as follows:

1 Base Speed. The base speed is the nominal or rated speed of the motor, also referred to as the *nameplate* speed. In a dc shunt motor, it is the speed for rated armature voltage and rated field current. In an ac motor, it is the speed at rated frequency. The motor can be operated above and below the base speed.

2 Speed Regulation. Regulation generally refers to the ability of the drive system to maintain the preset speed under varying loads. For a full-load speed ω_m and a speed drop $\Delta\omega_m$ from no load to full load, the percent speed regulation is

$$\text{Reg.} = \frac{\Delta\omega_m}{\omega_m} \times 100 \qquad (8\text{-}1)$$

The percent speed regulation is usually given at the base speed of the motor. At a lower speed, the motor may have the same speed drop $\Delta\omega_m$, but the percent speed regulation will be higher.

3 Speed Range. The speed range is the ratio of the maximum-to-minimum speed of the drive system. The speed range can be given only with the motor loading specified, e.g., no-load or full-load torque. The limitation of the speed range may lie with the inherent design of the control equipment, or it may lie with the ability of the motor to cool itself at low speeds or avoid flying apart at high speeds.

4 Torque Limit. Circuits are incorporated in control equipment to limit the torque of the motor during acceleration or overload periods. The purposes are numerous: to obtain timed acceleration of high-inertia loads, to protect the motor and control equip-

ment from overcurrent, and to prevent the drive from getting into unstable operating regions.

5 Constant-horsepower Drive.   The drive system has the inherent capability of supplying a constant horsepower (torque × speed) over a given speed range.   The highest torque at the lowest speed generally sets the size of the motor and the control equipment.

6 Constant-torque Drive.   The drive system has the inherent capability of supplying a specific torque over the given speed range.

7 Thyristor.   A three-terminal solid-state device which conducts current from its anode terminal to its cathode terminal after a voltage signal is applied to its gate terminal.   The thyristor, also termed *SCR*, is employed to obtain controlled dc voltages from ac supply lines.

b. DC Motor Systems

As described in Chap. 5, the torque-speed characteristics of a dc motor can be controlled by adjusting the armature voltage, by adjusting the field current, and by inserting resistance into the armature circuit.   Solid-state motor controls are designed to utilize each of these modes for particular purposes.

Adjustable-armature-voltage systems utilize phase-controlled thyristor rectifier circuits to provide dc power for the dc motors.   The rectifier circuits operate from single-phase and three-phase ac lines, and in various configurations depending upon horsepower rating, reversibility, and braking requirements.   A commercial system is shown in Fig. 8-1.   The armature voltage is adjusted by controlling the electrical angle within the ac wave at which the gate signal is applied to each thyristor.

Field current is supplied from the main rectifier when the field current is not controlled.   When the field current must be reduced to obtain speeds above the base speed, the current is supplied from an auxiliary controlled-rectifier circuit, which is actuated by a crossover circuit in the speed-control reference circuit.

The equivalent of armature resistance is used for speed control of dc series motors operating from dc sources, such as a battery in an electric-drive vehicle or a third rail in a rapid-transit system.   A thyristor is made to switch on and off at a fast rate so that the applied voltage divides between the armature and the switch, resulting in a controllable average armature voltage.   The thyristor switch circuit is called a *chopper* and acts as a lossless armature resistance.   A system for rapid-transit cars is shown in Fig. 8-2.

Fig. 8-1. Commercial thyristor dc motor drive system. *(Electric Regulator Corporation.)*

c. Adjustable-frequency AC Systems

The synchronous speed of induction and synchronous motors is directly proportional to the applied frequency. Speed control of these motors is obtained by supplying the stator electrical power from an adjustable-frequency, solid-state control unit. The synchronous motor will follow the frequency directly; the induction motor will follow within the slip frequency. Such control is efficient and, with synchronous motors, can be made highly accurate. However, the control equipment is more complicated and expensive than the controlled rectifiers for the dc drive systems.

The adjustable-frequency power is generated by a thyristor circuit called an *inverter*. To form a set of three-phase voltage waves in an inverter, six thyristors are connected in a bridge configuration to a dc bus.

By switching the thyristors with externally formed signals in a prescribed sequence, a rectangular or stepped approximation to a set of
three-phase sine waves is formed at the ac terminals of the bridge. By
speeding up or slowing down the switching rate, the frequency and motor
speed are controlled. A typical drive system consists of a rectifier to
convert power from the ac supply line to dc form, an inverter to form the
adjustable-frequency ac waves from the dc bus, an induction or synchronous motor, and a control system to set the frequency, adjust the motor
voltage, and insure that the maximum torque of the motor is not exceeded
during speed changes. Typical commercial drive systems are shown in
Figs. 8-3 and 8-4.

d. Adjustable-voltage Induction-motor Systems

At a given value of slip, the air-gap torque of an induction motor is nearly
proportional to the square of stator voltage. Depending upon the shape
of the torque-speed curve and the load characteristic, a degree of speed
control can be obtained with a relatively simple control system. The
rotor power loss, which is proportional to torque and slip, can become
large as the motor slows down. This method of control is particularly
suitable for loads such as pumps and fans whose torque requirements
drop rapidly as speed is reduced.

Fig. 8-2. Transit vehicle thyristor control assembly. *(Westinghouse Electric Corporation.)*

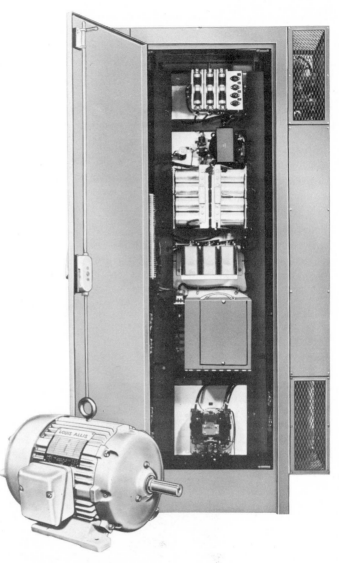

Fig. 8-3. Commercial ac adjustable-frequency drive system. *(The Louis Allis Company.)*

A typical adjustable-voltage control system uses thyristors connected *back-to-back* in each stator supply line. As the electrical angle at which the gate signals are applied to the thyristors is advanced toward the start of the ac wave, more of the line voltage is applied to the motor and its speed rises. The resultant motor voltage is distorted by the switching action of the thyristors, so that the motor losses tend to be higher than normal.

e. Wound-rotor Motor Systems

As described in Chap. 7, the speed-torque curve of an induction motor can be shifted so as to obtain speed control by inserting resistance into the rotor circuit. The method is wasteful of power because the rotor power loss must be $s/(1 - s)$ of the mechanical power. For example, at half speed, $s = 0.5$, the rotor power loss is equal to the mechanical power. The bulk of the rotor power loss can be recovered in a solid-state motor control system consisting of a rectifier and an inverter. The rotor voltage from the slip rings is rectified to obtain a dc voltage whose magnitude is directly proportional to slip s; a thyristor inverter connected between the dc bus and the stator terminals returns the rotor power to the supply line. The inverter operates with the fixed output frequency of the motor supply line, e.g., 60 Hz, and from any dc input voltage level from nearly zero at full motor speed to the maximum value when the motor is at standstill. The system is relatively simple because the thyristor gate signals are formed from the supply line without the complicated logic circuits and regulators of the adjustable-frequency system.

Fig. 8-4. Ac drive system consisting of 6 inverters and 1 dc rectifier for synthetic fiber production. The inverters are rated from 40 to 84 kva and all drive synchronous reluctance motors over a 6-to-1 speed range. *(Westinghouse Electric Corporation.)*

f. Series Universal Motors

Series universal motors are seldom built over one-horsepower rating and are used principally for portable tools and appliances to obtain high horsepower per unit weight. Their speed at fixed torque is proportional to applied voltage. A single bidirectional ac solid-state controlled switch, like two thyristors connected back-to-back, in series with the motor offers a simple and inexpensive means for speed control. The electrical angle at which the gate signal is applied is controlled in various ways: manually from a trigger; from preset buttons or knobs; or from a speed-regulating circuit. The system will be described in Chap. 11.

8-2
INTRODUCTION TO RECTIFIER CIRCUITS

The rectifier is the principal item in the solid-state control system for a dc motor. The rectifier is a circuit assembled from diodes and thyristors which supplies direct current to the armature and field of the motor from the ac supply line. The dc voltage of the rectifier, which is controlled by the thyristors, controls in turn the speed or torque of the motor. Several circuits with passive loads will be described in this article; motor loads will be discussed in Art. 8-3.[1]

The diode and the thyristor used in rectifier circuits are highly non-linear devices. In order to analyze circuits containing these devices, we require circuit models and the technique for using these models in the repetitive operation that occurs.[2]

The actual volt-ampere characteristic of a diode is shown in Fig. 8-5a. The diode is shown to conduct its rated forward current of 160 amps with a forward drop of 1.0 volts, and can block 1,200 volts in the reverse direction. The circuit model for the diode is merely a switch as shown in Fig. 8-5b; the characteristic for the model is shown in Fig. 8-5c. When forward voltage is impressed on the model, the switch is closed and the diode carries current with zero voltage drop. When reverse voltage is impressed, the switch is opened and the diode carries zero reverse current.

The actual volt-ampere characteristic of a thyristor is shown in Fig. 8-6a. This thyristor is rated to block the applied voltage up to

---

[1] The reader is referred to J. Schaefer, "Rectifier Circuits, Theory and Design," John Wiley & Sons, Inc., New York, 1965.

[2] Descriptions and applications of these devices are given in the "SCR Manual," 4th ed., General Electric Co., Auburn, New York, 1967.

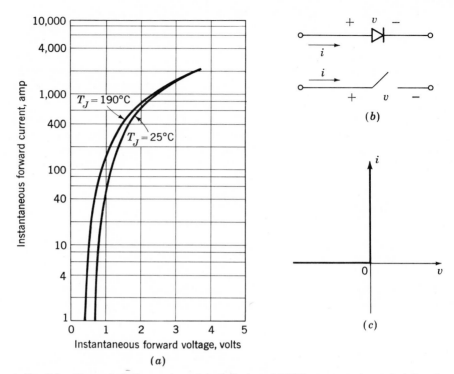

Fig. 8-5. Characteristics and model of diode. *(a)* Volt-ampere characteristic of a 160-amp diode in terms of junction temperature T_J; *(b)* symbol for diode *(upper)* and circuit model switch *(lower)*; *(c)* volt-ampere characteristic of circuit model.

1,800 volts in either direction in the absence of a gate signal. When a gate signal, which is typically a pulse of a few volts amplitude, is applied, the thyristor immediately becomes conducting in the forward direction like a diode. The forward voltage drop is 1.5 volts for a current of 110 amps. The circuit model is a switch as shown in Fig. 8-6*b* and the characteristic for the model is shown in Fig. 8-6*c*. The switch is open for either direction of current flow before a gate signal is applied. When forward voltage and a gate signal are applied, the switch is closed and the thyristor model carries forward current with no voltage drop.

a. Single-phase, Half-wave, Diode Rectifier

The simplest rectifier circuit consists of a single diode operating into a resistance load as shown in Fig. 8-7*a*. The line voltage is a sine wave as shown in Fig. 8-7b. According to the diode model of Fig. 8-5, the diode acts like an open switch when the diode voltage is negative and conducts zero current; this occurs typically from $\omega t = \pi$ to 2π. The diode acts like a closed switch when the diode voltage is positive, typically from $\omega t = 0$ to π, and conducts current i_n. The load voltage v_n is equal to the line voltage v_0 during the conducting interval because the diode model is lossless.

b. Single-phase, Half-wave Thyristor

The simplest controlled-rectifier circuit consists of a single thyristor operating into a resistance load as shown in Fig. 8-8a. The circuit for applying gating pulses to the thyristor is not shown. According to the thyristor model of Fig. 8-6, the thyristor can only conduct current when its voltage is positive, typically during the intervals $\omega t = 0$ to π, 2π to 3π, etc. In addition, the thyristor must also receive a gating pulse, which we assume is applied at the firing angle α. By convention, the

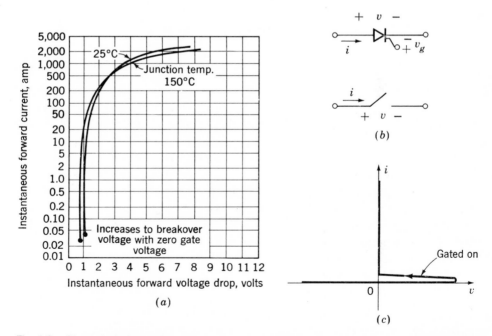

Fig. 8-6. Characteristics and model of a thyristor (SCR). *(a)* Volt-ampere characteristics of a 110-amp thyristor in terms of junction temperature; *(b)* symbol for thyristor *(upper)* and circuit model switch *(lower)*; *(c)* volt-ampere characteristic of circuit model.

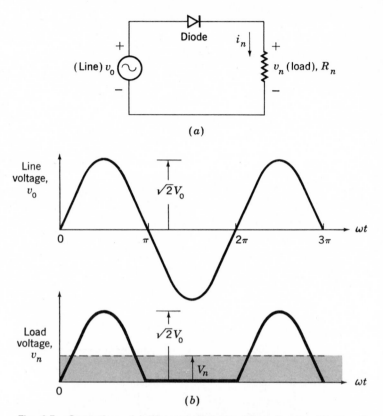

Fig. 8-7. Operation of half-wave diode-rectifier circuit. *(a)* Circuit; *(b)* waveforms of line voltage v_0 and load voltage v_n, where V_0 is the rms value of the sine-wave line voltage and V_n is the average value of the load voltage, shown shaded.

firing angle α is measured from the angle that produces the largest load voltage, in this case, from $\omega t = 0$, 2π, etc. As shown in Fig. 8-8b, the thyristor conducts from $\omega t = \alpha$ to π, $2\pi + \alpha$ to 3π, etc. During the conduction intervals the load voltage v_n is equal to the line voltage v_0. As the firing angle α is shifted by the control circuit from zero to π, the average value V_n of the load voltage decreases as well.

EXAMPLE 8-1

Find the expression for the average load voltage V_n as a function of the firing angle α for the circuit of Fig. 8-8a.

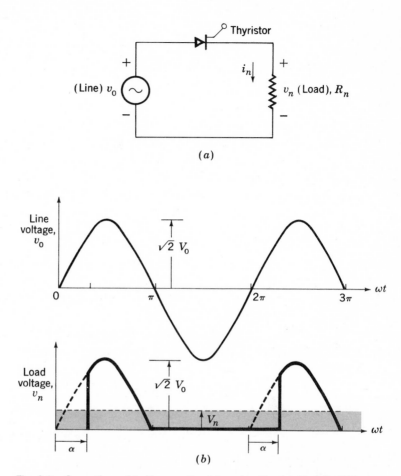

(a)

(b)

Fig. 8-8.   Operation of half-wave thyristor circuit.   (a) Circuit; (b) wave-forms of line voltage v_0 and load voltage v_n, where V_0 is the rms value of the sine-wave line voltage and V_n is the average value of the load voltage, shown shaded.

Solution

The average voltage is given by

$$V_n = \frac{1}{2\pi} \int_\alpha^\pi v_n \, d(\omega t) \qquad (8\text{-}2)$$

$$= \frac{1}{2\pi} \int_\alpha^\pi \sqrt{2} \, V_0 \sin \omega t \, d(\omega t)$$

$$= \frac{-1}{\sqrt{2}\,\pi} [V_0 \cos \omega t]_\alpha^\pi$$

$$= 0.225 V_0 (1 + \cos \alpha) \qquad (8\text{-}3)$$

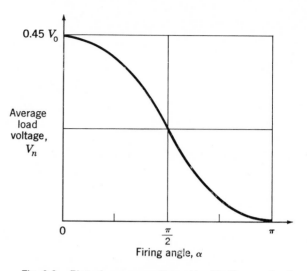

Fig. 8-9. Plot of average voltage V_n of half-wave thyristor circuit as a function of firing angle α.

The relationship is shown in Fig. 8-9. The maximum value of V_n, corresponding to the half-wave diode rectifier operation of Fig. 8-7, occurs for $\alpha = 0$ and is

$$V_n = 0.450 V_0 \qquad (8\text{-}4)$$

c. Single-phase Rectifier, Reactive Load

Actual dc motor loads on rectifiers do not act as resistance alone. Armature circuits have resistance and inductance; the armature and load inertia act like capacitance. Field circuits are highly inductive. The operation of the diode and thyristor elements is different from that with resistance load, but the elements must still behave in accordance with the constraints of their models.

A half-wave diode rectifier with RL load is shown in Fig. 8-10a. The waveforms of line voltage v_0, resistance voltage v_r, and inductance voltage v_l are shown in Fig. 8-10b. At $\omega t = 0$, the line voltage v_0 becomes positive and the diode starts to conduct the current i_n. The inductance forces the current to lag the voltage until the current reaches its peak value at angle ωt_1. During this period, the flux linkage λ in the inductance increases by

$$\Delta\lambda = \int_0^{t_1} v_l \, dt = \frac{1}{\omega} \int_0^{\omega t_1} v_l \, d(\omega t) \qquad (8\text{-}5)$$

From angle ωt_1 to β, the current declines until the flux linkage λ returns to its value at $\omega t = 0$, or

$$\Delta\lambda = \int_{t_1}^{\beta/\omega} v_l \, dt = \frac{1}{\omega} \int_{\omega t_1}^{\beta} v_l \, d(\omega t) \qquad (8\text{-}6)$$

The integrals of Eqs. 8-5 and 8-6 are represented by the areas shown shaded in Fig. 8-10b. The shaded areas, frequently termed *volt-time areas*, on both sides of the current peak at ωt_1 must be equal.

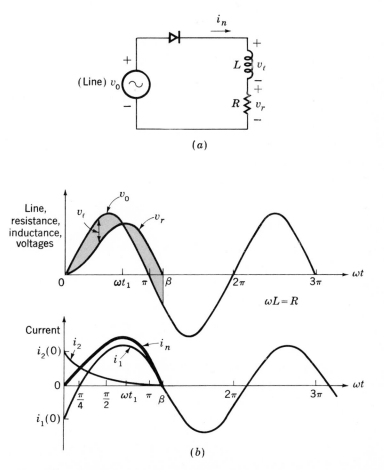

Fig. 8-10.   Operation of half-wave diode rectifier with reactive *RL* load. (a) Circuit; (b) waveforms of load-voltage components v_r and v_l, and load current i_n.

EXAMPLE 8-2

For the circuit of Fig. 8-10a, find the current i_n for the conduction interval $\omega t = 0$ to β.   Assume that $R = \omega L$.

Solution

The current i_n consists of a steady-state component i_1 and a transient component i_2.   The boundary condition at $\omega t = 0$ is

$$i_n(0) = i_1(0) + i_2(0) = 0$$

The steady-state component i_1 is given by

$$i_1 = \frac{\sqrt{2}\,V_0}{[R^2 + (\omega L)^2]^{\frac{1}{2}}} \sin(\omega t - \phi) \qquad (8\text{-}7)$$

where $\phi = \arctan \omega L/R$.   The transient component is

$$i_2 = I_2 \epsilon^{-t/\tau} \qquad (8\text{-}8)$$

where $\tau = L/R$.   The boundary condition requires that

$$I_2 = \frac{\sqrt{2}\,V_0}{[R^2 + (\omega L)^2]^{\frac{1}{2}}} \sin \phi \qquad (8\text{-}9)$$

For $R = \omega L$, the components are

$$i_1 = \frac{V_0}{R} \sin\left(\omega t - \frac{\pi}{4}\right) \qquad (8\text{-}10)$$

$$i_2 = 0.707 \frac{V_0}{R} \epsilon^{-\omega t} \qquad (8\text{-}11)$$

The components and total current i_n for the example are shown in the waveforms of Fig. 8-10b.   The peak current occurs at approximately $\omega t_1 = 3\pi/4$, (135°) and the extinction angle at $\beta = 5\pi/4$, (225°).

During the period $\omega t = 0$ to ωt_1 in Fig. 8-10b, the line supplies energy to the resistance R at the rate $v_r i_n$ and to the inductance L at the rate $v_l i_n$. From $\omega t = \omega t_1$ to π, the line continues to supply energy to the resistance R in addition to the energy extracted from the inductance L at the rate

$(-)\, v_{Ln}$.   From $\omega t = \pi$ to β, the energy extracted from the inductance L is not only supplied to the resistance R, but is returned to the line as well, forcing the conduction period of the current to extend beyond π to the extinction angle β.

d. Three-phase, Half-wave Rectifier

The simplest three-phase rectifier circuit is the half-wave configuration shown in Fig. 8-11a.   The circuit consists of three diodes connected to a common resistance load R; the load current is returned to the neutral n

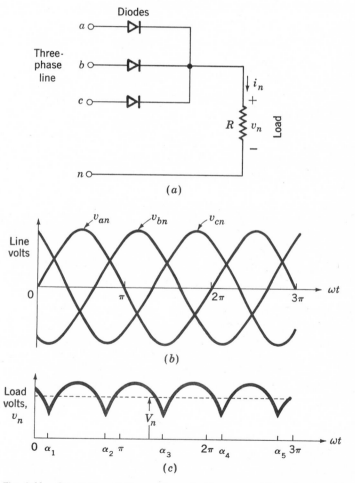

Fig. 8-11. Operation of three-phase half-wave rectifier circuit. *(a)* Circuit; *(b)* three-phase line-to-neutral voltages; *(c)* load voltage v_n.

of the three-phase supply line. The three-phase supply voltages, v_{an}, v_{bn}, and v_{cn} and the load voltage are shown in Fig. 8-11b and c.

The diodes conduct one at a time for 120° each, in sequence. The cathode ends of the three diodes are connected to a common point on the load resistance, while the anode ends are connected to the respective supply voltages. In accordance with the model of Fig. 8-5, each diode will conduct only when its voltage is positive and block when it is negative. At any time, the conducting diode is the one connected to the highest instantaneous supply voltage. The common terminal of the cathodes is also raised to the same voltage, so that the blocking diodes are *reverse biased.* As shown in Fig. 8-11c, the load voltage follows the envelope of the highest instantaneous supply voltages.

The load voltage v_n is continuous and has an average value V_n given by

$$V_n = \frac{3}{2\pi} \int_{\alpha_1}^{\alpha_2} \sqrt{2}\, V_0 \sin\,(\omega t)d(\omega t) \tag{8-12}$$

$$= \frac{-3}{\sqrt{2}\,\pi} \,[V_0 \cos\,(\omega t)]_{\alpha_1}^{\alpha_2}$$

For $\alpha_1 = \pi/6$ and $\alpha_2 = 5\pi/6$,

$$V_n = 1.17V_0 \tag{8-13}$$

There are numerous three-phase rectifier circuits using both diodes and thyristors. In the circuits using diodes, the dc load voltage is directly proportional to the ac source voltage. In the circuits using thyristors, the load voltage can be controlled independently of the ac source voltage. Rectifier circuits are selected to meet the requirements of the application and to minimize the cost. The cost of diodes and thyristors increases with both current and voltage rating; the use of a lesser number of costly elements must be balanced against a greater number of less costly elements to achieve the best rectifier design.

8-3
DC MOTOR DRIVE SYSTEMS

We will consider three dc motor drive circuits corresponding to the three rectifier circuits described in Art. 8-2 for resistance load.[1]   The rectifiers

---

[1] See A. Kusko, "Solid State DC Motor Drives," M.I.T. Press, Cambridge, Mass., 1969.

will be used to supply the armature circuit of a dc motor.   For simplicity, the armature circuit will be modeled as an armature-generated voltage e_a and an armature-circuit inductance L_a; the resistance R_a and the effect of the brushes will be neglected.   The field flux Φ_f is assumed constant. The armature will be assumed to have an inertia J and a mechanical load torque proportional to velocity, $B\omega_m$.

a. Single-phase, Half-wave Diode Motor Drive System

The simplest dc motor drive circuit operating from an ac line consists of a single diode supplying the armature circuit.   The field winding requires a separate source of excitation, or the motor can use a permanent-magnet field.   The drive system has no means for armature-voltage control; unless the field current is adjusted, it will operate on one torque-speed characteristic.   However, the circuit will be used to show the basic principles of armature-circuit operation from a rectifier.   In the next section, we will replace the diode with a thyristor and consider a con-trolled drive system.

The waveforms of voltage, current, and speed are shown in Fig. 8-12 for two levels of torque at the same average speed.   The armature receives pulses of current i_a once per cycle of the line voltage v_0.   As shown in Fig. 8-12a, the current i_a starts at $\omega t = \alpha$, when the line voltage v_0 attempts to rise above the armature emf e_a, and the diode conducts. The current i_a continues until $\omega t = \beta$ for a conduction angle γ.   Just as for the passive RL circuit of Fig. 8-10, the conduction period ends at $\omega t = \beta$, when the shaded volt-time areas of the inductance voltage bal-ance.   The end of current conduction indicates that the armature induc-tance has returned its stored energy to the circuit, or the armature-circuit flux linkages have returned to their starting point at $\omega t = \alpha$.

During the armature-current conduction period, electrical energy flows into the armature circuit; the interaction of the current and field flux results in positive electromagnetic torque.   The motor accelerates dur-ing the conduction period by the speed rise $\Delta\omega_m$, as shown in Fig. 8-12a. From the end of the conduction period at $\omega t = \beta$ of one cycle and the start of the next conduction period at $\omega t = 2\pi + \alpha$, the motor coasts down by $\Delta\omega_m$.   The motor supplies the load energy during the coasting period from its own kinetic energy; the torque is $T_m = J \, d\omega_m/dt$.   The armature inductance acts as a reservoir of electrical energy during the conduction period; the armature and load inertia act as a reservoir of mechanical energy during the coasting period.

The waveforms of Fig. 8-12b show the operation when the load torque is increased but the speed readjusted to the same average value.

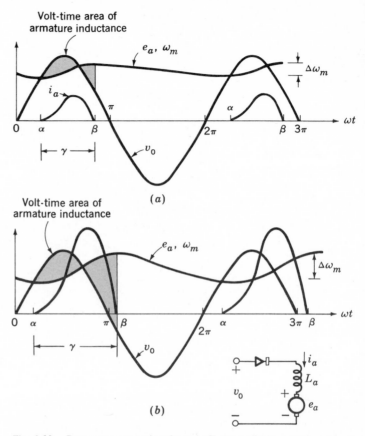

Fig. 8-12.   Dc motor operating from half-wave diode-rectifier circuit.
(a) Waveforms of speed ω_m, current i_a, and emf e_a at light load; (b)
same waveforms at heavy load and same average speed.

The speed dip $\Delta\omega_m$ during the coasting period increases because more
torque must be supplied than for Fig. 8-12a and the diode starts conducting earlier during its half cycle.   Furthermore, the amplitude and average
value of the current pulse i_a is greater, the volt-time areas are greater,
and the conduction period γ increases.   At low speeds and high torques,
the conduction period will exceed the angle π and will approach 2π.

EXAMPLE 8-3

A 1-hp, 500-rpm, dc motor having armature inertia $J = 0.050$ lb-ft-
sec² and magnetic torque of 15 lb-ft is driven by a 60-Hz half-wave
diode circuit.   Assume that the coasting period of the motor is

π rad between conduction periods, as shown in Fig. 8-12b.   Find the speed dip between conduction periods when the motor is delivering rated torque at rated speed.

Solution

During the coasting period, the air-gap torque is zero; the differential equation describing the velocity of the motor is given by

$$J \frac{d\omega_m}{dt} + B\omega_m = 0 \qquad (8\text{-}14)$$

where we have assumed that the load torque is proportional to velocity.   The motor velocity in terms of the initial velocity ω_{m0} is thus

$$\omega_m = \omega_{m0}\epsilon^{-t/\tau_L} \qquad (8\text{-}15)$$

The mechanical time constant of the motor is given by

$$\tau_L = \frac{J}{B} \qquad (8\text{-}16)$$

The values of J and B are

$$J = 0.050 \text{ lb-ft-sec}^2 \times 1/0.738 = 0.0678 \text{ kg-m}^2$$

$$T = 15 \text{ lb-ft} \times 1/0.738 = 20.3 \text{ newton-meters}$$

$$B = 20.3/(500 \times 2\pi \times \tfrac{1}{60}) = 0.388 \text{ newton-meter-sec}$$

$$\tau_L = \frac{0.0678}{0.388} = 0.174 \text{ sec}$$

The mechanical time constant τ_L is large compared to the time Δt of π/ω, 8.3 ms.   Hence, the speed dip is given by

$$\Delta\omega_m = \frac{d\omega_m}{dt} \Delta t \qquad (8\text{-}17)$$

From Eq. 8-15

$$\frac{d\omega_m}{dt} = -\omega_{m0} \frac{1}{\tau_L}$$

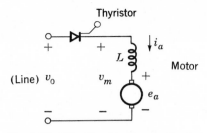

Fig. 8-13. Circuit of half-wave thyristor dc motor drive.

Substituting in Eq. 8-17

$$\text{Speed dip} = -\,\frac{8.3}{174} \times 500 \text{ rpm} = -23.8 \text{ rpm}$$

The motor-speed fluctuation is almost 5 percent of the base speed, so
that the motor can supply the load torque between conduction
periods without stalling.

___

b. Single-phase, Half-wave Thyristor Drive System

The half-wave diode drive system shown in the previous article 8-3a is
suitable for field control over a typical 3-to-1 range of speed.   The circuit
and waveforms for a thyristor drive system are shown in Figs. 8-13 and
8-14.   The diode of Fig. 8-12 has been replaced by a thyristor; by control of the firing angle the voltage applied to the armature circuit and
the speed of the motor can be adjusted over a wide range.   The operation

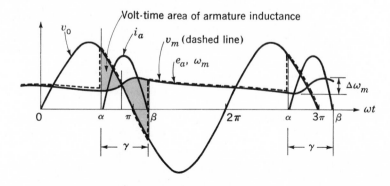

Fig. 8-14   Waveforms of speed ω_m, current i_a, motor voltage v_m, and
emf e_a for circuit of Fig. 8-13.

of the circuit is analogous to the RL load case of Art. 8-2c and also to the previous case of the half-wave diode circuit.

As shown in Fig. 8-14 the pulse of armature current i_a starts at firing angle α when the line voltage v_0 exceeds the motor voltage v_m and a gate pulse is applied to the thyristor. The current pulse continues for conduction angle γ until it reaches zero at $\omega t = \beta$; the thyristor then blocks until it receives its next gate pulse at $\omega t = 2\pi + \alpha$. The volt-time area of the armature inductance voltage v_l is shown by the shaded area. The positive and negative areas balance when the current i_a reaches zero, designating the return of the flux linkage to its initial value and the return of the magnetic-field energy to the circuit. For larger torques at the same speed, the angle α must be advanced; the current pulse increases in amplitude and conduction angle and the speed dip is more pronounced. For higher speed at the same torque, the angle α must be advanced so that the conduction period takes place closer to the peak of the line voltage v_0 wave.

The equation for the armature circuit during the conduction period is

$$v_0 = i_a R_a + L_a \frac{di_a}{dt} + e_a \tag{8-18}$$

The equation can be integrated over the conduction period

$$\int_{\alpha/\omega}^{\beta/\omega} v_0 \, dt = R_a \int_{\alpha/\omega}^{\beta/\omega} i_a \, dt + L_a \int di_a + \int_{\alpha/\omega}^{\beta/\omega} e_a \, dt \tag{8-19}$$

The interpretation of the terms of Eq. 8-19 must be done carefully. The inductance term is zero because the current i_a returns to its initial value at the end of the period of integration. The left-hand term can only be evaluated during the conduction period when the thyristor is conducting and the voltage v_0 is a segment of the line voltage. Each term can be divided by the conduction period γ/ω and interpreted as

$$V'_m = I'_a R_a + E'_a \tag{8-20}$$

where V'_m = average voltage applied to the motor over the conduction period, a segment of the line voltage.

I'_a = average current over the conduction period.

E'_a = average armature emf over the conduction period.

If we wish to consider the average values over the period of a line-voltage cycle, then we can write the unprimed expression

$$V_m = I_a R_a + E_a \qquad (8\text{-}21)$$

where V_m = average voltage at the motor terminals
I_a = average armature current = $(\gamma/2\pi)I_a'$
E_a = average armature-generated voltage = $K_m \Omega_m = E_a'$
Ω_m = average motor speed

The relationships given in Art. 4-4 for the magnetic torque T and generated voltage e_a can be expressed in terms of an electromechanical constant K_m for the condition of constant field flux, such that $T = K_m i_a$; $e_a = K_m \omega_m$.

The equation for the mechanical system is

$$T = K_m i_a = T_L + J \frac{d\omega_m}{dt} \qquad (8\text{-}22)$$

The equation can be integrated over a period of the line voltage

$$K_m \int_0^{2\pi/\omega} i_a \, dt = \int_0^{2\pi/\omega} T_L \, dt + J \int d\omega_m \qquad (8\text{-}23)$$

to yield $$K_m I_a = T_L \qquad (8\text{-}24)$$

Equations 8-21 and 8-24 show that the relationship between the average speed Ω_m and average load torque T_L is given in terms of average motor voltage V_m and average current I_a, just as for a dc motor operated from a fixed dc voltage source. One must be careful in using the equations to predict performance because the average motor voltage V_m' is not an independent variable; the firing angle α is usually the independent variable and the average motor voltage is a function of the angle α and the conduction angle γ. The conduction angle γ changes with the armature current. Example 8-4 will show how the equations are used.

EXAMPLE 8-4

The 1-hp motor of Example 8-3 is operating at 38.4 percent magnetic torque in a half-wave thyristor circuit. The firing angle α is observed to be 90° and the extinction angle β, 210°. The motor constants are R_a = 7.56 ohms; L_a = 0.55 henry; K_m = 4.23 newton-meters/amp or 4.23 volt-sec/rad. The rms line voltage is 120 volts. Find the average speed Ω_m.

Solution

The average motor voltage over the conduction period $\gamma = 2\pi/3$ is

$$V'_m = \frac{3\sqrt{2}\ V_0}{2\pi} \int_{\pi/2}^{7\pi/6} \sin \omega t\ d(\omega t) = \frac{-3\sqrt{2}\ V_0}{2\pi}\ [\cos \omega t]_{\pi/2}^{7\pi/6}$$

(8-25)

$$= \frac{3\sqrt{2} \times 120 \times 0.866}{2\pi} = 69.6 \text{ volts}$$

The average armature current for 38.4 percent of rated torque is

$$I_a = \frac{T_L}{K_m} = \frac{7.80}{4.23} = 1.84 \text{ amp} \tag{8-26}$$

The average armature current over the conduction period is

$$I'_a = 3I_a = 5.52 \text{ amp}$$

The average speed Ω_m is given by Eq. 8-20 as

$$\Omega_m = \frac{V'_m - I'_a R_a}{K_m}$$

$$= \frac{69.6 - 5.52 \times 7.56}{4.23}$$

$$= 6.6 \text{ rad/sec} \tag{8-27}$$

$$\text{Average speed} = 6.6 \text{ rad/sec} \times \frac{1 \text{ rev}}{2\pi \text{ rad}} \times \frac{60 \text{ sec}}{1 \text{ min}} = 63 \text{ rpm}$$

Using the results of Example 8-3, we see that the speed dip for a coasting period of about $3\pi/2$ rad and 38.4 percent torque is $0.384 \times 23.8 \times 1.5 = 13.7$ rpm. This operating point at 63-rpm and 13.7-rpm speed dip is obviously not one that yields smooth steady-speed operation.

The half-wave drive system is inexpensive because it uses only a single thyristor. However, it has several disadvantages. First, the armature current flows in relatively short pulses, one pulse per cycle of line voltage. The current has a high rms-to-average ratio, so that the armature heating for the same torque is higher than for continuous current. The motor must be either force-cooled to obtain its rated horse-power or derated. Second, the motor coasts between conduction periods; at high torques and low speeds the speed fluctuation is very pronounced. Third, the half-wave circuits all introduce a dc component into the supply line which can saturate supply transformers and cause other difficulties. One solution is to utilize single-phase full-wave circuits, which produce two current pulses per cycle, thus reducing the coasting time and speed dip, and also reducing the rms-to-average ratio of the current. Another solution is to utilize three-phase circuits.

c. Three-phase, Half-wave Thyristor Drive System

Three-phase supply voltages are usually used for dc motor drive systems of about 5 hp and larger. Three-phase rectifier circuits provide more voltage pulses per cycle of line frequency, thus assuring armature current over a larger portion of the cycle, increasing the ratio of average-to-rms current and thereby reducing the heating of the armature. Furthermore, the power is taken from a three-phase system which generally has more capability of supplying the power than a single-phase system. Three configurations of thyristors and diodes are used for such drives: three-phase bridge using six thyristors; three-phase incomplete bridge using three thyristors and three diodes; and three-phase half-wave circuit using three thyristors. The half-wave circuit is described in Art. 8-2*d* for resistive load; it will now be described for dc motor load.

The diagram for the half-wave drive circuit is shown in Fig. 8-15.

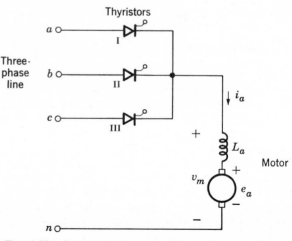

Fig. 8-15. Circuit diagram for three-phase half-wave drive system.

It consists of three thyristors connected so that when I is fired, a segment of voltage v_{an} is applied to the motor; when II is fired, a segment of voltage v_{bn} is applied; and when III is fired, a segment of voltage v_{cn} is applied. Two thyristors cannot conduct simultaneously because the one connected to the instantaneously highest line voltage will apply a negative anode-to-cathode voltage to the other and turn it off. At the instant that a thyristor is gated on in normal operation, there is a short interval when two thyristors conduct simultaneously to allow the current in the inductance associated with each thyristor branch to decline to zero. This interval is called the *commutation* interval and is treated in detail in a text on rectifier circuits.

The waveforms of motor voltage v_m, armature inductance voltage v_l, armature emf e_a, and armature current i_a are shown in Fig. 8-16 for three values of firing angle α. Zero firing angle is conventionally taken for maximum rectifier output voltage. In this case, $\alpha = 0$ occurs at $\omega t = \pi/6$ on the wave for voltage v_{an}. Hence, the rectifier voltage would become zero when α is retarded by 150° for a resistance load. The waveforms shown in Fig. 8-16b and c for $\alpha = 0$ represent operation of the dc motor at its maximum armature voltage and speed. Each thyristor conducts for 120° and the armature current i_a is continuous. The voltage v_m applied to the motor is identical with that under resistance load shown in Fig. 8-11. Both the motor voltage and current have the characteristic three pulses of ripple per cycle of line voltage, which is primarily triple-line-frequency ripple. The instantaneous speed variation is small so that the armature emf e_a is basically constant and equal to the average value of v_m. The instantaneous difference between the motor voltage v_m and the armature emf e_a is the voltage v_l of the armature-circuit inductance, represented by the shaded area. It must average to zero over the time of either the ripple period or the line period. The magnitude of the armature-current ripple depends upon the armature-circuit inductance.

The waveforms shown in Fig. 8-16e and f for $\alpha = 60°$ and for sufficient armature inductance to maintain continuous conduction of the armature current represent partial speed operation. The average motor voltage V_m for continuous conduction is given by

$$V_m = \frac{3V_0}{\sqrt{2}\,\pi}\int_{\alpha+\pi/6}^{\alpha+5\pi/6} \sin \omega t\; d(\omega t)$$

$$= \frac{3\sqrt{2}}{\pi} 0.866 V_0 \cos \alpha \qquad (8\text{-}28)$$

where V_0 is the rms line-to-neutral voltage and the value of V_m for

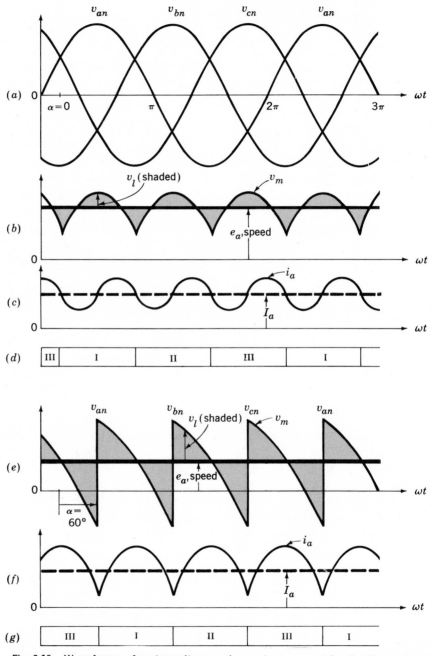

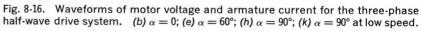

Fig. 8-16. Waveforms of motor voltage and armature current for the three-phase half-wave drive system.   *(b)* $\alpha = 0$; *(e)* $\alpha = 60°$; *(h)* $\alpha = 90°$; *(k)* $\alpha = 90°$ at low speed.

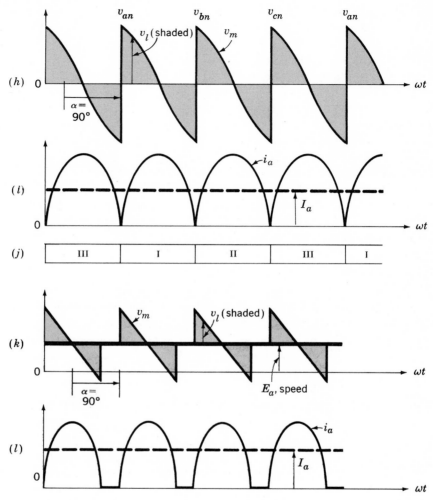

For legend, see opposite page.

$\alpha = 0$ is $1.17 V_0$. For $\alpha = 60°$, the average motor voltage and speed are proportional to cos α, or 0.5 of the value for $\alpha = 0°$. The ripple of the armature current is greater than for $\alpha = 0°$, reflecting the larger armature inductance volt-time areas that must be absorbed each ripple cycle.

The waveforms shown in Fig. 8-16h and i for $\alpha = 90°$ and sufficient armature-circuit inductance to just maintain continuous armature-current conduction represent the motor at zero speed but delivering load torque proportional to armature current i_a. The motor voltage v_m is basically a triple-line-frequency wave and the current has a peak value equal to twice its average value. A greater armature-circuit inductance would tend to reduce the ripple amplitude; a lesser value would allow the current to become discontinuous. The discontinuous case is shown in Fig. 8-16k for $\alpha = 90°$ at low speed, part load.

EXAMPLE 8-5

A 100-hp, 1,750-rpm, dc shunt motor has an armature inductance $L_a = 1.1 \times 10^{-3}$ henry, resistance $R_a = 14.4 \times 10^{-3}$ ohm, and a voltage constant $K_m = 1.27$ volt-sec/rad. The motor is operated from a three-phase line rated at 480 volts line to line through a half-wave thyristor rectifier at the rated armature current of $I_a = 340$ amp.

a. Find the firing angle α of the thyristors to obtain the rated speed of 1,750 rpm. Consider that the thyristors have a forward voltage drop of 1 volt; assume continuous conduction.

b. Determine the ripple amplitude of the armature current at $\dot\alpha = 90°$. The motor is carrying rated armature current and is at standstill. Neglect the resistance R_a for the calculation.

Solution

a. The rms line-to-neutral voltage V_0 is

$$V_0 = \frac{480}{\sqrt{3}} = 277 \text{ volts}$$

The average motor voltage is given by Eq. 8-28 as

$$V_m = \frac{3\sqrt{2}}{\pi} 0.866 \times 277 \cos \alpha \text{ volts}$$

The armature emf for 1,750 rpm is

$$E_a = \frac{1,750}{60} \times 2\pi \text{ rad/sec} \times 1.27 \text{ volt-sec/rad} = 232 \text{ volts}$$

To this value must be added 1 volt for the thyristors and 4.9 volts for the $I_a R_a$ drop for a total of 238 volts.   The firing angle is thus

$$\cos \alpha = \frac{238}{325} = 0.735$$

$$\alpha = 43°$$

b. The voltage applied to the armature-circuit inductance at $\alpha = 90°$ is the shaded area of Fig. 8-16h.   The armature-current rise ΔI_a from $\omega t = 2\pi/3$ to π for the inductance L_a is given by

$$\Delta I_a = \frac{1}{L_a} \int_{2\pi/3\omega}^{\pi/\omega} \sqrt{2} \, V_0 \sin \omega t \, dt \qquad (8\text{-}29)$$

$$= \frac{-392}{\omega L_a} [\cos \omega t]_{2\pi/3\omega}^{\pi/\omega}$$

$$= \frac{196}{\omega L_a} \text{ amp}$$

For $\omega = 377$ rad/sec and $L_a = 1.1 \times 10^{-3}$ henry, the peak-to-peak ripple amplitude is

$$\Delta I_a = 472 \text{ amp}$$

At $\alpha = 90°$, the peak ripple current is $\frac{1}{2}\Delta I_a$ or 236 amp.   Since the average current is 340 amp, the current never reaches zero and is therefore continuous.

A pair of three-phase half-wave circuits is frequently used for a reversing drive; one set of thyristors is connected to apply positive voltage to the motor; the other set is connected to apply negative voltage.   By allocating the gating signals for the thyristors, speed control of the motor in either direction is obtained.

In order to obtain larger armature-current conduction angles in single-direction drives, a free-wheeling diode is placed across the arma-

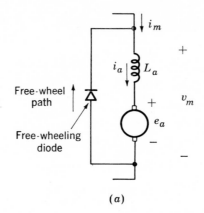

(a)

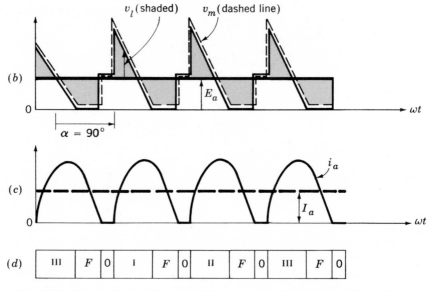

(b)

v_l (shaded) v_m (dashed line)

E_a

$\alpha = 90°$

ωt

(c)

i_a

I_a

ωt

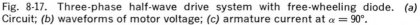

(d)

| III | F | 0 | I | F | 0 | II | F | 0 | III | F | 0 |

Fig. 8-17. Three-phase half-wave drive system with free-wheeling diode. *(a)* Circuit; *(b)* waveforms of motor voltage; *(c)* armature current at $\alpha = 90°$.

ture terminals as shown in Fig. 8-17*a*. The diode provides a path for the armature current when the thyristors are not conducting. During the time that the diode is conducting, the armature terminals are short-circuited and the armature circuit inductance discharges its energy into the armature. The waveforms for one case, $\alpha = 90°$, are shown in Fig. 8-17*b*. The motor voltage v_m follows the line voltage when a thy-

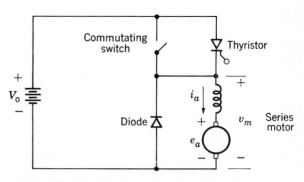

Fig. 8-18.   Simplified circuit of chopper speed control
system.

ristor is conducting; the motor voltage is zero when the free-wheeling
diode is conducting; the motor voltage v_m is the same as the armature
emf e_a when neither the thyristors nor the diode is conducting.   Now,
zero motor voltage is obtained when $\alpha = 150°$, rather than $\alpha = 90°$ in
Fig. 8-16h.

d. Choppers

A special class of solid-state circuits has been developed for controlling dc
motors which are supplied from fixed voltage sources.   These circuits are
used where the source is a battery, as in industrial electrically powered
trucks, and where it is a third rail or overhead trolley wire, as in rapid-
transit cars.   The circuits are called *choppers* and are used to replace
switched, series, armature-circuit resistors; the advantage is higher efficiency, continuous control, and the ability to operate the motor in a
regenerative braking mode.[1]   A simplified diagram of the chopper circuit
with a series motor is shown in Fig. 8-18.   The chopper operates from a
fixed dc voltage V_0 and controls the average motor voltage V_m from zero
to V_0.   The thyristor acts as a switch which is closed and opened at the
rate of several hundred cycles per second.   The relative on-to-off time
of the thyristor determines the average motor voltage.   The thyristor is
unable to turn itself off when carrying current; it requires a commutating
circuit which impresses a negative voltage on the thyristor for a short
period of typically 40 microseconds to turn it off.   The commutating
circuit is represented by a switch.

The waveforms of the circuit are shown in Figs. 8-19 and 8-20.   In
Fig. 8-19, the chopper is operating at about $0.2V_0$ motor voltage.   When

[1] K. Heumann, Pulse Control of DC and AC Motors by Silicon Controlled Rectifiers,
IEEE Transactions on Communication and Electronics, **83**:390–399 (1964).

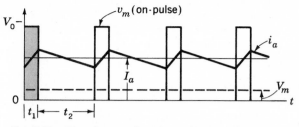

Fig. 8-19. Waveforms of motor voltage and current at low speed.

the thyristor is turned on by a gating signal at $t = 0$, the armature current i_a is delivered from the battery and rises as the circuit inductance absorbs the volt-time area of the difference between V_0 and the armature emf e_a. When the thyristor is turned off after the time t_1, the armature current declines through the free-wheeling diode as the energy stored in the circuit inductance is applied to the armature. As the ratio of on-to-off time t_1/t_2 is increased, the average motor voltage V_m rises, as shown in Fig. 8-20.

The chopper circuit requires inductance to store energy. Usually, a series motor is employed and the field winding serves as the inductance. However, shunt motors can be employed with external inductors and the normal field winding supplied from the dc source. The chopper can be controlled by pulse width or pulse frequency. The motor operates as though it were subjected to voltage control, not to resistance control, because the average motor voltage V_m is independent of the armature current.

EXAMPLE 8-6

A 100-hp series motor rated 180 amp is operating in a chopper circuit from a 500-volt dc source. The armature and field inductance is

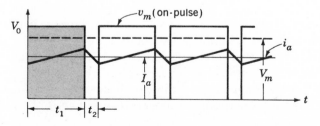

Fig. 8-20. Waveforms of motor voltage and current at high speed.

0.060 henry.   At the minimum ratio $t_1/(t_1 + t_2)$ of 0.20, as shown in Fig. 8-19, find the pulse frequency to limit the amplitude of armature-current excursion to 10 amp.

Solution

For a pulse ratio of 0.20, the average armature voltage is

$$0.2 \times 500 = 100 \text{ volts}$$

The volt-time area applied to the inductance is $(500 - 100)t_1$ volt sec. Thus the rise of current is

$$\Delta i_a = \frac{400t_1}{0.060} = 10 \text{ amp}$$

$$t_1 = \frac{0.6}{400} = 1.5 \times 10^{-3} \text{ sec}$$

$$t_1 + t_2 = \frac{1.5 \times 10^{-3}}{0.2} = 7.5 \times 10^{-3} \text{ sec}$$

$$\text{Pulse frequency} = \frac{1}{7.5 \times 10^{-3}} = 133 \text{ pps}$$

The chopper equipment for a rapid-transit car is shown in Fig. 8-2.

8-4
INTRODUCTION TO INVERTERS

Inverters are used to transfer energy from a dc source to an ac load of arbitrary frequency and phase.   More specifically, they are used typically in drive systems to provide power for adjustable frequency ac motors, to regenerate energy back to the ac line from decelerating dc motors, and to pump rotor-circuit power back to the ac line from wound-rotor induction motors.   In nonmotor systems, they are used to supply uninterruptible ac power to computers and to convert energy between ac and dc form at the terminals of high-voltage dc power-transmission systems.   Motor-drive-system inverters employ thyristors as controlled switching elements to form the desired ac waveforms.[1]

[1] B. D. Bedford and R. G. Hoft, "Principles of Inverter Circuits," John Wiley & Sons, Inc., New York, 1964.

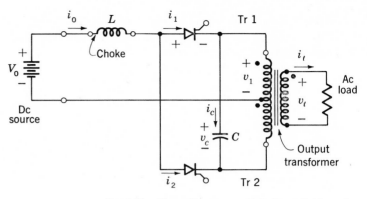

Fig. 8-21.　Capacitor commutated parallel inverter.

a. Single-phase Parallel Inverter

The basic inverter circuit that utilizes the principles of more complicated inverters is the single-phase parallel inverter circuit shown in Fig. 8-21. The inverter converts electrical energy from the dc source V_0 to ac form as voltage v_l applied to the load impedance. The frequency of the ac voltage is governed entirely by the gating pulses applied to the thyristors $Tr\ 1$ and $Tr\ 2$. In addition to the thyristors, the circuit includes an inductor L, an output transformer, and a commutating capacitor C. The means for generating the gating pulses is not shown but it usually consists of an oscillator or multivibrator, a pulse-shaping circuit, and pulse transformers for applying the gating pulses to the thyristors.

　　The inverter circuit of Fig. 8-21 functions by alternately turning on the thyristors; when one thyristor is conducting, the other is blocking and the source voltage V_0 is directly applied to one-half of the transformer winding forming one-half cycle of the output voltage. When the other thyristor is turned on, the commutating capacitor forces the first thyristor to become nonconducting and the source voltage is applied to the other half of the transformer winding to form the second-half cycle of the output voltage. The inductor serves to buffer the action of the inverter elements from the fixed-voltage source. The output waveform is essentially a square wave which can be filtered to a sine wave if necessary.

b. Operation of the Single-phase Parallel Inverter

The operation of the inverter will be described for a typical cycle after it has settled into a steady-state condition. The inductor L will be assumed to have infinite inductance. The parameters are shown in Fig. 8-21. The key waveforms are shown in Fig. 8-22. Let us assume

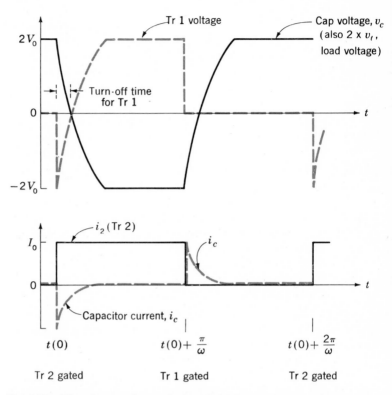

Fig. 8-22.   Waveforms of capacitor and thyristor voltages and currents over one cycle of operation with resistive load.

that $Tr\ 1$ is conducting, $Tr\ 2$ is blocking, and the load current and load voltage are I_0 and V_0, respectively.   Assume that the transformer turns ratio from each primary half to secondary winding is unity.   The capacitor C is charged to voltage $2V_0$; the capacitor current is zero because the capacitor is fully charged.   Thyristor $Tr\ 2$ is blocking voltage $2V_0$ in the forward direction.

At time $t(0)$, we apply a gating pulse to $Tr\ 2$; it turns "on" and forces the capacitor voltage $2V_0$ across $Tr\ 1$ in the inverse direction, turning it off.   The circuit now appears as in Fig. 8-23.   The choke L forces the current I_0 to pass through $Tr\ 2$ and the lower half of the transformer. However, the load current cannot be reversed instantaneously in the transformer both because of the leakage inductances, any load inductance, and because the capacitor voltage impressed across the primary winding maintains the load voltage at its $t(0-)$ polarity.   The current I_0 carried by $Tr\ 2$ develops ampere-turns of the wrong polarity in the transformer. Hence, a current of amplitude I_0 must flow out of the capacitor, as shown in Fig. 8-23, to cancel the $Tr\,2$-current ampere-turns and supply the required ampere-turns to balance the ampere-turns of the transformer. This capacitor current then is supplying the load current and is also discharging the capacitor; the current will continue until the capacitor reaches the voltage $-2V_0$.

The capacitor current is shown in Fig. 8-22 jumping to amplitude I_0 and dying down exponentially as the capacitor voltage reaches $-2V_0$. The load voltage has the same waveform as the capacitor voltage, but half its amplitude. The voltage applied to $Tr\,1$ is shown in Fig. 8-22. At $t(0+)$, the inverse voltage is $-2V_0$; the voltage passes through zero and rises to $+2V_0$ in the forward direction. The time that the $Tr\,1$ voltage is negative is the time available to turn off the thyristor. The interval must be larger than the rated turn-off time of the thyristor; otherwise the inverter will not commutate.

The problem of design and operation of the capacitor-commutated parallel inverter lies in the selection of the commutating capacitor and then accepting its effect over the operating range of frequency and load. The capacitor has to be large enough to insure adequate turn-off time for the off-going thyristor, while it is discharging into the load resistor. Hence, the capacitor must be selected for the lowest value of load resistance, that is, the largest load power. However, as the load resistance may be increased toward the light-load condition, the capacitor becomes too large; with a fixed inductor the capacitor tends to charge each half cycle to a voltage higher than $2V_0$. The consequence is that the load voltage can vary over a wide range as a function of loading, for example, 2-to-1 range of load voltage for 10-to-1 range of load resistance.

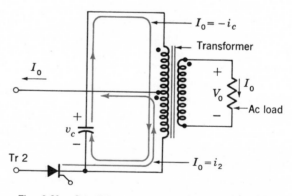

Fig. 8-23. Conditions at instant of commutation in parallel inverter.

The higher capacitor voltage at light load also appears across the thyristors as higher forward and inverse voltages.   If the load has an inductive component, such as a motor winding, then the commutating capacitor has to be large enough to, in effect, correct the power factor and also be large enough to commutate the equivalent resistive load.

EXAMPLE 8-7

A single-phase inverter as shown in Fig. 8-21 operates from a 120-volt battery.   The output transformer has a turns ratio from each half of the primary winding to the secondary winding of 2:1.   For the lowest load resistance of 10 ohms, find the necessary value of commutating capacitance C to obtain 20-microsecond turn-off time on the thyristor.   Assume the inductor L is infinite and the transformer is ideal.

Solution

The discharge path of the capacitor C is shown in Fig. 8-23.   The resistance reflected into the primary circuit is 10 ohms $\times$ 16 = 160 ohms.   The capacitor discharge curve of Fig. 8-22 is described by

$$v_c = 4V_0\epsilon^{-t/160C} - 2V_0 \qquad (8\text{-}30)$$

At $t = 20 \times 10^{-6}$ sec, the voltage $v_c = 0$; for C in μf,

$$\epsilon^{-20/160C} = 0.5$$

$$C = 0.18 \ \mu f$$

---

c. Three-phase Inverters

Inverters for providing power at adjustable frequency for ac motor drive systems are usually three-phase inverters.   Like the single-phase inverter, the three-phase inverter takes its energy from a dc source which may be a battery or dc distribution system, but usually is a rectifier.   A battery may be floated on the dc bus to ensure a supply of energy for critical applications when the ac supply to the rectifier is interrupted. The three-phase inverter has a minimum of six thyristors arranged in a bridge configuration; these thyristors are switched sequentially to synthesize at the ac terminals of the inverter bridge a set of three-phase voltages which are applied to the motor.   The thyristors are turned off by commutating capacitors either by the on-coming thyristors as in the previously explained single-phase inverter, called self-commutation, or by

auxiliary thyristors in a forced commutation mode.   With forced commutation, the conduction time of each thyristor can be controlled independently of frequency for purposes of waveform or voltage control.

There is a wide variety of three-phase inverter circuits.   A basic
circuit is shown in Fig. 8-24 in which the thyristors each conduct for 180°
of the cycle and are turned off by self-commutation.   Each single-phase
section of the inverter consisting of two opposing bridge arms operates
as a single-phase inverter.   Section a consists of two thyristors Tr 1 and
Tr 4, two commutating capacitors C 1 and C 4, a center-tapped inductor
L 1-4 and two diodes D 1 and D 4.   The inductor performs the function
of turning off one thyristor of the section when the other one is turned on
just as the center-tapped winding of the output transformer does for the
single-phase inverter.   The diodes D 1 and D 4 are called *feedback*
diodes and provide paths for the load current to return energy to the
dc source from reactive loads.

The pattern in which the thyristors in the circuit of Fig. 8-24 are
turned *on* and *off* to synthesize the desired ac waveforms will be described
in Art. 8-4d.   The manner in which the thyristor Tr 1 is turned *off* when
Tr 4 is turned *on* and assumes the conduction of load current i_a is shown
by the sequence in Fig. 8-25.   The same sequence occurs for the thyristor

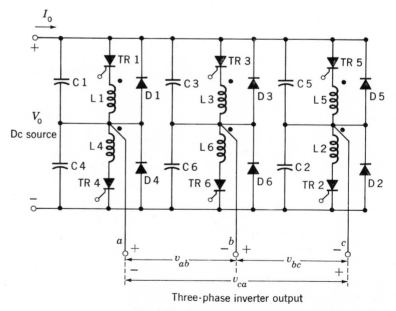

Three-phase inverter output

Fig. 8-24.   Three-phase self-commutated inverter.

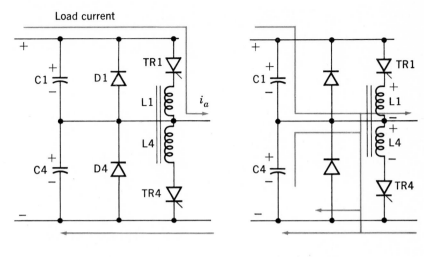

(a)

(b)

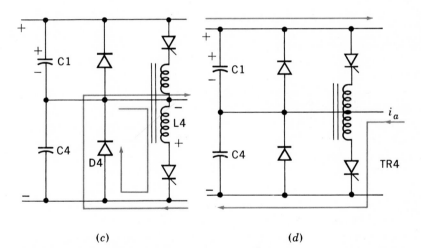

(c)

(d)

Fig. 8-25. Commutation sequence of three-phase inverter. *(a)* Thyristor *Tr* 1 carrying load current i_a; *(b)* thyristor *Tr* 4 is fired; *(c)* load current i_a carried by diode *D* 4; *(d)* thyristor *Tr* 4 carrying load current i_a.

pair *Tr* 3-*Tr* 6, 120° later in the cycle, and for *Tr* 5-*Tr* 2, 240° later than *Tr* 1-*Tr* 4.   While the commutating sequence for one section of the bridge is occurring, one thyristor is conducting in each of the other two sections of the bridge; these thyristors provide the return path for the load current.

The commutating sequence starts as shown in Fig. 8-25a with $Tr\ 1$ carrying load current i_a. Capacitor $C\ 4$ is charged as shown and capacitor $C\ 1$ is zero. Thyristor $Tr\ 4$ is now turned on in Fig. 8-25b; capacitor $C\ 4$ discharges through $L\ 4$ and the upper end of winding $L\ 1$ applies the positive capacitor voltage to the cathode end of $Tr\ 1$. Thyristor $Tr\ 1$ turns off and the load current transfers to the path through capacitor $C\ 1$, charging it as shown as $C\ 4$ discharges. In Fig. 8-25c, capacitor $C\ 1$ is fully charged; the reactive load current continues to circulate through diode $D\ 4$ and the trapped energy in $L\ 4$ is dissipated by the circulating current. Finally, in Fig. 8-25d, the load current i_a reverses and is carried by $Tr\ 4$ in a steady-state condition. The capacitor $C\ 1$ is charged to the source voltage and is waiting for the next commutation operation.[1]

d. Three-phase Inverter Waveforms

The sequence in which the thyristors of the circuit of Fig. 8-24 are operated to obtain three-phase output waveforms is shown in Fig. 8-26. The thyristors are represented by switches and the load is represented by a Y-connected set of impedances. Each step corresponds to 60° on the output waveform. During step I, thyristors 1, 6, and 5 are conducting. The a and c ends of the load are connected to the positive dc supply bus while terminal b is connected to the negative supply bus. The line voltages v_{ab} and v_{bc} are each V_0 in magnitude. The line-to-neutral voltages v_{a0} and v_{c0} are each $V_0/3$ while v_{b0} is $-2V_0/3$, as shown in Fig. 8-26. At each 60° interval, one of the thyristors switches in the sequence 1-2-3-4-5-6. The thyristors can be operated in a different mode of 120° conduction, depending upon the commutating circuit, to obtain a different set of output waveforms.

The waveforms for the line-to-neutral voltages v_{a0} and v_{b0} and the line voltage v_{ab} are shown in Fig. 8-27. The line-to-neutral voltages are three-step-per-half-cycle approximations to sine waves while the line-to-line voltages are 120° wide positive and negative pulses. When the inverter voltages are applied to a motor, the motor responds to the fundamental and the harmonics of the waveforms to produce positive and braking torques and both normal and additional losses. Inverters can be built that produce output waveforms with less harmonic content by various techniques. For example, the outputs of two inverters whose waveforms are phase displaced can be added.

---

[1] For a description of various commutation circuits for inverters, see A. J. Humphrey, Inverter Commutation Circuits, *IEEE Trans.*, *I.G.A.*, vol. IGA-4, no. 1, pp. 104–110, January–February 1968.

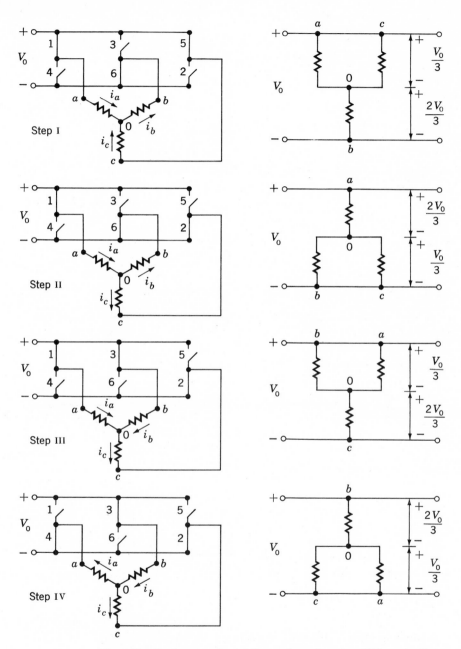

Step I

Step II

Step III

Step IV

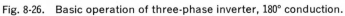

Fig. 8-26.   Basic operation of three-phase inverter, 180° conduction.

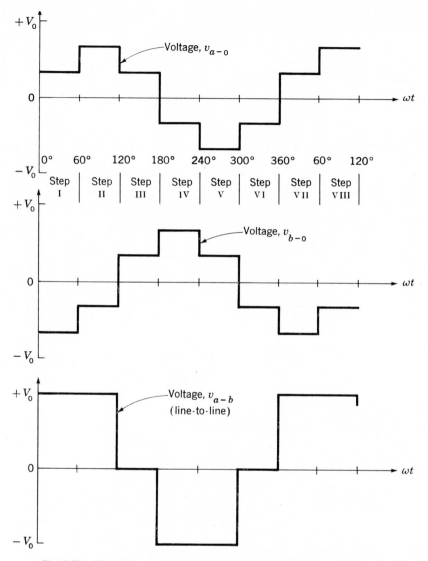

Fig. 8-27. Waveforms generated by three-phase inverter, 180° conduction.

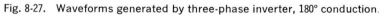

EXAMPLE 8-8

The set of three-phase line-to-line voltages of the form shown in Fig. 8-27 are to be applied to the stator terminals of an induction motor. Find the amplitude of the fifth-harmonic component relative to the fundamental.

Solution

The waveform of the voltage v_{ab} is shown in Fig. 8-27.   Assume that the positive half cycle is centered on $\omega t = 0$; it can be represented by Fourier series of the form

$$v_{ab} = \sqrt{2}\, V_1 \cos \omega t + \sqrt{2}\, V_5 \cos 5\omega t + \cdots \qquad (8\text{-}31)$$

The even harmonic components are excluded because the waveform is symmetrical about $\omega t = 0$.   The triplen harmonics (3d, 9th, . . .) are excluded because they would be in phase in the three line voltages and could not add to zero unless they are of zero amplitude themselves.   The amplitude V_1 of the fundamental is given by

$$V_1 = \frac{1}{\sqrt{2}\,\pi} \int_0^{2\pi} v_{ab} \cos \omega t \, d(\omega t)$$

$$= \frac{4}{\sqrt{2}\,\pi} \int_0^{\pi/3} V_0 \cos \omega t \, d(\omega t)$$

$$= \frac{\sqrt{6}}{\pi} V_0 \qquad\qquad\qquad (8\text{-}32)$$

The amplitude V_5 of the fifth-harmonic component is given by

$$V_5 = \frac{4}{\sqrt{2}\,\pi} \int_0^{\pi/3} v_{ab} \cos 5\omega t \, d(\omega t)$$

$$= \frac{-1}{5} \frac{\sqrt{6}}{\pi} V_0 \qquad\qquad (8\text{-}33)$$

The amplitude of the fifth-harmonic component is $\frac{1}{5}$ the fundamental.   The phase sequence of the fifth harmonic is negative, or opposite to the fundamental.   The torque produced by the fifth-harmonic components will thus produce a braking torque.

8-5
ADJUSTABLE–FREQUENCY AC MOTOR DRIVE SYSTEMS

The output power from inverters of the type described in Art. 8-4 is applied to synchronous and induction motors to obtain adjustable-speed operation by means of frequency control.[1]   Where closely regulated speed is required from one or more motors, such as in textile applications, synchronous motors are used; they will operate in synchronism with the oscillator that controls the inverter independently of their loading.   On the other hand, where nominally regulated speed is required, as in high-speed grinders or special vehicle drives, induction motors are used. Special types of motors are frequently employed because the operation is different from the usual fixed-frequency application.   The synchronous motors are usually hybrid induction-synchronous motors which synchronize by reluctance torque or by using permanent-magnet fields. The induction motors are either squirrel-cage or solid-iron rotor types. Examples are shown in Fig. 8-28.   Operation is frequently carried both above and below the nominal frequency of the motor, for example, 10 to 120 Hz for a 60-Hz motor.

The synchronous speed is directly proportional to frequency.   What is of primary interest is the shape of the torque-speed curve of the induction motor at each frequency in the operating range, including the starting and breakdown torques.   The torques are determined by how the

---

[1] See A. Kusko, "Solid-State AC Motor Drives," The M.I.T. Press, Cambridge, Mass., 1971.

Fig. 8-28. Cutaway views of a rotor of a Synchrospeed (trademark) motor. *Left:* The magnetic laminations are shown dark, the conductors light. *Right:* Only the rotor conductors are shown; the iron has been etched away. *(The Louis Allis Company.)*

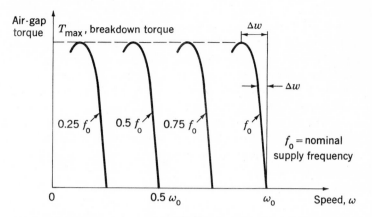

Fig. 8-29.   Idealized torque-speed curves of induction motor under adjustable-frequency control.

terminal voltage of the motor is managed as the frequency is adjusted. It is clear that to secure operation of a given motor at the maximum air-gap flux density over its speed range, the air-gap voltage must be adjusted proportional to frequency f. By so doing, we obtain the maximum available torque from the induction motor at each frequency setting.   In a practical system, the terminal voltage V_1 is controlled so that V_1/f is constant; the type of control is termed *constant volts per cycle*. For a voltage waveform that changes shape as a function of frequency, the half-wave average voltage rather than the rms voltage is controlled in the above-described manner. Present practice is to maintain V_1 constant below a specific frequency to overcome the $I_1 r_1$ drop; in some drives, the voltage V_1 is held constant for operation above the nominal motor frequency and for obtaining constant-horsepower operation.

The torque-speed curve of an induction motor for a given frequency can be calculated using the methods of Chap. 7 within the accuracy of the motor parameters at that frequency.   The torque-speed curves for the motor operated over a range of frequency, but at constant air-gap flux density are practically the same on a slip-speed scale as shown in Fig. 8-29.   This can be confirmed by viewing the air gap from the rotor; the rotor behavior is determined only by the slip speed, which sets the rotor induced voltage, frequency, current, and the torque produced.   Note that a speed regulation of 3 percent at nominal frequency ω_0 becomes a speed regulation of 30 percent at $\omega_s = 0.1\omega_0$.

An adjustable-frequency drive can be started in several ways.   It can be started at the lowest frequency and be brought up to speed by raising the frequency.   It can be started at any frequency in the range as a constant-frequency induction motor is started. Or, it can be started from the ac supply line and transferred to the inverter after it is up to speed.   The problems of starting are the dual ones of obtaining sufficient starting torque and of not exceeding the current rating of the inverter, which is usually based on the ability of the inverter to commutate the current, rather than the thyristor ratings.

The measured torque-speed curves on an actual adjustable-frequency drive system correspond closely with the curves of Fig. 8-29 for the ideal motor.   Commercial drive systems are shown in Figs. 8-3 and 8-4.   The

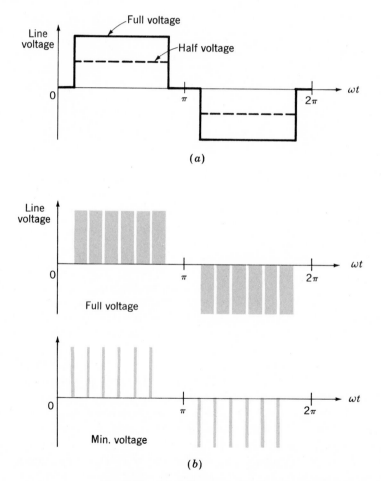

Fig. 8-30. *(a)* Voltage control by amplitude; *(b)* v pulse width.

cabinet for the solid-state equipment is several times larger than a dc drive system of the same horsepower.

The terminal voltage of the motor is usually held constant for operation above nominal frequency. Hence, the inverter-rectifier system requires no modification other than to provide thyristor gating pulses for such operation. The breakdown torque, instead of remaining constant as in Fig. 8-29, declines. The drive operates in a constant-horsepower mode. The operation is directly analogous to field-weakening control of a dc drive with fixed armature voltage above the nominal speed.

The waveform of each half cycle of the motor voltage must be controlled below nominal speed to maintain constant air-gap flux density. The method of such control should minimize the harmonic content to avoid undue losses and braking torques. It should also be done without complicating or penalizing the inverter. The waveforms for two typical methods for such control are shown in Fig. 8-30.

Figure 8-30a shows control of a line-to-line voltage by amplitude variation of a 120° wide pulse. Such control is obtained by varying the dc voltage from the rectifier portion of the drive. If self-commutation is employed in the inverter, it may not be possible to charge the commutation capacitors from the dc bus to properly commutate the full motor current at low frequencies. The inverter may have to be derated or a separate fixed-voltage dc bus set up just to supply the commutating capacitors.

Figure 8-30b shows waveform control by pulse-width control of each half cycle of the ac voltage. The 120° width of the waveform is divided typically into six 20° wide pulses where maximum voltage is obtained at 20° and minimum voltage at, say, 2° for a 10-to-1 speed range. The inverter dc voltage level remains unchanged but auxiliary commutation must be used to turn off the thyristors at prescribed points in the half cycle.

8-6
RÉSUMÉ

Solid-state devices are employed for the control of every type of ac and dc electric machine. In this chapter we have selected two important systems to discuss: the rectifier dc motor drive system and the adjustable-frequency ac drive system. In addition, solid-state devices such as thyristors are used with induction motors to control speed by primary voltage, with wound-rotor motors to control speed by pumping rotor power back into the line, and with series universal motors in portable tools and appliances. Thyristors are also used as choppers to control the average voltage applied to dc series motors and to control the field current of dc shunt motors. Other types of solid-state devices such as diodes and transistors are used at signal levels to process command and sensor (feedback) signals to determine how the power-level semiconductors will operate to produce the prescribed characteristics.

Various types of rectifier circuits are described in Art. 8-2. These circuits use thyristors to obtain firing angle or phase control of the output voltages of these rectifiers. The many types of rectifier circuits are categorized by the number of supply phases and whether half-wave or bridge connection. The selection of a rectifier circuit depends upon power level and its function.

Operation of dc motors from rectifiers is discussed in Art. 8-3. The waveforms of armature voltage and current influence the operation of the motor and vary with speed and load. Fractional-horsepower and low-integral-horsepower motors are usually operated from single-phase, half-wave, and full-wave controlled rectifiers. Motors of larger rating are supplied from three-phase rectifiers.

The basic principles of single-phase and three-phase inverters are treated in Art. 8-4. The inverters are supplied from dc sources and deliver ac power at adjustable frequency to ac motor loads. The inverters can be relatively simple self-commutated units or more complicated auxiliary commutated systems which are capable of pulse-width voltage control. The inverters are capable of working over wide frequency ranges but generally have limited overload capability, limited by the commutation circuits.

The principles of frequency control of induction motors are developed in Art. 8-5. For operation below nominal frequency, the terminal voltage is adjusted proportional to frequency to insure constant air-gap flux density and constant breakdown torque. For operation above nominal frequency, the voltage is usually held fixed and the motor operates in a constant-horsepower mode. Starting is a problem because of the limited overcurrent capability of the inverter and the variation of starting torque with frequency.

The field of solid-state motor control is developing rapidly. As larger thyristors are made available, the ratings of dc and ac drives will rise. As less expensive smaller-rating thyristors are placed on the market, their use will spread for small motors. Rapid-transit cars and electric-drive vehicles will use large quantities of devices for control. Reliable and inexpensive integrated-circuit packages are allowing designers more freedom in the functions that the motor-control system can accomplish. More control systems will utilize signals and sensors that can be coupled directly to computers and to motor-drive elements.

PROBLEMS

8-1. A small vehicle is driven by an electric motor rated at 15 hp at 1,800 rpm. The motor is required to operate at constant horsepower from rated speed to one-third rated speed, and at constant torque from one-third rated speed to standstill.

Sketch the profiles of motor torque and horsepower from zero to rated speed. Give the torque at standstill and at rated speed.

8-2. An induction motor is being selected to drive a pump. The speed of the motor and pump will be controlled by adjustment of the primary stator voltage. The motor slip is 0.05 pu for rated torque.

Sketch the profile of the allowable motor torque as a function of slip from $s = 0$ to 1.0 pu for the following levels of rotor power dissipation: *a.* nominal, corresponding to dissipation at $s = 0.05$ pu; *b.* twice the value for *a*; *c.* four times the value for *a*.

8-3. The power source in Fig. 8-8a is an inverter which delivers a square-wave voltage of amplitude 100 volts at 200 Hz.

Find the relationship of average load voltage V_n to firing angle α.

8-4. The load in the circuit of Prob. 8-3 is replaced with a series-connected combination of $R = 10$ ohms and $L = 25$ mh. The thyristor is fired at $\alpha = 10°$.

Find an expression for the current $i_n(t)$ over one cycle.

8-5. A three-phase, half-wave rectifier as shown in Fig. 8-11a is charging a battery through a filter inductor. The inductor is $L = 60$ mh and $R = 1$ ohm. The battery voltage is 24 volts and the charging current is 2 amp.

Find the required rms line-to-neutral voltage at the input to the rectifier. Assume that the diodes have a forward voltage drop of 1 volt each.

8-6. The motor of Example 8-3 is operated with a mechanical load that imposes rated torque at rated speed and a load inertia of 0.10 lb-ft-sec². The motor is supplied from a 60-Hz single-phase, full-wave rectifier. The coasting period is estimated to be $\pi/2$ rad between conduction periods.

Find the speed dip between conduction periods.

8-7. A 5-hp, dc motor is operating in a single-phase, half-wave thyristor circuit, as shown in Fig. 8-13. The motor is operating at 50 percent of its rated electromagnetic torque at an average speed of 600 rpm. The firing angle of the thyristor is $\alpha = 45°$ and the conduction angle is observed to be $\gamma = 180°$. The rated motor armature current is 21.4 amp; $K_m = 1.21$ volt-sec/rad; $R_a = 0.615$ ohm.

Find the rms value V_0 of the applied line voltage $v_0 = \sqrt{2}\, V_0 \sin \omega t$. Neglect the forward voltage drop of the thyristor.

8-8. For the three-phase, half-wave circuit of Fig. 8-15 and the conditions shown in Fig. 8-16, namely $\alpha = 0°$, $\alpha = 60°$, $\alpha = 90°$, sketch the waveform of the voltage across thyristor I over the period of one cycle.

8-9. The 100-hp motor of Example 8-5 is operated at a firing angle $\alpha = 0$. Find the average motor speed at no load and full load for a line voltage of 480 volts line to line.

8-10. Refer to Example 8-5, part *b*. What is the effect on the peak ripple current in the armature of loading the motor beyond the condition that just produces continuous conduction?

8-11. A 500-hp motor used in a high-speed train is controlled by a chopper circuit as shown in Fig. 8-18. The inductance of the armature and series field winding is augmented by an external inductor L_e. The dc source voltage for the train is 1,000 volts. The pulse ratio $t_1/(t_1 + t_2)$ varies from 0.15 to 1.0 in operation. What total inductance is required in the armature circuit to limit the amplitude of the armature-current excursion to 25 amp?

8-12. The single-phase inverter circuit of Fig. 8-21 can be built in a bridge configuration using four thyristors and a non-center-tapped transformer. Sketch the bridge circuit and calculate the value of C for the conditions of Example 8-7.

8-13. A common type of output waveform from a three-phase inverter is a rectangular wave having 120° wide positive and negative half cycles as shown in Fig. 8-27. Show that this waveform does not contain any third-harmonic voltage components.

dc machine dynamics

The outstanding characteristic of dc machines is their versatility. By means of various combinations of shunt-, series-, and separately excited field windings they can be designed to have a wide variety of built-in volt-ampere or speed-torque characteristics for both dynamic and steady-state operation. Because of the ease with which they can be controlled, dc motors with solid-state controls are often used in applications requiring a wide range of motor speeds or precise control of motor output. Often the inherent characteristics of the machines are modified by the addition of feedback circuits. The purpose of this chapter is to study dc machines with emphasis on their dynamic characteristics as electromechanical system components.

9-1
THE IDEAL DC MACHINE

Because of the complexity of dynamic-system problems, idealizing assumptions will be made. The usual assumptions are:

1 The brushes are narrow, and commutation is linear, as in Fig. 4-6. The brushes are located so that commutation occurs when the coil sides are in the neutral zone midway between the field poles. The axis of the armature-mmf wave then is fixed in space and lies along the *quadrature axis*.

2 The armature mmf is assumed to have no effect on the total direct-axis flux because the armature-mmf wave is perpendicular to the field axis. This assumption neglects the demagnetizing effect of armature reaction discussed in Chap. 5.

3 For most of the problems considered in this chapter the effects of magnetic saturation will be neglected. Superposition of magnetic fields can then be used, and inductances can be considered to be independent of the currents.

The schematic representation of the model is shown in Fig. 9-1. The convention will be adopted that the arrows represent the reference directions for both current and magnetic field. A consistent set of reference directions for all the other variables can then be adopted to conform with the flux-current reference directions. For example, in Fig. 9-1a the reference direction for magnetic torque is counterclockwise, tending to align the stator and rotor fields, as shown by the arrow labeled T_{fld}. If the machine is a motor, it will turn in the counterclockwise direction against the opposing torque T_L applied to the motor by the driven mechanical load, as shown by the arrows ω_m and T_L. The upper brush will be $+$, because electrical power must be supplied to the motor.

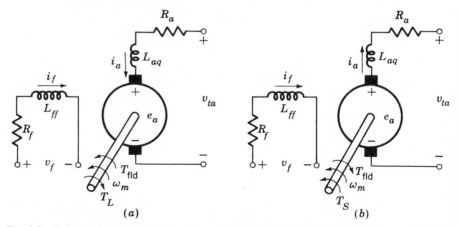

Fig. 9-1. Schematic representation of a dc machine showing (a) motor and (b) generator reference directions.

Figure 9-1*b* shows the reference directions for a generator, where T_S is the torque applied by the mechanical source. A consistent system of reference directions is especially important in dealing with the more complex cross-field machines discussed later in this chapter.

The electromechanical coupling terms are the magnetic torque T and the generated voltage e_a, which have already been derived in Art. 4-4. From Eqs. 4-8 and 4-12,

$$T = K_a \Phi_d i_a \tag{9-1}$$

$$e_a = K_a \Phi_d \omega_m \tag{9-2}$$

where

$$K_a = \frac{P Z_a}{2 \pi a} \tag{9-3}$$

The symbols have been defined in Eqs. 4-8 and 4-12. These equations together with the differential equation of motion of the mechanical system, the volt-ampere equations for the armature and field circuits, and the magnetization curve describe the system performance.

Consider first the ideal dc machine shown in Fig. 9-1 with 1 field winding and negligible magnetic saturation. The direct-axis air-gap flux Φ_d is then linearly proportional to the field current i_f, and Eqs. 9-1 and 9-2 can be expressed as

$$T = k_f i_f i_a \tag{9-4}$$

$$e_a = k_f i_f \omega_m \tag{9-5}$$

where k_f is a constant. With the brushes in the quadrature axis the mutual inductance between the field and armature circuits is zero, just as it would be for 2 coils whose axes are perpendicular. The voltage equation for the field circuit then is

$$v_f = L_{ff} p i_f + R_f i_f \tag{9-6}$$

where v_f, i_f, R_f, and L_{ff} are the terminal voltage, current, resistance, and self-inductance of the field circuit, respectively, and p is the derivative operator d/dt.

For motor reference directions (Fig. 9-1a) the voltage equation for the armature circuit is

$$v_{\text{ta}} = e_a + L_{\text{aq}}pi_a + R_a i_a \tag{9-7}$$

$$v_{\text{ta}} = k_f i_f \omega_m + L_{\text{aq}}pi_a + R_a i_a \tag{9-8}$$

where v_{ta}, i_a, R_a, and L_{aq} are the terminal voltage, current, resistance, and self-inductance of the armature circuit, respectively. The subscript q is used with the inductance because the axis of the armature mmf is along the quadrature axis. The inductance L_{aq} includes the effect of any quadrature-axis stator windings in series with the armature, such as interpoles and pole-face compensating windings used on large machines to improve commutation, as described in Chap. 5. For a motor the dynamic equation for the mechanical system is

$$T = k_f i_f i_a = Jp\omega_m + T_L \tag{9-9}$$

where J is the moment of inertia and T_L is the mechanical load torque opposing rotation.

For generator reference directions (Fig. 9-1b), the armature voltage and torque equations become

$$v_{\text{ta}} = e_a - L_{\text{aq}}pi_a - R_a i_a \tag{9-10}$$

$$v_{\text{ta}} = k_f i_f \omega_m - L_{\text{aq}}pi_a - R_a i_a \tag{9-11}$$

and $\qquad\qquad T_S = Jp\omega_m + T = Jp\omega_m + k_f i_f i_a \tag{9-12}$

where T_S now is the mechanical driving torque applied to the shaft in the direction of rotation.

Energy storage is associated with the magnetic fields produced by the field and armature currents and with the kinetic energy of the rotating parts. The state of a physical system can be described in terms of its stored energy. Accordingly, the field and armature currents and the speed are *state variables*. Equations 9-6 through 9-12 are first-order differential equations containing product nonlinearities $i_f \omega_m$ and $i_f i_a$ of these state variables. These equations, together with the Kirchhoff-law equations for the circuits connected to the field and armature terminals and the torque-speed characteristics of the mechanical system connected to the shaft, determine the system performance. Their application to specific cases will be illustrated in Arts. 9-3 and 9-4.

TRANSFER FUNCTIONS AND BLOCK DIAGRAMS OF DC MACHINES

The most difficult obstacle to overcome in analysis of dc machines is the inclusion of magnetic saturation. Linear analyses omitting saturation serve two useful purposes, however. First, by virtue of the relatively simple linear differential equations which may then be written, a fuller appreciation of other factors affecting transient performance is made possible, and an approximate picture of the events is gained. Second, for those system problems involving complex combinations of machines and other equipment, dynamic system studies are made possible which otherwise would be practically prohibitive without resort to a computer.

In this article we shall consider separately excited dc machines. In part a we shall be concerned primarily with the electrical transients in dc generators resulting from changes in excitation. The analysis will be made on a linear basis, with discussion of the effects of saturation postponed until Art. 9-6. In part b attention will be focused on the dynamics of dc motors with constant field excitation.

a. DC Generators.   Linear Analysis

Consider the dc generator of Fig. 9-1b, and assume that operation is restricted to the linear portion of the magnetization curve of Fig. 9-2. The inductance of the field winding then is constant, and the voltage equation for the field circuit is

$$v_f = R_f i_f + L_{ff} p i_f = R_f(1 + \tau_f p)i_f \qquad (9\text{-}13)$$

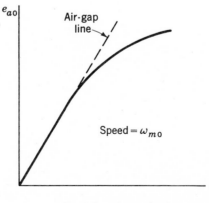

Fig. 9-2.   Magnetization curve.

where $\tau_f = L_{ff}/R_f$ is the time constant of the field circuit. At magnetization-curve speed ω_{m0} and with operation restricted to the linear range, the armature emf e_{a0} is

$$e_{a0} = K_g i_f \qquad (9\text{-}14)$$

where K_g is the slope of the air-gap line at speed ω_{m0}.

Rearrangement of Eq. 9-13 in state-variable form gives

$$p i_f = \frac{1}{\tau_f}\left(\frac{v_f}{R_f} - i_f\right) \qquad (9\text{-}15)$$

The block diagram with an integrator $1/p$ in the forward path is shown in Fig. 9-3a. Multiplication of the output i_f by K_g then gives the generated emf e_{a0} at magnetization-curve speed. Since generated emf is proportional to speed, the emf e_a at any other speed ω_m is

$$e_a = e_{a0}\frac{\omega_m}{\omega_{m0}} \qquad (9\text{-}16)$$

as shown by the multiplier in the output, Fig. 9-3a. The corresponding transfer function in complex variables is obtained by replacing the derivative operator p in Eq. 9-13 by the complex frequency s. The variables $E_{a0}(s)$ and $V_f(s)$ then are the complex amplitudes of the corresponding time variables, and the equations become algebraic equations in s; thus

$$\frac{E_{a0}(s)}{V_f(s)} = \frac{K_g I_f}{V_f} = \frac{K_g/R_f}{1 + \tau_f s} \qquad (9\text{-}17)$$

as shown in Fig. 9-3b.

The armature current i_a is determined by the generated emf e_a and the electrical circuits connected to the armature terminals. The magnetic torque T is then determined by the direct-axis flux and armature

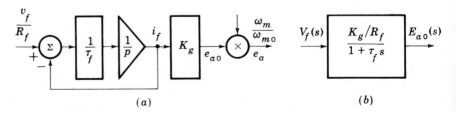

(a)   (b)

Fig. 9-3.   Block diagram of Eqs. 9-15 and 9-17.

current, as in Eq. 9-1. From Eq. 4-14, the torque can be expressed in terms of the magnetization curve as

$$T = \frac{e_{a0}}{\omega_{m0}} i_a \tag{9-18}$$

The relation between speed and torque is then given by Eq. 9-12 for a generator. The complete performance depends on the electrical and mechanical systems connected to the machine.

EXAMPLE 9-1

A 200-kw 250-volt dc generator has the following constants:

$$R_f = 33.7 \text{ ohms} \qquad R_a = 0.0125 \text{ ohm}$$

$$L_{ff} = 25 \text{ henrys} \qquad L_{aq} = 0.008 \text{ henry}$$

The slope of the air-gap line drawn on its magnetization curve at rated speed is

$$K_g = 38 \text{ volts/field amp}$$

The armature circuit is connected to a load having a resistance $R_L = 0.313$ ohm and an inductance $L_L = 1.62$ henrys.

The generator is initially unexcited but rotating at rated speed. A 230-volt dc source of negligible impedance is suddenly connected to the field terminals. Assume that as the terminal voltage builds up and the generator takes on load its speed does not change appreciably.

Compute and plot a curve of the armature current $i_a(t)$, and investigate the possibilities of simplifying approximations.

Solution

The block diagram is shown in Fig. 9-4 in terms of complex-frequency variables. The first block represents the build-up of generated emf, the second the build-up of armature current, where R_a and

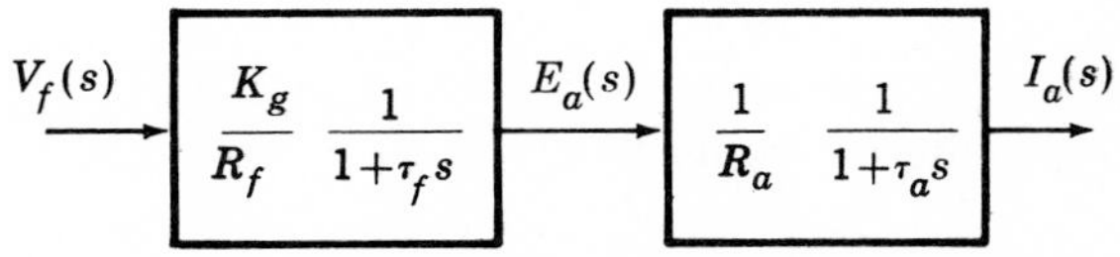

Fig. 9-4. Block diagram, Example 9-1.

L_a are the total resistance and inductance of the armature and load in series. Because the mutual inductance between armature and field is zero, the output of the second block does not influence the behavior of the first block. Each block represents an exponential term of the form $\epsilon^{-t/\tau}$, where

$$\tau_f = \frac{L_{ff}}{R_f} = \frac{25}{33.7} = 0.74 \text{ sec} \qquad \frac{1}{\tau_f} = 1.35$$

$$\tau_a = \frac{L_a}{R_a} = \frac{L_{aq} + L_L}{R_a + R_L}$$

$$= \frac{1.63}{0.326} = 5 \text{ sec} \qquad \frac{1}{\tau_a} = 0.2$$

The final steady-state value of the generated emf is

$$E_a = \frac{(230)(38)}{33.7} = 260 \text{ volts}$$

and the final steady-state value of the armature current is

$$I_a = \frac{E_a}{R_a} = \frac{260}{0.326} = 800 \text{ amp}$$

The equation for the armature current as a function of time is

$$i_a(t) = 800 + A\epsilon^{-1.35t} + B\epsilon^{-0.2t}$$

with initial conditions

$$i_a(0) = 0 \qquad \text{and} \qquad \frac{di_a}{dt}(0) = 0$$

whence $i_a(t) = 800 + 139\epsilon^{-1.35t} - 939\epsilon^{-0.2t}$

A plot of the current build-up is shown by the solid curve in Fig. 9-5. If the smaller of the two time constants were ignored, the current build-up would be given by

$$i_a(t) = 800 - 800\epsilon^{-0.2t}$$

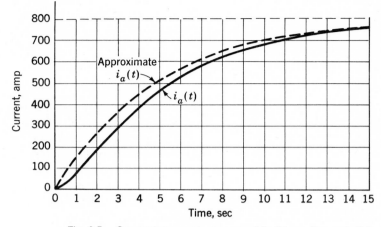

Fig. 9-5.   Generator armature-current build-up, Example 9-1.

which is shown by the dashed curve in Fig. 9-5.   Comparison of these two curves shows that the shorter of two time lags in series often may be ignored when its time constant is less than about one-quarter of the longer one.

The terminal voltage $v_{\text{ta}}(t)$ is

$$v_{\text{ta}}(t) = (R_L + L_L p)i_a(t) = 250 - 260\epsilon^{-1.35t} + 10\epsilon^{-0.2t}$$

Compare with the equation for generated emf given by

$$e_a(t) = 260 - 260\epsilon^{-1.35t}$$

An idea of the influence of armature resistance and inductance of this machine, which is typical of dc generators used as exciters for synchronous generators, can now be gained.   Obviously the influence of armature inductance is very small.   The principal effect of armature resistance is to reduce the final voltage from 260 volts to the 250 volts actually obtained.   When this effect is taken into account, it is evidently possible to base many engineering analyses on the assumption of negligible armature inductance.

In Example 9-1 the generator speed is assumed to be constant.   The dynamics of the mechanical drive do not enter into the problem; it is assumed that the drive is capable of delivering whatever mechanical

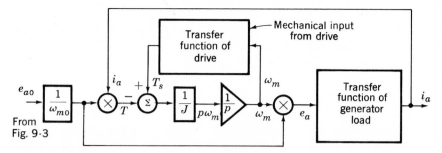

Fig. 9-6. Schematic block diagram for a dc generator and its mechanical drive.

torque is required to hold constant speed.   Figure 9-6 shows schematically the components which would have to be added to the block diagram of Fig. 9-3a or b to take into account the dynamics of the mechanical drive.   In Fig. 9-6, the first multiplier (reading from left to right) represents Eq. 9-18 for the magnetic torque T.   Rearrangement of Eq. 9-12 gives

$$p\omega_m = \frac{T_S - T}{J} \qquad (9\text{-}19)$$

as shown by the summation and the coefficient multiplier $1/J$, where J is the combined inertia of the generator and drive and T_S is obtained from the transfer function of the drive.   Integration then gives ω_m, which becomes an input to the transfer function of the drive and also to the multiplier representing Eq. 9-16 for e_a.   Finally i_a, obtained from the transfer function of the electrical load on the generator, is fed back into the first multiplier.

As more and more refinements are added, it can readily be appreciated that a complete system problem can soon become too complex for an analytical solution.   For example, in Fig. 9-6 the transfer function of the mechanical drive may involve the transfer function of a mechanical speed governor.   Fortunately, simplifying approximations usually can be made.   Otherwise a computer solution is necessary.   The laying out of a block diagram to take into account as many refinements as desired is not especially difficult, however, because it can be constructed piece by piece.

b. Separately Excited DC Motors

Direct-current motors are often used in applications requiring precise control of speed and torque output over a wide range.   One of the common ways of control is the use of a separately excited motor with constant

field excitation.   The speed is controlled by variation of the voltage
applied to the armature terminals.   The analysis then involves the
electrical transients in the armature circuit and the dynamics of the
mechanical load driven by the motor.

A separately excited motor is shown in Fig. 9-7.   The source may
be either a solid-state controlled rectifier, one of the types described
in Chap. 8, or a separately excited dc generator, as in the Ward Leonard
system described in Chap. 5.   At constant motor field current I_f the
magnetic torque and generated emf are

$$T = K_m i_a \qquad \text{newton-meters} \qquad (9\text{-}20)$$

$$e_a = K_m \omega_m \qquad \text{volts} \qquad (9\text{-}21)$$

where $K_m = k_f I_f$ is a constant.   In terms of the magnetization curve

$$K_m = \frac{e_{a0}}{\omega_{m0}} \qquad (9\text{-}22)$$

with e_{a0} the generated emf corresponding to the field current I_f at the
speed ω_{m0} rad/sec.   In mks units the constant K_m in newton-meters per
ampere (Eq. 9-20) equals the constant K_m in volt-seconds per radian
(Eq. 9-21).   The response of the motor to changes in source voltage and
the effects of load torque will now be investigated.

From Eq. 9-7, after rearrangement of the terms and division by R_a,
the differential equation for the armature current i_a is

$$\frac{L_a}{R_a} p i_a = \tau_a p i_a = \frac{v_s - e_a}{R_a} - i_a \qquad (9\text{-}23)$$

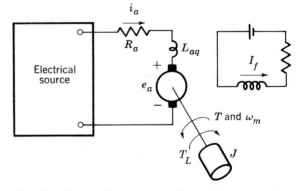

Fig. 9-7.   Schematic diagram of a separately excited
dc motor.

where v_s is the source voltage, e_a is the back emf (Eq. 9-21), R_a and L_a include the series resistance and inductance of the source and armature circuit, and $\tau_a = L_a/R_a$ is the *electrical time constant* of the armature circuit. The magnetic torque T is given by Eq. 9-20, and from Eq. 9-9 the acceleration is

$$p\omega_m = \frac{T}{J} - \frac{T_L}{J} = \frac{K_m i_a}{J} - \frac{T_L}{J} \tag{9-24}$$

where J is the moment of inertia including that of the load, and T_L is the load torque opposing rotation.

The block diagram representing Eqs. 9-20 to 9-24 is shown in Fig. 9-8a in terms of the state variables i_a and ω_m with v_s and T_L/J as inputs. Division of the input v_s by K_m and combination of the constants in the forward path yields the simpler form shown in Fig. 9-8b, where

$$\tau_m = \frac{JR_a}{K_m^2} \tag{9-25}$$

is the *inertial time constant*. Physically interpreted, v_s/K_m is the steady-state no-load speed corresponding to a constant dc input voltage V_{dc}.

In general, the load torque is a function of speed. It is sometimes

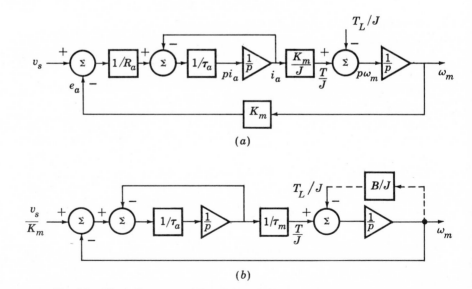

(a)

(b)

Fig. 9-8. Block diagrams of Eqs. 9-20 to 9-24 for a separately excited dc motor.

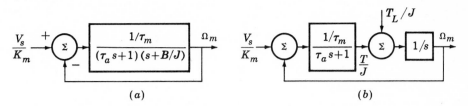

Fig. 9-9. Block diagrams of a separately excited dc motor in terms of complex variables.

assumed that the load torque is proportional to the speed; thus

$$T_L = B\omega_m \qquad \text{or} \qquad \frac{T_L}{J} = \frac{B\omega_m}{J} \tag{9-26}$$

where B is the slope of the torque-speed curve at the operating point and may be assumed to be constant for small changes.   The parameter J/B is the *load time constant* τ_L describing the rate at which the motor coasts when its armature circuit is open (see Eq. 8-16), and B/J is the corresponding damping factor.   It varies over a wide range from no load to full load but its effect usually is not very important with integral-horsepower motors.   The effect of the load damping is shown in Fig. 9-8b by the feedback B/J around the second integrator.

The block diagram in terms of complex frequency s is shown in Fig. 9-9, where now Ω_m, V_s, and T_L are the complex amplitudes of the corresponding time variables.   The functional notation "of s" will be omitted for the sake of simplicity in the rest of this chapter with the understanding that capital letters such as Ω_m and V_s are complex amplitudes. (Although the same symbol is used for torque $T(t)$ and torque $T(s)$, the meaning should be clear from the context.)   The first integrator in Fig. 9-8b becomes the algebraic term $1/(\tau_a s + 1)$ in Fig. 9-9.   The second integrator in Fig. 9-8b with damping B/J becomes the algebraic term $1/(s + B/J)$ in Fig. 9-9a.   The block diagram with mechanical damping neglected and T_L assumed to be an independent variable is shown in Fig. 9-9b.   The transfer function relating speed to input voltage, found by elimination of the negative feedback in Fig. 9-9a, is

$$\frac{\Omega_m}{V_s/K_m} = \frac{1}{\tau_m(\tau_a s + 1)\left(s + \dfrac{B}{J}\right) + 1} \tag{9-27}$$

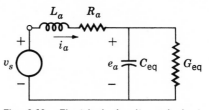

Fig. 9-10. Electrical circuit equivalent of a separately excited dc motor.

With mechanical damping neglected

$$\frac{\Omega_m}{V_s/K_m} = \frac{1}{\tau_m s(\tau_a s + 1) + 1} \tag{9-28}$$

and the transfer function relating speed to load torque then is

$$\frac{\Omega_m}{T_L} = -\frac{R_a}{K_m^2} \frac{1}{\tau_m s(\tau_a s + 1) + 1} \tag{9-29}$$

An analog electrical circuit is shown in Fig. 9-10 in which the inertia is represented by an equivalent capacitance C_{eq} and the damping by a shunt conductance G_{eq}. The torque equation for the mechanical system can be expressed as

$$T = K_m i_a = Jp\omega_m + B\omega_m \tag{9-30}$$

Division by K_m and substitution of $\omega_m = e_a/K_m$ in the result gives

$$i_a = \frac{J}{K_m^2} \frac{de_a}{dt} + \frac{B}{K_m^2} e_a \tag{9-31}$$

which is identical to the node equation for the $C_{eq}G_{eq}$ circuit provided that

$$C_{eq} = \frac{J}{K_m^2} \quad \text{and} \quad G_{eq} = \frac{B}{K_m^2} \tag{9-32}$$

In terms of the analog circuit the time constants τ_m and τ_L are

$$\tau_m = R_a C_{eq} \quad \text{and} \quad \tau_L = C_{eq}/G_{eq} \tag{9-33}$$

The natural frequencies s of the system are given by the poles of the transfer function, Eq. 9-27, or the roots of the equation

$$\left(s + \frac{1}{\tau_a}\right)\left(s + \frac{B}{J}\right) + \frac{1}{\tau_a \tau_m} = 0 \tag{9-34}$$

$$s^2 + \left(\frac{1}{\tau_a} + \frac{B}{J}\right)s + \frac{1}{\tau_a}\left(\frac{1}{\tau_m} + \frac{B}{J}\right) = 0 \tag{9-35}$$

Comparison with the standard form for a second-order equation,

$$s^2 + 2\alpha s + \omega_n^2 = 0$$

shows that the undamped natural frequency ω_n is

$$\omega_n = \sqrt{\frac{1}{\tau_a}\left(\frac{1}{\tau_m} + \frac{B}{J}\right)} \tag{9-36}$$

and the damping factor α is

$$\alpha = \frac{1}{2}\left(\frac{1}{\tau_a} + \frac{B}{J}\right) \tag{9-37}$$

The relative damping factor or damping ratio ζ is

$$\zeta = \alpha/\omega_n \tag{9-38}$$

The roots are given by the well-known solution

$$s_1, s_2 = -\zeta\omega_n \pm \omega_n \sqrt{\zeta^2 - 1} = -\zeta\omega_n \pm j\omega_n \sqrt{1 - \zeta^2} \tag{9-39}$$

where the first form, $\zeta > 1$, gives two exponential terms with negative real exponents, and the second form, $\zeta < 1$, gives a damped sinusoid. The mechanical load usually has only a small effect on ω_n and α, although of course it does affect the steady-state speed.   If B/J is neglected

$$\omega_n = \sqrt{\frac{1}{\tau_a \tau_m}} \tag{9-40}$$

$$\alpha = \frac{1}{2\tau_a} \tag{9-41}$$

$$\zeta = \frac{1}{2}\sqrt{\frac{\tau_m}{\tau_a}} \tag{9-42}$$

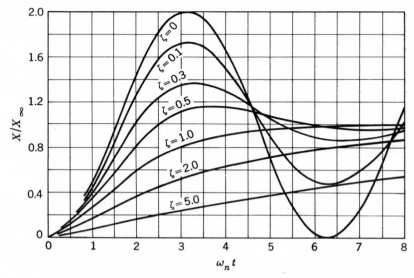

Fig. 9-11.   Normalized solutions of second-order linear differential equation for initial-rest conditions.

The solutions of a second-order system with a step input and initial rest conditions are shown in normalized form by the family of curves in Fig. 9-11.   The ordinates are the ratio of the output χ to its final steady-state value χ_∞.   These curves can be used to find the change in speed $\Delta\omega_m(t)$ resulting from a step change Δv_s in source voltage.   The ordinates are then interpreted as the ratio $\Delta\omega_m(t)/\Delta\omega_m(\infty)$, where $\Delta\omega_m(\infty)$ is the final value of the change in speed.   From Eq. 9-27, with $s = 0$

$$\frac{\Delta\omega_m(\infty)}{\Delta v_s/K_m} = \frac{1}{(\tau_m B/J) + 1} = \frac{1}{(R_a B/K_m^2) + 1} \tag{9-43}$$

From Eq. 9-29, with the load torque assumed to be independent of speed

$$\frac{\Delta\omega_m(\infty)}{\Delta T_L} = - \frac{R_a}{K_m^2} \tag{9-44}$$

EXAMPLE 9-2

The following constants are given for two typical high-performance compensated dc motors.[1]

---

[1] A table of dc motor constants is given by A. Kusko, "Solid-State DC Motor Drives," The M.I.T. Press, Cambridge, Mass., 1969, pp. 22–23.

Motor No. 1	*Motor No. 2*
1 hp, 500 rpm, 240 volts	100 hp, 1,750 rpm, 240 volts
$R_a = 7.56$ ohms	$R_a = 0.0144$ ohm
$L_a = 0.055$ henry	$L_a = 0.0011$ henry
$K_m = 4.23$ volt-sec/rad	$K_m = 1.27$ volt-sec/rad
$J = 0.050$ lb-ft-sec²	$J = 1.34$ lb-ft-sec²

Assume that the resistance and inductance of the source equal the resistance and inductance of the motor, and that the load inertia equals the motor inertia. Also assume that the torque-speed characteristic of the load is a straight line through the origin and the rated-load point. Neglect the rotational losses in the motors.

Find the undamped natural frequency ω_n and the damping factor ζ for each motor. Discuss the effect of approximations.

Solution

Convert the motor rating and inertia to mks units.

Motor No. 1	*Motor No. 2*
1 hp = 746 watts	100 hp = 74,600 watts
500 rpm = 52.3 rad/sec	1,750 rpm = 183 rad/sec
Rated torque T	
$= \dfrac{746}{52.3} = 14.3$ newton-meters	$T = \dfrac{74,600}{183} = 407$
$B = T/\omega = 0.273$	$B = 2.23$
$J = \dfrac{2(0.050)}{0.738} = 0.136,\ \text{kg-m}^2$	$J = \dfrac{2(1.34)}{0.738} = 3.64$
$B/J = 2.0$	$B/J = 0.61$
$\tau_a = \dfrac{0.110}{15.1} = 0.0073$	$\tau_a = \dfrac{0.0022}{0.0288} = 0.0765$
$1/\tau_a = 137$	$1/\tau_a = 13.1$
$\tau_m = \dfrac{(0.136)(15.1)}{(4.23)^2} = 0.115$	$\tau_m = \dfrac{(3.64)(0.0288)}{(1.27)^2} = 0.065$
$1/\tau_m = 8.7$	$1/\tau_m = 15.4$

For the assumed source and load parameters the effect of the load damping B/J is small, especially with the larger motor. From Eqs. 9-40 to 9-42 the results are:

$$Motor\ No.\ 1 \qquad\qquad\qquad Motor\ No.\ 2$$

$$\omega_n = \sqrt{(137)(8.7)} = 34.5 \qquad \omega_n = \sqrt{(13.1)(15.4)} = 14.2$$

$$\zeta = \frac{1}{2}\sqrt{\frac{0.115}{0.0073}} = 1.98 \qquad \zeta = \frac{1}{2}\sqrt{\frac{0.065}{0.0765}} = 0.46$$

From the curves of Fig. 9-11 the transient response of the 100-hp motor to a step change in source voltage is a damped sinusoid with an overshoot to about 1.2 at $\omega_n t = 3.5$, or $t = 3.5/14.2 = 0.25$ sec. The response of the 1-hp motor is overdamped. Its transient is 0.8 of its final value at $\omega_n t = 6.5$, or $t = 6.5/34.5 = 0.19$ sec. If its armature-circuit time constant is neglected, the expression for its response reduces to

$$\frac{\Delta\omega_m(t)}{\Delta\omega_m(\infty)} = 1 - \varepsilon^{-t/\tau_m}$$

The approximate expression gives a value of 0.8 when $\varepsilon^{-t/\tau_m} = 0.2$, or $t/\tau_m = 1.61$, $t = (1.61)(0.115) = 0.185$ sec. This approximate value is substantially the same as the value of 0.19 sec from the curves. The armature-circuit time constant then has very little effect on the transient behavior of the 1-hp motor, but a significant effect on the 100-hp motor.

In general, armature-circuit inductance may be neglected for damping ratios ζ greater than about 1.5, corresponding to time-constant ratios τ_a/τ_m less than about 1/9. Critical damping corresponds to the ratio $\tau_a/\tau_m = 1/4$. As a general trend for the motor alone, τ_a increases with increasing frame size and τ_m decreases slightly. Large low-speed motors, above 10 hp at speeds of 1,150 rpm and below, are underdamped. Of course, the source impedance affects τ_a and the load inertia affects τ_m. The damping B/J usually is negligible.

9-4

AN ELEMENTARY MOTOR–SPEED REGULATOR

When high precision and freedom from the effects of external disturbances are important specifications, a feedback control system must be employed. The theory of feedback control has been developed to a high degree of accuracy, and numerous textbooks and technical papers are

available covering the broad subject.[1]   We shall assume that the reader
is familiar with the basic concepts.   An example of a speed-control system will be discussed in this article.

The schematic diagram of a speed-control system using a separately
excited dc motor is shown in Fig. 9-12a.   The motor speed is measured
by means of a dc tachometer generator and its voltage e_t compared with
a reference voltage E_R.   The error voltage ε is amplified and controls the
output voltage of the power-conversion equipment, so as to maintain
substantially constant speed at the value set by the reference voltage.

[1] For example, see H. Chestnut and R. W. Mayer, "Servomechanisms and Regulating
System Design," John Wiley & Sons, New York, vol. I, 2d ed., 1959, vol. II, 1955;
G. C. Newton, Jr., L. A. Gould, and J. F. Kaiser, "Analytical Design of Linear Feed-
back Controls," John Wiley & Sons, Inc., New York, 1957; G. J. Thaler and R. G.
Brown, "Analysis and Design of Feedback Control Systems," McGraw-Hill Book
Company, New York, 2d ed., 1961; R. N. Clark, "Introduction to Automatic Control
Systems," John Wiley & Sons, New York, 1962; J. J. D'Azzo and C. H. Houpis,
"Feedback Control System Analysis and Synthesis," McGraw-Hill Book Company,
New York, 2d ed., 1966; A. E. Fitzgerald, D. E. Higginbotham, and A. Grabel, "Basic
Electrical Engineering," McGraw-Hill Book Company, New York, 3d ed., 1967,
chap. 17.

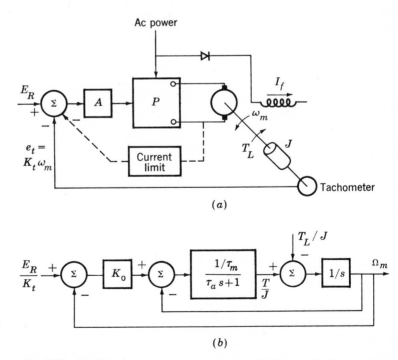

(a)

(b)

Fig. 9-12. Feedback speed-control system. *(a)* Schematic diagram and
(b) block diagram.

Usually, a current-limiting feedback is added to the controller to protect the equipment from overloads when the motor is called upon to respond to large changes in reference voltage or load torque. A current-limiting feedback is indicated schematically by the dotted line in Fig. 9-12a.

The details of the error-processing amplifier A and the power-conversion equipment depend on the type of system. For example, P may be a solid-state controlled rectifier, one of the types described in Chap. 8, and A may be a phase-shifting circuit for controlling the firing angle of the rectifiers. Or P may be a motor-generator set with A an amplifier controlling the field current of the dc generator, as in the Ward Leonard system described in Chap. 5. We shall assume that the combination of A and P is equivalent to a linear controlled voltage source $v_s = K_A \varepsilon$ with negligible time lag and gain K_A. (With solid-state rectifiers the time lags are about $\frac{1}{2}$ cycle of the ac supply and will be neglected. With the Ward Leonard system the time lag in the dc generator field may be significant, however.) We shall assume that the load torque T_L is independent of the speed, that is, the damping B/J is zero.

The block diagram is shown in Fig. 9-12b, where E_R/K_t is the steady-state no-load speed setting, K_t is the tachometer speed-voltage constant, and

$$K_0 = \frac{K_t K_A}{K_m} \tag{9-44a}$$

This block diagram is found by addition of the tachometer feedback to the block diagram of the motor, Fig. 9-9b. The response to changes in reference voltage E_R and load torque T_L will now be investigated.

With $T_L = 0$, reduction of the block diagram gives the transfer function

$$\frac{\Omega_m}{E_R} = \frac{K_0}{K_t} \frac{1}{\tau_m s(\tau_a s + 1) + 1 + K_0} \tag{9-45}$$

Similarly, with $E_R = 0$

$$\frac{\Omega_m}{T_L} = -\frac{R_a}{K_m^2} \frac{\tau_a s + 1}{\tau_m s(\tau_a s + 1) + 1 + K_0} \tag{9-46}$$

The natural frequencies s_1, s_2 of the closed-loop system are given by the poles of the transfer function. The undamped natural frequency ω_n is

$$\omega_n = \sqrt{\frac{1 + K_0}{\tau_a \tau_m}} \tag{9-47}$$

and the damping factor α is

$$\alpha = \frac{1}{2\tau_a} \tag{9-48}$$

The damping ratio ζ is

$$\zeta = \frac{\alpha}{\omega_n} = \frac{1}{2} \sqrt{\frac{\tau_m}{\tau_a} \frac{1}{1 + K_0}} \tag{9-49}$$

The natural frequencies s_1, s_2 are then given by Eq. 9-39. If the armature-circuit inductance is neglected, the response reduces to a single natural frequency $s_1 = -(1 + K_0)/\tau_m$ describing a single exponential term with a time constant $\tau_m' = \tau_m/(1 + K_0)$.

Examination of the above equations shows several important facts concerning the transient behavior of the system and its design for satisfactory performance. For a step input ΔE_R the final steady-state response $\Delta\omega_m(\infty)$ is, from Eq. 9-45 with $s = 0$,

$$\frac{\Delta\omega_m(\infty)}{\Delta E_R} = \frac{1}{K_t} \frac{K_0}{1 + K_0} \tag{9-50}$$

For initial rest conditions the step-input response is shown in normalized form by the curves in Fig. 9-11. For a step input ΔT_L of load torque, from Eq. 9-46,

$$\frac{\Delta\omega_m(\infty)}{\Delta T_L} = -\frac{R_a}{K_m^2} \frac{1}{1 + K_0} \tag{9-51}$$

For best performance, the system should be insensitive to load-torque disturbances. This requirement means that R_a should be as small as possible and K_m as large as possible; i.e., the motor should be operated at the maximum permissible flux. Also, K_0 should be as large as possible. Recall that K_0 is proportional to the amplifier gain. Increasing the amplifier gain therefore stiffens the system against the effect of load disturbances.

Increasing the amplifier gain has undesirable effects on the dynamic behavior, however. From Eqs. 9-47 and 9-49, increasing the amplifier gain increases the natural frequency ω_n and decreases the relative damping factor ζ; that is, the system oscillates rapidly and through wide extremes. The relative stability is poor. The components may wear out rapidly. In more complicated systems with three or more time lags, too much gain may lead to absolute instability, i.e., exponentially increas-

ing oscillations.   In practice it has been found that amplifier gains resulting in damping ratios lying in the range between 0.4 and 0.7 usually give
satisfactory results.   If straight amplification will not give a satisfactory
system, various forms of compensation can be added.   The techniques of
compensation by use of corrective networks are discussed in texts devoted
to feedback theory.

EXAMPLE 9-3

A 5-hp, 240-volt, 1,750-rpm dc motor is used in the speed-control
system shown in Fig. 9-12a.   The armature is supplied from a solid-
state controlled rectifier.   The armature-circuit resistance and
inductance including the rectifier are

$$R_a = 1.20 \text{ ohms} \qquad L_a = 0.010 \text{ henry}$$

The motor field is supplied with constant field current provided by a
separate rectifier.   The speed-voltage constant of the motor is

$$K_m = 1.21 \text{ volt-sec/rad}$$

The motor + load inertia is

$$J = 0.068 + 0.140 = 0.208 \text{ kg-m}^2$$

The tachometer speed-voltage constant is

$$K_t = 0.1 \text{ volt/rpm} \qquad \text{or} \qquad 0.96 \text{ volt-sec/rad}$$

The voltage gain of the error-detector amplifier and rectifier is

$$K_A = 10 \text{ volts/volt}$$

The reference voltage E_R is adjusted for a no-load speed of 1,800 rpm.

a. Solve for E_R.
b. Find the steady-state speed drop resulting from an applied torque
   of 20 newton-meters (approximately rated load).
c. Find ω_n, α, and ζ.   Comment on the system performance.

Solution

Work in mks units.   Neglect no-load rotational losses.

a. 1,800 rpm = 188 rad/sec

$\qquad$ Motor counter emf $E_a = K_m \omega_m = (1.21)(188) = 228$ volts

$\qquad$ Input to amplifier $= E_a/K_A = 22.8$ volts

$\qquad$ Reference voltage $E_R = 180 + 22.8 = 202.8$ volts

b. From Eq. 9-44a,

$$K_0 = \frac{(0.96)(10)}{1.21} = 7.9$$

and from Eq. 9-51,

$$\frac{\Delta \omega_m(\infty)}{\Delta T_L} = -\frac{1.20}{(1.21)^2} \frac{1}{8.9} = -0.092$$

$$\text{Speed drop} = (20)(0.092) = 1.84 \text{ rad/sec}$$

$$= \text{less than 1.0 percent of no-load speed}$$

c. The time constants are

$$\tau_a = \frac{0.010}{1.20} = 0.00833 \text{ sec}$$

$$\frac{R_a}{K_m^2} = \frac{1.20}{(1.21)^2} = 0.82 \qquad \tau_m = (0.208)(0.82) = 0.171 \text{ sec}$$

From Eqs. 9-47, 9-48, and 9-49

$$\omega_n = \sqrt{\frac{8.9}{(0.0083)(0.171)}} = 79 \text{ rad/sec}$$

$$\alpha = \frac{1}{(2)(0.00833)} = 60 \text{ rad/sec}$$

$$\zeta = \frac{60}{79} = 0.76$$

The steady-state speed regulation and the damping are within satisfactory limits for most industrial applications.

Many industrial applications of adjustable-speed drives require the coordinated control of several motors in tandem drives for processing continuous strips of material passed through a succession of rollers. Paper-making machines and some steel-mill drives are examples. These drives require careful coordination of the motor controls to avoid a disastrous tug-of-war between the individual drive motors.[1]

The transient analysis of dynamic systems in terms of the time response to disturbances shows what may happen to a system when it is subjected to disturbances such as step changes in inputs. However, the transient analysis becomes unwieldy when the system contains more than two energy-storage elements. Dynamic analysis by frequency-response methods is an important complement to transient analysis and is treated thoroughly in texts devoted to feedback theory. A brief summary is given here.

Any closed-cycle system can be represented by the block diagram of Fig. 9-13 in which $G(s)$ is the transfer function relating the complex amplitude of the output C to the complex amplitude of the error signal ε, and $H(s)$ is the transfer function of the feedback link. For steady-state sinusoidal excitation, $s = j\omega$, these functions become phasor functions of the frequency ω. The transfer function

$$\frac{C}{R}(j\omega) = \frac{G(j\omega)}{1 + G(j\omega)H(j\omega)} = |A(j\omega)|\underline{/\theta(j\omega)} \qquad (9\text{-}52)$$

is a phasor function of frequency having an amplitude $|A|$ and angle θ. The values of $|A|$ and θ can be plotted as functions of frequency ω. High peaks in the amplitude-response curve evidently correspond to undesirable oscillations in the transient response. On the other hand, a rapid falling off of the amplitude response curve as frequency increases corre-

---

[1] W. K. Boice, Controlling Speed in Multidrive Systems, *Machine Design*, **42**(2):130–134 (1970).

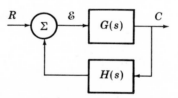

Fig. 9-13. Generalized block diagram of a feedback system.

sponds to sluggish transient response. The plots usually are made as Nyquist or Bode diagrams of the open-loop transfer function $G(j\omega)H(j\omega)$ as described in texts devoted to feedback theory. Compensating networks may be added to improve the system performance.

In any system involving feedback from the output to the input, there is always the possibility of self-excited operation, i.e., the system may undergo sustained oscillations even with no input. In the system of Fig. 9-13, a sufficient number of energy-storage elements, with their associated time lags and phase shifts between output and input, will cause the output C to be 180° out of phase with error ε at some finite frequency. In other words, the transfer function $G(j\omega)H(j\omega)$ will be a negative real number at this frequency. If that number is unity, self-excitation may take place; i.e., even with $R = 0$, feedback of the output around the closed loop supplies the proper input, and the system oscillates steadily at the frequency corresponding to the 180° phase shift. If that number is greater than unity, oscillations may take place at an amplitude which increases until it is limited by nonlinearities in the system. Compensating networks must then be added. The system will not undergo sustained oscillations if the magnitude of the transfer function is less than unity at the frequency producing a 180° phase shift.

It should be noted that these conditions furnish a criterion for absolute stability. The normal closed-cycle system operates well within these limits.

9-5
METADYNES AND AMPLIDYNES

So far, we have considered dc machines with brushes located only in the quadrature axis. The purpose of this article is to examine the effects of additional brushes located in the direct axis. By these means the armature mmf can be used to provide most of the excitation and high-power gains can be achieved. Machines with more than two brush sets per pair of poles are called *metadynes*. This article is concerned with metadyne generators, with emphasis on the most commonly used form, the amplidyne.[1]

[1] For a discussion of the steady-state theory and descriptions of a number of applications, see J. M. Pestarini, "Metadyne Statics," Technology Press and John Wiley & Sons, Inc., New York, 1952. For discussions of the transient theory, see M. Riaz, Transient Analysis of the Metadyne Generator, *Trans. AIEE*, **72**(III):52–62 (1953); K. A. Fegley, Metadyne Transients, *Trans. AIEE*, **74**(III):1179–1188 (1955).

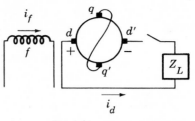

Fig. 9-14.   Basic metadyne.

a. Basic Metadyne Generators

A modification of the basic dc machine is shown in Fig. 9-14.   The stator
has a control-field winding f on the direct axis.   Brushes qq' are located
on the commutator so that commutation takes place along the quadrature axis, as in the normal dc generator.   With the generator driven at
constant speed ω_{m0} and with magnetic saturation neglected, the voltage
e_{aq} generated in the armature between the quadrature-axis brushes is

$$e_{\mathrm{aq}} = K_{\mathrm{qf}}i_f \tag{9-53}$$

where K_{qf} is a constant and i_f is the field current.

Now reduce the field current to a small value and short-circuit the
quadrature-axis brushes, as shown in Fig. 9-14.   Since the impedance
of the short-circuited armature is small, a weak control-field current will
produce a relatively much larger quadrature-axis armature current and a
corresponding flux-density wave centered on the quadrature axis.   By
commutator action this magnetic field is stationary in space.   Its effect
is similar to that of a fictitious stator winding on the quadrature axis.

If brushes dd' are now placed on the commutator in the direct axis,
as shown in Fig. 9-14, the emf e_{ad} generated in the armature by its rotation in the quadrature-axis flux will appear across these brushes.   With
the continued assumption of constant speed and negligible saturation

$$e_{\mathrm{ad}} = K_{\mathrm{dq}}i_q \tag{9-54}$$

where i_q is the quadrature-axis armature current and K_{dq} is a constant.

Now connect a load Z_L to the direct-axis brushes.   The direct-axis
armature current i_d produces an mmf which *opposes* the control-field mmf.
Each stage of voltage generation results in a current whose magnetic field
is spatially 90° ahead of the flux wave producing the voltage.   With two
stages of voltage generation the mmf of the direct-axis output current is

shifted 90° twice and therefore opposes the original field excitation. The quadrature-axis generated emf now is

$$e_{aq} = K_{qf}i_f - K_{qd}i_d \tag{9-55}$$

where K_{qd} is a constant under the assumed conditions of constant speed and negligible saturation.

The metadyne generator of Fig. 9-14, therefore, is a two-stage power amplifier with strong negative current feedback from the final output stage to the input.   For a fixed value of field current, it maintains very nearly constant output current i_d over a wide range of load impedance. Its power amplification, however, is reduced by the effect of the negative feedback.

b. Amplidynes

The commonest version of the metadyne is the *amplidyne*. It consists of the basic metadyne generator plus a cumulative winding on the direct axis connected in series with the direct-axis load current, as shown by the winding labeled *Comp* in the schematic diagram of Fig. 9-15. This winding, called a *compensating winding*, is very carefully designed to provide a flux as nearly as possible equal and opposite to the flux pro- duced by the direct-axis armature current.   The negative-feedback effect of the load current is thereby canceled, and the control-field winding has almost complete control over the direct-axis flux.   Very little control-field power input is required to produce a large current in the short-circuited quadrature axis of the armature.   The quadrature-axis current then produces the principal magnetic field.   The power required to sustain the quadrature-axis current and the load is supplied mechanically by the motor driving the amplidyne. Power amplification of the order of 20,000:1 can easily be obtained.   This power amplification may be compared with values in the range from about 20:1 and 100:1 for con- ventional generators.

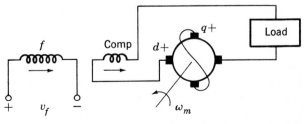

Fig. 9-15.   Basic amplidyne.

If we assume perfect compensation, negligible saturation, and constant speed, the transfer function relating direct-axis generated emf E_{ad} to control-field applied voltage V_f is

$$\frac{E_{ad}}{V_f} = \frac{K_{qf}/R_f}{\tau_f s + 1} \frac{K_{dq}/R_{aq}}{\tau_{aq} s + 1} \tag{9-56}$$

where R_f and τ_f are the resistance and time constant of the control field, and R_{aq} and τ_{aq} are the resistance and time constant of the quadrature-axis armature circuit. The resistance of the direct-axis armature circuit is considered to be lumped with the load. The direct-axis armature inductance usually is neglected because the compensating winding very nearly eliminates the flux produced by the direct-axis armature current. The principal time lag is that produced by the quadrature-axis time constant τ_{aq} and is in the range from 0.02 to 0.25 sec.

Various auxiliary or control-field windings may be added to either axis of an amplidyne to improve performance characteristics. For example, a cumulative series field may be wound on the quadrature axis and connected in series with the quadrature-axis current. This field decreases the quadrature-axis current for a specified direct-axis voltage output. Quadrature-axis commutation is thereby improved.

Amplidynes are used to provide the power amplification in a variety of feedback control systems requiring controlled power output in the range from about 1 to 50 kw. For example, they are used as the voltage-regulating unit in the excitation systems of large ac generators to insert a buck-or-boost voltage in series with the field winding of the main exciter.[1] Or the main exciter may be an amplidyne when the excitation requirements are within the range where amplidynes are competitive with other types of excitation systems. An amplidyne may be used as the generator in a Ward Leonard speed or position control system if the power requirements of the regulated motor do not exceed a few kilowatts. Position-control servomechanisms are treated in textbooks devoted to feedback theory.

9-6

EFFECTS OF SATURATION. SELF-EXCITED GENERATORS

In Art. 9-3a the transient response of a dc generator was studied on a linear basis with magnetic saturation neglected. Often the critical por-

---

[1] H. C. Barnes, J. A. Oliver, A. S. Rubenstein, and M. Temoshok, Alternator-Rectifier Exciter for Cardinal Plant 724-MVA Generator, *IEEE Trans.*, *Power Apparatus and Systems*, vol. PAS 87, no. 4, pp. 1189–1198, April 1968.

tion of the transient response takes place in substantially the linear region, or the response to small disturbances can be treated on an incrementally linear basis.   For more comprehensive studies, however, transient investigations may require a nonlinear analysis because of saturation.   For example, the response of an exciter driven into saturation by the demands of the voltage-regulating system during major system disturbances or the analysis of self-excited shunt generators very definitely require that saturation be taken into account.

The purpose of this article is to show how magnetic saturation can be included by a modification of the block diagram of Fig. 9-3a in a form adaptable to computer solution of system problems.   We shall also investigate the voltage build-up of a shunt generator as an example of a nonlinear problem that can be analyzed by relatively simple graphical means.

a. The Block Diagram with Saturation

Consider a dc generator driven at constant speed ω_{m0} with a voltage v_f applied to its field terminals.   With saturation included, the relation between generated voltage e_{a0} and field current i_f is the magnetization curve.   Furthermore, the inductance of the field winding is no longer constant.   It is more convenient then to express the voltage induced in the field winding in terms of the field flux linkages; thus

$$v_f - R_f i_f = N_f p \Phi_f \qquad (9\text{-}57)$$

where R_f is the field-circuit resistance, i_f is the field current, Φ_f is the field flux per pole, N_f is the total number of turns in the field winding (all poles assumed to be connected in series), and p is the derivative operator d/dt.

The field flux Φ_f is somewhat greater than the direct-axis air-gap flux Φ_d because of field leakage flux.   The increase may be included approximately by the use of a *coefficient of dispersion* σ, usually about 1.15, which we shall assume to be constant.   Thus

$$\Phi_f = \sigma \Phi_d \qquad (9\text{-}58)$$

The air-gap flux Φ_d is related to the generated voltage e_{a0} by Eq. 9-2; thus

$$\Phi_d = \frac{e_{a0}}{K_a \omega_{m0}} \qquad (9\text{-}59)$$

Substitution of Eq. 9-59 in Eq. 9-58, differentiation, and substitution of the result in Eq. 9-57 yields

$$\frac{N_f\sigma}{K_a\omega_{m0}} pe_{a0} = v_f - R_f i_f \qquad (9\text{-}60)$$

The coefficient of the left-hand side of Eq. 9-60 can now be expressed in terms of recognizable and easily determinable constants if its numerator and denominator are multiplied by $N_f\mathcal{P}_{ag}$, where $\mathcal{P}_{ag}$ is the permeance of the air gap; thus

$$\frac{N_f\sigma}{K_a\omega_{m0}} = \frac{N_f^2\sigma\mathcal{P}_{ag}}{K_a\omega_{m0}\mathcal{P}_{ag}N_f} \qquad (9\text{-}61)$$

Now, $N_f^2\sigma\mathcal{P}_{ag}$ is the unsaturated value of the field inductance L_{ff}, and $K_a\omega_{m0}\mathcal{P}_{ag}N_f$ is the slope K_g of the air-gap line in generated volts per field ampere. Both of these quantities are *constants*. They can easily be determined by tests taken under conditions of negligible saturation. Substitution of these constants in Eq. 9-60 gives

$$\frac{L_{ff}}{K_g} pe_{a0} = v_f - R_f i_f \qquad (9\text{-}62)$$

or, after division by R_f

$$pe_{a0} = \frac{K_g}{\tau_f}\left(\frac{v_f}{R_f} - i_f\right) \qquad (9\text{-}63)$$

where $\tau_f = L_{ff}/R_f$ is the unsaturated value of the field-circuit time constant.

The block diagram representing Eq. 9-63 is shown in Fig. 9-16. It is a relatively simple modification of the linearized block diagram of Fig. 9-3a. Saturation is taken care of by feeding back the output e_{a0} through the magnetization curve to obtain i_f. This operation can be

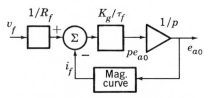

Fig. 9-16. Block diagram with saturation included.

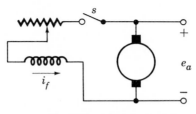

Fig. 9-17. Shunt generator.

carried out on a computer.   Based on Eq. 9-63, a computer can be pro-
grammed to solve relatively complex system problems involving saturated dc machines as system components.

b. Shunt Generator Voltage Build-up

The build-up of voltage of a self-excited shunt generator, described qualitatively in Art. 5-6b, is obviously a process inherently dependent on saturation.   A shunt generator driven at constant speed ω_{m0} is shown in Fig.
9-17, and its magnetization curve in Fig. 9-18.   A small residual flux is
assumed, corresponding to a small generated voltage at zero excitation.
The straight line Oa in Fig. 9-18, called the *field-resistance line*, is a plot
of the relation

$$v_f = R_f i_f \tag{9-64}$$

where v_f, i_f, and R_f are the voltage, current, and resistance of the field
circuit.   The slope of line Oa is adjustable by means of the field rheostat.
In Fig. 9-18 the slope of Oa is less than that of the air-gap line, and point a
is the intersection of the field-resistance line with the magnetization curve.

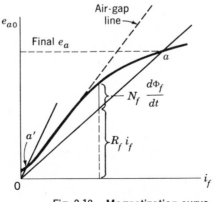

Fig. 9-18. Magnetization curve.

Fig. 9-19. Voltage build-up of a shunt generator.

Now let the field switch S be closed at $t = 0$. The small voltage generated in the armature by the residual flux is then applied to the shunt-field circuit, and the shunt-field current starts to build up. If the connections of the shunt field to the armature terminals are such that the resulting field current increases the flux, positive feedback results and the voltage continues to build up until it is finally limited by magnetic saturation at point a, where the generated voltage supplies just enough field current to sustain itself. This statement ignores the extremely small voltage drop caused by the shunt-field current in the armature-circuit resistance. The normal operating point of a shunt generator is well up on the magnetization curve.

Figure 9-19 shows a curve of the armature terminal voltage as a function of time following closure of the field switch. This curve can be calculated by a graphical method. Since the generated emf e_a is applied to the field circuit, the voltage equation for the field circuit is

$$N_f \frac{d\Phi_f}{dt} = e_a - R_f i_f \qquad (9\text{-}65)$$

where Φ_f is the field flux per pole and N_f is the number of turns in the field winding. Thus, the rate of change of field flux, and consequently of generated voltage, is proportional to the vertical difference between the magnetization curve and the field-resistance line.

The curve of e_a as a function of time can be computed by separation of the variables and graphical integration. From Eq. 9-62 with e_{a0} substituted for v_f and the variables separated

$$dt = \frac{L_{ff}}{K_g} \frac{de_{a0}}{e_{a0} - R_f i_f} \qquad (9\text{-}66)$$

The time required for the voltage to change from an initial value e_r to the value e_a is then

$$t = \frac{L_{ff}}{K_g} \int_{e_r}^{e_a} \frac{1}{e_{a0} - R_f i_f} \, de_{a0} \qquad (9\text{-}67)$$

This integral can be evaluated graphically by finding the areas on a plot of $1/(e_{a0} - R_f i_f)$ as a function of e_{a0}. The response is rather slow, because only relatively small voltage differences act to build up the flux.

The graphical process may be applied to many first-order equations where the variables can be separated in this manner. Example 10-5 shows an application of this method to the starting transient of an induction motor.

9-7
RÉSUMÉ

Our primary objective in this chapter is to construct and analyze mathematical models of the idealized dc machine. From these models we wish to find the dynamic performance not only of the machine itself as either a generator or motor but also of combinations of dc machines and simple control elements.

For the basic dc machine, the dynamic equations are relatively simple and are easily established. They are given in Art. 9-2. Accordingly, that article is the basic one in the chapter. The articles which follow it are devoted to examples of applying the equations, to the development of techniques for solving them, and to illustrating the adaptability and versatility of dc machines as control devices.

When the dc machine comprises a single field winding and an armature circuit, the basic equations show that there are three sources of time lag in the dynamic response of the machine. One is created by the field inductance, one by the armature-circuit inductance, and one by the mechanical equipment connected to the machine shaft, including the inertia of the armature itself. Analysis may often be simplified, however, by virtue of one or more time constants being small compared with others. Such simplification is of especial value when one is concerned with analysis of a system of machines rather than a single machine.

Since the more important and more interesting problems do deal with systems of machines, we have illustrated the use of several techniques for expediting their analysis. A common system of machines for wide-range and precise control of speed is based on a dc motor whose armature is supplied from an adjustable-voltage solid-state rectifier or dc generator (the so-called Ward Leonard system). Among the techniques of analysis, in addition to the classical solution of differential equations, are the use of block diagrams, transfer functions, equivalent circuits, and frequency-response methods. These techniques, of course, are broadly applicable to system problems in general.

One result of using these techniques is to view the machine as more than simply a brute-force energy-conversion device. The separately excited generator, for example, is looked upon as a power amplifier—a gain and one or more time constants, together with an increase in power level. The shunt generator is seen to be similar to a feedback oscillator. From such broader viewpoints one can more fully assess the control possibilities of more complex dc machines. The additions to the basic machine may include brushes in the direct as well as in the quadrature axis (as in the amplidyne and metadyne). When combined with feedback through external circuits, much can be done to "tailor" the system characteristics to meet performance specifications. Ideally, the objectives are to increase the sensitivity or gain, to decrease the effective time constants or response time, and to decrease the sensitivity of the system to uncontrolled external disturbances. Not all these ideals are compatible. For example, the addition of negative feedback to a system has the beneficial effect of decreasing the response time and stiffening the system against the effects of disturbances; but the gain is decreased. To obtain the same power output with the same control power input, the power amplification must accordingly be increased. The cost of equipment with the increased power amplification must be balanced against the improvements in system performance.

Throughout all these analyses, then, it should always be borne in mind that the machines must have adequate capability to handle the voltage, current, and power surges demanded by the control signals. In other words, the signal-flow characteristics of the ideal machine must be coordinated with the limitations imposed by the properties of the materials making up the realistic machine and the cost of equipment to give the required power amplification.

PROBLEMS

9-1. A separately excited dc generator has the following constants:

Field-winding resistance $R_f = 100$ ohms
Field-winding inductance $L_{ff} = 50$ henrys

Armature resistance R_a = 0.05 ohm

Armature inductance L_a = 0.5 mh

Generated emf constant K_g = 100 volts/field amp at 1,200 rpm

The generator is driven at a constant speed of 1,200 rpm. Its field and armature circuits are initially open.

a. At $t = 0$ a constant-voltage source of 250 volts is suddenly applied to the terminals of its field winding. Find the equation for the armature terminal voltage as a function of time, and sketch the curve.

b. After steady-state conditions have been established in (a), the armature is suddenly connected to a load of resistance 1.20 ohms and inductance 1.5 mh in series. Find the equations for (1) the armature current and (2) the armature terminal voltage as functions of time. Include the effect of the armature inductance and resistance. Sketch the curves.

c. Find the magnetic torque as a function of time.

9-2. A dc motor M has its armature permanently connected to a source S as shown in Fig. 9-20a. The volt-ampere characteristic of the source is shown in Fig. 9-20b. The motor field f is separately excited from a voltage source E_f as shown.

Motor armature resistance R_a = 0.5 ohm

Motor armature inductance negligible

Motor field resistance R_f = 50 ohms

Motor field inductance L_{ff} = 50 henrys

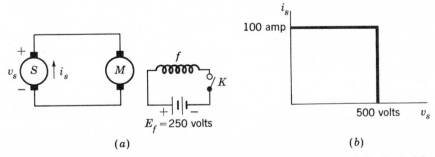

(a)　　　　　　　　　　　　　　　　(b)

Fig. 9-20. (a) Circuit diagram and (b) idealized source characteristic, Prob. 9-2.

Motor torque at 5-amp field current and 100-amp armature current = 200 newton-meters

Motor load = pure inertia.   Total moment of inertia J of load and armature = 10 kg-m²

Neglect magnetic-saturation effects in the motor.

The motor field switch K is closed at $t = 0$.

Derive an expression, with numerical values, for the speed in radians per second as a function of time in seconds.   Sketch this curve roughly to scale.   Indicate on it the final steady-state speed and the speed and time at which the break point in the source characteristic is reached.   Compute the time required for the motor to reach approximately 96 percent of its final speed.

9-3. A separately excited dc motor drives a pure-inertia load.   The combined inertia of motor and load is J kg-m².   The armature resistance is R_a ohms.   Neglect armature inductance.   The motor torque constant is K_m newton-meters/amp.

With the motor initially at standstill and the field excited in the steady state, a constant voltage V_t is suddenly applied to the armature. Find the total heat dissipated in the motor armature in bringing the motor up to its final steady-state speed.   Compare with the kinetic energy stored in the rotating masses.

9-4. A dc shunt motor is driving a pure-inertia load.   The armature and field are supplied from a source of constant direct voltage.   The motor is initially operating in the steady state.   Neglect all rotational losses and armature reaction.   Assume that the flux is directly proportional to the field current and that armature inductance is negligible.

The field rheostat is suddenly short-circuited.   Develop the differential equation for the speed of the motor following this disturbance, using the ordinary symbols for the various quantities.   Indicate all initial conditions.   It is not necessary to solve the equation.

9-5. In continuous rolling mills the stands, or rolls, through which the bar passes in the rolling process are arranged in tandem, with the majority of the stands driven by separate motors.   It is common to use dc motors supplied with power from one or several generators.   The transient changes of the motor speed under suddenly applied loads as the bar enters one stand after another may seriously affect the quality of the product.   In particular, the *impact speed drop* which occurs at the maximum of the transient oscillation is of major importance.

Consider a single motor M supplied by a generator G, each with separate and constant field excitation. The internal voltage E of the generator may be considered constant, and the armature reaction of both machines may be considered negligible. With the motor running without external load and the system in the steady state, a bar enters the stand at $t = 0$, causing the load torque to be increased suddenly from zero to T. The following numerical values apply:

Internal voltage of G, $E = 387$ volts

Motor-plus-generator armature inductance, $L = 0.00768$ henry

Motor-plus-generator armature resistance, $R = 0.0353$ ohm

Moment of inertia of motor armature and connected rolls, all referred to motor speed, $J = 42.2$ kg-m²

Electromechanical conversion constant for motor, $K_m = 4.23$ newton-meters/amp

No-load armature current, $i_o = 35$ amp

Suddenly applied torque, $T = 2,040$ newton-meters

Determine the following quantities:

a. The undamped angular frequency of the transient speed oscillations
b. The damping ratio of the system
c. The initial speed, in rpm
d. The initial acceleration, in rpm per second
e. The ultimate speed drop, in rpm
f. The impact speed drop, in rpm

9-6. Figure 9-21 shows a dc generator whose field current is supplied

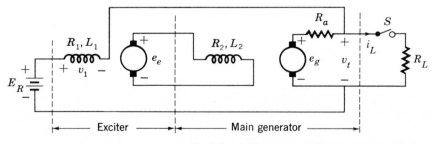

Fig. 9-21. Voltage regulating system, Prob. 9-6.

from an exciter.   The generator and exciter are driven at constant speed.
The machine constants are:

Exciter:
   Field inductance L_1 = 125 henrys
   Field resistance R_1 = 250 ohms
   Generated voltage = 1,000 volts/field amp
   Armature resistance negligible

Main generator:
   Field inductance L_2 = 100 henrys
   Field resistance R_2 = 100 ohms
   Generated voltage = 250 volts/field amp
   Armature resistance R_a = 1.0 ohm
   Armature inductance negligible

Neglect the effect of the exciter-field current on the voltage drop in the
armature of the main generator.

   a. With switch *S* closed the reference voltage E_R is adjusted until
   v_t = 250 volts with R_L = 10 ohms.   Find E_R.
   b. With the system in the steady state as in part *a*, switch *S* is opened
   at *t* = 0.   Find $v_t(t)$.

9-7. This problem concerns one of the two motor-generator sets in
the Francis Bitter National Magnet Laboratory.   Each set consists of
the following machines, all mechanically coupled on one shaft:

   One 6,000-hp 360-rpm synchronous motor
   One 600-hp 18-pole wound-rotor induction motor
   One 84-ton flywheel
   Two dc generators each with the following capability:

| continuous | 250 volts, 10,000 amp, 360 rpm |
| 5-sec pulse | 200 volts, 40,000 amp, 385/300 rpm |

The dc generators are used to supply direct current to water-cooled air-
core magnets for experimental work in very high magnetic fields.   The
dc generators can be connected in parallel to supply 20,000 amp at
250 volts continuously, or a maximum current of 80,000 amp at 200 volts
for 5 sec.

Three-phase power is supplied to the drive motors at 4,160 volts, 60 Hz. The induction motor is used to start the set. The starting current is held constant at 90 amp by means of an amplidyne servo-system which adjusts liquid resistors in the rotor circuits. It takes about 15 min to get up to 360 rpm. For continuous loads the synchronous motor is then synchronized and the induction motor is disconnected. For pulsed loads the synchronous motor is disconnected, the induction motor drives the set at 385 rpm initially, and most of the energy is supplied by the flywheel. The dc generators have interpoles, pole-face compensating windings, and series fields to improve commutation, speed of response, and load sharing.

The excitation system consists of an exciter motor-generator set comprising a 200-hp induction motor driving two 75-kw 250-volt fast-response amplidynes for excitation of the two main dc generator fields. The amplidynes are controlled by feedback amplifiers so as to regulate the main generator outputs.

With the main generators driven at 360 rpm and no load, the response of the control system is as shown in Fig. 9-22, where v_f is the voltage applied to the main generator field terminals in per unit. Here 1.0 per unit is the field voltage that will result in 250 volts generated emf in the main generator at 360 rpm, no load.

Estimated data:
    Moment of inertia $J = 4 \times 10^5$ kg-m^2.
    Main generator field time constant $\tau_f = 1.3$ sec.
    Main generator armature-circuit resistance (including all series fields) = 0.02 per unit on continuous rating.

Results required:
    a. Estimate the time t_1, Fig. 9-22, for the generator output voltage to reach 250 volts at 360 rpm, no load. Sketch the curve.
    b. With the two generators in parallel supplying the magnet, estimate the speed at the end of a 5-sec 80,000-amp pulse. Sketch the speed-time curve.
    c. With the load of b, estimate the net field excitation in per unit required to maintain 200 volts at 80,000-amp output. Sketch the curve.

9-8. The speed-control system described in Art. 9-4 is used with a 10-hp 1,750-rpm 240-volt separately excited motor.

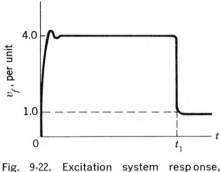

Fig. 9-22. Excitation system response, Prob. 9-7.

Motor + source resistance R_a = 1.20 ohms
Motor + source inductance L_a = 0.008 henry
Motor + load inertia J = 0.20 kg-m²
Motor speed-voltage constant K_m = 1.27 volt-sec/rad
Tachometer constant = 0.1 volt/rpm

The reference voltage E_R is adjusted for 1,800 rpm at no load. The voltage gain is adjusted so that the damping ratio ζ is 0.50. Compute the corresponding values of the amplifier gain K_A, the undamped natural frequency ω_n, and the damping factor α.

With the system initially in the steady state at no load, a step of load torque T_L = 40 newton-meters (approximately full load) is suddenly applied to the motor shaft.

   a. Find the final steady-state values of (1) the motor speed, and (2) the motor armature current.
   b. Find the initial values (t = 0+) of (1) the speed, and (2) the armature current.
   c. Find the initial rates of change of (1) the speed, and (2) the armature current.
   d. Find the equations for (1) the speed, and (2) the armature current as functions of time. Sketch the curves. Are the normalized curves of Fig. 9-11 applicable? Estimate the minimum speed and the maximum armature current.

   9-9. A 2-kw 200-volt metadyne generator of the type shown in Fig. 9-14 is driven by a synchronous motor at 1,800 rpm and has the following constants:

Control-field resistance $R_f = 20$ ohms
Control-field inductance $L_{ff} = 2$ henrys
Voltage constant $K_{qf} = 240$ volts/field amp
Armature resistances $R_{aq} = R_{ad} = 4$ ohms
Armature inductance $L_{aq} = 1.0$ henry
Voltage constants $K_{dq} = K_{qd} = 60$ volts/amp

The metadyne supplies a 20-ohm resistive load at a voltage of 200 volts. Find the power input to the control field and the power amplification.

9-10. The voltage applied to the field winding of the metadyne generator described in Prob. 9-9 is held constant at 60 volts, and the load resistance is varied from 0 to 25 ohms. Plot the steady-state output volt-ampere characteristic.

9-11. A compensating winding is added to the metadyne of Prob. 9-9, thereby converting it into an amplidyne. The amplidyne supplies 200 volts to a 20-ohm load. Find the power input to the control field, and compare with the result of Prob. 9-9. Neglect the resistance of the compensating winding.

9-12. An amplidyne exciter supplies field current to a 10-kw 125-volt dc generator.

Amplidyne data:
    Control-field turns $N_c = 400$
    Control-field resistance $R_c = 40$ ohms
    Control-field inductance $L_c = 1.6$ henrys
    Quadrature-axis time constant $\tau_q = 0.10$ sec
    Direct-axis generated voltage $= 5.0$ volts/control-field amp-turn
    Armature resistance $= 5.0$ ohms

Generator data:
    Field resistance $R_{fg} = 35$ ohms
    Field inductance $L_{fg} = 10$ henrys
    Generated voltage constant $= 50$ volts/field amp
    Armature resistance $= 0.075$ ohm

a. Compute the open-loop transfer function $G(j\omega)H(j\omega)$ relating generator output voltage at no load and amplidyne control-field voltage.

b. Compute the complex value of $G(j\omega)H(j\omega)$ at $\omega = 20$ rad/sec. Sketch the complete locus for positive values of ω.

c. The system is now connected as a closed-loop voltage-regulating system. Will the system be stable without any antihunt feedback?

d. Assume that the system is inherently stable or has been stabilized by an antihunt feedback circuit. What must be the value of the constant reference voltage to give a generator output voltage of 125 volts at no load?

e. With the reference voltage held constant as in d, what will be the generator terminal voltage when it is delivering an armature current of 80 amp?

9-13. A metadyne having no stator windings is driven at a constant speed. Its armature resistance $R_a = 0.10$ ohm measured between either pair of brushes. Its armature inductance $L_a = 0.01$ henry measured between either pair of brushes. A test taken with the direct axis open and a constant voltage of 20 volts applied to the quadrature-axis brushes gives a steady-state direct-axis open-circuit voltage of 600 volts. A similar test with the direct and quadrature axes interchanged gives similar results. Magnetic saturation is negligible.

This machine is used as a constant-voltage to constant-current transformer to supply substantially constant current to a variable-resistance load. The constant source voltage of 600 volts is applied to the quadrature-axis brushes, and the resistive load is connected to the direct-axis brushes.

Compute the steady-state load current and source current when the load voltage is 600 volts.

10
ac machines, transients and dynamics

Energy conversion by electromagnetic methods is, as we have seen, associated with energy storage in magnetic fields. When changes in operating conditions take place, the accompanying changes in stored magnetic energy cannot occur instantaneously. Instead, a transient period of readjustment must be interposed between the initial and final operating conditions. Very often, consideration must be given to energy storage and energy flows into and out of moving masses, and in some cases elastic compliances, as well as in magnetic fields; that is, electromechanical transients rather than simply electric transients are involved.

Because the general subject is so broad, this chapter can provide only an introduction to the transients and dynamics of ac machines. The development is first carried out for electrical transients in synchronous machines, largely on a physical or semi-intuitive basis. The thought process is then applied to transients in induction machines and finally to the dynamics of both types.

10-1
SYNCHRONOUS-MACHINE TRANSIENTS:
COUPLED-CIRCUIT VIEWPOINT

The inherent complexity of synchronous-machine transient phenomena can be appreciated by inspecting the main structural details of the machine with the object of pointing out the significant circuits when transient rather than steady-state conditions prevail.   For this purpose, the schematic diagram of Fig. 10-1 is presented.   The rotor damping circuits are included because they may be of determining importance.   A salient-pole structure is shown to emphasize the differences between the polar, or direct, axis and the interpolar, or quadrature, axis; similar differences exist between the two axes of cylindrical-rotor machines when transients are considered, so that the treatments of the two structural classes become essentially alike.

Under balanced steady-state conditions, the component-mmf wave of the stator winding and the associated component-flux wave revolve at the same speed as the rotor and are of essentially fixed waveform.   The flux linkages with the rotor circuits then do not change with time, and no voltages are induced in these circuits.   In effect, the main-field winding is the only rotor circuit which need be considered, and its excitation is determined by a simple Ohm's law relationship.

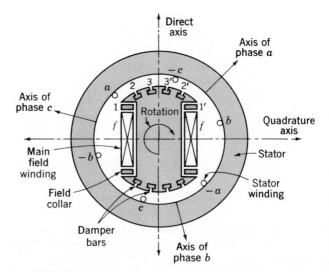

Fig. 10-1.   Schematic diagram for synchronous machine showing significant circuits for transients.

Under transient conditions time-varying currents may exist in all of the rotor circuits. The stator currents will no longer be constant-amplitude sinusoids, and the stator component-mmf and -flux waves will likewise not be constant-amplitude sinusoids in space. Moreover, the possibility of the mechanical speed of the rotor changing with time may have to be considered. As a consequence, the flux linkages with all of the rotor circuits will change with time, and induced currents will exist in all the circuits; the main-field current, for example, is no longer determined by a simple Ohm's law relation. All of these rotor currents are, of course, effective in providing excitation for the machine, for they contribute to the air-gap flux and thus influence the instantaneous values of stator currents. In short, the machine must be regarded as an inconveniently large group of mutually coupled circuits in order that a systematic approach to the transient problem may be devised.

In Fig. 10-1 there is a stator circuit for each of the three phases, a, b, and c, and there are rotor circuits corresponding to the field winding and to bars 2-2′ and 3-3′ and to the conducting field collars 1-1′. An additional equivalent rotor circuit may be formed by the bolts and iron of the rotor structure. Cylindrical-rotor machines may likewise have rotor circuits other than the main-field winding, especially in the quadrature axis, where the rotor iron may form an equivalent circuit almost as effective as the main field for induced currents.

All of these circuits have their own resistance and their own self-inductance and mutual inductances with respect to every other circuit. And to make a complex situation still more complicated, the self- and mutual inductances associated with the stator circuits are functions of rotor position, varying periodically as the rotor revolves. Fortunately, the self-inductances of the rotor circuits and the mutual inductances between rotor circuits can be considered constant as long as the effect of stator teeth and slots is ignored. In view of this complexity, no attempts have been made to include saturation, hysteresis, or eddy currents in a basic transient analysis of synchronous-machine performance. To reduce the complexity of the problem somewhat, analyses are usually confined to machines having but one effective rotor circuit other than the main field in the direct axis and one effective rotor circuit in the quadrature axis—in other words, one equivalent damper circuit in each axis.

Basic analysis of synchronous-machine transient performance accordingly involves the solution of a set of simultaneous coupled-circuit differential equations. Because of idealization of the machine, the relations between flux linkages and currents are linear, and, because of symmetry of machine geometry, certain of the self- and mutual-inductance coefficients can be recognized as having equal values. Nevertheless, the solution of the equations, even with only one damper circuit in each axis, is a formidable task, not because of any profundity, but because of the complexity of details. The solution is expedited appreciably by a linear transformation of variables in which the three stator phase currents i_a, i_b, and i_c are replaced by three component currents, the *direct-axis component* i_d, the *quadrature-axis component* i_q, and a single-phase component i_0 known as the *zero-sequence component* or *zero-axis component*. Similar transformations are made for voltages and flux linkages. For steady-state balanced operation, i_0 is zero (the significance of i_0 will not be discussed in this chapter because only balanced conditions are discussed in any detail), and the physical significance of the direct- and quadrature-axis components of current is that given in Chap. 6; in fact, the idea of making this transformation of variables arose from an extension of the physical picture corresponding to steady-state two-reaction theory. The changes of variables permit the simultaneous equations to be written in reasonably compact form with each equation relating only variables in the same axis. Furthermore, they enable recognition of certain frequently recurring combinations of machine constants, which become the reactances and time constants discussed in the next four articles. Analyses based on these differential equations are available for problems involving a single machine, a single machine and an infinite bus, and simple two-machine systems.

For engineering application of machine-transient theory, it must be recognized that a single synchronous machine is by no means the only element in the usual system problem. There are in general a goodly number of other such machines with different constants interconnected by a complicated network with each other and with numerous loads. In addition, the main field of each generator is supplied by an exciter in whose field is usually placed an automatic voltage regulator which recognizes the presence of a disturbance and alters conditions in the field. Moreover, action of the prime-mover governors may have a profound influence on system performance. All of these considerations clearly indicate the need for simple methods of characterizing the transient performance of a synchronous machine. Representation of the machine as a system element by means of simple equivalent circuits is desirable—representation not unlike that afforded by synchronous reactance for steady-state performance. The method actually used is to characterize the machine by a relatively few reactances and time constants. The reactances permit the computation of the initial magnitudes of transient currents, and the time constants characterize their decay. The coupled-circuit viewpoint is basic to the method, but because approximations and simplifications are necessary, correlation with test results and with a good physical picture of the phenomena is necessary.

One of the important details in the evolution of a physical picture to interpret transient test data is the handling of circuit resistances. Complete neglect of resistances greatly simplifies the problem. Then, in the absence of capacitance, the total flux linkages with any closed circuit on the rotor cannot change when a disturbance occurs but must remain constant at the initial value. This constancy is caused by the fact that if the linkages with such a circuit did change, an induced *voltage* would necessarily appear in the circuit in violation of Kirchhoff's voltage law. Any impetus toward a change of linkages, such as might be caused by a rapid increase of stator current following a shortcircuit, is therefore counteracted by an induced *current* of an appropriate magnitude to maintain constancy of linkages in spite of the impetus. Computation of currents following a disturbance is then simply a matter of finding the various values of flux linkages prior to the disturbance from the specified initial conditions, and solving the coupled-circuit equations which state that these values must remain the same after the disturbance. Algebraic rather than differential equations are involved, and, with zero resistance, the resulting induced currents do not die out with time.

Actually, of course, neither test results nor common sense permits adoption of the notion that the induced currents do not decay with time. The method used practically is justified by the fact that, while all circuits have some resistance so that the flux eventually will be changed through any closed circuit, yet most of the circuits dealt with in synchronous machines have low enough resistance so that the behavior at the first instant after a sudden change is very nearly the same as if the resistances were zero. On this basis the initial currents and flux linkages following a sudden change are determined from a group of reactances set up as if no resistance were present; and the decay of these currents and linkages is handled by means of a group of time constants determined from the resistances as well as the reactances of the circuits. The method is thus approximate in that resistance is only indirectly included in the solution. It is of great practical importance, however, because only by this means can the problems arising in complicated multimachine systems be handled.

This approach leads to a physical picture of happenings inside the machine which ties in directly with the results of experiments. Consider a synchronous generator operating at synchronous speed with a constant dc slip-ring voltage. One effective rotor circuit in the direct axis in addition to the main-field winding is formed by the amortisseur bars. The machine is operating initially unloaded, and a 3-phase shortcircuit sud-

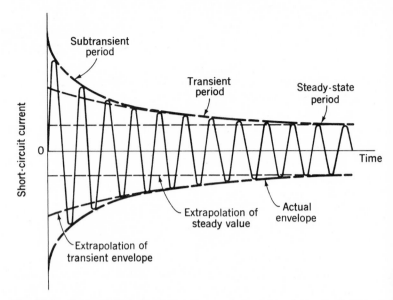

Fig. 10-2. Symmetrical short-circuit armature current in synchronous machine.

denly appears at its terminals. A symmetrical trace of a short-circuit stator-current wave such as might be obtained oscillographically is given in Fig. 10-2. The wave, whose envelope is shown in Fig. 10-3, may be divided into three periods or time regimes: the *subtransient period*, lasting only for the first few cycles during which the current decrement is very rapid; the *transient period*, covering a relatively longer time during which the current decrement is more moderate; and finally the *steady-state period*, during which the current is determined by the principles of Chaps. 4 and 6. That the three successive periods merge through the medium of nearly exponential envelope decays can be shown by appropriate semilog plots. The difference $\Delta i'$ (Fig. 10-3) between the transient envelope and the steady-state amplitude is plotted to a logarithmic scale as a function of time in Fig. 10-4. In similar fashion the difference $\Delta i''$ between the subtransient envelope and an extrapolation of the transient envelope is also plotted in Fig. 10-4. When the work is done carefully, both plots closely approximate straight lines, illustrating the essentially exponential nature of the decrement.

A physical picture of the happenings during these periods can be constructed by recognizing that under short-circuit conditions with zero stator-circuit resistance the stator-mmf wave is in the direct axis of the machine. The direct-axis rotor circuits and their flux linkages thus

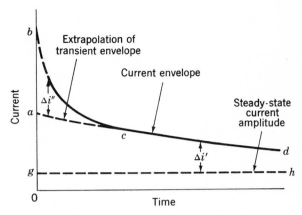

Fig. 10-3. Envelope of synchronous-machine symmetrical short-circuit current.

become of determining importance. The flux linkages with the main-field winding must remain constant at their initial value as determined by the field inductance and prefault field current. And this constancy must be maintained in the face of the demagnetizing stator mmf accompanying the short-circuit current of Fig. 10-2. An induced component of field current, like that shown in Fig. 10-5, must therefore appear in order to counteract the demagnetizing mmf. The induced component of field current determines the behavior of stator current during the transient period; it simply represents greater excitation on the machine than is present in the steady state, and consequently the stator currents during the transient period are greater than in the steady state. The induced field current, not being supported by an applied voltage in the field circuit,

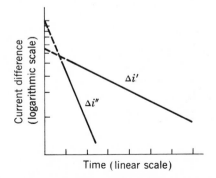

Fig. 10-4. Current differences plotted to semilog coordinates.

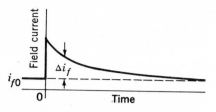

Fig. 10-5. Approximate form of synchronous-machine field current following armature shortcircuit.

dies away at a rate determined by the field-circuit resistance and equivalent inductance, and the corresponding increment of stator current dies away at the same rate. The associated transient time constant can be determined from the envelopes of Figs. 10-2 to 10-4.

The main field is not the only direct-axis rotor circuit, however. Flux linkages with the damper winding must remain constant at their initial value as determined by the mutual inductance between the field and damper circuits and the prefault field current. And again this constancy must be maintained in the face of the demagnetizing stator mmf accompanying the short-circuit current of Fig. 10-2. A suddenly induced damper current must therefore appear. It determines the behavior of the stator current during the subtransient period; it simply represents a still greater rotor excitation than is present in the transient period, and consequently the stator currents during the subtransient period are greater than in the transient period. The induced damper current dies away at a rate determined by the damper-circuit resistance and equivalent inductance, and the corresponding increment of stator current dies away at the same rate. Because the damper resistance-to-equivalent-inductance ratio is relatively higher than that of the field circuit, the subtransient decrement is much faster than the transient decrement. The associated subtransient time constant can be determined from the envelopes of Figs. 10-2 to 10-4.

But the oscillogram of Fig. 10-2 is a special rather than a general case in that a symmetrical current wave is shown. The more usual short-circuit oscillograms have the general appearance illustrated in Fig. 10-6. These traces are not symmetrical about the zero-current axis but exhibit definite dc components which result in offset waves. A symmetrical wave like that of Fig. 10-2 can be obtained either by replotting the offset waves with the dc component subtracted or by taking a series of oscillograms until a symmetrical wave is obtained for one of the three phases.

The dc component of stator current fits into the physical picture when it is recognized that constancy of flux linkages applies as well to

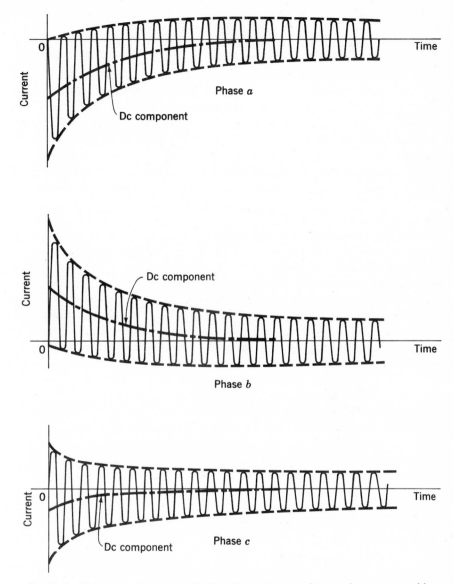

Fig. 10-6.   Short-circuit currents in the three phases of a synchronous machine.

each of the three phases of the resistanceless stator.   If the shortcircuit appears at an instant when the linkages with a stator phase are zero, no dc component is required to maintain them constant at that value and the short-circuit-current wave for that phase is symmetrical.   If, however, the shortcircuit appears at an instant when the phase linkages have a nonzero value, a dc component must appear in that phase in order to keep the linkages constant. The dc component, in effect, fills the same need as the dc component of transient current in a simple RL circuit with a suddenly impressed ac voltage. As in the RL circuit, the largest possible value of dc component is equal to the greatest instantaneous value of short-circuit current during the subtransient period. This largest dc component occurs when the shortcircuit appears at the instant of maximum linkages for a stator phase, and the corresponding short-circuit-current wave is then fully offset from the zero axis. The dc component, being unsupported by a voltage in the stator circuit, dies away at a rate determined by the stator-circuit resistance and equivalent inductance.

The dc component of stator current establishes a component field in the air gap which is stationary in space and which therefore induces a fundamental-frequency voltage and current in the synchronously revolving rotor circuits. Figure 10-7 shows the superimposed ac component in the field current immediately following a 3-phase shortcircuit at the stator terminals; also illustrated in this figure is the fact that, in contrast to the approximate sketch of Fig. 10-5, the field current cannot change suddenly at the first instant. As shown in Chap. 11, the pulsating fields produced by these single-phase currents can be resolved into oppositely rotating components. One component is stationary with respect to the stator and reacts back upon the dc component of stator current. The other component travels at twice synchronous speed with respect to the stator winding and induces a second harmonic in it. If the stator circuit is unbalanced (by a line-to-line or line-to-neutral instead of a balanced 3-phase shortcircuit, for example), higher harmonics may be caused in both the rotor and stator currents by successive reflections back and forth across the air gap. Harmonics are also introduced by the alternating component of stator current under unbalanced conditions. The result is that for an unbalanced shortcircuit at the machine terminals the waveform may be decidedly different from those given in Figs. 10-2 and 10-5. When shortcircuits occur at points removed from the machine terminals, the harmonics may be greatly decreased by the

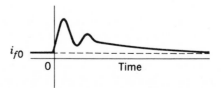

Fig. 10-7.   Field current following armature shortcircuit.

intervening reactance. For many engineering purposes, the harmonics are ignored; they will not be included in any subsequent analysis here. Unbalance of the stator circuit will also be omitted from consideration.

10-3

SYNCHRONOUS–MACHINE REACTANCES AND TIME CONSTANTS

On the basis of these physical considerations, it becomes possible to characterize the performance of an unloaded machine following a 3-phase shortcircuit directly at its terminals by three reactances and three time constants. The direct-axis synchronous reactance x_d determines the steady-state short-circuit current. The *direct-axis transient reactance* x'_d is so defined that it determines the initial value Oa of the symmetrical transient envelope acd (Fig. 10-3); it is equal to the rms value of prefault open-circuit phase voltage divided by $Oa/\sqrt{2}$, the factor $\sqrt{2}$ appearing because acd is the envelope of peak current values. The *direct-axis short-circuit transient time constant* T'_d is so defined that it determines the decay of the transient envelope acd; it is equal to the time required for the envelope to decay to the point where the difference between it and the steady-state envelope gh is $1/\epsilon$, or 0.368 of the initial difference ga. The *direct-axis subtransient reactance* x''_d is so defined that it determines the initial value Ob of the symmetrical subtransient envelope bc (Fig. 10-3); it is equal to the rms value of prefault open-circuit phase voltage divided by $Ob/\sqrt{2}$, the factor $\sqrt{2}$ appearing because bc is the envelope of peak current values. The *direct-axis short-circuit subtransient time constant* T''_d is so defined that it determines the decay of the subtransient envelope bc; it is equal to the time required for the envelope to decay to the point where the difference between it and the transient envelope acd is $1/\epsilon$, or 0.368 of the initial difference ab. The initial value of the dc component or offset of the current wave, evident in Fig. 10-6, is determined by the point in the cycle at which the fault appears; its largest possible value is equal to the peak amplitude Ob (Fig. 10-3) of the symmetrical subtransient current. The *armature time constant* T_a is so defined that it determines the decay of the dc component; it is equal to the time required for the dc component to decay to $1/\epsilon$, or 0.368 of its initial value. (The symbol T is used for time constant in this chapter to conform with standard notation. Unlike Chap. 9, where τ is used, there is practically no possibility of confusion with torque here.)

These reactances are appropriate machine constants for use not only for a 3-phase shortcircuit directly at the terminals of an unloaded machine but also for any application involving a sudden change in direct-axis current. This fact is emphasized by the conventional definitions of the reactances, which are formulated in terms of a suddenly impressed current. These definitions are:

▲ The direct-axis transient reactance is the ratio of the fundamental component of reactive armature voltage, due to the fundamental direct-axis ac component of the armature current, to this component of current under suddenly applied load conditions and at rated frequency, the value of current to be determined by the extrapolation of the envelope of the ac component of the current wave to the instant of the sudden application of load, neglecting the high-decrement currents during the first few cycles.

▲ The direct-axis subtransient reactance is the ratio of the fundamental component of reactive armature voltage, due to the initial value of the fundamental direct-axis ac component of armature current, to this component of current under suddenly applied load conditions and at rated frequency.

The time constants T'_d, T''_d, and T_a, on the other hand, can be used only for 3-phase shortcircuits at the machine terminals. When external impedance is present, the decay of the induced currents is influenced not only by the self- and mutual inductances of the machine itself, but also by the constants of the external circuit. Simple methods are available, however, for the appropriate adjustment of time constants for the presence of external reactance.

Only direct-axis events have been considered up to this point because only direct-axis quantities are involved in 3-phase shortcircuits on purely reactive networks under the assumptions adopted. When the machine has an active-power loading before the disturbance, quadrature-axis quantities are also involved because changes in quadrature-axis current i_q are to be expected, and constant flux linkages must be maintained with the quadrature-axis rotor circuits in the face of these changes. When these aspects are to be included, the *quadrature-axis transient reactance* x'_q, *subtransient reactance* x''_q, *transient short-circuit time constant* T'_q, and *subtransient short-circuit time constant* T''_q must be considered. These reactances and time constants bear the same relation to quadrature-axis events as the corresponding direct-axis quantities do to direct-axis events. In fact, the definitions of x'_q and x''_q are the same as those quoted for x'_d and x''_d, respectively, except that the words *direct axis* are replaced by the words *quadrature axis*.

No field winding exists in the quadrature axis of a normal machine, so that the quadrature-axis rotor circuits must be composed of damper bars or of the rotor iron in the interpolar axis of a cylindrical-rotor machine. If there are no effective rotor circuits in the quadrature axis, $x_q = x_q' = x_q''$. If the quadrature-axis circuit is composed of damper bars in a salient-pole machine, the induced currents are of subtransient order and $x_q' = x_q$, with x_q'' having a lower value. If the quadrature-axis circuit is formed by the interpolar iron in a solid cylindrical-rotor machine, the induced currents are usually of transient rather than sub-transient order, and x_q' almost equals x_q'', with x_q having a higher value. For a solid-cylindrical-rotor turboalternator, the interpolar iron forms just as effective an induced-current path as the main-field winding in the direct axis, and x_q' is approximately equal to x_d'.

Table 10-1 presents typical values of the constants for different types of synchronous machines. Reactances are given in per unit with the machine rating as a base. Reasonable variation either side of these values may be expected for any particular machine.

TABLE 10-1
TYPICAL VALUES OF MACHINE CONSTANTS
(Reactances are per-unit values based on the machine rating; time constants are in seconds)

Machine constant	Cylindrical-rotor generators		Salient-pole generators	Salient-pole motors (low speed)	Syn-chronous condensers
	Solid rotor	Laminated rotor			
x_d	1.10	1.10	1.00	1.10	1.60
x_d'	0.20	0.20	0.35	0.50	0.60
x_d''	0.10	0.10	0.23	0.35	0.25
x_q	1.00	1.00	0.65	0.80	1.00
x_q'	0.20	1.00	0.65	0.80	1.00
x_q''	0.15	0.25	0.65	0.40	0.30
T_d'	1.0	1.0	1.8	1.4	2.0
T_d''	0.035	0.035	0.035	0.035	0.035
T_a	0.15	0.15	0.15	0.15	0.15

10-4
EQUIVALENT CIRCUIT FOR TRANSIENT CONDITIONS

By use of the concept of constant flux linkages, coupled-circuit equations can be written corresponding to the conventional definition of direct-axis transient reactance x_d', and from these equations x_d' may be evaluated in terms of mutual and leakage inductances. From this evaluation, a simple equivalent circuit for x_d' may be obtained, and transient time constants may be interpreted with respect to the equivalent circuit. On the basis of these results, the concepts may readily be extended to apply to a short-circuit with external reactance between it and the machine terminals.

To reproduce the conditions for the conventional definition of x_d', consider an unloaded synchronous machine operating at synchronous speed with zero initial field current but with the slip rings short-circuited so that the field circuit is closed. Neglect of the high decrement during the first few cycles is equivalent to ignoring all direct-axis rotor circuits other than the main field. A direct-axis current is to be suddenly impressed, and the accompanying terminal voltage is to be evaluated.

Let L_f be the leakage inductance of the field winding, L_a the leakage inductance of the armature winding, and M_d the mutual inductance between the armature and the field winding in the direct axis. The terms L_a and M_d correspond to the armature leakage reactance x_a and direct-axis magnetizing reactance $x_{\varphi d}$ for fundamental-frequency steady-state armature currents. Note that $x_{\varphi d}$ and hence M_d include the effect of all three stator phases when the stator currents are balanced. Note also that the symbol x_a is used here for armature leakage reactance, instead of x_l as in Chaps. 4 and 6. This change is made in order that one may distinguish between armature leakage, designated by the subscript a, and field leakage, designated by the subscript f.

At an instant when the field current and direct-axis stator current have the values i_f and i_d, respectively, the flux linkages with the field winding are

$$\lambda_f = (L_f + M_d)i_f - M_d i_d \qquad (10\text{-}1)$$

The sign convention is that positive values of i_d give rise to linkages in the opposite direction from positive values of i_f. Also, the phenomena are viewed from the stator, and all quantities are referred to the stator winding.

But λ_f is initially zero and, in the absence of resistance, must remain zero at all subsequent instants. Hence, when a nonzero value of i_d appears, it must be accompanied by an induced field current given by

$$i_f = \frac{M_d}{L_f + M_d} i_d \qquad (10\text{-}2)$$

Thus, if i_d varies sinusoidally with time, i_f *as viewed from the stator* must vary sinusoidally with time. This sinusoidally varying i_f corresponds to a direct current in the physical field winding, for a direct field current plus rotation of the rotor looks like an alternating current when viewed from the stator.

In a similar manner, the direct-axis stator linkages are

$$\lambda_d = M_d i_f - (L_a + M_d)i_d \qquad (10\text{-}3)$$

which, upon substitution of Eq. 10-2, becomes

$$\lambda_d = -\left(L_a + M_d - \frac{M_d^2}{L_f + M_d}\right)i_d \tag{10-4}$$

For sinusoidal variation of i_d with the rms value I_d, the rms linkage magnitude is

$$\Lambda_d = \left(L_a + M_d - \frac{M_d^2}{L_f + M_d}\right)I_d \tag{10-5}$$

and the corresponding rms magnitude of direct-axis voltage is

$$\omega\Lambda_d = \left(\omega L_a + \omega M_d - \frac{\omega^2 M_d^2}{\omega L_f + \omega M_d}\right)I_d \tag{10-6}$$

From the conventional definition, x_d' is then

$$x_d' = \frac{\omega\Lambda_d}{I_d} = \omega L_a + \omega M_d - \frac{\omega^2 M_d^2}{\omega L_f + \omega M_d} \tag{10-7}$$

$$= x_a + x_{\varphi d} - \frac{x_{\varphi d}^2}{x_f + x_{\varphi d}} \tag{10-8}$$

$$= x_d - \frac{x_{\varphi d}^2}{x_f + x_{\varphi d}} \tag{10-9}$$

where $x_f = \omega L_f$ is the field leakage reactance referred to the stator.

Equation 10-8 can be put in alternate form by algebraic modification. Thus,

$$x_d' = \frac{x_a x_f + x_a x_{\varphi d} + x_{\varphi d} x_f + x_{\varphi d}^2 - x_{\varphi d}^2}{x_f + x_{\varphi d}}$$

$$= x_a + \frac{x_{\varphi d} x_f}{x_{\varphi d} + x_f} \tag{10-10}$$

Equation 10-10 shows that x_d' is composed of armature leakage reactance plus modified field leakage reactance. It is the general form commonly used for computation of x_d' from design data.

With x_d' evaluated in terms of mutual and leakage reactances, it can now be shown that it is the reactance which determines the initial value of short-circuit current for a machine with no amortisseur winding.

Under prefault conditions with the machine unloaded and having the field current i_{f0}, the field linkages are, from Eq. 10-1, with $i_d = 0$,

$$\lambda_{f0} = (L_f + M_d)i_{f0} \tag{10-11}$$

When the shortcircuit appears, the field linkages must remain at this value, so that

$$(L_f + M_d)i_f - M_d i_d = (L_f + M_d)i_{f0} \tag{10-12}$$

or

$$i_f = i_{f0} + \frac{M_d}{L_f + M_d} i_d \tag{10-13}$$

The second term on the right-hand side of Eq. 10-13 is the induced component of field current required to maintain constant field linkages.

Substitution of Eq. 10-13 in 10-3 gives for direct-axis linkages

$$\lambda_d = M_d i_{f0} - \left(L_a + M_d - \frac{M_d^2}{L_f + M_d}\right) i_d \tag{10-14}$$

Under short-circuit conditions with the dc component and all harmonics ignored, λ_d must be zero in order that the terminal voltage may be zero. Equation 10-14 then yields

$$i_d = \frac{M_d i_{f0}}{L_a + M_d - \left(\dfrac{M_d^2}{L_f + M_d}\right)} = \frac{x_{\varphi d} i_{f0}}{x_d'} \tag{10-15}$$

But the rms value of $x_{\varphi d} i_{f0}$ is also the rms value of the prefault terminal voltage E_i', as can be seen by multiplying Eq. 10-14 by ω, putting $i_d = 0$, and writing it for rms values.   Accordingly, the rms short-circuit current is

$$I_d = \frac{E_i'}{x_d'} \tag{10-16}$$

and its value is limited by the direct-axis transient reactance x_d'.

From Eq. 10-10 for transient reactance x_d' and by recalling that synchronous reactance x_d is the sum of x_a and $x_{\varphi d}$, it may be recognized that the transformer-type equivalent circuit of Fig. 10-8 will serve as an equivalent circuit for both x_d and x_d'.   With the machine operating balanced and in the steady state, the stator current has no effect in the rotor circuit, the rotor terminals are regarded as open-circuited, and the reactance

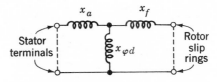

Fig. 10-8. Transformer-type equivalent circuit for synchronous machine.

viewed from the stator terminals is the synchronous reactance x_d. With the machine in the transient state, the change in stator current is accompanied by an induced current in the field, the rotor terminals are regarded as short-circuited, and the reactance viewed from the stator terminals is the transient reactance x_d'.

The factors influencing the transient time constant may be investigated with the aid of Fig. 10-8. First consider a machine with a closed field circuit but an open armature circuit; the time constant T_{do}' describing the decay of a field transient under these conditions is known as the *direct-axis open-circuit transient time constant*. The stator terminals (Fig. 10-8) are open-circuited, and the field terminals are short-circuited. The field transient is then affected by the self-inductance of the field, $(x_f + x_{\varphi d})/2\pi f$. If r_f is the field resistance, the time constant is

$$T_{do}' = \frac{x_f + x_{\varphi d}}{2\pi f r_f} = \frac{L_f + M_d}{r_f} \qquad (10\text{-}17)$$

The time constant is a characteristic of the field circuit alone and is simply the usual self-inductance-to-resistance ratio which normally constitutes a time constant for a simple series circuit.

For short-circuit conditions, both the stator and rotor terminals are short-circuited. The equivalent field inductance affecting the decay of a field transient is then that of x_f plus $x_{\varphi d}$ and x_a in parallel, or

$$\text{Equivalent field inductance} = \frac{1}{2\pi_f}\left(x_f + \frac{x_a x_{\varphi d}}{x_a + x_{\varphi d}}\right) \qquad (10\text{-}18)$$

$$= \frac{x_f + x_{\varphi d}}{2\pi f} \frac{x_a + \dfrac{x_{\varphi d} x_f}{x_{\varphi d} + x_f}}{x_a + x_{\varphi d}}$$

$$= \frac{x_f + x_{\varphi d}}{2\pi f} \frac{x_d'}{x_d}$$

The short-circuit time constant is therefore

$$T'_d = \frac{x_f + x_{\varphi d}}{2\pi f r_f} \frac{x'_d}{x_d} = T'_{do} \frac{x'_d}{x_d} \tag{10-19}$$

The reactance and time constant characterizing the transient-current component for a 3-phase shortcircuit with external reactance x_e between it and the machine terminals can now be indicated. As far as the internal phenomena of the machine are concerned, external reactance is equivalent to increasing the armature leakage reactance from x_a to $x_a + x_e$, and the *direct-axis short-circuit transient time constant adjusted for external reactance*, T'_{de}, becomes

$$T'_{de} = T'_{do} \frac{x'_d + x_e}{x_d + x_e} = T'_d \frac{x_d}{x'_d} \frac{x'_d + x_e}{x_d + x_e} \tag{10-20}$$

The reactance determining the initial magnitude of transient alternating current is, of course, $x'_d + x_e$.

10-5

APPLICATION TO SYSTEM TRANSIENTS

When the machine constants x_d, x'_d, x''_d, T'_d, T''_d, and T_a are known, the stator currents can be predicted with reasonable accuracy for a 3-phase shortcircuit separated from the terminals of an initially unloaded machine by the external reactance x_e. Thus, if the internal prefault machine voltages behind synchronous reactance, transient reactance, and subtransient reactance are E_i, E'_i, and E''_i, respectively, then the symmetrical, or ac, component of short-circuit current is

$$I_{ac} = \frac{E_i}{x_d + x_e} + \left(\frac{E'_i}{x'_d + x_e} - \frac{E_i}{x_d + x_e}\right)\epsilon^{-t/T_{de}'}$$

$$+ \left(\frac{E''_i}{x''_d + x_e} - \frac{E'_i}{x'_d + x_e}\right)\epsilon^{-t/T_{de}''} \tag{10-21}$$

T'_{de} and T''_{de} being the appropriate transient and subtransient time constants. Very commonly, rms values of voltage are used in this expression and an rms value of current obtained. Accordingly, the approximate viewpoint adopted is that of the quasi-steady state in which rms values change exponentially with time. Instantaneous values can, of course, readily be obtained when the point in the cycle at which the fault occurs is known.

Superimposed on the symmetrical component of current in each phase is a dc component given by

$$I_{dc} = I_{dc0}\epsilon^{-t/T_{ae}} \tag{10-22}$$

where I_{dc0} is the initial value, equal and opposite to the instantaneous value of I_{ac} for that phase at $t = 0$, and T_{ae} is the appropriately adjusted armature time constant. The maximum possible dc component, corresponding to a completely offset wave, is

$$I_{dcm} = \sqrt{2}\,\frac{E_i''}{x_d'' + x_e}\,\epsilon^{-t/T_{ae}} \tag{10-23}$$

The total rms value of the dissymmetrical wave at any instant is

$$I_{sc} = \sqrt{I_{ac}^2 + I_{dc}^2} \tag{10-24}$$

when the rms value is used for I_{ac}. On the basis of conservatism, the condition to be studied is commonly taken as that corresponding to the largest dc component. By a process generally similar to that leading to Eq. 10-20, it can be shown that an approximate but sufficiently accurate value of T_{de}'' is

$$T_{de}'' = T_d''\,\frac{x_d'\,x_d'' + x_e}{x_d''\,x_d' + x_e} \tag{10-25}$$

The armature time constant T_a with a 3-phase shortcircuit directly at the terminals depends on the armature resistance r_a and the equivalent inductance of the armature circuit to direct current. This equivalent inductance depends not only on the armature circuit but also on both the field and damper circuits, for the dc component induces fundamental-frequency currents in both of these closed circuits. As the rotor revolves, the stationary mmf distribution which the dc component creates in space reacts on direct-axis rotor circuits to one extent and on quadrature-axis rotor circuits to another because of the different permeances and different rotor circuits in the two axes. It is reasonable on an intuitive basis, therefore, to conclude that the equivalent armature inductance is somewhere between the inductance corresponding to x_d'' and the inductance corresponding to x_q''; usually it is taken as the arithmetic mean of the two-

The corresponding armature time constant is

$$T_a = \frac{1}{2\pi f r_a} \frac{x_d'' + x_q''}{2} \tag{10-26}$$

The reasoning leading up to Eq. 10-20 may be used to adjust the time constant for external reactance, but it is important in this case to include also the external resistance r_e between the machine terminals and the fault. The adjusted time constant is

$$T_{ae} = \frac{1}{2\pi f(r_a + r_e)} \left(\frac{x_d'' + x_q''}{2} + x_e \right) \tag{10-27}$$

For many engineering applications, the subtransient component may be ignored because of its rapid decrement. The third term in Eq. 10-21 is then omitted. Not infrequently, however, the possibility of a large dc component must be accounted for, especially for fault-current values shortly after the fault has occurred. At least the initial value of subtransient current must be found to permit determination of the maximum possible dc component.

This summary leads up to the representation of the machine as a system element in system-transient problems. When the highly decremented response during the first few cycles can be ignored, the machine is commonly represented by its direct-axis transient reactance x_d'. To correspond approximately to constant flux linkages with the main-field winding, the voltage behind transient reactance may be kept constant. Flux-linkage decrement may be included by decrementing the machine currents in accordance with the appropriate time constant. Such representation of each individual machine by a single constant reactance makes it possible to perform transient analyses almost as expeditiously as steady-state analyses and at the same time yields results to engineering accuracy for many problems. For computation of power-network short-circuit currents, such as are required for determining circuit-breaker rupturing duties, protective-relay settings, and bus-bar stresses, the power-system loads are usually ignored except insofar as they affect the prefault internal machine voltages. Also, it is common to assume all machine internal emfs to be in phase.

For studies involving machine torques and power outputs under transient conditions, such as are required for investigating maintenance of synchronism during disturbances, the dc component of fault current is usually ignored because it has negligible influence on synchronous power and torque. Also, the voltages behind transient reactance may frequently be considered constant for periods up to about 1 sec after fault occurrence, an assumption which is equivalent to ignoring field-flux-linkage decrement or to implying that the machine exciters, acted upon by the voltage regulators, build up field current at a rate approximately sufficient to compensate the demagnetizing influence of the stator short-circuit currents. There are, of course, numerous problems not susceptible to such simplified treatment and requiring inclusion of quadrature-axis events and external system resistance. To mention one example, studies of the terminal-voltage dip of an alternator following sudden load application rather obviously do not permit neglect of either the load or its resistive component. Such studies demand more comprehensive methods of analysis.

EXAMPLE 10-1

A hydroelectric station furnishes power to a large metropolitan area over a double-circuit transmission line with transformer banks at the sending and receiving ends, as shown in Fig. 10-9. Because of the large generating capacity within the metropolitan area, the receiving-end low-tension bus may be considered infinite. The system constants indicated in Fig. 10-9 are per-unit values based on the kva rating of the hydrogenerators. Except for their influence on time constants, resistances are to be neglected.

The receiving-end low-tension bus voltage E_b has its normal 100 percent value. The generators are initially so loaded that the power delivered is 80 percent of their kva rating, and the receiving-end low-tension power factor is unity. A solid 3-phase shortcircuit occurs on one transmission circuit just outside the sending-end high-tension bus.

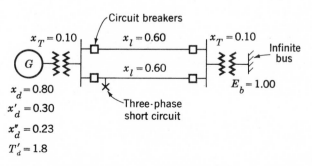

Fig. 10-9. Simplified power system for Example 10-1.

a. Find the rms total current in one phase of the fault just after its occurrence. Consider that the dc components contributed by both the hydrogenerators and the infinite bus to this phase have their largest possible values.

b. The sending-end circuit breaker on the faulted circuit opens after 0.10 sec. Find the current which one phase of this breaker may have to interrupt in clearing the 3-phase fault. For this purpose, consider that both the dc and subtransient components have reached negligible proportions in 0.1 sec.

Solution

a. The prefault voltage behind x_d'' is

$$E_i'' = 1.00 + j0.80(0.50 + 0.23) = 1.16\underline{/30.3°}$$

The initial symmetrical current from the generators is then

$$\frac{1.16}{0.23 + 0.10} = 3.52$$

and from the infinite bus is

$$\frac{1.00}{0.10 + 0.30} = 2.50$$

making the initial symmetrical fault current $3.52 + 2.50 = 6.02$.

The dc component from the generators is, for a completely offset wave, $\sqrt{2}\,(3.52) = 4.97$, and from the infinite bus is $\sqrt{2}\,(2.50) = 3.53$, making the dc component in the fault

$$4.97 + 3.53 = 8.50$$

The largest rms total fault current is therefore

$$\sqrt{(6.02)^2 + (8.50)^2} = 10.4$$

b. The prefault voltage behind x_d' is

$$E_i' = 1.00 + j0.80(0.50 + 0.30) = 1.19\underline{/32.6°}$$

and behind x_d

$$E_i = 1.00 + j0.80(0.50 + 0.80) = 1.44\underline{/46.1°}$$

The initial symmetrical generator current with the subtransient ignored is

$$\frac{1.19}{0.30 + 0.10} = 2.97$$

and the final steady-state generator current when the fault remains on the system is

$$\frac{1.44}{0.80 + 0.10} = 1.60$$

The symmetrical generator current passes from the first of these values to the second exponentially with a time constant (Eq. 10-20) equal to

$$T'_{de} = 1.8 \frac{0.80}{0.30} \frac{0.30 + 0.10}{0.80 + 0.10} = 2.13 \text{ sec}$$

The value of this current at $t = 0.10$ sec is accordingly

$$1.60 + (2.97 - 1.60)\epsilon^{-0.10/2.13} = 2.91$$

Only half of the infinite-bus current, or 1.25, passes through the breaker, so that the symmetrical breaker current is

$$2.91 + 1.25 = 4.16$$

10-6
ELECTRICAL TRANSIENTS IN INDUCTION MACHINES

The transient behavior of induction machines can be examined by following substantially the same approach as we have just been through for synchronous machines.   The results will be very similar to those in the synchronous case.

Consider that the induction machine is acting as either a motor or a generator when a 3-phase shortcircuit takes place at its terminals.

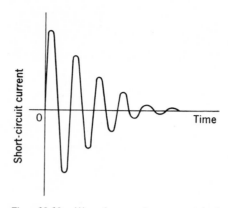

Fig. 10-10. Waveform of symmetrical short-circuit current in induction machine.

In either case, the machine will feed current into the fault because of the "trapped" flux linkages with the rotor circuits. This current will, in time, decay to zero. In addition to an ac component, it will in general have a decaying dc component in order to keep the flux linkages with the associated phase initially constant. The ac component of the short-circuit current as a function of time appears as in Fig. 10-10.

The initial magnitude of the ac component can be determined in terms of a *transient reactance* x' and a voltage E_1' behind that reactance assumed equal to its prefault value. The decay of the ac component can be characterized in terms of a *short-circuit transient time constant T'*. Underlying this thought process is the assumption that the machine is operating at negligible slip in order that speed voltages induced in the rotor shall have insignificant influence on the short-circuit current. In a practical case, the assumption is usually justifiable for two reasons: (1) the normal machine usually operates at small values of slip, and (2) the short-circuit current dies out rapidly enough (as we shall see) so that the machine does not have time to change speed appreciably.

Much of the approximate analytical development of Art. 10-4 can be applied by making the appropriate changes in notation. The currents i_d and i_f, for example, become the stator and rotor currents i_1 and i_2. The reactances x_a, x_f, and $x_{\varphi d}$ become the stator and rotor leakage reactances x_1 and x_2 and the magnetizing reactance x_φ, respectively (changes which may also be applied to the transformer-type equivalent circuit of Fig. 10-8). Because of the cylindrical structure of the induction machine rotor, the distinction between direct and quadrature axis need not be maintained. The results corresponding to Eqs. 10-9, 10-10, 10-16, 10-17, and 10-19 are, respectively,

$$x' = x_1 + x_\varphi - \frac{x_\varphi^2}{x_2 + x_\varphi} \tag{10-28}$$

$$= x_1 + \frac{x_\varphi x_2}{x_\varphi + x_2} \tag{10-29}$$

$$I_1 = \frac{E_1'}{x'} \tag{10-30}$$

$$T_o' = \frac{x_2 + x_\varphi}{2\pi f r_2} \tag{10-31}$$

and $$T' = T_o' \frac{x'}{x_1 + x_\varphi} \tag{10-32}$$

where T_o' is the *open-circuit transient time constant* of the induction machine and r_2 is the rotor-circuit resistance.

The induction machine can then be represented by the simple transient equivalent circuit of Fig. 10-11. The reactance is the transient reactance x', defined by Eqs. 10-28 and 10-29. Although it has been ignored so far, the stator resistance r_1 can also be added if somewhat greater precision is desired. The voltage E_1' behind transient reactance is a voltage proportional to rotor linkages. It changes with these linkages and, for a 3-phase shortcircuit, decreases to zero at a rate determined by the time constant T' (Eq. 10-32). Adjustment of the time constant for external reactance between the machine terminals and the fault can readily be carried out by adding the reactance to the numerator and denominator of the fraction in Eq. 10-32.

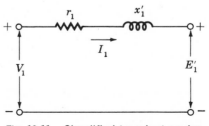

Fig. 10-11. Simplified transient equivalent circuit of induction machine.

EXAMPLE 10-2

A 400-hp 440-volt (line-to-line) 60-Hz Y-connected 6-pole squirrel-cage induction motor has a full-load efficiency of 93 percent and power factor of 90 percent. The motor constants in ohms per phase referred to the stator are as follows:

$$x_1 = 0.060 \qquad x_2 = 0.060 \qquad x_\varphi = 2.50$$

$$r_1 = 0.0073 \qquad r_2 = 0.0064$$

While the motor is operating in the steady state under rated conditions, a 3-phase shortcircuit occurs on its supply line near the motor terminals. Determine the motor rms short-circuit current.

Solution

From Eq. 10-28, the motor transient reactance is seen to be

$$x' = 0.060 + 2.50 - \frac{(2.50)^2}{0.06 + 2.50} = 0.12 \text{ ohm per phase}$$

The prefault stator current is

$$I_1 = \frac{400 \times 746}{0.90 \times 0.93 \times 440 \sqrt{3}} = 467 \text{ amp}$$

With terminal voltage as the reference phasor, the prefault voltage behind transient reactance is then

$$E_1' = \frac{440}{\sqrt{3}} - (0.0073 + j0.12)(467\underline{/- \cos^{-1} 0.90})$$

$$= 232\underline{/-12.2°}$$

From Eq. 10-30, the initial rms short-circuit current is

$$\frac{232}{0.12} = 1,940 \text{ amp}$$

The open-circuit time constant (Eq. 10-31) is

$$T_o' = \frac{2.50 + 0.060}{2\pi \times 60 \times 0.0064} = 1.06 \text{ sec}$$

and the short-circuit time constant (Eq. 10-32) is

$$T' = 1.06 \times \frac{0.12}{2.56} = 0.050 \text{ sec}$$

The rms short-circuit current is therefore

$$I_1 = 1,940\epsilon^{-t/0.050} \qquad \text{amp}$$

The short-circuit time constant is 3 cycles on a 60-cycle base. Accordingly, in 3 cycles the short-circuit current decreases to 36.8 percent of its initial value. It has substantially disappeared in about 10 cycles. It is thus seen that, while the initial short-circuit current of an induction machine is relatively high compared with its normal current, the transient usually disappears rapidly. Because of this fact, the electrical transients in induction machines are not infrequently neglected.

It can now be clearly recognized that the subtransient effects of Art. 10-2 are very like the superposition of induction-machine action on synchronous-machine action. The usual damper circuits in synchronous machines are essentially equivalent to squirrel-cage rotors. Damper torques are induction-machine torques; to a fair approximation, both vary linearly with slip in the neighborhood of synchronous speed. Under short-circuit conditions, the induced damper currents in the synchronous machine are comparable with the induced rotor currents required to maintain constant rotor linkages at the first instant in the induction machine; both die away very rapidly. In effect, then, during the short subtransient period following a synchronous-machine disturbance, we have induction and synchronous effects combined; when the subtransient period is over, synchronous transient effects remain.

10-7
SYNCHRONOUS–MACHINE DYNAMICS

Important dynamic problems arise in synchronous-machine systems because successful operation of the machines demands equality of the mechanical speed of the rotor and the speed of the stator field, and because synchronizing forces tending to maintain this equality are brought into play whenever the relationship is disturbed. If the instantaneous speed of a synchronous machine in a system containing other synchronous equipment should decrease slightly, the decrease would be associated with a decrease in torque angle if the machine were a generator or an increase if it were a motor. For example, if a large load is suddenly applied to the shaft of a synchronous motor, the motor must slow down at least momentarily in order that the torque angle may assume the increased value necessary to supply the added load. In fact, until the new angle is reached, an appreciable portion of the energy furnished to the load comes from stored energy in the rotating mass as it slows down. When the newly required value of angle is first reached, equilibrium is not yet attained, for the mechanical speed is then below synchronous speed. The angle must momentarily increase further in order to permit replacing the deficit of stored energy in the rotating mass. The ensuing processes involve a series of oscillations about the final position even when equilibrium is ultimately restored. Exact description of such events can be given only in terms of the associated electromechanical differential equation, and decisions on restoration of equilibrium can be based only on the solution of the equation.

Similar oscillations or hunting, with the accompanying power and current pulsations, may be particularly troublesome in synchronous motors driving loads whose torque requirements vary cyclically at a fairly rapid frequency, as in motors driving reciprocating air or ammonia compressors. If the natural frequency of mechanical oscillation of the synchronous motor approximates the frequency of an important torque harmonic in the compressor cycle, intolerable oscillations result. Electrodynamic transients of a very complicated form but of the same basic nature occur in electric power systems. Unless they are carefully investigated during system planning, they may result in complete shutdowns over wide areas.

a. The Basic Electromechanical Equation

As in all other types of machines, the electromechanical equation for a synchronous machine follows from recognition of the three classes of torque acting on the rotating members. They are an inertia torque, an electromagnetic torque T_e resulting from energy conversion, and a mechanical shaft torque T_{sh} representing input from the prime mover or output to turn the load. Thus,

$$T_{\text{inertia}} + T_e = T_{sh} \qquad (10\text{-}33)$$

This equation, like those to follow, is written with a generator specifically in mind. The same equations may be applied to a motor by following an appropriate sign convention. The convention used will be stated below after the final form of the equation is given.

The electromagnetic term in Eq. 10-33, and often the shaft input or output as well, can be obtained more conveniently as a power than as a torque. Hence the equation is usually rewritten in power form as

$$P_{\text{inertia}} + P_e = P_{\text{sh}} \tag{10-34}$$

The inertia power is, of course, found from the angular acceleration. The angular position of the shaft at any instant is taken as the electrical angle δ between a point on it and a reference which is rotating at synchronous speed. Often in simple problems the angle δ is taken identical with the power angle of the machine. The inertia power is then

$$P_{\text{inertia}} = P_j \frac{d^2\delta}{dt^2} \tag{10-35}$$

where P_j is the inertia power per unit acceleration. When δ is measured in electrical radians,

$$P_j = J \frac{2}{\text{poles}} \frac{2\pi n}{60} \tag{10-36}$$

where J is the moment of inertia and n the speed in rpm. The factor $2/\text{poles}$ changes electrical to mechanical angle, and the factor $2\pi n/60$ changes torque to power. The coefficient P_j is usually considered a constant and evaluated with n equal to synchronous speed. This procedure is justified by the fact that the angular oscillations become intolerable before the speed departs more than 1 or 2 percent from synchronous speed.

The electromagnetic power P_e has two components. One component is the *damping power*, mentioned at the end of the preceding article. As stated there, it is often considered to vary linearly with the departure $d\delta/dt$ from synchronous speed. The second component is the *synchronous power* resulting from synchronous-machine action and characterized by Eqs. 6-18, 6-19, and 6-30 for a single machine on an infinite bus. The electromechanical equation 10-34 then becomes

$$P_j \frac{d^2\delta}{dt^2} + P_d \frac{d\delta}{dt} + P(\delta) = P_{\text{sh}} \tag{10-37}$$

where P_d is the damping power per unit departure in speed from synchronous and the term $P(\delta)$ indicates that synchronous power is a function of
angle δ.

The specific nature of the external network must be known before the
function $P(\delta)$ can be identified. When consideration is restricted to one
machine connected directly to the terminals of a very large system (see
Art. 6-5) and saliency is ignored, the function is $P_m \sin \delta$, where
P_m is the amplitude of the sinusoidal power-angle curve. Equation 10-37
becomes

$$P_j \frac{d^2\delta}{dt^2} + P_d \frac{d\delta}{dt} + P_m \sin \delta = P_{sh} \qquad (10\text{-}38)$$

Positive values of δ denote generator action and therefore energy conversion from mechanical to electrical form, positive values of P_{sh} denote
mechanical power input to the shaft, positive values of $d\delta/dt$ denote
speeds above synchronous speed, and positive values of $d^2\delta/dt^2$ denote
acceleration. Alternatively, the reverse convention may be used. That
is, positive values of δ may denote motor action, positive values of P_{sh}
mechanical power output from the shaft, positive values of $d\delta/dt$ speeds
below synchronous speed, and positive values of $d^2\delta/dt^2$ deceleration.

Both Eqs. 10-37 and 10-38 are nonlinear. One method of analysis,
applicable for small oscillations and illustrated below, is to linearize the
power-angle expression. It is simply a special case of linearization about
an operating point.

b. Linearized Analysis

When a single machine connected to a large system is under study, only a
single differential equation rather than a group of such equations is
involved. If, in addition, the variations of δ are small, the term $P(\delta)$ in
Eq. 10-37 may be replaced by the equation for the slope of the power-
angle curve at the operating point.

When δ varies between about $+\pi/6$ and $-\pi/6$ electrical radians, the
sine of the angle is closely equal to the angle in radians. The term
$P_m \sin \delta$ in Eq. 10-38 may then be replaced by the term $P_s \delta$, where P_s is
the *synchronizing power*, or slope, of the power-angle curve evaluated at
the origin. Equation 10-38 becomes

$$P_j \frac{d^2\delta}{dt^2} + P_d \frac{d\delta}{dt} + P_s \delta = P_{sh} \qquad (10\text{-}39)$$

Since the equation is now linear, its solution in a particular case may be
obtained by conventional methods.

EXAMPLE 10-3

A 200-hp 2,300-volt 3-phase 60-Hz 28-pole 257-rpm synchronous motor is directly connected to a large power system. The motor has the following characteristics:

$Wk^2 = 10,500$ lb-ft² (motor plus load)

Synchronizing power $P_s = 11.0$ kw/elec deg

Damping torque $= 1,770$ lb-ft/mech rad/sec

a. Investigate the mode of electrodynamic oscillation of the machine.
b. Rated mechanical load is suddenly thrown on the motor shaft at a time when it is operating in the steady state, but unloaded. Study the electrodynamic transient which will ensue.

Solution

a. Throughout this solution, the angle δ will be measured in electrical degrees rather than radians. This fact must be recognized in obtaining P_j, P_d, and P_s from the given data.

The inertia is given as Wk^2 (weight times square of radius of gyration) in English units, a common practice for large machines. In mks units (see table of conversion factors in Appendix C).

$$J = \frac{Wk^2}{23.7} = \frac{10,500}{23.7} = 444 \text{ kg-m}^2$$

From Eq. 10-36 with the factor $\pi/180$ inserted to convert angular measurement to degrees,

$$P_j = 444 \times \frac{2}{28} \times \frac{2\pi(257)}{60} \times \frac{\pi}{180} = 14.9 \text{ watts/elec deg/sec}^2$$

The remaining motor constants in the appropriate units are

$$P_d = 2\pi \times 257 \times 1,770 \times \frac{746}{33,000} \times \frac{\pi}{180} \times \frac{2}{28}$$

$$= 80.6 \text{ watts/elec deg/sec}$$

$P_s = 11.0 \times 1,000 = 11,000$ watts/elec deg

The force-free equation which determines the mode of oscillation is then

$$14.9 \, \frac{d^2\delta}{dt^2} + 80.6 \, \frac{d\delta}{dt} + 11{,}000\delta = 0$$

The undamped angular frequency and damping ratio are, respectively,

$$\omega_n = \sqrt{\frac{11{,}000}{14.9}} = 27 \text{ rad/sec}$$

$$\zeta = \frac{80.6}{2\sqrt{14.9 \times 11{,}000}} = 0.10$$

The magnitude of ζ places the transient response decidedly in the oscillatory region, as it does for all synchronous machines. Any operating disturbance will be followed by a relatively slowly damped oscillation, or swing, of the rotor before steady operation at synchronous speed is resumed. A large disturbance may, of course, be followed by complete loss of synchronism. The damped angular velocity of the motor is

$$\omega_d = 27 \sqrt{1 - (0.10)^2} = 26.9 \text{ rad/sec}$$

corresponding to a damped oscillation frequency of

$$f_d = \frac{26.9}{2\pi} = 4.3 \text{ Hz}$$

b. The full load of 200 hp is equivalent to $200 \times 746 = 149{,}200$ watts. The steady-state operating angle is

$$\delta_\infty = \frac{149{,}200}{11{,}000} = 13.6 \text{ elec deg}$$

The angular excursions are characterized by the equation

$$\delta = 13.6°[1 - 1.004\epsilon^{-2.7t} \sin{(26.9t + 84.3°)}]$$

---

c. Nonlinear Analysis. Equal-area Methods

In most of the serious dynamic problems, the oscillations are of such magnitude that the foregoing linearization is not permissible. The equations of motion must be retained in nonlinear form.   Analog or digital computers are then often used to aid the analysis.   For programming the study on a digital computer, use is made of numerical methods of solving sets of differential equations.[1]   The object of the study is usually to find whether or not synchronism is maintained, i.e., whether or not the angle δ settles down to a steady operating value after the machine has been subjected to a sizeable disturbance.

For simple synchronous-machine systems with negligible damping, use may be made of graphical interpretation of the energy stored in the rotating mass as an aid to determining the maximum angle of swing and to settling the question of maintenance of synchronism.   Because of the physical insight which it gives to the dynamic process, application of the method to analysis of a single machine connected to a large system will be discussed.

Consider specifically a synchronous motor having the power-angle curve of Fig. 10-12.   With the motor initially unloaded, the operating point is at the origin of the curve.   When a mechanical load P_{sh} is suddenly applied, the operating point travels along the sinusoid ABC and, if synchronism is maintained, finally comes to rest at point B with a new torque angle δ_∞.   To reach this new operating point, the motor must decelerate at least momentarily under the influence of the difference $P_{sh} - P_m \sin \delta$ between the power required by the load and that resulting from electromechanical energy conversion.   Now recall from the thought process leading to Eq. 10-38 that both P_{sh} and $P_m \sin \delta$ are proportional to the corresponding torques.   Recall also that the integral $\int T \, d\delta$ of torque with respect to angle is energy.   The area OAB in Fig. 10-12 is then seen to be proportional to the energy abstracted from the rotating mass during the initial period when electromagnetic energy conversion is insufficient to supply the shaft load.   When point B is reached on the first excursion, therefore, the rotor has a momentum in the direction of deceleration.   Acting under this momentum, the rotor must swing past point B until an equal amount of energy is recovered by the rotating mass.   The result is that the rotor swings to point C and the angle δ_{max}, at which point

$$\text{Area } BCD = \text{area } OAB \tag{10-40}$$

Thereafter, in the absence of damping, the rotor continues to oscillate between points O and C at its natural frequency.   The damping present in any physical machine causes successive oscillations to be of decreasing

[1] See, for instance, Glenn W. Stagg and Ahmed H. El-Abiad, "Computer Methods in Power System Analysis," McGraw-Hill Book Company, New York, 1968.

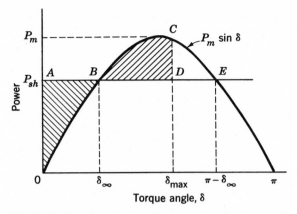

Fig. 10-12. Synchronous-motor power-angle curve and power required by load.

amplitude and finally results in dynamic equilibrium at point B. The analogy to the oscillations of a pendulum may be noted.

This *equal-area method* provides a ready means of finding the maximum angle of swing. It also provides a simple indication of whether synchronism is maintained and a rough measure of the margin of stability. Thus, if area $BCED$ in Fig. 10-12 is less than area OAB, the decelerating momentum can never be overcome, the angle-time curve follows the course of curve A in Fig. 10-13, and synchronism is lost. On the other hand, if area $BCED$ is greater than area OAB, synchronism is maintained with a margin indicated by the difference in areas and the angle-time curve

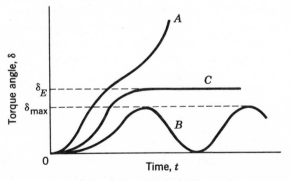

Fig. 10-13. Simple synchronous-machine swing curves showing instability (curve *A*), stability (curve *B*), and marginal or critical case (curve *C*).

follows curve B of Fig. 10-13. Equality of areas $BCED$ and OAB yields a borderline solution of unstable equilibrium for which curve C is followed.

EXAMPLE 10-4

Determine the maximum shaft load which may be suddenly applied to the motor of Example 10-3 when it is initially operating unloaded. The synchronizing power of 11.0 kw per electrical degree quoted in that example is the initial slope of the power-angle curve followed under these conditions. Damping is to be ignored.

Solution

The initial slope of the sinusoidal power-angle curve expressed in kilowatts per radian is equal to the amplitude of the curve in kilowatts. Hence

$$P_m = 11.0 \times \frac{180}{\pi} = 630 \text{ kw}$$

The load P_{sh} (Fig. 10-14) must be adjusted so that

$$\text{Area } OAB = \text{area } BCD$$

or $$P_{\text{sh}}\delta_\infty - \int_0^{\delta_\infty} 630 \sin \delta \, d\delta = \int_{\delta_\infty}^{\pi - \delta_\infty} 630 \sin \delta \, d\delta - P_{\text{sh}}(\pi - 2\delta_\infty)$$

Also, $$P_{\text{sh}} = 630 \sin \delta_\infty$$

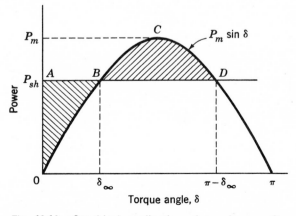

Fig. 10-14. Graphical application of equal-area criterion, Example 10-4.

Trial-and-error solution yields

$$P_{sh} = 455 \text{ kw} = 610 \text{ hp}$$

Notice that the result is independent of inertia when damping is neglected. Under these circumstances, inertia determines the period of oscillation but does not influence its amplitude.

10-8

INDUCTION-MACHINE DYNAMICS

Among integral-horsepower induction motors (i.e., those used primarily for power purposes), the most common dynamic problems are associated with starting and stopping and with the ability of the motor to continue operation during serious disturbances of the supply system. For example, a typical problem in an industrial plant may concern the ability to start a large motor without causing other parallel motors to cease normal operation because of the voltage reduction caused by the heavy inrush current to the motor being started.

The methods of induction-motor representation in dynamic analyses depend to a considerable extent on the nature and complexity of the problem and the associated precision requirements. When the electrical transients in the motor are to be included as well as the motional transients (and especially when the motor is an important element in a complex network), the equivalent circuit of Fig. 10-11 may be used. In many problems, however, the electrical transients in the induction machine may be ignored. This simplification is possible, because as illustrated in Example 10-2, the electrical transient subsides so rapidly—most commonly in a time short compared with the duration of the motional transient. (Among the principal exceptions to this statement are large 3,600-rpm motors.) We shall confine ourselves here to this type of problem.

Representation of the machine under these conditions may be based on steady-state theory, including the equivalent circuits of Figs. 7-6 and 7-9, the torque-slip curve of Fig. 7-10, and the torque-slip relation (Eq. 7-23). The problem then becomes one of arriving at a sufficiently simple yet reasonably realistic representation which will not unduly complicate the dynamic analysis, particularly through the introduction of non-linearities. One approach applicable to relatively simple problems is a graphical one. Both the torque produced by the motor and the torque required to turn the load are considered to be nonlinear functions of speed

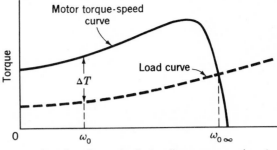

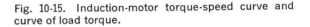

Fig. 10-15. Induction-motor torque-speed curve and curve of load torque.

for which data are given in the form of curves.   The procedure is illustrated in the following example.

EXAMPLE 10-5

A polyphase induction motor has the torque-speed curve for rated impressed voltage shown in Fig. 10-15. A curve of the torque required to maintain rotation of the load is also given in Fig. 10-15. The inertia of the load and rotor is J mks units.

Consider that across-the-line starting at rated voltage is used and that the steady-state torque-speed curve represents the performance under transient conditions with sufficient accuracy.   Show how a curve of speed as a function of time may be obtained.

Solution

At any motor speed ω_o mechanical radians per second, the torque differential ΔT between that produced by the motor and that required to turn the load is available to accelerate the rotating mass. Consequently,

$$J\frac{d\omega_o}{dt} = \Delta T \tag{10-41}$$

The time required to attain the speed ω_o is therefore

$$t = J\int_0^{\omega_o}\frac{1}{\Delta T}\,d\omega_o \tag{10-42}$$

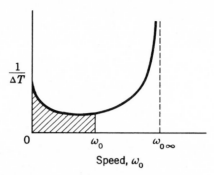

Fig. 10-16.   Graphical analysis of induction-motor starting.

The integral in Eq. 10-42 can be evaluated graphically by plot-
ting a curve of $1/\Delta T$ as a function of ω_o and finding the area between
the curve and the ω_o axis up to the value corresponding to the upper
limit of the integral, a procedure illustrated in Fig. 10-16.   The area
can be found by planimeter, by counting small squares on the curve
sheet, or by dividing it into uniform segments and using the average
ordinate for each section.   This area in units of radians per second
per newton-meter times the inertia J in mks units gives the time t in
seconds.   The computations can be carried out conveniently in
tabular form to yield a result of the type shown in Fig. 10-17.

Analytical methods are often based on piecewise linear representations of the torque-slip curve or the pertinent portion of it.   For example,
the curve is close to linear in the normal operating region of small slips.
In this region, the relation

$$T = ks \qquad\qquad (10\text{-}43)$$

may be used, where k is a constant.   Or if terminal voltage is variable,

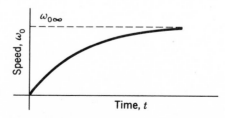

Fig. 10-17.   Speed-time curve of induction
motor during starting.

the relation

$$T = k'V_1^2 s \tag{10-44}$$

may be substituted to incorporate the fact that, according to Eq. 7-23, the electromagnetic torque varies as the square of the voltage. When a wider region of the torque-slip curve is involved, a number of straight lines may be required. Alternatively, it is shown in Art. 7-5 that a reasonably good representation of the torque-slip relation in accordance with Eq. 7-23 is given by the equation

$$\frac{T}{T_{\max}} = \frac{2}{s/s_{\max T} + s_{\max T}/s} \tag{10-45}$$

where $T_{\max}$ is the maximum torque and $s_{\max T}$ is the slip at maximum torque.

EXAMPLE 10-6

The period of heavy current inrush to an induction motor during starting usually lasts for about the time required to reach the slip $s_{\max T}$ at maximum torque, after which it decreases to the normal running value. Develop an expression for the time t required to reach the speed corresponding to $s_{\max T}$ for an induction motor being started at full voltage.

The combined inertia of the rotor and the connected mechanical equipment is J. The motor is unloaded at starting, and rotational losses are to be neglected. Consider that the torque-slip curve may be represented by Eq. 10-45.

Solution

Let all quantities except time t be expressed in per unit. The base for torque T may be the rated torque of the motor. Unit speed is synchronous speed for the motor, and s, of course, is given in per unit by the usual expression 4-3. The inertia J will then have the dimensions

$$\frac{\text{Per-unit torque}}{\text{Per-unit change in speed per second}}$$

With n representing motor speed, the basic differential equation is then

$$J \frac{dn}{dt} = -J \frac{ds}{dt} = \frac{2T_{max}}{s/s_{max\,T} + s_{max\,T}/s}$$

Upon integration,

$$t = -\frac{J}{2T_{max}} \left(\frac{1}{s_{max\,T}} \int_1^{s_{max\,T}} s\,ds + s_{max\,T} \int_1^{s_{max\,T}} \frac{ds}{s} \right)$$

$$= \frac{J}{2T_{max}} \left(\frac{1 - s^2_{max\,T}}{2s_{max\,T}} + s_{max\,T} \ln \frac{1}{s_{max\,T}} \right)$$

Neglect of rotational losses tends to make this result optimistic. A significant breakaway torque may be required to turn the motor over at very low speeds, for example, especially if the bearings are cold.

10-9
RÉSUMÉ

The study of electrical transients in ac machines is necessarily a compromise between the detailed analyses which may be made when only conditions within a single machine are of concern and the more approximate artifices which must be adopted for simplified representation of the machine as one element of a dynamic system. The basic problem is essentially that of transients in nonlinear coupled circuits. The most important question from the systems viewpoint is that of deciding on admissible approximations.

From the partly intuitive and empirical viewpoint adopted in this chapter, a simple approximate picture of the synchronous machine as a circuit element emerges. The machine is essentially a reactance behind which a readily determined voltage exists. In the steady state, the reactance is the synchronous reactance, and the internal voltage is the excitation voltage. Under transient conditions, the reactance decreases and becomes the transient reactance; the associated internal voltage is one dependent on the field flux linkages. In some problems, one may need to take account of induced damper-winding currents by using the still lower subtransient reactance. Decrement of the current from one state to another is handled by means of the appropriate time constant. The concept of a transient reactance is similarly applicable to the induction machine.

The viewpoint of simple equivalents is an essential one for the analyses of dynamic problems concerned with systems of synchronous machines.   Moreover, as shown in Art. 10-6, the electrical transients in the typical induction machine often subside rapidly in comparison with the duration of mechanical transients.   Hence, for many purposes the steady-state equivalent circuit can be used as well for dynamic analysis.

PROBLEMS

10-1. The following are the constants of a 35,000-kva 13.8-kv 60-Hz water-wheel generator: $x_d = 1.00$, $x_d' = 0.35$, $x_d'' = 0.25$, $T_{do}' = 5.0$ sec, $T_d'' = 0.04$ sec.   Reactances are in per unit on the generator rating as a base.   This generator supplies a load over a line whose reactance is 0.50 per unit and whose resistance is negligible.   Under normal conditions, the generator is fully loaded with a terminal power factor of 0.80 lagging and rated terminal voltage.   A 3-phase short circuit is considered to occur at the receiving end of the line.

    a. Compute the prefault values of the voltages behind synchronous and transient reactances.

    b. Compute the largest possible initial value of the dc component of short-circuit current in the machine.

    c. Give the numerical equation for the generator rms short-circuit current as a function of time after fault occurrence.   Ignore the dc component.

    d. The fault is cleared after 0.15 sec.   From the results of *c*, give the rms value of generator short-circuit current just before clearing.

10-2. A low-speed salient-pole synchronous motor has constants which are indicated in Table 10-1 as typical for such a machine.   It is connected to an infinite bus through a transformer whose reactance is 0.07 on the rated kva input of the motor as a base (the motor reactances are on the same base).   The infinite-bus voltage is unity.   The motor excitation is so adjusted that, with no shaft load, the transformer takes 0.50 per unit of leading reactive power from the bus at the transformer input.   Motor losses are negligible.

    a. For purposes of analyzing the response of the motor to a suddenly applied shaft load, the motor would usually be represented by its direct-axis transient reactance with a constant voltage back of that reactance equal to the value before the load was applied.

Give the numerical equation for the power-angle curve with respect to the infinite bus under these conditions. Sketch this curve approximately to scale.

b. When the load is applied very slowly, the pertinent power-angle curve is determined by the steady-state principles of Chaps. 4 and 6. Consider that cylindrical-rotor theory is to be used, with the motor represented by its direct-axis synchronous reactance. As an approximate adjustment for saturation, the unsaturated value from Table 10-1 is to be multiplied by 0.8 and assumed to be constant at this value. Give the numerical equation for the power-angle curve under these conditions, and sketch it on the same plot as in a.

c. Repeat b, but use salient-pole theory. Consider that the quadrature-axis synchronous reactance is unaffected by saturation.

d. Compute the initial symmetrical value of per-unit short-circuit current in the transformer for a 3-phase shortcircuit at its input terminals.

10-3. While an alternator is undergoing a standard short-circuit test on a factory test floor, the shortcircuit is suddenly removed. Prior to this removal, rated steady-state short-circuit current was flowing. The machine constants are $x_d = 1.20$, $x_q = 0.80$, $x'_d = 0.40$, $x'_q = 0.80$, $T'_{do} = 5$ sec, $T'_d = 1.67$ sec. There are no effective damper circuits in either the direct or the quadrature axis, and the resistance and inductance of the exciter may be assumed negligible.

Give the numerical expression for the per-unit field current as a function of time after removal of the short circuit.

10-4. Particularly severe dips in generator terminal voltage are produced by the application of inductive loads at or close to zero power factor. The starting inrush to a large motor is one such type of load.

Consider that a synchronous generator is operating initially unloaded at normal terminal voltage. A balanced inductive load x_L is suddenly applied to its terminals. The field voltage is not changed, and the generator continues to operate at synchronous speed. Ignore saturation.

a. Show that the variation of terminal voltage with time after load application is given by

$$V_t = E_{f0} \frac{x_L}{x_d + x_L} + E_{f0} \left(\frac{x_L}{x'_d + x_L} - \frac{x_L}{x_d + x_L} \right) \varepsilon^{-t/T'_d}$$

where E_{f0} is the preload excitation voltage and

$$T'_d = T'_{do} \frac{x'_d + x_L}{x_d + x_L}$$

b. Give per-unit voltage magnitudes when $x_d = 1.10$, $x'_d = 0.20$, $x_L = 1.25$, and $T'_{do} = 5.0$ sec. The reactances are per-unit values. Those for the machine are typical of large turbine generators. That for x_L will, at normal voltage, load the generator to about its continuous reactive capability at zero power factor.

10-5. A 3-phase turbine generator is rated 13.8 kv (line to line), 110,000 kva. Its constants, with reactances expressed in per unit on the machine rating as a base, are

$$x_d = 1.10 \qquad x'_d = 0.20 \qquad T'_d = 1.0 \text{ sec}$$

It is operating unloaded at a terminal voltage of 1.00 per unit when a 3-phase shortcircuit occurs at its terminals. Except in part g, ignore the dc component in the short-circuit current. Express numerical answers both in per unit and in amperes.

a. What is the rms steady-state short-circuit current? Does it make sense physically for the steady-state short-circuit current to be lower than rated current as it is here? Explain.

b. Write the numerical equation for the instantaneous phase a current as a function of time. Consider the fault to occur when the angle between the axis of phase a and the direct axis is 90°. Because of the neglect of the dc component, this is the *symmetrical* short-circuit current.

c. Write the numerical equation for the envelope of the short-circuit current wave as a function of time.

d. Using the results of part c, write the numerical equation showing how the rms value of short-circuit current varies with time.

e. What value is given by the expression in d at $t = 0$? This is known as the *initial symmetrical* rms short-circuit current.

f. Generalize the result in d by writing the equation for rms symmetrical short-circuit current as a function of time, initial voltage behind transient reactance, and the machine constants.

g. In part b, suppose the fault occurs when the magnitude of the initial angle is other than 90°. The value of i_a at $t = 0$ would then

be nonzero.   But since the phase *a* winding is a resistance-induc-
tance circuit, the complete phase *a* current cannot change instantaneously from zero.   Hence a dc component must be present in i_a
to reconcile the situation.   This component dies away rapidly.
Give the maximum possible initial magnitude of the dc component.

10-6. The machine in Prob. 10-5 has the constants

$$x_d'' = 0.10 \qquad T_d'' = 0.035 \text{ sec}$$

in addition to those given there.   Work Prob. 10-5, except for part *b*.
with the subtransient effects included as well as the transient effects,
In part *f*, the initial voltage behind subtransient reactance must also be
reflected.   In part *g*, the principle still holds that the dc component must
preserve continuity of instantaneous phase current just before and just
after the shortcircuit.

10-7. A 4-pole 440-volt 400-hp 3-phase 60-Hz Y-connected wound-
rotor induction motor has the following constants in ohms per phase
referred to the stator: $x_1 = x_2 = 0.055$, $x_\varphi = 2.23$, $r_1 = 0.0054$, $r_2 =$
0.0071.   The motor is supplied at normal terminal voltage through a
series reactance of 0.03 ohm per phase representing a step-down transformer bank.   It is fully loaded, the slip rings are short-circuited, and
the efficiency and power factor are 90.5 and 90.0 percent, respectively.
     A 3-phase shortcircuit occurs at the high-tension terminals of the
transformer bank.   Determine the initial symmetrical short-circuit current in the motor, and show how it is decremented.

10-8. Reciprocating air and ammonia compressors require a torque
which fluctuates periodically about a steady average value.   For a
2-cycle unit, the torque harmonics have frequencies in cycles per second
which are multiples of the speed in revolutions per second.   When, as is
commonly the case, the compressors are driven by synchronous motors,
the torque harmonics cause periodic fluctuation of the torque angle δ and
may result in undesirably high pulsations of power and current to the
motor.   It is therefore essential that, for the significant harmonics, the
electrodynamic response of the motor be held to a minimum.

    a. To investigate the response of the motor to torque harmonics, use
    a linearized analysis.   Let

$$P_{sh} = P_{shm} \sin \omega t$$

where P_{shm} corresponds to the amplitude of the harmonic-torque pulsation whose angular frequency is ω. Then show that the differential equation can be written

$$\frac{d^2\delta}{dt^2} + 2\zeta\omega_n \frac{d\delta}{dt} + \omega_n^2\delta = \frac{P_{shm}}{P_j} \sin \omega t$$

Identify these quantities in terms of P_s, P_d, and P_j.

b. Show that the phasor expression for the steady-state solution is

$$\Delta = \frac{P_{shm}/P_j}{(\omega_n^2 - \omega^2) + j2\zeta\omega_n\omega}$$

c. The motor driving the compressor is that of Example 10-3. The compressor has a first-order torque harmonic with an amplitude of 580 lb-ft and a frequency of 4.3 Hz. Determine the maximum deviation in power angle δ and the corresponding pulsation of synchronous power.

d. Consider that a flywheel is added to bring the total Wk^2 up to 16,500 lb-ft². Repeat the computation of c, and compare the results.

10-9. The ideal conditions for synchronizing an alternator with an electric power system are that the alternator voltage be the same as that of the system bus in magnitude, phase, and frequency. Departure from these conditions results in undesirable current and power surges accompanying electromechanical oscillation of the alternator rotor. As long as the oscillations are not too violent, they may be investigated by a linearized analysis.

Consider that a 2,500-kw 0.80-power-factor 25-Hz 26-pole oil-engine-driven alternator is to be synchronized with a 25-Hz system large enough to be considered an infinite bus. The Wk^2 of the alternator, engine, and associated flywheel is 750,000 lb-ft². The damping-power coefficient P_d is 3,600 watts per electrical degree per second, and the synchronizing-power coefficient P_s is 1.21×10^5 watts per electrical degree. Both P_d and P_s may be assumed to remain constant. In all cases below, the terminal voltage is adjusted to its correct magnitude. The engine governor is sufficiently insensitive so that it does not act during the synchronizing period.

a. Consider that the alternator is initially adjusted to the correct speed but that it is synchronized out of phase by 20 electrical

degrees, with the alternator leading the bus.   Obtain a numerical
expression for the ensuing electromechanical oscillations.   Also
give the largest value of torque exerted on the rotor during the
synchronizing period.   Ignore losses, and express this torque as a
percentage of that corresponding to the nameplate rating.

b. Repeat *a* with the alternator synchronized at the proper angle
   but with its speed initially adjusted 1.0 Hz fast.

c. Repeat *a* with the alternator initially leading the bus by 20
   electrical degrees and its speed initially adjusted 1.0 Hz fast.

10-10. A synchronous motor whose input under rated operating conditions is 10,000 kva is connected to an infinite bus over a short feeder
whose impedance is purely reactive.   The motor is rated at 60 Hz, 600
rpm, and has a total Wk^2 of 500,000 lb-ft^2 (including the shaft load).
The power-angle curve under transient conditions is 2.00 sin δ, where the
amplitude is in per unit on a 10,000-kva base.

a. With the motor operating initially unloaded, a 10,000-kw shaft
   load is suddenly applied.   Does the motor remain in synchronism?

b. How large a shaft load may be suddenly applied without loss of
   synchronism?

c. Consider now that the suddenly applied load is on for only 0.2 sec,
   after which a comparatively long time elapses before any load is
   again applied.   Determine the maximum value of such a load
   which will still allow synchronism to be maintained.   Use the
   equal-area criterion as an aid in the process.   For purposes of
   computing the angle δ at 0.2 sec, ignore damping, and use a linearized analysis in which the power-angle curve is approximated by a
   straight line through the origin and the 60° point.

10-11. *a.* A polyphase synchronous motor is initially operating at no
   load.   A mechanical load requiring rated torque is suddenly applied to the shaft.   Sketch the curve of instantaneous speed as a function of time immediately after load
   application.

   b. Repeat *a* with the synchronous motor replaced by a polyphase induction motor.

   c. If your two curves in *a* and *b* are different in any respect,
   give basic reasons for these differences.

10-12. A polyphase induction motor has negligible rotor rotational losses and is driving a pure-inertia load. The moment of inertia of the rotor plus load is J mks units.

a. Obtain an expression for the rotor energy loss during starting. Express the result in terms of J and the synchronous angular velocity ω_s.

b. Obtain an expression for the rotor energy loss associated with reversal from full speed forward by reversing the phase sequence of the voltage supply (a process known as *plugging*). Express the result in terms of J and ω_s.

c. State and discuss the degree of dependence of the results in a and b on the current-limiting scheme which may be used during starting and reversal.

d. A 5-hp 3-phase 4-pole 60-Hz squirrel-cage induction motor has a full-load efficiency of 85.0 percent. The total Wk^2 of rotor plus load is 1.5 lb-ft². The total motor losses for a reversal may be assumed to be 2.25 times the rotor losses. The impairment of ventilation arising from the lower average speed during reversing is to be ignored.

Using the result in b, compute the number of times per minute that the motor can be reversed without its allowable temperature rise being exceeded.

e. Discuss the optimism or pessimism of the result in d.

10-13. Following are points on the torque-speed curve of a 3-phase squirrel-cage induction motor with balanced rated voltage impressed:

Torque, per unit	0	1.00	2.00	3.00	3.50	3.25	3.00
Speed, per unit	1.00	0.97	0.93	0.80	0.47	0.20	0

Unit speed is synchronous speed; unit torque is rated torque. The motor is coupled to a machine tool which requires rated torque regardless of speed. The inertia of the motor plus load is such that it requires 1.2 sec to bring them to rated speed with a constant *accelerating* torque equal to rated torque.

With the motor driving the load under normal steady conditions, the voltage at its terminals suddenly drops to 50 percent of rated value because of a shortcircuit in the neighborhood. It remains at this reduced value for 0.6 sec and is then restored to its full value by clearing of the shortcircuit. The undervoltage release on the motor does not operate. Will the motor stop? If not, what is its lowest speed? Neglect any effects of secondary importance.

10-14. A 230-volt 3-phase Y-connected 6-pole 60-Hz wound-rotor induction motor has a stator-plus-rotor leakage reactance of 0.50 ohm per phase referred to the stator, a rotor-plus-load moment of inertia of 1.0 kg-m², negligible losses (except for rotor copper loss), and negligible exciting current. It is connected to a balanced 230-volt source and drives a pure-inertia load. Across-the-line starting is used, and the rotor-circuit resistance is to be adjusted so that the motor brings its load from rest to one-half synchronous speed in the shortest possible time.

Determine the value of the rotor resistance referred to the stator and the minimum time to reach one-half of synchronous speed.

10-15. A 3-phase induction motor is operating in the steady state at the slip s_1. By suddenly interchanging two stator leads, the motor is to be plugged to a quick stop. Use the notation and assumptions of Example 10-6 and consider the motor to be unloaded.

a. Develop an expression for the braking time of the motor.
b. Suppose that the motor is to be reversed instead of simply brought to a stop. Develop an expression for the reversing time.

10-16. A 4,160-volt 2,500-hp 2-pole 3-phase 60-Hz squirrel-cage induction motor is driving a boiler-feed pump in an electric generating plant. The table below lists points on the motor torque-speed curve at rated voltage, speed being in percent of synchronous speed and torque in percent of rated torque. Also listed are the torque requirements of the boiler-feed pump expressed in percent of motor rated torque. The drive is started at rated voltage with the discharge valve open but working against a check valve until the pump head equals the system head. The straight-line portion of the pump characteristic between 0 and 10 percent speed represents the breakaway-torque requirements. At 92 percent speed the check valve opens, and there is a discontinuity in the slope of the pump curve. At 98 percent speed the motor produces its maximum torque.

Percent speed	0	10	30	50	70	90	92	98	99.5
Percent motor torque	75	75	75	75	80	125	155	240	100
Percent pump torque	15	0+	4	12	26	42	44	87	100

The inertia of the drive is such that 5,400 kw-sec of energy is stored in the rotating mass at synchronous speed.

a. Determine the time required for the check valve to open after rated voltage is applied to the motor.
b. Determine the time required to reach the motor maximum-torque point.

10-17. For the boiler-feed-pump drive in Prob. 10-16, consider that the motor must occasionally be started with the motor voltage at 80 percent of its rated value. The motor torque may be assumed to be proportional to the square of the voltage.

a. Determine the time required for the check valve to open.
b. To what value may the motor starting voltage be reduced without making it impossible to reach a speed at which water is delivered to the system?

11

fractional-horsepower ac motors

Fractional-horsepower motors supply the motive power for all types of equipment in the home, office, and factory. They differ from integral-horsepower motors in the wide variety of designs and characteristics available, prompted by the economics and special requirements of their application. In addition, fractional-horsepower ac motors are nearly all designed for and utilized on single-phase lines.

Although fractional-horsepower ac motors are relatively simple in construction, they are considerably more difficult to analyze than larger 3-phase motors. Much design is carried out by building and testing prototype motors until the required design is achieved. Computer design programs are being used to realize more accurate paper designs and to reduce the extent of cut-and-try to obtain required performance. In this chapter, we will describe the various types of motors qualitatively in terms of the rotating-field theory, and finally develop a method of analysis appropriate for 2-phase motors from which the starting and running characteristics of single-phase induction motors can be calculated.[1]

---

[1] For an extensive treatment of fractional-horsepower motors, see C. G. Veinott, "Fractional- and Subfractional Horsepower Electric Motors," McGraw-Hill Book Company, New York, 1970.

The series ac motor, also termed the universal motor, is widely used in portable tools, vacuum cleaners, and kitchen equipment, where its high speed results in high horsepower per unit motor size. These motors are frequently used with solid-state devices for speed control and will be treated in this chapter.

SINGLE-PHASE INDUCTION MOTORS—QUALITATIVE EXAMINATION

Structurally, the commonest types of single-phase induction motors resemble polyphase squirrel-cage motors except for the arrangement of the stator windings. An induction motor with a cage rotor and a single-phase stator winding is represented schematically in Fig. 11-1. Instead of being a concentrated coil, the actual stator winding is distributed in slots so as to produce an approximately sinusoidal space distribution of mmf. Such a motor inherently has no starting torque, but if started by auxiliary means, it will continue to run. Before considering auxiliary starting methods, the basic properties of the elementary motor of Fig. 11-1 will be described.

Consideration of conditions with the rotor at rest readily shows that no starting torque is produced. From Fig. 11-1, it is evident that the axis of the stator field remains fixed in position along the coil axis. With alternating current in the stator coil, the stator-mmf wave is stationary in space but pulsates in magnitude, the stator-field strength alternating in polarity and varying sinusoidally with time. Currents are induced in the rotor by transformer action, these currents being in such a direction as to produce an mmf opposing the stator mmf. The axis of the rotor-mmf wave coincides with that of the stator field, the torque angle there-

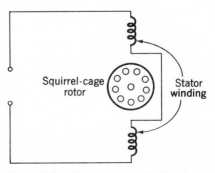

Fig. 11-1. Elementary single-phase induction motor.

fore is zero, and no starting torque is produced. The motor is merely a single-phase static transformer with a short-circuited secondary.

Conditions are not so simple, however, when the rotor is made to revolve. Two different viewpoints may then be adopted in explaining the operation of the motor: the first is to derive the conditions from those already established for polyphase motors; the second is to start afresh and show that, under certain circumstances, the necessary conditions for the production of motor torque are satisfied. Both viewpoints, of course, lead to the same results, and both can be presented in quantitative terms. The resulting analytical methods are known as the *revolving-field theory* and the *cross-field theory*, respectively. Both viewpoints have their advantages, but on the whole there is little choice between them for computational purposes.

According to the cross-field theory, when the rotor is made to revolve, there is, in addition to the transformer voltage, a voltage generated in the rotor by virtue of its rotation in the stationary stator field. In Fig. 11-1, for example, the rotational voltages in the rotor conductors are all in one direction in the upper half of the rotor and all in the other direction in the lower half. The rotational voltage produces a component rotor current and a component rotor-mmf wave whose axis is displaced 90 electrical degrees from the stator axis. The torque angle for this component of the rotor mmf is 90°, and a torque is obtained. Further detailed analysis will show that this torque is in the direction of rotation and that the necessary conditions for the continued production of torque are satisfied.[1]

The argument in the revolving-field theory is that, if a rotating magnetic field is produced, then an induction-motor torque results. Moreover, this torque will be quantitatively similar to that of the polyphase motor treated in Chap. 7, and approximately the same type of performance can be expected. The treatment of single-phase induction motors in the rest of this chapter will be from the revolving-field point of view.

Consider the elementary motor of Fig. 11-1, whose developed stator winding for 1 pole is represented schematically by the concentrated coil sides in Fig. 11-2a. Remember, however, that the stator winding actually is distributed in a number of slots so as to produce approximately a sinusoidal space distribution of mmf centered on the coil axis. If space harmonics are neglected, the space wave of stator mmf F_1 can then be expressed as

$$F_1 = F_{1(\text{peak})} \cos \theta \qquad (11\text{-}1)$$

[1] For discussion and application of the cross-field theory, see *ibid.*, chap. 2.

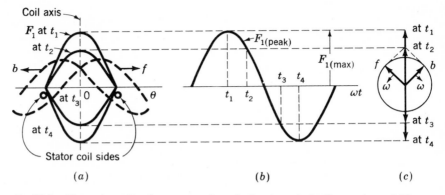

Fig. 11-2. Magnetomotive-force waves in a single-phase induction motor. *(a)* Space waves. *(b)* Time variations. *(c)* Representation by space phasors.

where θ is the electrical space angle measured from the stator coil axis and $F_{1(peak)}$ is the instantaneous value of the mmf wave at the coil axis and is proportional to the instantaneous stator current. If the stator current varies sinusoidally, then $F_{1(peak)}$ varies sinusoidally with time, as shown in Fig. 11-2b. The space distributions of stator mmf F_1 corresponding to several instants of time are shown in Fig. 11-2a. The stator-mmf wave is stationary, and its amplitude varies sinusoidally with time.

For analytical purposes, this pulsating, stationary wave can be resolved into two constant-amplitude traveling waves. Consider that the pulsating stator mmf may be represented by a space phasor of varying length, pointing up half the time, down the other half, and having a magnitude and direction determined by the instantaneous magnitude and direction of the stator current. This space phasor is shown by the vertical arrows in Fig. 11-2c for the same instants of time indicated in Fig. 11-2a and b. But it can be seen from Fig. 11-2c that such a phasor may be considered as the sum of two equal phasors rotating in opposite directions, each component phasor having a constant length equal to half the maximum length of the original pulsating phasor. Consequently the pulsating stator-mmf wave can be divided into two rotating waves of equal magnitudes. These component waves rotate in opposite directions at synchronous speed. The forward- and backward-rotating mmf waves f and b are shown dashed in Fig. 11-2a for the instant of time t_2, and the corresponding rotating vectors representing them are f and b in Fig. 11-2c.

The same conclusions can be reached by analytical methods. The analysis is essentially the same as the rotating-field theory of Art. 3-4 except that we are now concerned with only 1 stator phase, whereas in Art. 3-4 the mmfs of 3 stator phases were involved. Thus, if the stator current is a cosine function of time, the instantaneous value of the spatial peak of the pulsating mmf wave is

$$F_{1(\text{peak})} = F_{1(\text{max})} \cos \omega t \qquad (11\text{-}2)$$

where $F_{1(\text{max})}$ is the peak value corresponding to maximum instantaneous current. Consequently, by substitution of Eq. 11-2 in Eq. 11-1, the mmf wave as a function of both time and space is

$$F_1 = F_{1(\text{max})} \cos \omega t \cos \theta \qquad (11\text{-}3)$$

and, from the relation for the product of two cosines,

$$F_1 = \tfrac{1}{2}F_{1(\text{max})} \cos (\theta - \omega t) + \tfrac{1}{2}F_{1(\text{max})} \cos (\theta + \omega t) \qquad (11\text{-}4)$$

Each of the cosine terms in Eq. 11-4 describes a sinusoidal function of the space angle θ. Each has a peak value of half the maximum amplitude of the pulsating wave, and a space-phase angle ωt. Both waves are centered on the axis of the stator winding at the instant when the stator mmf has its maximum value. The angle ωt provides rotation of each wave around the air gap at the constant angular velocity ω electrical radians per second, the waves traveling in opposite directions. The first wave, whose argument is $\theta - \omega t$, travels in the forward direction of θ; the second wave, whose argument is $\theta + \omega t$, travels in the backward direction of θ. With a balanced polyphase winding the backward-rotating components cancel, leaving only the forward components, as in Eq. 3-36 for a 3-phase winding. For a single-phase winding, however, both forward and backward components are present. Thus Eq. 11-4 leads to the same conclusion as that reached by means of the phasor diagram of Fig. 11-2c.

Each of these component-mmf waves produces induction-motor action, but the corresponding torques are in opposite directions. With the rotor at rest, the forward and backward air-gap flux waves created by the combined mmfs of stator and rotor currents are equal, the component torques are equal, and no starting torque is produced. If the forward and backward air-gap flux waves remained equal when the rotor is revolving, each of the component fields would produce a torque-speed characteristic similar to that of a polyphase motor with negligible stator leakage impedance, as illustrated by the dashed curves f and b in Fig. 11-3a. The

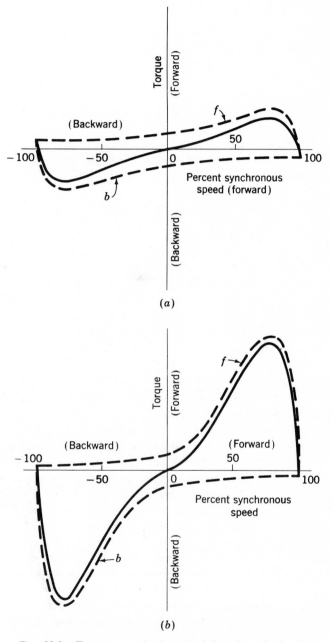

Fig. 11-3. Torque-speed characteristic of a single-phase induction motor. *(a)* On the basis of constant forward and backward flux waves. *(b)* Taking into account changes in the flux waves.

resultant torque-speed characteristic, which is the algebraic sum of the two component curves, shows that if the motor were started by auxiliary means it would produce torque in whatever direction it was started.

The assumption that the air-gap flux waves remain equal when the rotor is in motion is a rather drastic simplification of the actual state of affairs. In the first place, the effects of stator leakage impedance are ignored. Furthermore the effects of induced rotor currents are not properly accounted for. Both these effects will ultimately be included in the detailed quantitative theory of Art. 11-4. The following qualitative explanation shows that the performance of a single-phase induction motor is considerably better than would be predicted on the basis of equal forward and backward flux waves.

When the rotor is in motion, the component rotor currents induced by the backward field are greater than at standstill and their power factor is lower. Their mmf, which opposes that of the stator current, results in a reduction of the backward flux wave. Conversely, the magnetic effect of the component currents induced by the forward field is less than at standstill, because the rotor currents are less, and their power factor is higher. As speed increases, therefore, the forward flux wave increases while the backward flux wave decreases, their sum remaining roughly constant since it must induce the stator counter emf, which is approximately constant if the stator leakage-impedance voltage drop is small. Hence, with the rotor in motion, the torque of the forward field is greater and that of the backward field is less than in Fig. 11-3a, the true situation being about as shown in Fig. 11-3b. In the normal running region at a few percent slip, the forward field is several times greater than the backward field, and the flux wave does not differ greatly from the constant-amplitude revolving field in the air gap of a balanced polyphase motor. In the normal running region, therefore, the torque-speed characteristic of a single-phase motor is not too greatly inferior to that of a polyphase motor having the same rotor and operating with the same maximum air-gap flux density.

In addition to the torques shown in Fig. 11-3, double-stator-frequency torque pulsations are produced by the interactions of the oppositely rotating flux and mmf waves which glide past each other at twice synchronous speed. These interactions produce no average torque, but they tend to make the motor noisier than a polyphase motor. Such torque pulsations are unavoidable in a single-phase motor because of the pulsations in instantaneous power input inherent in a single-phase circuit. The effects of the pulsating torque can be minimized by using an elastic mounting for the motor. The torque referred to on the torque-speed curves is the time average of the instantaneous torque.

STARTING AND RUNNING PERFORMANCE OF SINGLE-PHASE INDUCTION AND SYNCHRONOUS MOTORS

Single-phase induction motors are classified in accordance with the methods of starting and are usually referred to by names descriptive of these methods.   Selection of the appropriate motor is based upon the starting and running torque requirements of the load, the duty cycle, and the limitations on starting and running current from the supply line for the motor.   The cost of single-phase motors increases with the horsepower and with the performance, such as starting-torque-to-current ratio; hence, the application engineer selects the lowest horsepower and performance motor that meets his requirement to minimize the cost.   Where a large volume of motors is to be used for a specific purpose, a special motor is designed for the application to ensure least cost.   In the fractional-horse-power motor business, differences of cost in pennies are important. Starting methods and resulting torque-speed characteristics will be considered qualitatively in this article.

a. Split-phase

Split-phase motors have 2 stator windings, a main winding m and an auxiliary winding a, with their axes displaced 90 electrical degrees in space.   They are connected as shown in Fig. 11-4a.   The auxiliary wind-

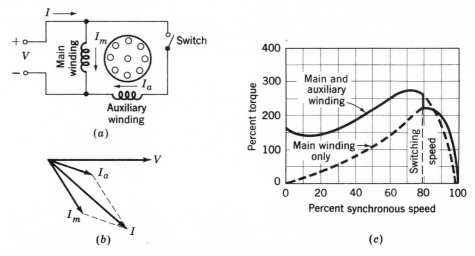

Fig. 11-4. Split-phase motor. (a) Connections. (b) Phasor diagram at starting. (c) Typical torque-speed characteristic.

ing has a higher resistance-to-reactance ratio than the main winding, so that the two currents are out of phase, as indicated in the phasor diagram of Fig. 11-4b, which is representative of conditions at starting. Since the auxiliary-winding current I_a leads the main-winding current I_m, the stator field first reaches a maximum along the axis of the auxiliary winding and then somewhat later in time reaches a maximum along the axis of the main winding. The winding currents are equivalent to unbalanced 2-phase currents, and the motor is equivalent to an unbalanced 2-phase motor. The result is a rotating stator field which causes the motor to start. After the motor starts, the auxiliary winding is disconnected, usually by means of a centrifugal switch that operates at about 75 percent of synchronous speed. The simple way to obtain the high resistance-to-reactance ratio for the auxiliary winding is to wind it with smaller wire than the main winding, a permissible procedure because this winding is in circuit only during starting. Its reactance can be reduced somewhat by placing it in the tops of the slots. A typical torque-speed characteristic is shown in Fig. 11-4c.

Split-phase motors have moderate starting torque with low starting current. Typical applications include fans, blowers, centrifugal pumps, and office equipment. Typical ratings are $\frac{1}{20}$-to-$\frac{1}{2}$ hp; in this range they are the lowest-cost motors available.

b. Capacitor-type Motors

Capacitors can be used to improve the starting performance, running performance, or both, of the motor depending upon the size and connection of the capacitor.

The capacitor-start motor is also a split-phase motor, but the time-phase displacement between the two currents is obtained by means of a capacitor in series with the auxiliary winding, as shown in Fig. 11-5a. Again the auxiliary winding is disconnected after the motor has started, and consequently the auxiliary winding and capacitor can be designed at minimum cost for intermittent service. By use of a starting capacitor of appropriate value, the auxiliary-winding current I_a at standstill could be made to lead the main-winding current I_m by 90 electrical degrees, as it would in a balanced 2-phase motor (see Fig. 11-5b). Actually, the best compromise among the factors of starting torque, starting current, and costs results with a phase angle somewhat less than 90°. A typical torque-speed characteristic is shown in Fig. 11-5c, high starting torque being an outstanding feature. These motors are used for compressors, pumps, refrigeration and air-conditioning equipment, and other hard-to-

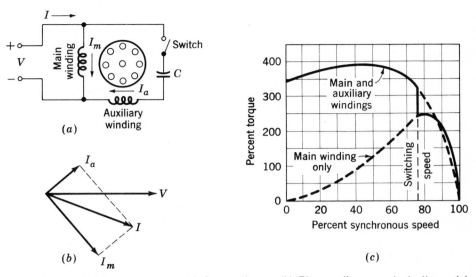

Fig. 11-5. Capacitor-start motor. *(a)* Connections. *(b)* Phasor diagram at starting. *(c)* Typical torque-speed characteristic.

start loads.   A cutaway view of a capacitor-start motor is shown in Fig. 11-6.

In the permanent-split-capacitor motor, the capacitor and auxiliary winding are not cut out after starting; the construction can be simplified by omission of the switch, and the power factor, efficiency, and torque

Fig. 11-6.   Cutaway view of a capacitor-start induction motor. The starting switch is at the right of the rotor.   The motor is of drip-proof construction.   *(General Electric Company.)*

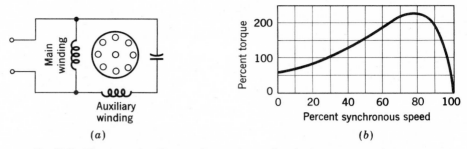

(a)   (b)

Fig. 11-7.   Permanent-split-capacitor motor and typical torque-speed characteristic.

pulsations improved. For example, the capacitor and auxiliary winding could be designed for perfect 2-phase operation at any one desired load. The backward field would then be eliminated with resulting improvement in efficiency. The double stator-frequency torque pulsations also would be eliminated, the capacitor serving as an energy-storage reservoir for smoothing out the pulsations in power input from the single-phase line. The result is a quiet motor. Starting torque must be sacrificed because the capacitance is necessarily a compromise between the best starting and running values. The resulting torque-speed characteristic, together with a schematic diagram, are given in Fig. 11-7.

If two capacitors are used, one for starting and one for running, theoretically optimum starting and running performance can both be obtained. One way of accomplishing this result is shown in Fig. 11-8a. The small value of capacitance required for optimum running conditions is permanently connected in series with the auxiliary winding, and the much larger value required for starting is obtained by a capacitor connected in parallel with the running capacitor. The starting capacitor is disconnected after the motor starts.

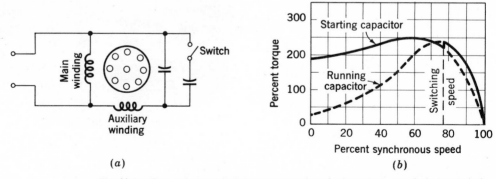

(a)   (b)

Fig. 11-8.   Two-value-capacitor motor and typical torque-speed characteristic.

The capacitor for a capacitor-start motor has a typical rating of 300 μf for a ½-hp motor.   Since it must carry current for just the starting time, the capacitor is a special compact ac electrolytic type made for motor-starting duty.   The capacitor for the same motor permanently connected has a typical rating of 40 μf; since it operates continuously, the capacitor is an ac paper, foil and oil type.   The cost of the motors is related to the performance: the permanent-capacitor motor is the lowest cost, the capacitor-start motor next, and the two-value-capacitor motor the highest cost.

EXAMPLE 11-1

A ⅓-hp 120-volt 60-Hz capacitor-start motor has the following constants for the main and auxiliary windings:

$$\text{Main winding, } Z_m = 4.5 + j3.7 \text{ ohms}$$

$$\text{Auxiliary winding, } Z_a = 9.5 + j3.5 \text{ ohms}$$

Find the value of starting capacitance that will place the main and auxiliary winding currents in quadrature at starting.

Solution

The currents I_m and I_a are as shown in Fig. 11-5a and b.   The impedance angle of the main winding is

$$\phi_m = \arctan \frac{3.7}{4.5} = 39.6°$$

The impedance angle of the auxiliary winding must be

$$\phi_a = 39.6° - 90.0° = -50.4°$$

The reactance X_c of the capacitor must satisfy the relationship

$$\arctan \frac{3.5 - X_c}{9.5} = -50.4°$$

$$\frac{3.5 - X_c}{9.5} = -1.21$$

$$X_c = 1.21 \times 9.5 + 3.5 = 15.0 \text{ ohms}$$

The capacitance C is

$$C = \frac{10^6}{15.0 \times 377} = 177 \ \mu f$$

c. Shaded-pole

As illustrated schematically in Fig. 11-9a, the shaded-pole motor usually has salient poles with one portion of each pole surrounded by a short-circuited turn of copper called a *shading coil*.   Induced currents in the shading coil cause the flux in the shaded portion of the pole to lag the flux in the other portion.   The result is like a rotating field moving in the direction from the unshaded to the shaded portion of the pole, and a low starting torque is produced.   A typical torque-speed characteristic is shown in Fig. 11-9b.   The efficiency is low.   Shaded-pole motors are the least expensive type of fractional-horsepower motor and are built up to about $\frac{1}{20}$ hp.

d. Self-starting Reluctance Motors

Any one of the induction-motor types described above can be made into a self-starting synchronous motor of the reluctance type.   Anything which makes the reluctance of the air gap a function of the angular position of the rotor with respect to the stator coil axis will produce reluctance torque when the rotor is revolving at synchronous speed.   For example, suppose some of the teeth are removed from a squirrel-cage rotor, leaving the bars and end rings intact as in an ordinary squirrel-cage induction motor. Figure 11-10a shows a lamination for such a rotor designed for use with a 4-pole stator.   The stator may be polyphase or any one of the single-phase

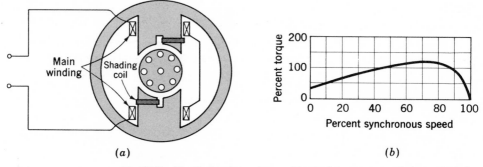

(a)   (b)

Fig. 11-9.   Shaded-pole motor and typical torque-speed characteristic.

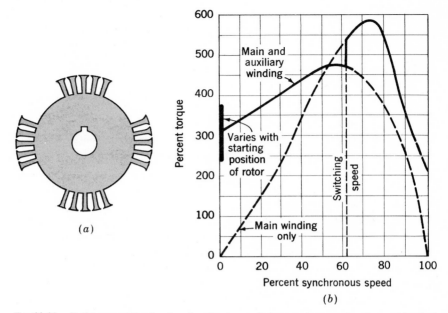

Fig. 11-10.   Rotor punching for 4-pole reluctance-type synchronous motor and typical starting characteristics.

types described above.   The motor will start as an induction motor and at light loads will speed up to a small value of slip.   The reluctance torque arises from the tendency of the rotor to try to align itself in the minimum-reluctance position with respect to the synchronously revolving forward air-gap flux wave, in accordance with the principles explained in Chap. 2. At a small slip, this torque alternates slowly in direction; the rotor is accelerated during a positive half cycle of the torque variation and decelerated during the succeeding negative half cycle.   If, however, the moment of inertia of the rotor and its mechanical load is sufficiently small, the rotor will be accelerated from slip speed up to synchronous speed during an accelerating half cycle of the reluctance torque.   The rotor will then pull into step and continue to run at synchronous speed. The torque of the backward-revolving field affects the synchronous-motor performance in the same way as an additional shaft load.

A typical torque-speed characteristic for a split-phase-start synchronous motor of the single-phase reluctance type is shown in Fig. 11-10b. Notice the high values of induction-motor torque.   The reason for this is that in order to obtain satisfactory synchronous-motor characteristics it has been found necessary to build reluctance-type synchronous motors on frames which would be suitable for induction motors of two or three

times the synchronous-motor rating.   Also notice that the principal
effect of the salient-pole rotor on the induction-motor characteristic is
at standstill, where considerable "cogging" is evident; i.e., the torque
varies considerably with rotor position.

e. Hysteresis Motors

The phenomenon of hysteresis can be used to produce mechanical torque.
In its simplest form, the rotor of a hysteresis motor is a smooth cylinder of
magnetically hard steel, without windings or teeth.   It is placed within a
slotted stator carrying distributed windings designed to produce as nearly
as possible a sinusoidal space distribution of flux, since undulations in the
flux wave greatly increase the losses.   In single-phase motors, the stator
windings usually are the permanent-split-capacitor type, as in Fig. 11-7.
The capacitor is chosen so as to result in approximately balanced 2-phase
conditions within the motor windings.   The stator then produces a rotating field, approximately constant in space waveform and revolving at
synchronous speed.

Instantaneous magnetic conditions in the air gap and rotor are indicated in Fig. 11-11a for a 2-pole stator.   The axis SS' of the stator-mmf
wave revolves at synchronous speed.   Because of hysteresis, the magnetization of the rotor lags behind the inducing mmf wave, and therefore
the axis RR' of the rotor flux wave lags behind the axis of the stator-mmf
wave by the hysteretic lag angle δ (Fig. 11-11a).   If the rotor is station-

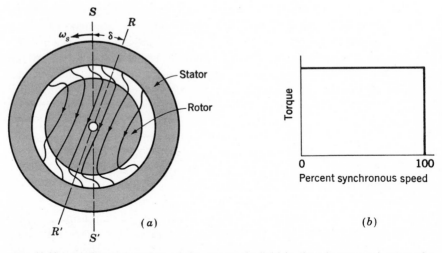

Fig. 11-11. *(a)* General nature of the magnetic field in the air gap and rotor of a
hysteresis motor; *(b)* idealized torque-speed characteristic.

ary, starting torque is produced proportional to the product of the fundamental components of the stator mmf and rotor flux and the sine of the torque angle δ. The rotor then accelerates if the counter torque of the load is less than the developed torque of the motor. So long as the rotor is turning at less than synchronous speed, each particle of the rotor is subjected to a repetitive hysteresis cycle at slip frequency. While the rotor accelerates, the lag angle δ remains constant if the flux is constant, since the angle δ depends merely on the hysteresis loop of the rotor and is independent of the rate at which the loop is traversed. The motor therefore develops constant torque right up to synchronous speed, as shown in the idealized torque-speed characteristic of Fig. 11-11b. This feature is one of the advantages of the hysteresis motor. In contrast with a reluctance motor, which must "snap" its load into synchronism from an induction-motor torque-speed characteristic, a hysteresis motor can synchronize any load which it can accelerate, no matter how great the inertia. After reaching synchronism, the motor continues to run at synchronous speed and adjusts its torque angle so as to develop the torque required by the load.

The hysteresis motor is inherently quiet and produces smooth rotation of its load. Furthermore, the rotor takes on the same number of poles as the stator field. The motor lends itself to multispeed synchronous-speed operation where the stator is wound with several sets of windings and utilizes pole-changing connections. The hysteresis motor can accelerate and synchronize high-inertia loads because its torque is uniform from standstill to synchronous speed.

11-3

SERIES MOTORS. UNIVERSAL MOTORS

A series motor has the convenient ability to run on either alternating or direct current and with similar characteristics, provided both stator and rotor cores are laminated. Such a single-phase series motor therefore is commonly called a *universal motor*. The torque angle is fixed by the brush position and is normally at its optimum value of 90°. If alternating current is supplied to a series motor, the stator and rotor field strengths will vary in exact time phase. Both will reverse at the same instant, and consequently the torque will always be in the same direction, though pulsating in magnitude at twice line frequency. Average torque will be produced, and the performance of the motor will be generally similar to that with direct current.

Small universal motors are used where light weight is important, as

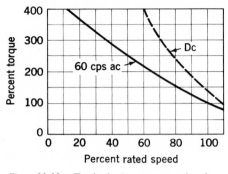

Fig. 11-12. Typical torque-speed characteristics of a universal series motor.

in vacuum cleaners and portable tools, and usually operate at high speeds (1,500 to 15,000 rpm). Typical characteristics are shown in Fig. 11-12. The ac and dc characteristics differ somewhat for two reasons: (1) with alternating current, reactance-voltage drops in the field and armature absorb part of the applied voltage, and therefore for a specified current and torque the rotational counter emf generated in the armature is less than with direct current and the speed tends to be lower; (2) with alternating current, the magnetic circuit may be appreciably saturated at the peaks of the current wave, and the rms value of the flux may thus be appreciably less with alternating current than with the same rms value of direct current; the torque therefore tends to be less and the speed higher with alternating than with direct current. The universal motor provides the highest horsepower per dollar in the fractional-horsepower range, at the expense of noise, relatively short life, and high speed.

To obtain control of the speed and torque of the series motor, the ac voltage is applied in series with a solid-state device which controls the voltage applied to the motor.[1] A thyristor as described in Chap. 8 can control alternate half cycles of the voltage so that the motor carries unidirectional current. A bidirectional switch (Triac) controls both half cycles so that the ac motor voltage is phase controlled. A circuit for ac control is shown in Fig. 11-13a. The waveforms of motor voltage and current for an arbitrary firing angle α are shown in Fig. 11-13b. The torque-speed characteristics for various firing angles are shown in Fig. 11-13c. The firing angle can be manually adjusted as in a trigger-controlled electric drill, or can be controlled by a speed-control circuit, as in some portable tools and appliances. The combination of solid-state

[1] See A. Kusko, "Solid-State DC Motor Drives," chap. 6, The M.I.T. Press, Cambridge, Mass., 1969.

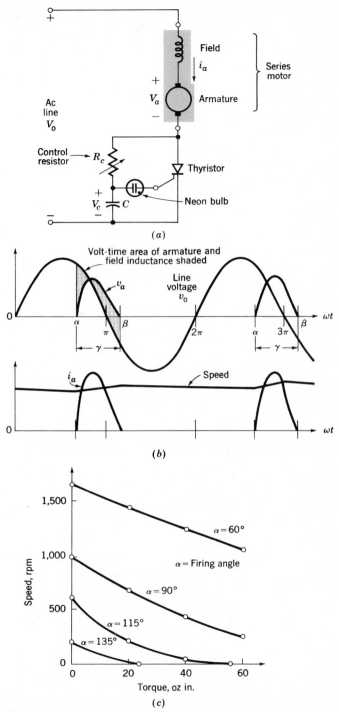

Fig. 11-13. Operation of series motor with thyristor speed-control circuit. (a) Circuit with manual control of speed; (b) waveforms of motor voltage, current, and speed; (c) speed-torque curves of a thyristor-controlled series motor as a function of firing angle α.

device and series motor provides an economical, controllable motor package.

EXAMPLE 11-2

Show that the time-average torque of an ac series motor is proportional to the square of the rms current when the magnetic structure is not saturated.

Solution

The instantaneous magnetic flux φ produced by the motor current i_a in the field poles is

$$\varphi = K_f i_a$$

The instantaneous torque is $K_t \varphi i_a$; hence the time-average torque T_{av} is

$$T_{av} = \frac{\omega}{\pi} \int_0^{\pi/\omega} K_t K_f i_a^2 \, dt$$

$$= K_t K_f I_a^2$$

where I_a is the rms value of the current i_a. This relationship is independent of the waveform of the current i_a.

11-4
REVOLVING–FIELD THEORY OF SINGLE–PHASE INDUCTION MOTORS

In Art. 11-1, the stator-mmf wave of a single-phase induction motor is shown to be equivalent to two constant-amplitude mmf waves revolving in opposite directions at synchronous speed. Each of these component stator-mmf waves induces its own component rotor currents and produces induction-motor action just as in a balanced polyphase motor. This double-revolving-field concept not only is useful for qualitative visualization but also can be developed into a quantitative theory applicable to a wide variety of induction-motor types. A simple and important case is that of the single-phase induction motor running on only its main winding.

First consider conditions with the rotor stationary and only the main stator winding m excited. The motor then is equivalent to a transformer

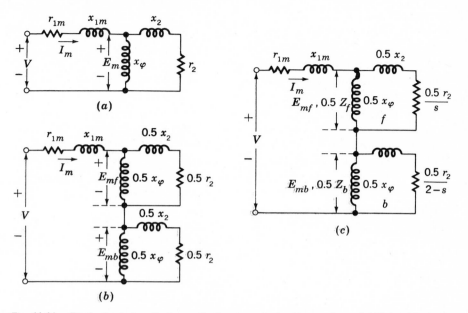

Fig. 11-14.   Equivalent circuits for a single-phase induction motor.   (a) Rotor blocked; (b) rotor blocked, showing effects of forward and backward fields; (c) running conditions.

with its secondary short-circuited.   The equivalent circuit is shown in Fig. 11-14a, where r_{1m} and x_{1m} are, respectively, the resistance and leakage reactance of the main winding, x_{φ} is the magnetizing reactance, and r_2 and x_2 are the standstill values of the rotor resistance and leakage reactance referred to the main stator winding by use of the appropriate turns ratio.   Core loss, which is omitted here, will be accounted for later as if it were a rotational loss.   The applied voltage is V, and the main-winding current is I_m.   The voltage E_m is the counter emf generated in the main winding by the stationary pulsating air-gap flux wave produced by the combined action of the stator and rotor currents.

In accordance with the double-revolving-field concept of Art. 11-1, the stator mmf can be resolved into half-amplitude forward- and backward rotating fields.   At standstill the amplitudes of the forward and backward resultant air-gap flux waves both equal half the amplitude of the pulsating field.   In Fig. 11-14b the portion of the equivalent circuit representing the effects of the air-gap flux is split into two equal portions representing the effects of the forward and backward fields, respectively.

Now consider conditions after the motor has been brought up to speed by some auxiliary means and is running on only its main winding in the direction of the forward field at a per-unit slip s.   The rotor currents induced by the forward field are of slip frequency sf, where f is the stator frequency. Just as in any polyphase motor with a symmetrical polyphase or cage rotor, these rotor currents produce an mmf wave traveling forward at slip speed with respect to the rotor and therefore at synchronous speed with respect to the stator. The resultant of the forward waves of stator and rotor mmf creates a resultant forward wave of air-gap flux which generates a counter emf E_{mf} in the main winding m of the stator. The reflected effect of the rotor as viewed from the stator is like that in a polyphase motor and can be represented by an impedance $0.5r_2/s + j0.5x_2$ in parallel with $j0.5x_\varphi$, as in the portion of the equivalent circuit (Fig. 11-14c) labeled f. The factors of 0.5 come from the resolution of the pulsating stator mmf into forward and backward components.

Now consider conditions with respect to the backward field. The rotor is still turning at a slip s with respect to the forward field, and its per-unit speed n in the direction of the forward field is

$$n = 1 - s \qquad (11\text{-}5)$$

The relative speed of the rotor with respect to the backward field is $1 + n$, or its slip with respect to the backward field is

$$1 + n = 2 - s \qquad (11\text{-}6)$$

The backward field then induces rotor currents whose frequency is $(2 - s)f$. For small slips, these rotor currents are of almost twice stator frequency. At a small slip, an oscillogram of rotor current therefore will show a high-frequency component from the backward field superposed on a low-frequency component from the forward field. As viewed from the stator, the rotor-mmf wave of the backward-field rotor currents travels at synchronous speed, but in the backward direction. The equivalent circuit representing these internal reactions from the viewpoint of the stator is like that of a polyphase motor whose slip is $2 - s$ and is shown in the portion of the equivalent circuit (Fig. 11-14c) labeled b. As with the forward field, the factors of 0.5 come from the resolution of the pulsating stator mmf into forward and backward components. The voltage E_{mb} across the parallel combination representing the backward field is the counter emf generated in the main winding m of the stator by the resultant backward field.

By use of the equivalent circuit of Fig. 11-14c, the stator current, power input, and power factor can be computed for any assumed value of slip when the applied voltage and the motor impedances are known. To simplify the notation, let

$$Z_f \equiv R_f + jX_f \equiv \left(\frac{r_2}{s} + jx_2\right) \text{ in parallel with } jx_\varphi \qquad (11\text{-}7)$$

$$Z_b \equiv R_b + jX_b \equiv \left(\frac{r_2}{2-s} + jx_2\right) \text{ in parallel with } jx_\varphi \qquad (11\text{-}8)$$

The impedances representing the reactions of the forward and backward fields from the viewpoint of the single-phase stator winding m are $0.5Z_f$ and $0.5Z_b$, respectively, in Fig. 11-14c.

Examination of the equivalent circuit (Fig. 11-14c) confirms the conclusion, reached by qualitative reasoning in Art. 11-1 (Fig. 11-3b), that the forward air-gap flux wave increases and the backward wave decreases when the rotor is set in motion. When the motor is running at a small slip, the reflected effect of the rotor resistance in the forward field, $0.5r_2/s$, is much larger than its standstill value, while the corresponding effect in the backward field, $0.5r_2/(2-s)$, is smaller. The forward-field impedance therefore is larger than its standstill value, while that of the backward field is smaller. The forward-field counter emf E_{mf} therefore is larger than its standstill value, while the backward-field counter emf E_{mb} is smaller; i.e., the forward air-gap flux wave increases, while the backward flux wave decreases.

Moreover, mechanical output conditions can be computed by application of the torque and power relations developed for polyphase motors in Chap. 7. The torques produced by the forward and backward fields can each be treated in this manner. The interactions of the oppositely rotating flux and mmf waves cause torque pulsations at twice stator frequency but produce no average torque.

As in Eq. 7-18, the internal torque T_f of the forward field in newton-meters equals $1/\omega_s$ times the power P_{gf} in watts delivered by the stator winding to the forward field, where ω_s is the synchronous angular velocity in mechanical radians per second; thus

$$T_f = \frac{1}{\omega_s} P_{gf} \qquad (11\text{-}9)$$

When the magnetizing impedance is treated as purely inductive, P_{gf} is the power absorbed by the impedance $0.5Z_f$; that is,

$$P_{gf} = I_m^2 0.5 R_f \qquad (11\text{-}10)$$

where R_f is the resistive component of the forward-field impedance defined in Eq. 11-7. Similarly, the internal torque T_b of the backward field is

$$T_b = \frac{1}{\omega_s} P_{gb} \tag{11-11}$$

where P_{gb} is the power delivered by the stator winding to the backward field, or

$$P_{gb} = I_m^2 0.5 R_b \tag{11-12}$$

where R_b is the resistive component of the backward-field impedance Z_b defined in Eq. 11-8. The torque of the backward field is in the opposite direction to that of the forward field, and therefore the net internal torque T is

$$T = T_f - T_b = \frac{1}{\omega_s} (P_{gf} - P_{gb}) \tag{11-13}$$

Since the rotor currents produced by the two component air-gap fields are of different frequencies, the total rotor I^2R loss is the numerical sum of the losses caused by each field. In general, as shown by comparison of Eqs. 7-12 and 7-13, the rotor copper loss caused by a rotating field equals the slip of the field times the power absorbed from the stator, whence

$$\text{Forward-field rotor } I^2R = sP_{gf} \tag{11-14}$$

$$\text{Backward-field rotor } I^2R = (2 - s)P_{gb} \tag{11-15}$$

$$\text{Total rotor } I^2R = sP_{gf} + (2 - s)P_{gb} \tag{11-16}$$

Since power is torque times angular velocity and the angular velocity of the rotor is $(1 - s)\omega_s$, the internal power P converted to mechanical form, in watts, is

$$P = (1 - s)\omega_s T = (1 - s)(P_{gf} - P_{gb}) \tag{11-17}$$

As in the polyphase motor, the internal torque T and internal power P are not the output values, because rotational losses remain to be accounted for. It is obviously correct to subtract friction and windage effects from T or P, and it is usually assumed that core losses may be treated in the same manner. For the small changes in speed encountered in normal operation, the rotational losses are often assumed to be constant.[1]

EXAMPLE 11-3

A $\frac{1}{4}$-hp 110-volt 60-Hz 4-pole capacitor-start motor has the following constants and losses:

$$r_{1m} = 2.02 \text{ ohms} \qquad x_{1m} = 2.79 \text{ ohms}$$

$$r_2 = 4.12 \qquad\qquad x_2 = 2.12$$

$$x_\varphi = 66.8$$

Core loss = 24 watts   Friction and windage = 13 watts

For a slip of 0.05, determine the stator current, power factor, power output, speed, torque, and efficiency when this motor is running as a single-phase motor at rated voltage and frequency with its starting winding open.

Solution

The first step is to determine the values of the forward- and backward-field impedances at the assigned value of slip.   The following relations, derived from Eq. 11-7, simplify the computations:.

$$R_f = \frac{x_\varphi^2}{x_{22}} \frac{1}{sQ_2 + (1/sQ_2)} \tag{11-18}$$

$$X_f = \frac{x_2 x_\varphi}{x_{22}} + \frac{R_f}{sQ_2} \tag{11-19}$$

where

$$x_{22} = x_2 + x_\varphi \tag{11-20}$$

and

$$Q_2 = \frac{x_{22}}{r_2} \tag{11-21}$$

Substitution of numerical values gives, for $s = 0.05$,

$$R_f + jX_f = 31.9 + j40.3 \text{ ohms}$$

Corresponding relations for the backward-field impedance Z_b are obtained by substituting $2 - s$ for s in Eqs. 11-18 and 11-19.   When

---

[1] For a treatment of the experimental determination of motor constants and losses, see *ibid.*, C. G. Veinott, chap. 18.

$(2 - s)Q_2$ is greater than 10, as is usually the case, less than 1 percent error results from use of the following approximate forms:

$$R_b = \frac{r_2}{2 - s}\left(\frac{x_\varphi}{x_{22}}\right)^2 \tag{11-22}$$

$$X_b = \frac{x_2 x_\varphi}{x_{22}} + \frac{R_b}{(2 - s)Q_2} \tag{11-23}$$

Substitution of numerical values gives, for $s = 0.05$,

$$R_b + jX_b = 1.98 + j2.12 \text{ ohms}$$

Addition of the series elements in the equivalent circuit of Fig. 11-14c gives

$$
\begin{aligned}
r_{1m} + jx_{1m} &= 2.02 + j2.79 \\
0.5(R_f + jX_f) &= 15.95 + j20.15 \\
0.5(R_b + jX_b) &= 0.99 + j1.06 \\
\hline
\text{Input } Z = \text{sum} &= 18.96 + j24.00 = 30.6\underline{/51.7^\circ}
\end{aligned}
$$

$$\text{Stator current } I_m = \frac{110}{30.6} = 3.59 \text{ amp}$$

$$\text{Power factor} = \cos 51.7^\circ = 0.620$$

$$\text{Power input} = (110)(3.59)(0.620) = 244 \text{ watts}$$

Power absorbed by forward field (Eq. 11-10)

$$P_{gf} = (3.59)^2(15.95) = 206 \text{ watts}$$

Power absorbed by backward field (Eq. 11-12)

$$P_{gb} = (3.59)^2(0.99) = 12.8 \text{ watts}$$

Internal mechanical power (Eq. 11-17)

$$
\begin{aligned}
P &= (0.95)(206 - 13) = 184 \\
\text{Rotational loss} &= 24 + 13 = \underline{37} \\
\text{Power output} &= \text{difference} = 147 \text{ watts, or } 0.197 \text{ hp}
\end{aligned}
$$

$$\text{Synchronous speed} = 1{,}800 \text{ rpm, or } 30 \text{ rev/sec}$$

$$\omega_s = 2\pi(30) = 188.5 \text{ rad/sec}$$

$$\text{Rotor speed} = (1 - s) \times (\text{synchronous speed})$$

$$= (0.95)(1{,}800) = 1{,}710 \text{ rpm}$$

$$= (0.95)(188.5) = 179 \text{ rad/sec}$$

$$\text{Torque} = \frac{\text{power}}{\text{angular velocity}}$$

$$= \frac{147}{179} = 0.821 \text{ newton-meter, or } 0.605 \text{ lb-ft}$$

$$\text{Efficiency} = \frac{\text{output}}{\text{input}} = \frac{147}{244} = 0.602$$

As a check on the power bookkeeping, compute the losses:

Stator $I_m^2 r_{1m} = (3.59)^2(2.02)$		= 26.0
Forward-field rotor I^2R, Eq. 11-14 = $(0.05)(206)$		= 10.3
Backward-field rotor I^2R, Eq. 11-15 = $(1.95)(12.8)$		= 25.0
Rotational losses		= 37.0
Sum		= 98.3
From input $-$ output, total losses		= 97
(Checks within accuracy of computations.)		

Examination of the order of magnitude of the numerical values in Example 11-3 suggests approximations which usually can be made. These approximations pertain particularly to the backward-field impedance.   Note that the impedance $0.5(R_b + jX_b)$ is only about 5 percent of the total motor impedance for a slip near full load.   Consequently, an approximation as large as 20 percent of this impedance would cause only about 1 percent error in the motor current.   Although, strictly speaking, the backward-field impedance is a function of slip, very little error usually results from computing its value at any convenient slip in the normal running region—say, 5 percent—and then assuming R_b and X_b to be constants.   With a slightly greater approximation, the shunting effect of jx_φ on the backward-field impedance can often be neglected, whence

$$Z_b \approx \frac{r_2}{2 - s} + jx_2 \tag{11-24}$$

This equation gives values of the backward-field resistance that are a few percent high, as can be seen by comparison with Eq. 11-22.   Neglecting

s in Eq. 11-24 would tend to give values of the backward-field resistance that would be too low, and therefore such an approximation would tend to counteract the error in Eq. 11-24. Consequently, for small slips

$$Z_b \approx \frac{r_2}{2} + jx_2 \tag{11-25}$$

In the polyphase motor (Art. 7-4) maximum internal torque and the slip at which it occurs can easily be expressed in terms of the motor constants; the maximum internal torque is independent of rotor resistance. No such simple relations exist for the single-phase motor. The single-phase problem is much more involved because of the presence of the backward field, the effect of which is twofold: first, it absorbs some of the applied voltage, thus reducing the voltage available for the forward field and decreasing the forward torque developed; and second, the backward field then absorbs some of the forward-field torque. Both these effects depend on rotor resistance as well as leakage reactance. Consequently, unlike the polyphase motor, the maximum internal torque of a single-phase motor is influenced by rotor resistance; increasing the rotor resistance decreases the maximum torque and increases the slip at which maximum torque occurs.

Principally because of the effects of the backward field, a single-phase induction motor is somewhat inferior to a polyphase motor using the same rotor and the same stator core. The single-phase motor has a lower maximum torque which occurs at a lower slip. For the same torque, the single-phase motor has a higher slip and greater losses, principally because of the backward-field rotor copper loss. The volt-ampere input to the single-phase motor is greater, principally because of the power and reactive volt-amperes consumed by the backward field. The stator copper loss also is somewhat higher in the single-phase motor, because 1 phase, rather than several, must carry all the current. Because of the greater losses, the efficiency is lower, and the temperature rise for the same torque is higher. A larger frame size must be used for a single-phase motor than for a polyphase motor of the same power and speed rating. Because of the larger frame size, the maximum torque can be made comparable with that of a physically smaller but equally rated polyphase motor. In spite of the larger frame size and the necessity for auxiliary starting arrangements, general-purpose single-phase motors in the standard fractional-horsepower ratings cost approximately the same as correspondingly rated polyphase motors, because of the much greater volume of production of the former.

UNBALANCED OPERATION OF SYMMETRICAL TWO-PHASE MACHINES. THE SYMMETRICAL-COMPONENT CONCEPT

Unbalanced operation of an induction motor occurs either when the voltages applied to the stator do not constitute a balanced polyphase set, or when the stator or rotor windings are not symmetrical with respect to the phases.   The single-phase motor is the extreme case of a motor operating under unbalanced stator-voltage conditions.   In some cases, unbalanced voltages are produced in the supply network to a motor, for example, when a line fuse is blown.   In other cases, unbalanced voltages are produced by the starting impedances of single-phase motors, as described in Art. 11-2.   An important use of unbalanced voltages is made in the control of 2-phase servomotors, as will be described in Art. 11-6. The purpose of this article is to develop the symmetrical-component theory of 2-phase induction motors from the double-revolving field concept and to show how the theory can be applied to a variety of problems involving induction motors having 2 stator windings in space quadrature.

First consider in review what happens when balanced 2-phase voltages are applied to the stator terminals of a 2-phase machine having a uniform air gap, a symmetrical polyphase or cage rotor, and 2 identical stator windings a and m in space quadrature.   The stator currents are equal in magnitude and in time quadrature.   When the current in winding a has its instantaneous maximum, the current in winding m is zero and the stator-mmf wave is centered on the axis of winding a.   Similarly, the stator-mmf wave is centered on the axis of winding m at the instant when the current in winding m has its instantaneous maximum.   The stator-mmf wave therefore travels 90 electrical degrees in space in an interval of 90° in time, the direction of its travel depending on the phase sequence of the currents.   A more complete analysis in the manner of Art. 3-4 or Fig. 11-2 proves that the traveling wave has constant amplitude and constant angular velocity.   This fact is, of course, the basis of the whole theory of balanced operation of induction machines.

The behavior of the motor for balanced 2-phase applied voltages of either phase sequence can readily be determined.   Thus, if the rotor is turning at a per-unit speed n in the direction from winding a toward winding m, the terminal impedance per phase is given by the equivalent circuit of Fig. 11-15a when the applied voltage v_a leads the applied voltage v_m by 90°.   Throughout the rest of this treatment, this phase sequence will be called *positive sequence* and will be designated by subscript f, since positive-sequence currents result in a forward field.   With the rotor still forced to run at the same speed and in the same direction, the terminal

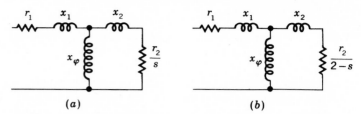

$$(a) \qquad\qquad\qquad\qquad (b)$$

Fig. 11-15.   Equivalent circuits for a 2-phase motor under unbalanced conditions.   *(a)* Forward field; *(b)* backward field.

impedance per phase is given by the equivalent circuit of Fig. 11-15*b* when v_a lags v_m by 90°.   This phase sequence will be called *negative sequence* and will be designated by subscript *b*, since negative-sequence currents produce a backward field.

Suppose now that *two* balanced 2-phase voltage sources *of opposite phase sequence* are connected in series and applied simultaneously to the motor, as indicated in Fig. 11-16*a*, where phasor voltages V_{mf} and jV_{mf} applied, respectively, to windings *m* and *a*, form a balanced system of positive sequence and phasor voltages V_{mb} and $-jV_{\mathrm{mb}}$ form another balanced system but of negative sequence.   The resultant voltage V_m applied to winding *m* is, as a phasor,

$$V_m = V_{\mathrm{mf}} + V_{\mathrm{mb}} \qquad\qquad (11\text{-}26a)$$

and that applied to winding *a* is

$$V_a = jV_{\mathrm{mf}} - jV_{\mathrm{mb}} \qquad\qquad (11\text{-}26b)$$

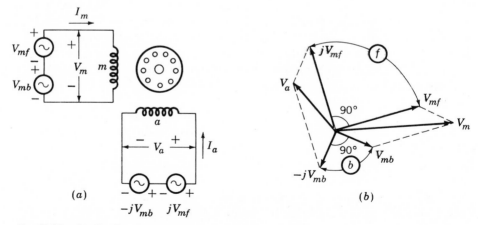

$$(a) \qquad\qquad\qquad\qquad\qquad\qquad (b)$$

Fig. 11-16.   Synthesis of an unbalanced 2-phase system from the sum of two balanced systems of opposite phase sequence.

If, for example, the forward, or positive-sequence, system is given by the phasors V_{mf} and jV_{mf} in Fig. 11-16b and the backward, or negative-sequence, system is given by the phasors V_{mb} and $-jV_{mb}$, then the resultant voltages are given by the phasors V_m and V_a. An unbalanced 2-phase system of applied voltages V_m and V_a has thus been synthesized by combining two symmetrical systems of opposite phase sequence.

The symmetrical component systems are, however, much easier to work with than is their unbalanced resultant system. Thus, it is easy to compute the component currents produced by each symmetrical component system of applied voltages because the induction motor operates as a balanced 2-phase motor for each component system. By superposition, the actual current in a winding then is the sum of its components. Thus, if I_{mf} and I_{mb} are, respectively, the positive- and negative-sequence component phasor currents in winding m, then the corresponding positive- and negative-sequence component phasor currents in winding a are, respectively, jI_{mf} and $-jI_{mb}$ and the actual winding currents I_m and I_a are

$$I_m = I_{mf} + I_{mb} \tag{11-27}$$

$$I_a = jI_{mf} - jI_{mb} \tag{11-28}$$

The inverse operation of finding the symmetrical components of specified voltages or currents must often be performed. Solution of Eqs. 11-26a and 11-26b for the phasor components V_{mf} and V_{mb} in terms of known phasor voltages V_m and V_a gives

$$V_{mf} = \tfrac{1}{2}(V_m - jV_a) \tag{11-29}$$

$$V_{mb} = \tfrac{1}{2}(V_m + jV_a) \tag{11-30}$$

These operations are illustrated in the phasor diagram of Fig. 11-17. Obviously, similar relations give the phasor symmetrical components I_{mf} and I_{mb} of the current in winding m in terms of specified phasor currents

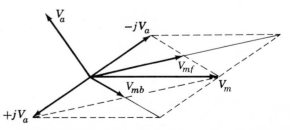

Fig. 11-17. Resolution of unbalanced 2-phase voltages into symmetrical components.

I_m and I_a in the 2 phases; thus

$$I_{mf} = \tfrac{1}{2}(I_m - jI_a) \tag{11-31}$$

$$I_{mb} = \tfrac{1}{2}(I_m + jI_a) \tag{11-32}$$

Resolution of the stator-mmf wave into its forward and backward components, as in Fig. 11-2c, may help to complete a physical picture of what is happening in the machine when one applies the symmetrical-component transformations of Eqs. 11-31 and 11-32. In Fig. 11-18a, I_m and I_a are rotating time phasors whose projections on the real axis are proportional to the instantaneous currents in the windings. Figure 11-18b is a space-phasor diagram in which the dash-dot lines m and a represent the winding axes and the phasors F_m and F_a represent the instantaneous values of the pulsating mmf wave produced by each winding. For simplicity, the phasors are shown in their positions at the moment when i_m, and therefore F_m, has its instantaneous maximum $F_{m(max)}$ Consequently, when F_m is resolved into its forward and backward half-amplitude revolving components (in the manner of Fig. 11-2c

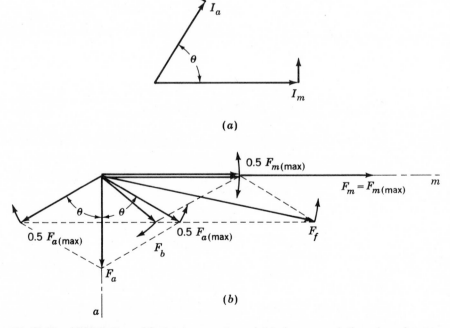

(a)

(b)

Fig. 11-18.   (a) Unbalanced 2-phase currents and (b) phasor resolution of mmf waves into forward and backward components.

and Eq. 11-4), the components at this instant are in line with the axis m, as shown by the two oppositely revolving components $0.5F_{m(\text{max})}$. At this same instant, however, i_a has passed beyond its maximum value by the time angle θ (Fig. 11-18a) and therefore the forward- and backward-revolving components $0.5F_{a(\text{max})}$ are in the two positions shown in Fig. 11-18b. The space phasors representing the resultant forward and backward fields F_f and F_b are the phasor sums of the components, as in Fig. 11-18b. Because of the space angle between the 2 windings, the angle between the two revolving-field components $0.5F_{m(\text{max})}$ and $0.5F_{a(\text{max})}$ for the forward field is 90° less than the time-phase angle θ between the currents and for the backward field is 90° greater. Consequently, the phasor summations by which the mmfs F_f and F_b are obtained in Fig. 11-18b are exactly similar to those of Eqs. 11-31 and 11-32 for obtaining the symmetrical-component currents. Thus, when the currents are resolved into symmetrical components, the stator mmf is thereby resolved into forward and backward components.

EXAMPLE 11-4

The equivalent-circuit constants of a 5-hp 220-volt 60-Hz 2-phase squirrel-cage induction motor are given below, in ohms per phase:

$$r_1 = 0.534 \qquad x_1 = 2.45 \qquad x_\varphi = 70.1$$
$$r_2 = 0.956 \qquad x_2 = 2.96$$

This motor is operated from an unbalanced 2-phase source whose phase voltages are, respectively, 230 volts and 210 volts, the smaller voltage leading the larger by 80°. For a slip of 0.05, find:

a. The positive- and negative-sequence components of the applied voltages

b. The positive- and negative-sequence components of the stator phase currents

c. The effective values of the phase currents

d. The internal mechanical power

Solution

a. Let V_m and V_a denote the voltages applied to the 2 phases, respectively. Then

$$V_m = 230\underline{/0°} = 230 + j0 \text{ volts}$$
$$V_a = 210\underline{/80°} = 36.4 + j207 \text{ volts}$$

From Eqs. 11-29 and 11-30, the forward and backward components of voltages are, respectively,

$$V_{mf} = \tfrac{1}{2}(230 + j0 + 207 - j36.4)$$
$$= 218.5 - j18.2 = 219.5\underline{/-4.8°} \text{ volts}$$
$$V_{mb} = \tfrac{1}{2}(230 + j0 - 207 + j36.4)$$
$$= 11.5 + j18.2 = 21.5\underline{/57.7°} \text{ volts}$$

b. From Eqs. 11-18 and 11-19, the forward-field impedance is, for a slip of 0.05,

$$Z_f = 16.46 + j7.15 \text{ ohms}$$
$$r_1 + jx_1 = \underline{0.53 + j2.45} \text{ ohms}$$
$$16.99 + j9.60 = 19.50\underline{/29.4°} \text{ ohms}$$

Hence, the forward component of stator current is

$$I_{mf} = \frac{219.5\underline{/-4.8°}}{19.50\underline{/29.4°}} = 11.26\underline{/-34.2°} \text{ amp}$$

For the same slip, from Eqs. 11-22 and 11-23 the backward-field impedance is

$$Z_b = 0.451 + j2.84 \text{ ohms}$$
$$r_1 + x_1 = \underline{0.534 + j2.45} \text{ ohms}$$
$$0.985 + j5.29 = 5.38\underline{/79.5°} \text{ ohms}$$

Hence, the backward component of stator current is

$$I_{mb} = \frac{21.5\underline{/57.7°}}{5.38\underline{/79.5°}} = 4.00\underline{/-21.8°} \text{ amp}$$

c. By Eqs. 11-27 and 11-28, the currents in the two phases are, respectively,

$$I_m = 13.06 - j7.79 = 15.2\underline{/-31°} \text{ amp}$$
$$I_a = 4.81 + j5.64 = 7.40\underline{/49.2°} \text{ amp}$$

Note that the currents are much more unbalanced than the applied voltages. Even though the motor is not overloaded insofar as shaft load is concerned, the losses are appreciably increased by the current unbalance and the stator winding with the greatest current may overheat.

d. The power delivered to the forward field by the 2 stator phases is

$$P_{\text{gf}} = 2I_{\text{mf}}^2 R_f = 2 \times 126.8 \times 16.46 = 4{,}175 \text{ watts}$$

and the power delivered to the backward field is

$$P_{\text{gb}} = 2I_{\text{mb}}^2 R_b = 2 \times 16.0 \times 0.451 = 15 \text{ watts}$$

Thus, according to Eq. 11-17, the internal mechanical power developed is

$$P = 0.95(4{,}175 - 15) = 3{,}950 \text{ watts}$$

If the core losses, friction and windage, and stray load losses are known, the shaft output can be found by subtracting them from the internal power. The friction and windage losses depend solely on the speed and are the same as they would be for balanced operation at the same speed. The core and stray load losses, however, are somewhat greater than they would be for balanced operation with the same positive-sequence voltage and current. The increase is caused principally by the $(2 - s)$-frequency core and stray losses in the rotor caused by the backward field.

11-6
TWO–PHASE CONTROL MOTORS

Control systems utilize fractional-horsepower size electromagnetic components as motors to drive the loads and as sensors to measure speed and position of the controlled elements. In this article we will describe the operation of a control motor, which operates as an unbalanced 2-phase induction motor, and is suitable for control systems up to a few hundred watts. In Art. 11-7, we will describe the operation of stepper motors which are suitable for digital control systems. In Arts. 11-8 and 11-9 we will describe the operation of speed and position sensors. More information on these and other control-system components can be found in the technical literature.

A schematic diagram of a 2-phase control motor is given in Fig. 11-19. Phase m of the motor is the *fixed*, or *reference*, *phase*. The voltage V_m is a

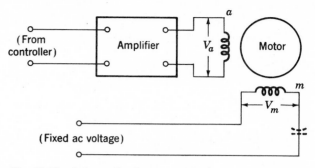

Fig. 11-19. Schematic diagram of 2-phase control motor.

fixed voltage applied from a constant-voltage constant-frequency source. Phase a is the *control phase*. The voltage V_a is supplied from an amplifier, usually of magnetic-amplifier or solid-state construction, and has an amplitude proportional to the input control signal. The voltages V_m and V_a must be in synchronism, which means that they must be derived from the same ultimate ac source. They must also be made to be approximately in time quadrature either by introducing a 90° phase shift in the amplifier or by connecting a suitable capacitor in series with the reference phase m. The amplifier derives its power from the same ac source that supplies the reference phase, so that the amplifier output voltage V_a is a modulated ac wave having its fundamental component at the same frequency as V_m. When V_a has a nonzero value and its phase leads V_m by approximately 90°, rotation in one direction is obtained; when V_a lags V_m, rotation in the other direction results. Since the torque is a function of both V_m and V_a, changing the magnitude of V_a changes the developed torque of the motor.

Typical torque-speed curves for a 2-phase control motor are given in Fig. 11-20 for a series of values of control voltage and unity reference voltage. They are based on a motor having identical 2-phase stator windings and on negligible source impedances.

The requirements of a control system very nearly specify the shape of the torque-speed characteristic of a 2-phase induction motor suitable for such use. As with any motor for such service, the torque should be high at speeds near zero, and the slope of the torque-speed characteristic should be negative in the normal operating range around zero speed in order to provide a stabilizing feature for the control system. Both these requirements can be met by use of a high-resistance rotor designed so that maximum torque is developed at a reverse speed of approximately one-half synchronous speed, as shown by the torque-speed characteristic labeled

$V_a = 1.0$ in Fig. 11-20.   Normal operation near zero speed is then in the stable region to the right of the maximum-torque points.   A further requirement is that the motor must not tend to run as a single-phase motor when the error signal is zero.   It can be shown that this requirement also is met by use of a high-resistance rotor.

Although the squirrel-cage induction motor prefers to run at a small slip and therefore is most readily adaptable to constant-speed drives, nevertheless this type of motor has other features which are sufficiently attractive so that it is used extensively in low-power control systems. The ruggedness and simplicity of the cage rotor are a great advantage, for both economic and technical reasons.   There are no brushes riding on sliding contacts and requiring inspection and maintenance as in other types of motors.   Because the rotor windings do not require insulation, the rotor temperature is limited only by mechanical considerations and indirectly by its effect on the stator-winding temperature.   If suitable means are provided for cooling the stator windings, higher rotor losses can be tolerated than in other types of motors.   Because there is relatively little inactive material, the inertia of a squirrel-cage motor can be made less than that of a correspondingly rated dc motor.   When the maximum power output is below a few watts, inertia can be minimized by using a thin metallic cup as the rotor.   As shown in the simplified sketch of Fig. 11-21, the rotating member is then like a can with one end removed.   A

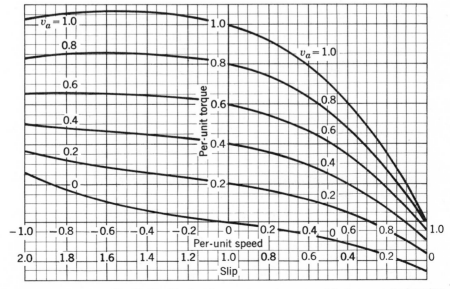

Fig. 11-20. Typical torque-speed curves of 2-phase control motor.

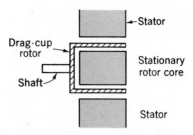

Fig. 11-21. Cross section of drag-cup rotor.

stationary iron core, like a plug inside the cup, completes the magnetic circuit. The construction is known as a *drag-cup rotor*.

The principal disadvantage of the 2-phase control motor is the inherent inefficiency of a squirrel-cage induction motor running at a large slip. As pointed out in Art. 7-3, the efficiency of a polyphase induction motor with short-circuited rotor windings is like that of a slipping mechanical clutch; the slip is a direct measure of the rotor losses. The inherent limit on efficiency will be even lower in a speed control scheme which involves unbalancing the applied voltages, because of the decrease in output and increase in rotor losses.

EXAMPLE 11-5

It may be noted from Fig. 11-20 that, in the low-speed range, the torque of a 2-phase control motor is linearly proportional to the rms control-winding voltage. In this range, the torque-speed curves have a negative slope which is approximately constant and substantially the same for the lower values of control voltage. Under these conditions, the torque-speed curves may be represented empirically by the relation

$$T = k_1 V_a - k_2 \omega_o$$

where ω_o is the shaft mechanical angular velocity and k_1 and k_2 are constants determined from the curves and the units involved.

The motor is driving a load requiring a viscous-friction torque $f_v \omega_o$ proportional to speed. The combined inertia of rotor and load is J. Determine the transfer functions relating:

a. Motor shaft position and control voltage

b. Motor shaft velocity and control voltage

Solution

a. The signal to the motor consists of variation of the rms value V_a
of the control-winding voltage. Let the value of this voltage at
any time t be v_s. The electrodynamic differential equation in
terms of the shaft position angle θ_o is then

$$J \frac{d^2\theta_o}{dt^2} + f_v \frac{d\theta_o}{dt} = k_1 v_s - k_2 \frac{d\theta_o}{dt}$$

In terms of the phasor amplitudes Θ_o and V_s of sinusoidal
variations at the angular frequency ω, this equation becomes

$$(j\omega)^2 J\Theta_o + j\omega(f_v + k_2)\Theta_o = k_1 V_s$$

which yields the transfer function

$$\frac{\Theta_o}{V_s} = \frac{k_1}{(j\omega)^2 J + j\omega(f_v + k_2)}$$

b. The differential equation in terms of shaft angular velocity ω_o is

$$J \frac{d\omega_o}{dt} + f_v \omega_o = k_1 v_s - k_2 \omega_o$$

Then $\qquad j\omega J\Omega_o + (f_v + k_2)\Omega_o = k_1 V_s$

or the transfer function is

$$\frac{\Omega_o}{V_s} = \frac{k_1}{f_v + k_2 + (j\omega)J}$$

The characteristics of the 2-phase control motor can be determined
by test on a dynamometer capable of driving it in the slip range from
1.0 to 2.0 and loading it from slip of 1.0 to zero. The characteristics
can also be calculated using the symmetrical-component method of Art.
11-5. The analysis is greatly simplified if the currents and voltages are
considered to be sinusoidal, if the effects of the source impedances are
neglected, and if the motor is assumed to have identical 2-phase stator
windings. The motor then is simply a symmetrical 2-phase motor operating from an unbalanced 2-phase source.

EXAMPLE 11-6

A symmetrical 2-phase induction motor develops a maximum internal torque at a reverse speed of 0.50 per unit when balanced 2-phase voltages are applied to its stator terminals.   The Q of its Thévenin equivalent circuit (see Art. 7-5) is 3.0.   This motor is to be used as a 2-phase control motor with constant voltage of 1.00 per unit applied to its reference phase and variable voltage applied to its control phase, these voltages being considered to be in time quadrature.

Plot a family of internal-torque-speed characteristics for per-unit values of the control-phase voltage of 1.00, 0.80, 0.60, 0.40, 0.20, and 0, covering a speed range from -1 to $+1$ per unit.   Express the torque in per unit, considering the unit of torque to be the internal torque developed at standstill when balanced 2-phase voltages of 1.00 per unit are applied to the two stator phases.

The torque-slip curves obtained in this example are those plotted in Fig. 11-20.

Solution

The speed for maximum torque, the Q, and the stalled torque for balanced 2-phase voltages fix the curve for $V_a = 1.00$.   This curve can readily be determined from the normalized torque-slip curves of Fig. 7-13.   The rest of the family then can be computed from this curve by resolving the applied voltages into 2-phase symmetrical components.

Since the speed for maximum torque is given as -0.50 per unit, the slip for maximum torque $s_{max\ T} = 1.50$.   At standstill, then,

$$\frac{s}{s_{max\ T}} = \frac{1.00}{1.50} = 0.667$$

From Fig. 7-13 for $Q = 3.0$, the corresponding torque ratio at standstill is

$$\frac{T_{stalled}}{T_{max}} = 0.938$$

But $T_{stalled} = 1.00$ per unit, by definition.   Hence, T_{max} is $1/0.938$, or 1.066 per unit.

Data for the torque-speed curve for $V_a = 1.00$ in Fig. 11-20 can now be obtained from Fig. 7-13.   The data are shown in Table 11-1. The first column gives the slip ratios, and the corresponding torque ratios are read from the curve for $Q = 3.0$ in Fig. 7-13. The actual slip s in column 3 is found by multiplying column 1 by $s_{\max\,T} = 1.50$. The corresponding torque T'_f in column 4 is found by multiplying column 2 by $T_{\max} = 1.066$. For balanced conditions $(V_a = 1.00)$ there is no backward torque, and column 4 gives the net torque from which the curve labeled $V_a = 1.0$ in Fig. 11-20 is plotted.

TABLE 11-1
COMPUTATIONS FOR EXAMPLE 11-6

$\dfrac{s}{s_{\max}\,T}$	$\dfrac{T}{T_{\max}}$	s	T_f'	$2 - s$	T_b'
0	0	0	0	2.0	1.03
0.133	0.32	0.2	0.34	1.8	1.055
0.267	0.565	0.4	0.60	1.6	1.06
0.40	0.745	0.6	0.795	1.4	1.06
0.533	0.86	0.8	0.92	1.2	1.045
0.667	0.94	1.0	1.00	1.0	1.00
0.80	0.98	1.2	1.045	0.8	0.92
0.933	0.995	1.4	1.06	0.6	0.795
1.067	0.995	1.6	1.06	0.4	0.60
1.20	0.99	1.8	1.055	0.2	0.34
1.33	0.965	2.0	1.03	0	0

When the voltages are unbalanced, they can be resolved into symmetrical components. Let the per-unit magnitude of the control-phase voltage be V_a, and assume that this voltage leads the reference voltage V_m by 90°. The phasor expression for the control-phase voltage then is jV_a, and Eqs. 11-29 and 11-30 reduce to

$$V_{mf} = \tfrac{1}{2}[1 - j(jV_a)] = \tfrac{1}{2}(1 + V_a) \tag{11-33}$$

$$V_{mb} = \tfrac{1}{2}[1 + j(jV_a)] = \tfrac{1}{2}(1 - V_a) \tag{11-34}$$

Both forward and backward fields are now present. The slip for the backward field is $2 - s$, as in column 5 of Table 11-1. Column 6 gives the values of backward torque T'_b that would be developed with negative-sequence voltages of 1.00 per unit and is obtained from column 4. (For example, the value of T'_b at a backward-field slip $2 - s = 1.8$ is the same as the value of T'_f at a forward-field slip $s = 1.8$.)

Now recall that the internal torque developed by a polyphase induction motor varies as the square of the voltage. The forward and backward torques therefore are

$$T_f = V_{mf}^2 T_f' \tag{11-35}$$

$$T_b = V_{mb}^2 T_b' \tag{11-36}$$

where V_{mf}, V_{mb} are the per-unit values of the positive- and negative-sequence components of the unbalanced applied voltages (Eqs. 11-33 and 11-34) and T_f', T_b' are the forward and backward torque corresponding to positive- and negative-sequence applied voltages, respectively, of 1.00 per unit, as given in columns 4 and 6. The net internal torque T is

$$T = T_f - T_b \tag{11-37}$$

The torque developed at any chosen values of V_a and slip can now be determined. For example, from Eqs. 11-33 to 11-36,

At $V_a = 0.60$:

$$V_{mf} = 0.80 \qquad T_f = 0.64T_f'$$

$$V_{mb} = 0.20 \qquad T_b = 0.04T_b'$$

Values of T_f' and T_b' can be read from Table 11-1. For example,

At $s = 0.20$:

$$T_f' = 0.34 \qquad T_f = (0.64)(0.34) \quad = 0.218$$
$$T_b' = 1.055 \qquad T_b = (0.04)(1.055) = 0.042$$
$$T = T_f - T_b \qquad = 0.176$$

Data for the family of curves in Fig. 11-20 can be computed by repeating these simple calculations for other assumed values of V_a and slip. The calculations can be arranged systematically in tabular form.

The nondimensional curves of Fig. 11-20 are approximately applicable to all 2-phase control motors since nearly all of them are designed to develop maximum torque at about the same per-unit speed and to have about the same Q. The slip $s_{max\ T}$ at maximum torque and the Q fix the shape of the characteristics, so long as the rotor resistance is constant and the effects of saturation are negligible. Fairly wide variations in the parameters $s_{max\ T}$ and Q have relatively little effect on the characteristics over the normal operating range.

11-7
STEPPER MOTORS

The stepper motor is a form of synchronous motor which is designed to rotate a specific number of degrees for each electrical pulse received by its control unit. Typical steps are 7.5° and 15° per pulse. The stepper motor is used in digital control systems, where the motor receives open-loop commands as a train of pulses to turn a shaft or move a plate by a specific distance. A typical application for the motor is positioning a work table in two dimensions for automatic drilling in accordance with hole-location instructions on tape. With a stepper motor a position sensor and feedback system is not normally required to make the output member follow input instructions. Stepper motors are built to follow signals as rapid as 1,200 pulses per second and with equivalent power ratings up to several horsepower.

Stepper motors are usually designed with a multipole, multiphase stator winding that is not unlike the windings of conventional machines. They typically use three-phase and four-phase windings with the number of poles determined by the required angular change per input pulse. The rotors are either of the variable-reluctance type or the permanent-magnet type. Stepper motors operate with an external drive logic circuit; as a train of pulses is applied to the input of the drive circuit, the circuit delivers appropriate currents to the stator windings of the motor to make the axis of the air-gap field step around in coincidence with the input pulses. Depending upon the pulse rate and the load torque, including inertia effects, the rotor follows the axis of the air-gap magnetic field by virtue of the reluctance torque and/or the permanent-magnet torque.

The elementary operation of a four-phase stepper motor with a two-pole rotor is shown in the sequence of Fig. 11-22. The rotor can be either a ferromagnetic element or a permanent magnet. The rotor assumes the angles $\theta = 0°, 45°, 90°, \ldots$, as the windings are excited in the sequence $N_a, N_a + N_b, N_b, \ldots$ The stepper motor of Fig. 11-22 can also be used for 90° steps by exciting the coils singly. In the latter case, only the permanent-magnet rotor can be used. The torque-angle curves for the two types of rotors are shown in Fig. 11-23; whereas the permanent-magnet rotor has its peak torque when the excitation is shifted 90°, the ferromagnetic rotor has zero torque and can move in either direction. The permanent-magnet rotor has the additional feature that the rotor position θ is defined by the winding currents with no ambiguity, whereas the ferromagnetic rotor has two possible positions for each winding-current pattern. Winding patterns can be visualized for steps of 22.5°, 11.25°, and smaller, per pulse to the input circuit.

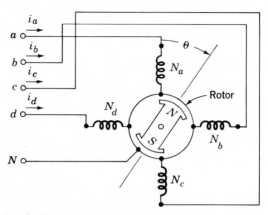

Fig. 11-22. Elementary diagram of 4-phase stepper motor.

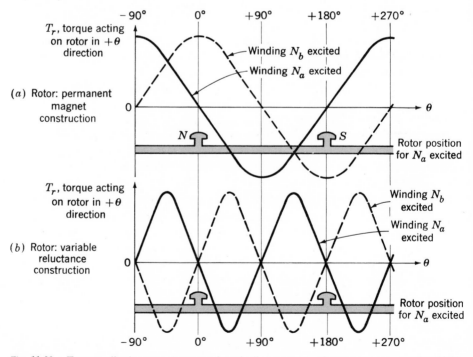

Fig. 11-23. Torque-displacement curves for winding pattern of stepper motor of Fig. 11-22: (a) Permanent-magnet rotor; (b) variable-reluctance rotor.

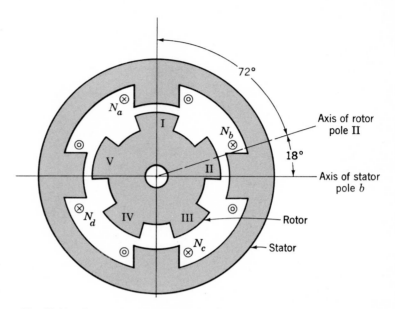

Fig. 11-24. Cross section of stepper motor for differential operation.

To obtain relatively small steps of angle, a differential construction as shown in Fig. 11-24 can be used. The stator has a 2-phase winding, while the rotor has five projecting poles. The position shown is for current in winding N_a. If the current is now transferred to winding N_b, the rotor will turn by $\theta = 90° - 72° = 18°$ to align pole 2 with the N_b axis. The rotor can be ferromagnetic or can utilize permanent-magnet construction. The permanent magnet is usually placed in the axial position as shown in Fig. 11-25 with ferromagnetic end plates to form the poles. Such construction tends to reduce the reluctance torque, and makes the

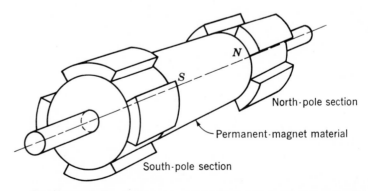

Fig. 11-25. Rotor of stepper motor with axial permanent magnet.

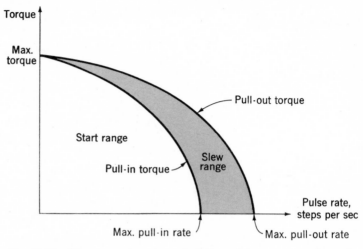

Fig. 11-26. Stepper-motor characteristics of torque vs. pulse rate.

motor smoothly responsive to the winding currents. It also provides a uniform magnetic loading on the permanent-magnet material.[1]

The characteristics of a stepper motor are frequently presented as the torque vs. stepping rate of the pulses applied to the drive unit, as shown in Fig. 11-26. As the stepping rate is increased the motor can provide less torque because the rotor has less time to drive the load from one position to the next as the stator-winding-current pattern is shifted. The start range of Fig. 11-26 is that in which the load position follows the pulses without losing steps. The slew range is that in which the load velocity follows the pulse rate without losing steps, but cannot start, stop, or reverse on command. The maximum torque point is the maximum holding torque of the excited motor to a steady load. At light loads, the maximum slew rate can be as much as ten times the position response rate.

The advantage of the stepper motor is the smaller size and lower cost of the motor-drive unit package compared to the corresponding parts of a proportional position or velocity servosystem. Typical variable-reluctance stepper motors operate at small steps, 15° or less, or at maximum position response rates up to 1,200 pps. Typical permanent-magnet types operate at larger steps, up to 90°, and at maximum response rates of 300 pps. Applications include table positioning for machine tools, tape drives, recorder pen drives, and X-Y plotters.[2]

[1] See A. E. Snowden and E. W. Madsen, Characteristics of a Synchronous Inductor Motor, *IEEE Trans.*, Part II, Applications and Industry, **81**:1–5 (1962).
[2] For a description of various types of stepper motors see the article Stepper Motor and Controls, *Electromechanical Design*, 107–119 (1969). An extensive bibliography is given by D. J. Robinson and C. K. Taft, A Dynamic Analysis of Magnetic Stepping Motors, *IEEE Trans. on Industrial Electronics and Control Instrumentation*, **IECI-16**(2):111–125 (1969).

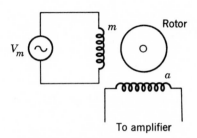

Fig. 11-27. Schematic diagram of a 2-phase tachometer.

11-8
AC TACHOMETERS

For feedback-control systems, it is frequently necessary to measure the angular velocity of a shaft, and it is often desirable that this measure be in the form of an alternating voltage of constant frequency. A small 2-phase induction motor may be used for this purpose.   The connections are shown in Fig. 11-27.   Winding m, often referred to as the *fixed field*, or *reference field*, is energized from a suitable alternating voltage of constant magnitude and frequency.   A voltage of the same frequency is then generated in the auxiliary winding, or *control field a*.   This voltage is applied to the high-impedance input circuit of an amplifier, and therefore winding a can be considered as open-circuited.   The electrical requirements are, ideally, that the magnitude of the signal voltage generated in winding a should be linearly proportional to the speed and that the phase of this voltage should be fixed with respect to the applied voltage V_m.

The operation of the ac tachometer may be visualized in terms of the double-revolving-field theory of Art. 11-4.[1]   As viewed from the reference winding m, the tachometer is equivalent to a small single-phase induction motor, and the equivalent circuit of Fig. 11-14c therefore applies to conditions as viewed from this winding.   The voltages across the impedances $0.5Z_f$ and $0.5Z_b$ in Fig. 11-14c are the voltages generated in winding m by the forward and backward flux waves, respectively.   These flux waves also generate voltages in the auxiliary winding a.   If the ratio of effective turns in winding a to effective turns in winding m is a, then the voltages generated in winding a are a times the corresponding voltages generated in winding m.   By effective turns is meant the number of turns corrected for the effects of winding distribution insofar as fundamental space distributions of flux and mmf are concerned.   If the direction of rotation is

[1] For a quantitative analysis, see R. H. Frazier, Analysis of the Drag-cup AC Tachometer, *Trans. AIEE*, **70** (1951).

such that the forward field revolves past winding a a quarter cycle in time before it passes winding m, then the voltage E_{af} generated by the forward field in winding a leads the corresponding voltage E_{mf} generated in winding m by 90°, or, as phasors,

$$E_{af} = jaE_{mf} = jaI_m0.5Z_f \qquad (11\text{-}38)$$

where I_m is the phasor current in winding m and is determined by the equivalent circuit of Fig. 11-14c. The backward field revolves in the opposite direction, and therefore the voltage E_{ab} generated by it in winding a lags the corresponding voltage E_{mb} generated in winding m by 90°, or

$$E_{ab} = -jaE_{mb} = -jaI_m0.5Z_b \qquad (11\text{-}39)$$

The total voltage E_a generated in winding a is the sum of the components generated by each field, or

$$E_a = jaI_m0.5(Z_f - Z_b) \qquad (11\text{-}40)$$

At standstill, the forward and backward fields are equal, and no voltage is generated in winding a. When the rotor is revolving, however, the impedance of the forward field increases while that of the backward field decreases, the difference between them being a function of the speed. The voltage generated in winding a is therefore a function of speed. Reversal of the direction of rotation reverses the phase of the auxiliary-winding voltage.

The shapes of the curves of voltage magnitude and phase angle as functions of speed depend on the speed range and tachometer constants— primarily on the rotor self-reactance-to-resistance ratio Q_2. It can be shown that either a low-Q_2 rotor (x_{22}/r_2 less than about 0.1) or a high-Q_2 rotor (x_{22}/r_2 greater than about 10) will provide nearly a constant phase angle and nearly a linear relation between the auxiliary-winding voltage and speed. The sensitivity in volts per rpm is sacrificed if a low-Q_2 rotor is used, but the linear speed range is wide. On the other hand, if a high-Q_2 rotor is used, the speed range around zero speed is limited to a fairly small fraction of synchronous speed when the requirements for linearity of voltage and constancy of phase angle are strict. These restrictions on rotor Q_2 should not be taken too literally, however, since satisfactory performance may be obtained with intermediate values of Q_2 if the requirements for linearity of voltage and constancy of phase angle are not too severe.

In common with other measuring instruments, the ac tachometer should have as little effect as possible on the system into which it is inserted. In other words, its torque should be small compared with other torques acting in the system, and its inertia should be small when rapid speed variations are encountered, as in automatic control systems. To minimize the inertia, ac tachometers are often built with a thin, metallic drag-cup rotor like that shown in the simplified sketch of Fig. 11-21. Because of the relatively long air gap, this construction inherently gives a fairly low Q_2, which can be made still lower, if desired, by making the drag cup of high-resistivity material.

Alternating-current tachometers require precise workmanship and care in design and assembly in order to maintain concentricity and to eliminate direct coupling through the leakage fluxes between the excited winding and the output winding. Such coupling would result in signal voltage at zero speed. Sometimes soft-iron shields are provided to minimize pickup from stray fields. Frequently ac tachometers are used in 400-Hz systems.

11-9

SYNCHROS AND CONTROL TRANSFORMERS

Synchros are used in control systems for transmitting shaft-position information, for maintaining synchronism between two or more shafts, and for performing arithmetic operations with angular information.   The designation *selsyn* is also used from the combination "self-synchronous." A control transformer is a particular type of synchro that provides an electrical error signal proportional to the deviation of its shaft from the synchronous position; it is widely used in position feedback-control systems.   Synchros are generally constructed like miniature versions of ac synchronous machines; they have a 3-phase distributed stator winding and a 2-pole wound single-phase rotor winding.

A basic arrangement of two synchros in which the two shafts maintain synchronism is shown in Fig. 11-28. In most respects, the construction of the *synchro generator*, or *transmitter*, is similar to that of the *synchro motor*, or *receiver*. Both have a single-phase winding (usually on the rotor) connected to a common ac voltage source. On the other member (usually the stator), both have 3 windings with axes 120° apart and connected in Y; these windings on the transmitter and motor have their corresponding terminals connected together. When the single-phase rotor windings are excited, voltages are induced by transformer action in the Y-connected stator windings. If the two rotors are in the

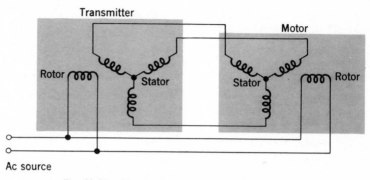

Fig. 11-28.   Single-phase synchrotransmitter-motor system.

same space position relative to their stator windings, the transmitter and
motor stator-winding voltages are equal, no current circulates in these
windings, and no torque is transmitted.   If, however, the two rotor space
positions do not correspond, the stator-winding voltages are unequal and
currents circulate in the stator windings.   These currents, in conjunction
with the air-gap magnetic fields, produce torques tending to align the two
rotors.

Mechanically, synchros have the same general construction features
as small motors.   The rotor structure of a synchro motor may be seen in
Fig. 11-29.   The rotor and stator are laminated, and ball bearings are
used to minimize friction.   Dampers are used to force the motors to
settle quickly at their angular positions.

The motor torque at standstill or for slow rotation may be shown to
depend closely upon the sine of the relative angular difference in position
of the transmitter and motor shafts.   Torque gradients developed by
electrically identical transmitters and motors interconnected as in Fig.
11-28 range from 0.07 in.-oz per degree for the smaller units to 1.75 in.-oz
per degree for the larger units.

A modification of the synchro system of Fig. 11-28 may be introduced
by including a *differential synchro*, thereby permitting the rotation of a
shaft to be a function of the sum or difference of the rotation of two other

Fig. 11-29.   Two dampers (front and rear) and wound rotor, with damper, for
synchro motor.   *(General Electric Company.)*

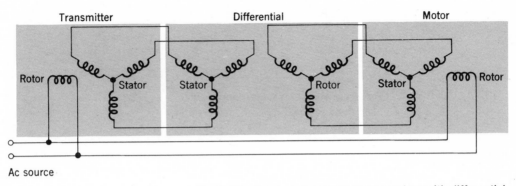

Fig. 11-30.   Synchro generator-motor system with differential.

shafts.   In Fig. 11-30, the differential synchro acts as a differential transmitter.   The voltages impressed on its stator windings induce corresponding voltages in the rotor windings.   The relative magnitudes of the
three rotor voltages are the same as would exist if the differential were
removed and the transmitter turned through an angle equal to the sum or
difference of the transmitter and differential angles.   Such differential
transmitters usually have a bank of three capacitors connected across the
primary terminals to improve power factor and hence minimize the possibility of overheating in the system.   Alternatively, the differential may be
used as a motor supplied from two separate synchro transmitters and producing a rotation dependent upon the sum or difference of the two
transmitter rotations.

The maximum static error for a system consisting of a transmitter
and a single motor of the same size is of the order of 1° and is caused
largely by friction in the motor bearings.   The error increases as additional synchros are added or as the line impedance between transmitter
and motor becomes appreciable.   Dynamic errors, created by mechanical
oscillation of the motor shaft about the correct position, may be two or
three times the static errors.   To minimize dynamic errors, mechanical
dampers are built into the rotors of motor units as shown in Fig. 11-29.
When the motor is called upon to supply significant torque, the error
increases because of the need of a definite angular displacement between
transmitter and motor shafts for torque transmission.   This fact, together
with heating of the synchro equipment, definitely limits the torque magnitudes and makes the use of a feedback-control system operated from
a control transformer more desirable.

The basic circuit by which a control transformer produces an error
signal proportional to rotor-angle deviation is shown in Fig. 11-31.   Two
interconnected units are again involved, one a transmitter and the other

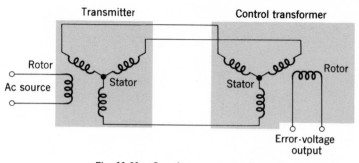

Fig. 11-31. Synchro generator-transformer system.

a very similar unit called *control transformer*. The rotor of the synchro
transmitter is excited from a single-phase source, producing a magnetic
field in the transmitter and voltages in the stator windings of both the
transmitter and the control transformer. If the voltage drops caused
by exciting current are neglected, the induced voltages in the 2 stator
windings must be equal. Therefore, the distribution of flux about the
control-transformer stator must be similar to that about the transmitter
stator. The effect is consequently the same as if the 2 rotor windings
were on the same magnetic circuit and arranged so that their axes could
be given any arbitrary displacement angle in space. The arrangement
is thus the equivalent of an adjustable mutual inductance between the
2 rotor windings, but with the added feature that geographical separation
of the 2 windings is possible. When the angle is 90 electrical degrees,
corresponding to a 90-electrical-degree displacement of the two shafts, no
voltage is induced in the transformer rotor; this displacement is the
equilibrium position of the two shafts. When the angle has any value
except 90° and 270°, a voltage is induced in the transformer rotor. The
magnitude of the voltage is a function of the angular discrepancy between
the two shafts, and the instantaneous polarity depends on the direction of
the displacement. In an actual control system, the control transformer
operates with an error angle of a fraction of a degree. In this range the
error voltage is proportional to error angle. Differential synchros may
also be incorporated between the transmitters and control transformers
of these systems.

    Figure 11-32, for example, illustrates an application to control of the
angular position of an output shaft in accordance with an input shaft.
The rotor of the synchro transmitter is mechanically connected to the
input shaft. The rotor of the control transformer is mechanically connected to the output shaft and electrically connected to the input of a
servoamplifier. Mechanical power to turn the output shaft and its

associated load is furnished by a 2-phase control motor. The input to the control winding of the motor is supplied by the amplifier, which includes phase-splitting capacitors in its circuitry. When the output shaft is in the correct position, 90° from the input shaft position, the voltage input to the amplifier, and hence the power input to the control winding of the motor, is zero, and the motor does not turn. When an angular discrepancy exists, a definite error voltage appears at the amplifier input. Its relative polarity is such that the motor is caused to turn in the direction to correct the angular discrepancy.

11-10
RÉSUMÉ

The main theme of this chapter is a continuation of the induction-machine theory of Chap. 7. This theory is expanded by a step-by-step reasoning process from the simple revolving-field theory of the symmetrical polyphase induction motor. The basic concept is the resolution of the stator-mmf wave into two constant-amplitude traveling waves revolving around the air gap at synchronous speed in opposite directions. If the slip for the forward field is s, then that for the backward field is $2 - s$. Each of these component fields produces induction-motor action,

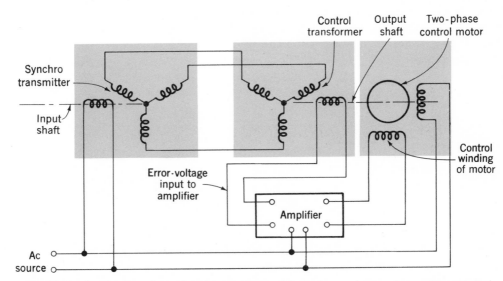

Fig. 11-32. Application of synchro transmitter and transformer to angular position control.

just as in a symmetrical polyphase motor. From the viewpoint of the stator, the reflected effects of the rotor can be visualized and expressed quantitatively in terms of simple equivalent circuits. The ease with which the internal reactions can be accounted for in this manner is the essential reason for the usefulness of the double-revolving-field theory.

For a single-phase winding, the forward- and backward-component mmf waves are equal, and their amplitude is half the maximum value of the peak of the stationary pulsating mmf produced by the winding. The resolution of the stator mmf into its forward and backward components then leads to the physical concept of the single-phase motor described in Art. 11-1 and finally to the quantitative theory developed in Art. 11-4 and to the equivalent circuits of Fig. 11-14.

The next step is investigation of the possibilities of applying the double-revolving-field resolution to a symmetrical 2-phase motor with unbalanced applied voltages, as in Art. 11-5. This investigation leads to the symmetrical-component concept, whereby an unbalanced 2-phase system of currents or voltages can be resolved into the sum of two balanced 2-phase component systems of opposite phase sequence. Resolution of the currents into symmetrical-component systems is equivalent to resolving the stator-mmf wave into its forward and backward components, and therefore the internal reactions of the rotor for each symmetrical-component system are the same as those which we have already investigated. A very similar reasoning process, not considered here, leads to the well-known 3-phase symmetrical-component method for treating problems involving unbalanced operation of 3-phase rotating machines. The ease with which the rotating machine can be analyzed in terms of revolving-field theory is the chief reason for the usefulness of the symmetrical-component method.

PROBLEMS

11-1. What type of motor would you use in the following applications? Give reasons. Vacuum cleaner. Refrigerator. Washing machine. Domestic oil burner. Desk fan. Sewing machine. Bench grinder. Clock. Food mixer. Record player. Portable electric drill. Water pump.

11-2. At standstill, the currents in the main and auxiliary windings of a capacitor-start induction motor are $I_m = 14.14$ amp and $I_a = 7.07$ amp. The auxiliary-winding current leads the main-winding current by $60°$. The effective turns per pole—i.e., the number of turns corrected for the effects of winding distribution—are $N_m = 80$ and $N_a = 100$. The windings are in space quadrature.

Determine the amplitudes of the forward and backward stator-mmf waves.

Suppose it were possible to adjust the magnitude and phase of the auxiliary-winding current. What should be its magnitude and phase to produce a pure forward mmf wave?

11-3. Find the mechanical power output of the ¼-hp 4-pole 110-volt 60-Hz single-phase induction motor, whose constants are given below, at a slip of 0.05:

$$r_{1m} = 1.86 \text{ ohms} \qquad x_{1m} = 2.56 \text{ ohms} \qquad x_\varphi = 53.5 \text{ ohms}$$

$$r_2 = 3.56 \text{ ohms} \qquad x_2 = 2.56 \text{ ohms}$$

Core loss = 35 watts   Friction and windage = 13.5 watts

11-4. For the single-phase induction motor of Example 11-3 running at a slip of 0.05, determine the ratio of the backward flux wave to the forward flux wave. Plot a half wave of the resultant flux distribution for instants of time corresponding to $\omega t = 0, 45°, 90°, 135°,$ and $180°$, zero time being chosen as the instant when the forward and backward flux waves are in space phase. If the forward and backward flux waves are represented by space phasors like the mmf phasors f and b in Fig. 11-2c, draw a diagram showing the components and the resultant for the same five instants of time. Sketch the locus of the tip of the phasor representing the resultant air-gap flux wave. What kind of curve do you think this locus is?

11-5. Derive an expression in terms of Q_2 for the nonzero speed of a single-phase induction motor at which the internal torque is zero (see Eq. 11-21).

11-6. A small 2-phase 2-pole induction motor has the following constants at 60 Hz:

$$r_{1m} = 375 \text{ ohms} \qquad r_2 = 255 \text{ ohms}$$

$$x_{1m} = x_2 = 50 \text{ ohms} \qquad x_\varphi = 920 \text{ ohms}$$

The main and auxiliary windings have the same number of turns. This motor is used as a tachometer with a 60-Hz reference voltage applied to its main winding, as in Fig. 11-27. Compute the speed voltage sensitivity in volts output per volt input per radian per second near zero speed.

Also compute the phase angle of the output voltage relative to the input voltage.

11-7. *a.* Find the starting torque of the motor given in Example 11-4 for the conditions specified.

b. Compare the result of *a* with the torque which the motor would develop at starting when balanced 2-phase voltages of 220 volts are applied.

c. Show, in general, that, if the stator voltages V_m and V_a of a 2-phase induction motor are in quadrature but unequal, the starting torque is the same as that developed when balanced 2-phase voltages of $\sqrt{V_m V_a}$ volts are applied.

11-8. The induction motor of Example 11-4 is supplied from an unbalanced 2-phase source by a four-wire feeder having an impedance of $1.0 + j3.0$ ohms per phase. The source voltages can be expressed as

$$V_m = 240\underline{/0^\circ} \text{ volts} \qquad V_a = 200\underline{/75^\circ} \text{ volts}$$

For a slip of 0.05, show that the induction-motor performance is such that the motor's terminal voltages correspond more nearly to those of a balanced 2-phase system than those at the source.

11-9. The equivalent-circuit constants in ohms per phase referred to the stator for a 2-phase 1.5-hp 220-volt 4-pole 60-Hz squirrel-cage induction motor are given below. The no-load rotational loss is 200 watts.

$$r_1 = 3.2 \qquad\qquad r_2 = 2.4$$

$$x_1 = x_2 = 3.2 \qquad x_\varphi = 100$$

a. The voltage applied to phase *m* is $220\underline{/0^\circ}$ volts, and the voltage applied to phase *a* is $220\underline{/60^\circ}$. At a slip $s = 0.04$, $Z_f = 41.9 + j27.2$ ohms, and $Z_b = 1.20 + j3.2$ ohms. What is the net air-gap torque?

b. What is the starting torque with the applied voltages of *a*?

c. The applied voltages are readjusted so that $V_m = 220\underline{/0^\circ}$ and $V_a = 220\underline{/90^\circ}$. Full load on the machine occurs at $s = 0.04$. At what value of slip does maximum torque occur? What is the value of maximum air-gap torque in newton-meters?

d. While the motor is running as in *c*, phase *a* is open-circuited. What is the horsepower developed by the machine at slip $s = 0.04$?

e. What voltage appears across the open phase a terminals under the conditions of d at $s = 0.04$?

11-10. Simplified dynamic considerations relating to 2-phase control motors are presented in Example 11-5 on the basis of linearity of the torque-speed curves of Fig. 11-20 at low speeds. A similar approximate investigation is called for here.[1]

The control motor has identical 2-phase stator windings and a high-resistance rotor and is supplied from a low-impedance source. When balanced sinusoidal voltages are applied, the torque-slip curve is linear over the range of interest, i.e.,

$$T = kV_1^2 s$$

where k is a constant and V_1 the balanced stator voltage.

a. For control-motor usage as in Art. 11-6, show that the forward and backward torques are given by

$$T_f = k \left(\frac{V_a + V_m}{2} \right)^2 \left(1 - \frac{\omega_o}{\omega_s} \right)$$

$$T_b = k \left(\frac{V_a - V_m}{2} \right)^2 \left(1 + \frac{\omega_o}{\omega_s} \right)$$

where ω_o is the shaft angular velocity and ω_s is the synchronous angular velocity.

b. The motor drives a load having moment of inertia J and a torque requirement $f_v \omega_o$ proportional to speed. Neglect motor losses. With the motor at rest and the voltage V_m on the reference field, the rms voltage V_a is suddenly applied to the control field. Determine the velocity ω_o as a function of time. Ignore electrical transients.

c. For small values of control-field voltage, V_a^2 may be neglected in comparison with V_m^2. What is the time constant in b under this assumption?

d. Using the assumption in c, determine the transfer functions relating shaft velocity and shaft position angle to the control-field signal.

[1] A more comprehensive examination along these lines can be found in A. M. Hopkin, *Transient Response of Small Two-phase Servomotors*, *Trans. AIEE*, **70**:881–886 (1951).

11-11. The motor of Prob. 11-6 is used as a 2-phase control motor. When the reference-field voltage is 100 volts and the control-field voltage is 70 volts and leads the reference voltage by 90° (both voltages at 60 Hz), compute:

    a. The ratio of the backward flux wave to the forward flux wave at standstill

    b. The ratio of the backward flux wave to the forward flux wave at a slip $s = 0.80$

    c. The internally developed mechanical power in watts at $s = 0.80$

11-12. For the 2-phase induction motor of Example 11-6, plot a family of curves of rotor power loss and internal power developed for per-unit values of control-phase voltage of 1.00, 0.50, and 0, covering a speed range from -1 to $+1$ per unit. The applied voltages are in quadrature. Express the power in per unit based on the power delivered to the air gap by the stator windings at standstill when balanced 2-phase voltages of 1.00 per unit are applied to the 2 stator phases.

11-13. A small 2-pole squirrel-cage induction motor for use in servo systems has symmetrical 2-phase stator windings. At standstill, the input impedance measured at the terminals of each stator winding at 60 Hz is $305 + j51$ ohms. For the purposes of this problem rotational and core losses may be neglected. Three points on the torque-slip characteristic of this motor with balanced 2-phase voltages of 100 volts at 60 Hz applied to its stator terminals are given below:

Torque, newton-meter	0.064	0.082	0.088
Slip, per unit	0.50	1.00	1.50

If the reference-phase voltage is held constant at 100 volts, 60 Hz, and the control-phase voltage is reduced to 50 volts (the two voltages being in time quadrature), compute:

    a. The standstill torque, in newton-meters

    b. The power input to the reference phase at standstill

    c. The power input to the control phase at standstill

    d. The total rotor I^2R loss at standstill

    e. The torque at $s = 0.50$

11-14. For the 2-phase control motor of Prob. 11-13 at standstill with 100 volts applied to the reference phase and variable voltage applied to the control phase, plot curves of the following variables as functions of the standstill torque:

   a. Total rotor copper loss
   b. Control-phase stator copper loss
   c. Reference-phase stator copper loss
   d. Power input to control phase
   e. Power input to reference phase

11-15. A symmetrical 2-phase control motor produces a torque of 1.25 lb-ft at standstill with balanced voltages of 100 volts applied.

If the motor is required to produce an acceleration at zero speed of 64.4 rad/sec² in a load having no friction but having an inertia of 0.5 lb-ft², what voltage must be supplied to the auxiliary winding when the main winding is supplied with 100 volts in time quadrature?

appendix A
3-phase circuits

Generation, transmission, and heavy-power utilization of ac electric
energy almost invariably involve a type of system or circuit called a
polyphase system or *polyphase circuit*. In such a system, each voltage
source consists of a group of voltages having related magnitudes and
phase angles. Thus, an *n*-phase system will employ voltage sources
which, conventionally, consist of *n* voltages substantially equal in magnitude and successively displaced by a phase angle of 360°/*n*. A 3-*phase*
system will employ voltage sources which, conventionally, consist of 3
voltages substantially equal in magnitude and displaced by phase angles
of 120°. Because it possesses definite economic and operating advantages, the 3-phase system is by far the most common, and consequently
predominating emphasis is given to 3-phase circuits in this appendix.

The 3 individual voltages of a 3-phase source may each be connected
to its own independent circuit. We would then have 3 separate *single-
phase systems*. Alternatively, as will be shown in the first Article, symmetrical electrical connections may be made among the 3 voltages and
the associated circuitry to form a 3-phase system. It is the latter alter-

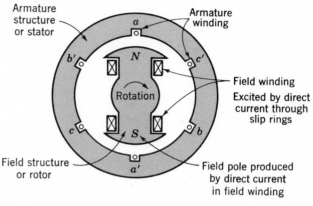

Fig. A-1.   Elementary 3-phase 2-pole generator.

native that we are concerned with in this appendix.   Note that the word *phase* now has two distinct meanings.   It may refer to a portion of a polyphase system or circuit, or, as in the familiar steady-state circuit theory, it may be used in reference to the angular displacement between voltage or current phasors.   There is very little possibility of confusing the two usages.

A-1

GENERATION OF 3-PHASE VOLTAGES

Consider the elementary 3-phase 2-pole generator of Fig. A-1.   On the armature are three coils, *aa'*, *bb'*, and *cc'*, whose axes are displaced 120° in space from each other.   This winding can be represented schematically as shown in Fig. A-2.   When the field is excited and rotated, voltages will

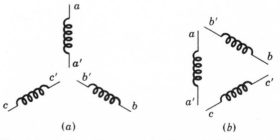

Fig. A-2.   Schematic representation of windings of Fig. A-1.

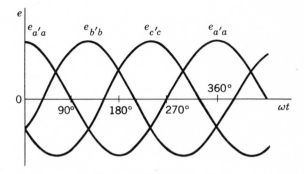

Fig. A-3.   Voltage waves generated in windings of Figs. A-1 and A-2.

be generated in the three phases in accordance with Faraday's law.   If the field structure is so designed that the flux is distributed sinusoidally over the poles, the flux linking any phase will vary sinusoidally with time and sinusoidal voltages will be induced in the three phases.   As shown in Fig. A-3, these three waves will be displaced 120 electrical degrees in time as a result of the phases being displaced 120° in space.   The corresponding phasor diagram is shown in Fig. A-4.   In general, the time origin and the reference axis in diagrams such as Figs. A-3 and A-4 are chosen on the basis of analytical convenience.

There are two possibilities for the utilization of voltages generated in this manner: The six terminals a, a', b, b', c, and c' of the winding may be connected to three independent single-phase systems, or the three phases of the winding may be interconnected and used to supply a 3-phase system.   The latter procedure is the one adopted almost universally.   The three phases of the winding may be interconnected in two possible ways, as shown in Fig. A-5: Terminals a', b', and c' may be joined to form the neutral o, yielding a Y *connection*, or terminals a and b', b and c', and c and a' may be joined individually, yielding a Δ *connection*.   In the Y connec-

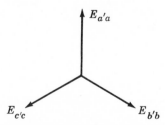

Fig. A-4.   Phasor diagram of generated voltages.

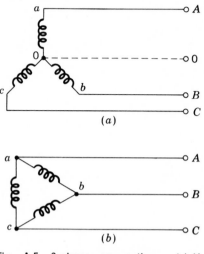

Fig. A-5. 3-phase connections. *(a)* Y connection; *(b)* Δ connection.

tion, a *neutral conductor*, shown dotted in Fig. A-5a, may or may not be brought out. If a neutral conductor exists, the system is a 4-*wire 3-phase system;* if not, it is a 3-*wire 3-phase system.* In the Δ connection (Fig. A-5b), no neutral exists and only a 3-wire 3-phase system can be formed.

The three phase voltages, Figs. A-3 and A-4, are equal and phase displaced by 120 electrical degrees, a general characteristic of a *balanced 3-phase system.* Furthermore, the impedance in any one phase is equal to that in either of the other two phases, so that the resulting phase currents are equal and phase displaced from each other by 120 electrical degrees. Likewise, equal power and equal reactive power flow in each phase. An *unbalanced 3-phase system,* on the other hand, may lack any or all of the equalities and 120° displacements. It is important to note that *only balanced systems are treated in this appendix and that none of the methods developed or conclusions reached apply to unbalanced systems.* Most practical problems are concerned with balanced systems. Many industrial loads are 3-phase loads and therefore inherently balanced, and in supplying single-phase loads from a 3-phase source, definite efforts are made to keep the 3-phase system balanced by assigning approximately equal single-phase loads to each of the three phases.

A-2
3–PHASE VOLTAGES, CURRENTS, AND POWER

When the three phases of the winding in Fig. A-1 are Y-connected, as in Fig. A-5a, the phasor diagram of voltages is that of Fig. A-6. The *phase*

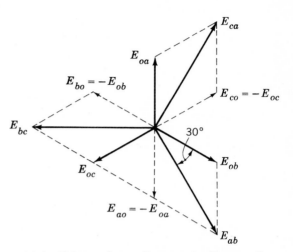

Fig. A-6.   Voltage phasor diagrams for Y connection.

order or *phase sequence* in Fig. A-6 is *abc*; that is, the voltage of phase *a* reaches its maximum 120° before that of phase *b*.   The use of *double-subscript notation* in Fig. A-6 greatly simplifies the task of drawing the complete diagram.   The subscripts indicate the points between which the voltage exists, and the order of subscripts indicates the direction in which the voltage rise is taken.   Thus, $\mathbf{E}_{ao} = -\mathbf{E}_{oa}$.

The 3-phase voltages are E_{oa}, E_{ob}, and E_{oc}.   They are also called *line-to-neutral voltages*.   The three voltages E_{ab}, E_{bc}, and E_{ca}, called *line voltages* or, more specifically, *line-to-line voltages*, are also important.   By Kirchhoff's voltage law, the line voltage $\mathbf{E}_{ab}$ is

$$\mathbf{E}_{ab} = \mathbf{E}_{ao} + \mathbf{E}_{ob} = -\mathbf{E}_{oa} + \mathbf{E}_{ob} \qquad\qquad (\text{A-1})$$
$$= \sqrt{3}\, \mathbf{E}_{ob} \underline{/-30°}$$

as shown in Fig. A-6.   Similarly,

$$\mathbf{E}_{bc} = \sqrt{3}\, \mathbf{E}_{oc} \underline{/-30°} \qquad\qquad (\text{A-2})$$

and
$$\mathbf{E}_{ca} = \sqrt{3}\, \mathbf{E}_{oa} \underline{/-30°} \qquad\qquad (\text{A-3})$$

Stated in words, these equations show that, *for a Y connection, the line voltage is $\sqrt{3}$ times the phase voltage, or the line-to-line voltage is $\sqrt{3}$ times the line-to-neutral voltage.*

The corresponding current phasors for the Y connection of Fig. A-5a

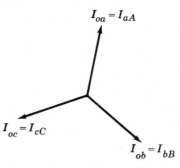

Fig. A-7. Current phasor diagram for Y connection.

are given in Fig. A-7. Obviously, *for a Y connection, the line currents and phase currents are equal.*

When the three phases are Δ-connected, as in Fig. A-5b, the phasor diagram of voltages is that of Fig. A-8. Obviously, *for a Δ connection, the line voltages and phase voltages are equal.*

The corresponding phasor diagram of currents is given in Fig. A-9. The 3-phase currents are I_{ab}, I_{bc}, and I_{ca}, the order of subscripts indicating the current directions. By Kirchhoff's current law, the line current $\mathbf{I}_{aA}$ is

$$\mathbf{I}_{aA} = \mathbf{I}_{ba} + \mathbf{I}_{ca} = -\mathbf{I}_{ab} + \mathbf{I}_{ca} \qquad \text{(A-4)}$$

$$= \sqrt{3}\,\mathbf{I}_{ca}\underline{/30°}$$

as shown in Fig. A-9. Similarly,

$$\mathbf{I}_{bB} = \sqrt{3}\,\mathbf{I}_{ab}\underline{/30°} \qquad \text{(A-5)}$$

and

$$\mathbf{I}_{cC} = \sqrt{3}\,\mathbf{I}_{bc}\underline{/30°} \qquad \text{(A-6)}$$

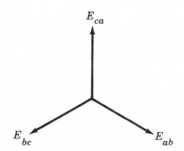

Fig. A-8. Voltage phasor diagram for Δ connection.

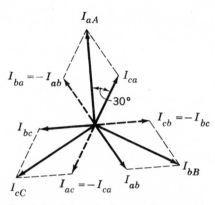

Fig. A-9. Current phasor diagram for
Δ connection.

Stated in words, Eqs. A-4 to A-6 show that *for a Δ connection, the line
current is* $\sqrt{3}$ *times the phase current.*   Evidently, the relations between
phase and line currents of a Δ connection are similar to those between
phase and line voltages of a Y connection.

   For both Y- and Δ-connected systems, it can be shown that the total
of the instantaneous power for all three phases of a balanced 3-phase
circuit does not pulsate with time.   Thus, with the time origin taken
at the maximum positive point of the phase-*a* voltage wave, the instantaneous voltages of the three phases are

$$e_a = \sqrt{2}\, E_p \cos \omega t \tag{A-7}$$

$$e_b = \sqrt{2}\, E_p \cos (\omega t - 120°) \tag{A-8}$$

$$e_c = \sqrt{2}\, E_p \cos (\omega t - 240°) \tag{A-9}$$

where E_p is the rms value of the phase voltage.   When the phase currents
are displaced from the corresponding phase voltages by the angle θ, the
instantaneous phase currents are

$$i_a = \sqrt{2}\, I_p \cos (\omega t + \theta) \tag{A-10}$$

$$i_b = \sqrt{2}\, I_p \cos (\omega t + \theta - 120°) \tag{A-11}$$

$$i_c = \sqrt{2}\, I_p \cos (\omega t + \theta - 240°) \tag{A-12}$$

where I_p is the rms value of the phase current.

The instantaneous power in each phase then becomes

$$p_a = e_a i_a = E_p I_p [\cos (2\omega t + \theta) + \cos \theta] \qquad \text{(A-13)}$$

$$p_b = e_b i_b = E_p I_p [\cos (2\omega t + \theta - 240°) + \cos \theta] \qquad \text{(A-14)}$$

$$p_c = e_c i_c = E_p I_p [\cos (2\omega t + \theta - 480°) + \cos \theta] \qquad \text{(A-15)}$$

The total instantaneous power for all three phases is

$$p = p_a + p_b + p_c = 3 E_p I_p \cos \theta \qquad \text{(A-16)}$$

Notice that the sum of the cosine terms which involve time in Eqs. A-13 to A-15 (i.e., the first terms in the brackets) is zero. The total instantaneous power is accordingly independent of time. This situation is depicted graphically in Fig. A-10. Instantaneous powers for the three phases are plotted, together with the total instantaneous power, which is the sum of the three individual waves. *The total instantaneous power for a balanced 3-phase system is constant and is equal to 3 times the average power per phase.*

In general, it can be shown that the total instantaneous power for any balanced polyphase system is constant. This is one of the outstanding advantages of polyphase systems. It is of particular advantage in the operation of polyphase motors, for example, for it means that the shaft power output is constant and that torque pulsations, with the consequent tendency toward vibration, do not result from pulsations inherent in the supply system.

On the basis of single-phase considerations, the average power per phase P_p for either a Y- or Δ-connected system is

$$P_p = E_p I_p \cos \theta = I_p^2 R_p \qquad \text{(A-17)}$$

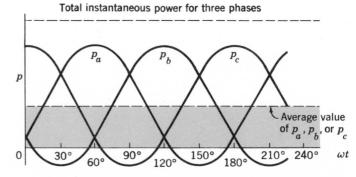

Fig. A-10.   Instantaneous power in a 3-phase system.

where E_p, I_p, and R_p are the voltage, current, and resistance, respectively, all per phase.   The total 3-phase power P is

$$P = 3P_p \tag{A-18}$$

Similarly, for reactive power per phase Q_p and total 3-phase reactive power Q,

$$Q_p = E_p I_p \sin \theta = I_p^2 X_p \tag{A-19}$$

and

$$Q = 3Q_p \tag{A-20}$$

The volt-amperes per phase $(VA)_p$ and total 3-phase volt-amperes VA are

$$(VA)_p = E_p I_p = I_p^2 Z_p \tag{A-21}$$

and

$$VA = 3(VA)_p \tag{A-22}$$

In Eqs. (A-17) and (A-19), θ is the angle between phase voltage and phase current.   As in the single-phase case, it is given by

$$\theta = \tan^{-1} \frac{X_p}{R_p} = \cos^{-1} \frac{R_p}{Z_p} = \sin^{-1} \frac{X_p}{Z_p} \tag{A-23}$$

The power factor of a balanced 3-phase system is therefore equal to that of any one phase.

**A-3
Y- AND Δ-CONNECTED CIRCUITS**

Three specific examples will be given to illustrate the computational details of Y- and Δ-connected circuits.   Explanatory remarks which are generally applicable are incorporated in the solutions.

EXAMPLE A-1

In Fig. A-11 is shown a 60-Hz transmission system consisting of a line having the impedance $Z_l = 0.05 + j0.20$ ohm, at the receiving end of which is a load of equivalent impedance $Z_L = 10.0 + j3.00$ ohms. The impedance of the return conductor should be considered zero.

a. Compute (1) the line current I; (2) the load voltage E_L; (3) the power, reactive power, and volt-amperes taken by the load; and (4) the power and reactive-power loss in the line.

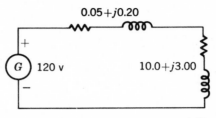

0.05+j0.20

G 120 v

10.0+j3.00

Fig. A-11.   Circuit for Example A-1a.

Suppose now that three such identical systems are to be constructed to supply three such identical loads.   Instead of drawing the diagrams one below the other, let them be drawn in the fashion shown in Fig. A-12, which is, of course, the same electrically.

b. Give, for Fig. A-12, (1) the current in each line; (2) the voltage at each load; (3) the power, reactive power, and volt-amperes taken by each load; (4) the power and reactive-power loss in each of the three transmission systems; (5) the total power, reactive power, and volt-amperes taken by the loads; and (6) the total power and reactive-power loss in the three transmission systems.

Next consider that the three return conductors are combined into one and that the phase relationship of the voltage sources is such that a balanced 4-wire 3-phase system results, as in Fig. A-13.

c. Give, for Fig. A-13, (1) the line current; (2) the load voltage, both line-to-line and line-to-neutral; (3) the power, reactive power, and volt-amperes taken by each phase of the load; (4) the power and reactive-power loss in each line; (5) the total 3-phase power, reac-

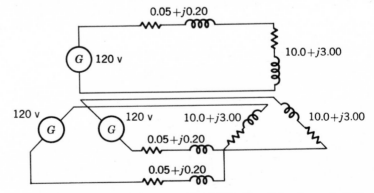

0.05+j0.20

G 120 v

10.0+j3.00

120 v   G   120 v   10.0+j3.00   10.0+j3.00

0.05+j0.20

0.05+j0.20

Fig. A-12.   Circuit for Example A-1b.

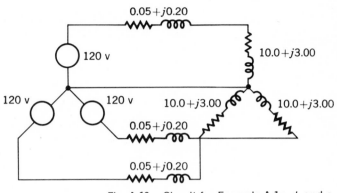

Fig. A-13. Circuit for Example A-1c, d, and e.

tive power, and volt-amperes taken by the load; and (6) the total power and reactive-power loss in the lines.

d. In Fig. A-13, what is the current in the combined return or neutral conductor?

e. May this conductor be dispensed with in Fig. A-13 if desired?
    Assume now that this neutral conductor is omitted. This results in the 3-wire 3-phase system of Fig. A-14.

f. Repeat part c for Fig. A-14.

g. On the basis of the results of this example, outline briefly the method of reducing a balanced 3-phase Y-connected circuit problem to its equivalent single-phase problem. Be careful to distinguish between the use of line-to-line and line-to-neutral voltages.

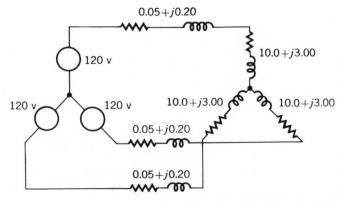

Fig. A-14. Circuit for Example A-1f.

Solution

a.

(1) $\qquad I = \dfrac{120}{\sqrt{(0.05 + 10.0)^2 + (0.20 + 3.00)^2}} = 11.4$ amp

(2) $\qquad E_L = IZ_L = 11.4 \sqrt{(10.0)^2 + (3.00)^2} = 119$ volts

(3) $\qquad P_L = I^2 R_L = (11.4)^2 (10.0) = 1{,}300$ watts

$\qquad\qquad Q_L = I^2 X_L = (11.4)^2 (3.00) = 390$ var

$\qquad (VA)_L = I^2 Z_L = (11.4)^2 \sqrt{(10.0)^2 + (3.00)^2} = 1{,}360$ va

(4) $\qquad P_l = I^2 R_l = (11.4)^2 (0.05) = 6.5$ watts

$\qquad\qquad Q_l = I^2 X_l = (11.4)^2 (0.20) = 26$ var

b. Parts (1) to (4) obviously have the same values as in *a.*

(5) $\quad$ Total power $= 3P_L = 3(1{,}300) = 3{,}900$ watts

$\qquad$ Total reactive power $= 3Q_L = 3(390) = 1{,}170$ var

$\qquad$ Total VA $= 3(VA)_L = 3(1{,}360) = 4{,}080$ va

(6) $\quad$ Total power loss $= 3P_l = 3(6.5) = 19.5$ watts

$\qquad$ Total reactive-power loss $= 3Q_l = 3(26) = 78$ var

c. The results obtained in *b* are unaffected by this change. The voltage in *b*(2) and *a*(2) is now the line-to-neutral voltage. The line-to-line voltage is

$$\sqrt{3}\,119 = 206 \text{ volts}$$

d. By Kirchhoff's current law, the neutral current is the phasor sum of the three line currents. These line currents are equal and phase-displaced 120°. Since the phasor sum of three equal phasors 120° apart is zero, the neutral current is zero.

e. The neutral current being zero, the neutral conductor may be dispensed with if desired.

f. Since the presence or absence of the neutral conductor does not affect conditions, the values are the same as in *c.*

g. A neutral conductor may be assumed, regardless of whether one is physically present. Since the neutral conductor in a balanced 3-phase circuit carries no current and hence has no voltage drop across it, the neutral conductor should be considered to have zero impedance.   Then one phase of the Y, together with the neutral conductor, may be removed for study.   Since this phase is uprooted at the neutral, *line-to-neutral voltages must be used.*   This procedure yields the equivalent single-phase circuit, in which all quantities correspond to those in one phase of the 3-phase circuit. Conditions in the other two phases being the same (except for the 120° phase displacements in the currents and voltages), there is no need for investigating them individually.   Line currents in the 3-phase system are the same as in the single-phase circuit, and total 3-phase power, reactive power, and volt-amperes are three times the corresponding quantities in the single-phase circuit.   If line-to-line voltages are desired, they must be obtained by multiplying voltages in the single-phase circuit by $\sqrt{3}$.

EXAMPLE A-2

Three impedances of value $Z_p = 4.00 + j3.00 = 5.00\underline{/36.9°}$ ohms are connected in Y, as shown in Fig. A-15.   For balanced line-to-line voltages of 208 volts, find the line current, the power factor, and the total power, reactive power, and volt-amperes.

Solution

The line-to-neutral voltage across any one phase, such as ao, is

$$E_p = \frac{208}{\sqrt{3}} = 120 \text{ volts}$$

Hence
$$I_l = I_p = \frac{E_p}{Z_p} = \frac{120}{5.00} = 24.0 \text{ amp}$$

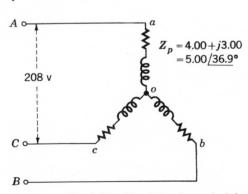

Fig. A-15.   Circuit for Example A-2.

Power factor $= \cos \theta = \cos 36.9° = 0.80$ lagging

$P = 3P_p = 3I_p^2 R_p = 3(24.0)^2(4.00) = 6,910$ watts

$Q = 3Q_p = 3I_p^2 X_p = 3(24.0)^2(3.00) = 5,180$ var

VA $= 3(\text{VA})_p = 3E_p I_p = 3(120)(24.0) = 8,640$ va

It should be noted that phases a and c (Fig. A-15) do not form a simple series circuit. Consequently, the current cannot be found by dividing 208 volts by the sum of the phase a and c impedances. To be sure, an equation can be written for voltage between points a and c by Kirchhoff's voltage law, but this must be a phasor equation taking account of the 120° phase displacement between the phase-a and phase-c currents. As a result, the method of thought outlined in Example A-1 leads to the simplest solution.

EXAMPLE A-3

Three impedances of value $Z_p = 12.00 + j9.00 = 15.00\underline{/36.9°}$ ohms are connected in Δ, as shown in Fig. A-16. For balanced line-to-line voltages of 208 volts, find the line current, the power factor, and the total power, reactive power, and volt-amperes.

Solution

The voltage across any one phase, such as ca, is evidently equal to the line-to-line voltage. Consequently,

$$E_p = 208 \text{ volts}$$

and
$$I_p = \frac{E_p}{Z_p} = \frac{208}{15.00} = 13.87 \text{ amp}$$

Power factor $= \cos \theta = \cos 36.9° = 0.80$ lagging

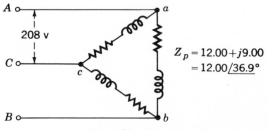

Fig. A-16.   Circuit for Example A-3.

From Eq. A-4,

$$I_l = \sqrt{3}\, I_p = \sqrt{3}\,(13.87) = 24.0 \text{ amp}$$

Also $P = 3P_p = 3I_p^2 R_p = 3(13.87)^2(12.00) = 6{,}910 \text{ watts}$

$Q = 3Q_p = 3I_p^2 X_p = 3(13.87)^2(9.00) = 5{,}180 \text{ var}$

and $\text{VA} = 3(\text{VA})_p = 3E_p I_p = 3(208)(13.87) = 8{,}640 \text{ va}$

It should be noted that phases ab and bc do not form a simple series circuit, nor does the path cba form a simple parallel combination with the direct path through the phase ca. Consequently, the line current cannot be found by dividing 208 volts by the equivalent impedance of $\mathbf{Z}_{ca}$ in parallel with $\mathbf{Z}_{ab} + \mathbf{Z}_{bc}$. Kirchhoff's-law equations involving quantities in more than one phase can be written, but they must be phasor quantities taking account of the 120° phase displacement between phase currents and between phase voltages. As a result, the method outlined above leads to the simplest solution.

Comparison of the results of Examples A-2 and A-3 leads to a valuable and interesting conclusion. It will be noted that the line-to-line voltage, line current, power factor, total power, reactive power, and volt-amperes are precisely equal in the two cases; in other words, conditions viewed from the terminals A, B, and C are identical, and one cannot distinguish between the two circuits from their terminal quantities. It will also be seen that the impedance, resistance, and reactance per phase of the Y connection (Fig. A-15) are exactly one-third of the corresponding values per phase of the Δ connection (Fig. A-16). Consequently, a balanced Δ connection may be replaced by a balanced Y connection providing that the circuit constants per phase obey the relation

$$\mathbf{Z}_Y = \tfrac{1}{3}\mathbf{Z}_\Delta \qquad\qquad\qquad (\text{A-24})$$

Conversely, a Y connection may be replaced by a Δ connection provided Eq. A-24 is satisfied. The concept of this Y-Δ equivalence stems from the general Y-Δ transformation and is not the accidental result of a specific numerical case.

Two important corollaries follow from this equivalence. First, a general computational scheme for balanced circuits may be based entirely upon Y-connected circuits or entirely on Δ-connected circuits, whichever one prefers. Since it is frequently more convenient to handle a Y connection, the former scheme is the one usually adopted. Second, in the frequently occurring problems in which the connection is not specified and is not pertinent to the solution, either a Y or a Δ connection may be assumed. Again the Y connection is more commonly selected. In analyzing 3-phase motor performance, for example, the actual winding connections need not be known unless the investigation is to include detailed conditions within the coils themselves. The entire analysis may be based on an assumed Y connection.

A-4

ANALYSIS OF BALANCED 3-PHASE CIRCUITS; SINGLE-LINE DIAGRAMS

By combining the principle of Δ-Y equivalence with the technique revealed by Example A-1, a simple method of reducing a balanced 3-phase-circuit problem to its corresponding single-phase problem may be developed. All the methods of single-phase-circuit analysis thus become available for its solution. The end results of the single-phase analysis are then translated back into 3-phase terms to give the final results.

In carrying out this procedure, phasor diagrams need be drawn for but one phase of the Y connection, the diagrams for the other two phases being unnecessary repetition. Furthermore, circuit diagrams may be simplified by drawing only one phase. Examples of such *single-line diagrams* are given in Fig. A-17, showing two 3-phase generators with their associated lines or cables supplying a common substation load. Specific connections of apparatus may be indicated if desired. Thus, Fig. A-17*b* shows that G_1 is Y-connected and G_2 is Δ-connected. Impedances are given in ohms per phase.

When dealing with power, reactive power, and volt-amperes, it is sometimes more convenient to deal with the entire 3-phase circuit at once instead of concentrating on one phase. This possibility arises because

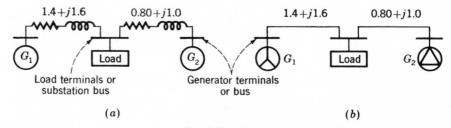

Fig. A-17.   Examples of single-line circuit diagrams.

simple expressions for 3-phase power, reactive power, and volt-amperes can be written in terms of line-to-line voltage and line current regardless of whether the circuit is Y- or Δ-connected.   Thus, from Eqs. A-17 and A-18, 3-phase power is

$$P = 3P_p = 3E_p I_p \cos \theta \tag{A-25}$$

For a Y connection, $I_p = I_{\text{line}}$ and $E_p = E_{\text{line}}/\sqrt{3}$.   For a Δ connection, $I_p = I_{\text{line}}/\sqrt{3}$ and $E_p = E_{\text{line}}$.   In either case, Eq. A-25 becomes

$$P = \sqrt{3}\, E_{\text{line}} I_{\text{line}} \cos \theta \tag{A-26}$$

Similarly, $$Q = \sqrt{3}\, E_{\text{line}} I_{\text{line}} \sin \theta \tag{A-27}$$

and $$\text{VA} = \sqrt{3}\, E_{\text{line}} I_{\text{line}} \tag{A-28}$$

It should be borne in mind, however, that the power-factor angle θ, given by Eq. A-23, is the angle between $\mathbf{E}_p$ and $\mathbf{I}_p$ and not that between $\mathbf{E}_{\text{line}}$ and $\mathbf{I}_{\text{line}}$.

EXAMPLE A-4

Figure A-17 is the equivalent circuit of a load supplied from two 3-phase generating stations over lines having the impedances per phase given on the diagram.   The load requires 30 kw at 0.80 power factor lagging.   Generator G_1 operates at a terminal voltage of 797 volts line to line and supplies 15 kw at 0.80 power factor lagging.

Find the load voltage and the terminal voltage and power and reactive power output of G_2.

Solution

Let I, P, and Q, respectively, denote line current and 3-phase active and reactive power.   The subscripts 1 and 2 denote the respective branches of the system; the subscript r denotes a quantity measured at the receiving end of the line.   We then have

$$I_1 = \frac{P_1}{\sqrt{3}\, E_1 \cos \theta_1} = \frac{15{,}000}{\sqrt{3}\,(797)(0.80)} = 13.6 \text{ amp}$$

$$P_{r1} = P_1 - 3I_1^2 R_1 = 15{,}000 - 3(13.6)^2(1.4) = 14{,}220 \text{ watts}$$

$$Q_{r1} = Q_1 - 3I_1^2 X_1 = 15{,}000 \tan(\cos^{-1} 0.80) - 3(13.6)^2(1.6)$$

$$= 10{,}350 \text{ var}$$

The factor 3 appears before $I_1^2 R_1$ and $I_1^2 X_1$ in the last two equations because the current I_1 exists in all three lines.   The load voltage is

$$E_L = \frac{\text{VA}}{\sqrt{3}\ (\text{current})} = \frac{\sqrt{(14{,}220)^2 + (10{,}350)^2}}{\sqrt{3}\ (13.6)}$$

$$= 748 \text{ volts line to line}$$

Since the load requires 30,000 watts and 30,000 tan (cos^{-1} 0.80) or 22,500 var,

$$P_{r2} = 30{,}000 - 14{,}220 = 15{,}780 \text{ watts}$$

and
$$Q_{r2} = 22{,}500 - 10{,}350 = 12{,}150 \text{ var}$$

$$I_2 = \frac{\text{VA}}{\sqrt{3}\ (\text{voltage})} = \frac{\sqrt{(15{,}780)^2 + (12{,}150)^2}}{\sqrt{3}\ (748)} = 15.4 \text{ amp}$$

$$P_2 = P_{r2} + 3I_2^2 R_2 = 15{,}780 + 3(15.4)^2(0.80) = 16{,}350 \text{ watts}$$

$$Q_2 = Q_{r2} + 3I_2^2 X_2 = 12{,}150 + 3(15.4)^2(1.0) = 12{,}870 \text{ var}$$

$$E_2 = \frac{\text{VA}}{\sqrt{3}\ (\text{current})} = \frac{\sqrt{(16{,}350)^2 + (12{,}870)^2}}{\sqrt{3}\ (15.4)}$$
$$= 780 \text{ volts line to line}$$

A-5
OTHER POLYPHASE SYSTEMS

Although 3-phase systems are by far the most common of all polyphase systems, other numbers of phases are used for specialized purposes.   The 5-wire 4-phase system (Fig. A-18) is sometimes used for low-voltage dis-

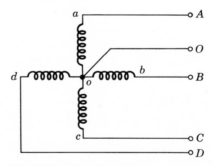

Fig. A-18.   Five-wire 4-phase system.

tribution.   It has the advantage that for a phase voltage of 115 volts, single-phase voltages of 115 (between a, b, c, or d and o, Fig. A-18) and 230 volts (between a and c or b and d) are available, as well as a system of polyphase voltages.   Essentially the same advantages are possessed by 4-wire 3-phase systems having a line-to-neutral voltage of 120 volts and a line-to-line voltage of 208 volts, however.

Four-phase systems are obtained from 3-phase systems by means of special transformer connections. Half of the 4-phase system—the part aob (Fig. A-18), for example—constitutes a 2-phase system.   In mercury-arc rectifiers, 6-, 12-, 18-, and 36-phase connections are used for the conversion of alternating to direct current.   These systems are also obtained by transformation from 3-phase systems.

When the loads and voltages are balanced, the methods of analysis for 3-phase systems may be adapted to any of the other polyphase systems by considering one phase of that polyphase system.   Of course, the basic voltage, current, and power relations must be modified to suit the particular polyphase system.

appendix B
voltages and magnetic fields
in distributed ac windings

Both amplitude and waveform of the generated voltages and armature mmfs in machines are determined by the winding arrangements and general machine geometry. These configurations in turn are dictated by economic use of space and materials in the machine and by suitability for the intended service. In this appendix we shall supplement the introductory discussion of these considerations in Chap. 3 by analytical treatment of ac voltages and mmfs in the balanced steady state. Attention will be confined to the time-fundamental component of voltages and the space-fundamental component of mmfs.

B-1
GENERATED VOLTAGES

In accordance with Eq. 3-15 the rms generated voltage per phase for a concentrated winding having N_{ph} turns per phase is

$$E = 4.44fN_{\mathrm{ph}}\Phi \tag{B-1}$$

f being the frequency and Φ the fundamental flux per pole.

A more complex and practical winding will have coil sides for each phase distributed in several slots per pole. Equation B-1 may then be used to compute the voltage distribution of individual coils. To determine the voltage of an entire phase group, the voltages of the component coils must be added as phasors. Such addition of fundamental-frequency voltages is the subject of this article.

a. Distributed Fractional-pitch Windings

A simple example of a distributed winding is illustrated in Fig. B-1 for a 3-phase 2-pole machine. This case retains all the features of a more general one with any integral number of phases, poles, and slots per pole per phase. At the same time, a *double-layer winding* is shown. Double-layer windings usually lead to simpler end connections and to a machine which is more economical to manufacture and are found in all machines except some small motors below 10 hp in size. Generally, one side of a coil, such as a_1, is placed in the bottom of a slot, and the other side, $-a_1$, is placed in the top of another slot. Coil sides such as a_1 and a_3 or a_2 and a_4, which are in adjacent slots and associated with the same phase, con-

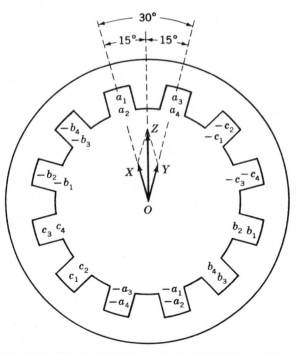

Fig. B-1. Distributed 3-phase 2-pole full-pitch armature winding with voltage phasor diagram.

stitute a *phase belt*. All phase belts are alike when an integral number of slots per pole per phase are used, and for the normal machine the peripheral angle subtended by a phase belt is 60 electrical degrees for a 3-phase machine and 90 electrical degrees for a 2-phase machine.

Individual coils in Fig. B-1 all span a full pole pitch, or 180 electrical degrees; accordingly, the winding is a *full-pitch winding*. Suppose now that all coil sides in the tops of the slots are shifted 1 slot counterclockwise, as in Fig. B-2. Any coil, such as a_1, $-a_1$, then spans only $\frac{5}{6}$ of a pole pitch or $(\frac{5}{6})(180) = 150$ electrical degrees, and the winding is a *fractional-pitch*, or *chorded, winding*. Similar shifting by 2 slots yields a $\frac{2}{3}$-pitch winding, and so forth. Phase groupings are now intermingled, for some slots contain coil sides in phases a and b, a and c, and b and c. Individual phase groups, such as that formed by a_1, a_2, a_3, a_4 on one side and $-a_1$, $-a_2$, $-a_3$, $-a_4$ on the other, are still displaced by 120 electrical degrees from the groups in other phases so that 3-phase voltages are produced. Besides the minor feature of shortening the end connections, fractional-pitch windings will be found to decrease the harmonic content of both the voltage and mmf waves.

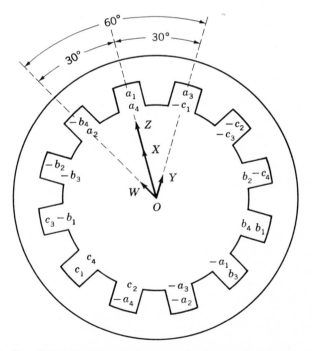

Fig. B-2. Distributed 3-phase 2-pole fractional-pitch armature winding with voltage phasor diagram.

The end connections between the coil sides are normally in a region of negligible flux density, and hence altering them does not significantly affect the mutual flux linkages of the winding.   Allocation of coil sides in slots is then the factor determining the generated voltages, and only that allocation need be specified in Figs. B-1 and B-2.   The only requisite is that all coil sides in a phase be included in the interconnection in such a manner that individual voltages shall make a positive contribution to the total.   The practical consequence is that end connections can be made according to the dictates of manufacturing simplicity; the theoretical consequence is that, when computational advantages result, the coil sides in a phase may be combined in an arbitrary fashion to form equivalent coils.

One sacrifice is made in using the distributed and fractional-pitch windings of Figs. B-1 and B-2 compared with a concentrated full-pitch winding: for the same number of turns per phase, the generated voltage is lower.   The harmonics are, in general, lowered by an appreciably greater factor, however, and the total number of turns which can be accommodated on a fixed iron geometry is increased.   The effect of distributing the winding in Fig. B-1 is that the voltages of coils a_1 and a_2 are not in phase with those of coils a_3 and a_4.   Thus, the voltage of coils a_1 and a_2 may be represented by phasor OX in Fig. B-1, and that of coils a_3 and a_4 by the phasor OY.   The time-phase displacement between these two voltages is the same as the electrical angle between adjacent slots, so that OX and OY coincide with the center lines of adjacent slots.   The resultant phasor OZ for phase a is obviously smaller than the arithmetic sum of OX and OY.

In addition, the effect of fractional pitch in Fig. B-2 is that a coil links a smaller portion of the total pole flux than if it were a full-pitch coil. The effect may be superimposed on that of distributing the winding by regarding coil sides a_2 and $-a_1$ as an equivalent coil with the phasor voltage OW (Fig. B-2), coil sides a_1, a_4, $-a_2$, and $-a_3$ as 2 equivalent coils with the phasor voltage OX (twice the length of OW), and coil sides a_3 and $-a_4$ as an equivalent coil with phasor voltage OY.   The resultant phasor OZ for phase a is obviously smaller than the arithmetic sum of OW, OX, and OY and is also smaller than OZ in Fig. B-1.

The combination of these two effects may be included in a *winding factor* k_w to be used as a reduction factor in Eq. B-1.   Thus, the generated voltage per phase is

$$E = 4.44k_w f N_{ph}\Phi \qquad\qquad (B-2)$$

where N_{ph} is the total turns in series per phase and k_w inserts the departure from the concentrated full-pitch case.   For a 3-phase machine, Eq. B-2 yields the line-to-line voltage for a Δ-connected winding and the line-to-neutral voltage for a Y-connected winding. As in any balanced Y connection, the line-to-line voltage of the latter winding is $\sqrt{3}$ times the line-to-neutral voltage.

b. Breadth and Pitch Factors

By considering separately the effects of distributing and of chording the winding, reduction factors may be obtained in generalized form convenient for quantitative analysis. The effect of distributing the winding in n slots per phase belt is to yield n voltage phasors phase-displaced by the electrical angle γ between slots, γ being equal to 180 electrical degrees divided by the number of slots per pole. Such a group of phasors is shown in Fig. B-3a and, in a more convenient form for addition, again in Fig. B-3b. Each phasor AB, BC, and CD is the chord of a circle with center at O and subtends the angle γ at the center. The phasor sum AD subtends the angle $n\gamma$, which, as noted previously, is 60 electrical degrees for the normal, uniformly distributed 3-phase machine and 90 electrical degrees for the corresponding 2-phase machine. From triangles OAa and OAd, respectively,

$$OA = \frac{Aa}{\sin(\gamma/2)} = \frac{AB}{2\sin(\gamma/2)} \tag{B-3}$$

$$OA = \frac{Ad}{\sin(n\gamma/2)} = \frac{AD}{2\sin(n\gamma/2)} \tag{B-4}$$

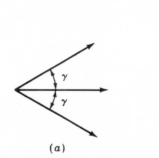

(a)

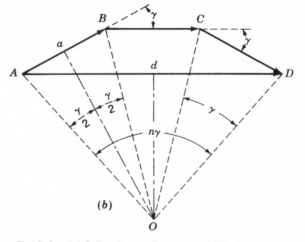

(b)

Fig. B-3.   (a) Coil voltage phasors and (b) phasor sum.

Equating these two values of OA yields

$$AD = AB \frac{\sin (n\gamma/2)}{\sin (\gamma/2)} \qquad \text{(B-5)}$$

But the arithmetic sum of the phasors is $n(AB)$.   Consequently, the reduction factor arising from distributing the winding is

$$k_b = \frac{AD}{n(AB)} = \frac{\sin (n\gamma/2)}{n \sin (\gamma/2)} \qquad \text{(B-6)}$$

The factor k_b is called the *breadth factor* of the winding.
   The effect of chording on the coil voltage may be obtained by first determining the flux linkages with the fractional-pitch coil.   Thus, in Fig. B-4 coil side $-a$ is only ρ electrical degrees from side a instead of the full 180°.   The flux linkages with the coil are

$$\lambda = NB_{\text{peak}}lr \frac{2}{P} \int_{\alpha}^{\rho+\alpha} \sin\theta \, d\theta \qquad \text{(B-7)}$$

$$\lambda = NB_{\text{peak}}lr \frac{2}{P} [\cos \alpha - \cos (\alpha + \rho)] \qquad \text{(B-8)}$$

where l is the axial length of the coil side, r the coil radius, and P the number of poles.   With α replaced by ωt to indicate rotation at ω elec-

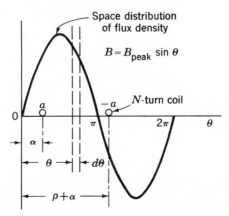

Fig. B-4.   Fractional-pitch coil in sinusoidal field.

trical radians per second, Eq. B-8 becomes

$$\lambda = N B_{\text{peak}} l r \, \frac{2}{P} \, [\cos \omega t - \cos (\omega t + \rho)] \qquad \text{(B-9)}$$

The addition of cosine waves required in the brackets of Eq. B-9 may be performed by a phasor diagram as indicated in Fig. B-5, from which it follows that

$$\cos \omega t - \cos (\omega t + \rho) = 2 \cos \frac{\pi - \rho}{2} \cos \left(\omega t - \frac{\pi - \rho}{2} \right) \qquad \text{(B-10)}$$

a result which may also be obtained directly from the terms in Eq. B-9 by the appropriate trigonometric transformations. The flux linkages are then

$$\lambda = N B_{\text{peak}} l r \, \frac{4}{P} \cos \frac{\pi - \rho}{2} \cos \left(\omega t - \frac{\pi - \rho}{2} \right) \qquad \text{(B-11)}$$

and the instantaneous voltage is

$$e = \omega N B_{\text{peak}} l r \, \frac{4}{P} \cos \frac{\pi - \rho}{2} \sin \left(\omega t - \frac{\pi - \rho}{2} \right) \qquad \text{(B-12)}$$

The phase angle $(\pi - \rho)/2$ in Eq. B-12 merely indicates that the instantaneous voltage is no longer zero when α in Fig. B-4 is zero. The factor $\cos (\pi - \rho)/2$ is an amplitude-reduction factor, however, so that the rms voltage of Eq. B-1 is modified to

$$E = 4.44 k_p f N_{\text{ph}} \Phi \qquad \text{(B-13)}$$

where the *pitch factor* k_p is

$$k_p = \cos \frac{\pi - \rho}{2} \qquad \text{(B-14)}$$

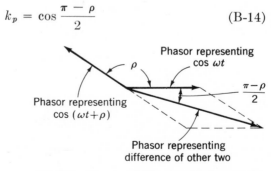

Fig. B-5. Phasor addition for fractional-pitch coil.

When both the breadth and pitch factors apply, the rms voltage is

$$E = 4.44 k_b k_p f N_{\text{ph}} \Phi \qquad \text{(B-15)}$$

which is an alternate form of Eq. B-2.

B-2
ARMATURE-MMF WAVES

Distribution of a winding in several slots per pole per phase and the use of fractional-pitch coils influence not only the emf generated in the winding but also the magnetic field produced by it.   Space-fundamental components of the mmf distributions will be examined in this article.

a. Concentrated Full-pitch Windings

We have seen in Art. 3-3 that a concentrated winding of N turns in a P-pole machine produces a rectangular magnetic-potential-difference wave around the air-gap circumference.   With excitation by a sinusoidal rms current I, the time-maximum height of the space-fundamental component of the wave is, in accordance with Eq. 3-21,

$$\frac{4}{\pi} \frac{N}{P} (\sqrt{2}\, I) \qquad \text{amp-turns per pole} \qquad \text{(B-16)}$$

For a polyphase concentrated winding, the amplitude for 1 phase becomes

$$\frac{4}{\pi} \frac{N_{\text{ph}}}{P} (\sqrt{2}\, I) \qquad \text{amp-turns per pole} \qquad \text{(B-17)}$$

where N_{ph} is the number of series turns per phase.   It is this last amplitude which is designated by the symbols $F_{a(\text{max})}$, $F_{b(\text{max})}$, or $F_{c(\text{max})}$ in Eqs. 3-30 to 3-32 and by the common symbol F_{max} in the balanced 3-phase case considered in Eqs. 3-33 to 3-36.

Each phase of a polyphase concentrated winding creates such a pulsating standing mmf wave in space.   This situation forms the basis of the analysis leading to Eq. 3-36.   Equation 3-36 may accordingly be rewritten

$$F_\theta = \frac{3}{2} \frac{4}{\pi} \frac{N_{\text{ph}}}{P} (\sqrt{2}\, I) \cos (\theta - \omega t) \qquad \text{(B-18)}$$

The amplitude of the resultant mmf wave in a 3-phase machine in ampere-turns per pole is then

$$F_A = \frac{3}{2}\frac{4}{\pi}\frac{N_{ph}}{P}(\sqrt{2}\,I) = 0.90\,\frac{3N_{ph}}{P}\,I \qquad (B\text{-}19)$$

Similarly, it may be shown that for a q-phase machine, the amplitude is

$$F_A = \frac{q}{2}\frac{4}{\pi}\frac{N_{ph}}{P}(\sqrt{2}\,I) = 0.90\,\frac{qN_{ph}}{P}\,I \qquad (B\text{-}20)$$

In Eqs. B-19 and B-20, I is the rms current per phase. The equations include only the fundamental component of the actual distribution and apply to concentrated full-pitch windings with balanced excitation.

b. Distributed Fractional-pitch Windings

When the coils in each phase of a winding are distributed among several slots per pole, the resultant space-fundamental mmf may be obtained by superposition from the preceding simpler considerations for a concentrated winding. The effect of distribution may be seen from Fig. B-6, which is a reproduction of the 3-phase 2-pole full-pitch winding with 2 slots per pole per phase given in Fig. B-1. Coils a_1 and a_2, b_1 and b_2, and c_1 and c_2 by themselves constitute the equivalent of a 3-phase 2-pole

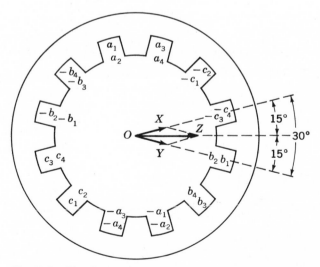

Fig. B-6. Distributed 3-phase 2-pole full-pitch armature winding with mmf phasor diagram.

concentrated winding because they form three sets of coils excited by polyphase currents and mechanically displaced 120° from each other. They therefore produce a rotating space-fundamental mmf; the amplitude of this contribution is given by Eq. B-19 when N_{ph} is taken as the sum of the series turns in coils a_1 and a_2 only. Similarly, coils a_3 and a_4, b_3 and b_4, and c_3 and c_4 produce another identical mmf wave, but one which is phase-displaced in space by the slot angle γ from the former wave. The resultant fundamental-mmf wave for the winding may be obtained by adding these two sinusoidal contributions.

The contribution from the $a_1a_2b_1b_2c_1c_2$ coils may be represented by the phasor OX in Fig. B-6. Such phasor representation is appropriate because the waveforms concerned are sinusoidal, and phasor diagrams are simply convenient means for adding sine waves. These are space sinusoids, however, not time sinusoids. Phasor OX is drawn in the space position of the mmf peak for an instant of time when the current in phase a is a maximum. The length of OX is proportional to the number of turns in the associated coils. Similarly, the contribution from the $a_3a_4b_3b_4c_3c_4$ coils may be represented by the phasor OY. Accordingly, the phasor OZ represents the resultant mmf wave. Just as in the corresponding voltage diagram, the resultant mmf is seen to be smaller than if the same number of turns per phase were concentrated in 1 slot per pole.

In like manner, mmf phasors can be drawn for fractional-pitch windings as illustrated in Fig. B-7, which is a reproduction of the 3-phase

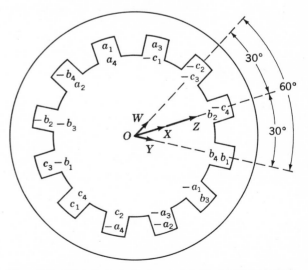

Fig. B-7. Distributed 3-phase 2-pole fractional-pitch armature winding with mmf phasor diagram.

2-pole $\frac{5}{6}$-pitch winding with 2 slots per pole per phase given in Fig. B-2. Phasor OW represents the contribution for the equivalent coils formed by conductors a_2 and $-a_1$, b_2 and $-b_1$, and c_2 and $-c_1$; OX for a_1a_4 and $-a_3 - a_2$, b_1b_4 and $-b_3 -b_2$, and c_1c_4 and $-c_3 -c_2$; and OY for a_3 and $-a_4$, b_3 and $-b_4$, and c_3 and $-c_4$. The resultant phasor OZ is, of course, smaller than the algebraic sum of the individual contributions and is also smaller than OZ in Fig. B-6.

By comparison with Figs. B-1 and B-2, these phasor diagrams may be seen to be identical with those for generated voltages. It therefore follows that the pitch and breadth factors previously developed may be applied directly to the determination of resultant mmf. Thus, for a distributed fractional-pitch polyphase winding, the amplitude of the space-fundamental component of mmf may be obtained by using $k_b k_p N_{\mathrm{ph}}$ instead of simply N_{ph} in Eqs. B-19 and B-20. These equations then become

$$F_A = \frac{3}{2} \frac{4}{\pi} \frac{k_b k_p N_{\mathrm{ph}}}{P} (\sqrt{2}\, I) = 0.90 \frac{3 k_b k_p N_{\mathrm{ph}}}{P} I \qquad (\text{B-21})$$

for a 3-phase machine and

$$F_A = \frac{q}{2} \frac{4}{\pi} \frac{k_b k_p N_{\mathrm{ph}}}{P} (\sqrt{2}\, I) = 0.90 \frac{q k_b k_p N_{\mathrm{ph}}}{P} I \qquad (\text{B-22})$$

for a q-phase machine, where F_A is in ampere-turns per pole.

appendix C
table of constants and conversion factors for rationalized mks units

Constants

Permeability of free space . . . . $\mu_0 = 4\pi \times 10^{-7}$ weber/amp-turn meter
Permittivity (capacitivity) of free space . . . $\epsilon_0 = 8.854 \times 10^{-12}$ coulomb^2/newton-meter^2
Acceleration of gravity . . . . $g = 9.807$ m/sec^2

Conversion Factors

Length . . . . 1 m = 3.281 ft
　　　　　　　　　　　　　　　　　　　= 39.37 in.
Mass . . . . 1 kg = 0.0685 slug
　　　　　　　　　　　　　　　　　　　= 2.205 lb (mass)
Force . . . . 1 newton = 0.225 lb
　　　　　　　　　　　　　　　　　　　= 7.23 poundals
Torque . . . . 1 newton-meter = 0.738 lb-ft
Energy . . . . 1 joule (watt-sec) = 0.738 ft-lb
Power . . . . 1 watt = 1.341 × 10^{-3} hp
Moment of inertia . . . . 1 kg-m^2 = 0.738 slug-ft^2
　　　　　　　　　　　　　　　　　　　= 23.7 lb-ft^2
Magnetic flux . . . . 1 weber = 10^8 maxwells (lines)
Magnetic flux density . . . . 1 weber/m^2 = 10,000 gauss
　　　　　　　　　　　　　　　　　　　= 64.5 kilolines/in.^{2}
Magnetizing force . . . . 1 amp-turn/m = 0.0254 amp-turn/in.
　　　　　　　　　　　　　　　　　　　= 0.0126 oersted

index